Ion Channels
and Disease

Ion Channels and Disease

Channelopathies

Frances M. Ashcroft
University Laboratory of Physiology
Oxford, United Kingdom

ACADEMIC PRESS
San Diego London Boston New York Sydney Tokyo Toronto

Copyright © 2000 by ACADEMIC PRESS

All Rights Reserved.
No part of this publication may be reproduced or transmitted in any form or by any
means, electronic or mechanical, including photocopy, recording, or any information
storage and retrieval system, without permission in writing from the publisher.

Requests for permission to make copies of any part of the work should be mailed to:
Permissions Department, Harcourt, Inc., 6277 Sea Harbor Drive,
Orlando, Florida 32887-6777

Academic Press
A Harcourt Science and Technology Company
525 B Street, Suite 1900, San Diego, California 92101-4495, USA
http://www.academicpress.com

Academic Press
Harcourt Place, 32 Jamestown Road, London NW1 7BY, UK
http://www.academicpress.com

Library of Congress Catalog Card Number: 98-85618

International Standard Book Number: 0-12-065310-9

PRINTED IN THE UNITED STATES OF AMERICA
01   02   03   04   05   06   MM   9   8   7   6   5   4   3   2

CN

*To Fiona and Graham,
who made it possible*

BRIEF CONTENTS

CONTENTS

PREFACE

In the last few years a new word has entered the medical and scientific vocabulary. This word, *channelopathy*, describes those human and animal diseases that result from defects in ion channel function. Ion channels are membrane proteins that act as gated pathways for the movement of ions across cell membranes. They play essential roles in the physiology and pathophysiology of all cells and it is therefore not very surprising that an ever increasing number of human and animal diseases have been found to be caused by defective ion channel function. This book aims to provide an informative and up to date account of our present understanding of ion channels and the molecular basis of ion channel diseases. It is intended for graduates and final year undergraduates in the medical and biological sciences. I hope it will also be of value to clinicians who wish to know more about the molecular basis of the channelopathies, and to research workers interested in the physiological role of ion channels and the diseases that result from their defective function.

To understand how ion channel mutations give rise to human and animal disease requires at least a basic understanding of the genetics, molecular structure, biophysical properties and physiological role of ion channels. It is difficult to give a comprehensive review of all these areas in a single book. However, I have provided a basic introduction to the relevant aspects of molecular biology and biophysics in the next two chapters, for the benefit of those unfamiliar with these topics. I have also included a brief description of the principal methods used to study channelopathies (Chapter 4). For more detailed descriptions of these topics readers are advised to consult one of the many excellent textbooks in the field, such as those by Hille (1992) or Aidley & Stanfield (1996). Chapters 5-23 each consider a single class of ion channel, while Chapter 24 includes a medley of channels which are distinct from those discussed in earlier chapters but about which too little is known to warrant an individual chapter. In each chapter, the initial section summarizes the physiological roles, subunit composition, molecular structure and chromosomal location of the channel in question. The relationship between channel structure and function is then considered. The third section describes those diseases associated with defective channel structure and/or regulation. For those channelopathies that result from mutations in ion channel genes, I consider how the mutation(s) affects channel function and to what extent this change in channel function can account for the clinical phenotype.

Identification of new ion channel disorders is progressing at an ever increasing rate. Many were identified during the course of writing this book, so that at times I felt like the Red Queen in Alice in Wonderland, who had to run ever faster in order to remain in the same place. I am confident that many more channelopathies will be found after this book has finally gone to press. For this reason, I have included a brief account of those ion channels—like the P2X receptors—which have been characterized at the molecular level but not (yet) been linked to any human disease.

Frances Ashcroft
Oxford, UK
April 1999

ACKNOWLEDGMENTS

I am grateful to a large number of people for help with this book. First, to my research group and my students who have been very patient and understanding throughout the long time the book took to write, and whose critical advice both on its content and style has been invaluable. In particular, Fiona Gribble, Trude Haug, Frank Reimann, Stefan Trapp and Emma Wells read almost all the chapters, Rebecca Ashfield and Stephen Tucker checked the molecular biology sections and Stephen also kept me up to date with the latest references. Many of my colleagues have read individual chapters concerned with their own areas of expertise. My special thanks are due to: John Adelman, Stephen Ashcroft, Peter Agre, Heinrich Betz, Richard Boyd, Buzz Brown, Stephen Cannon, Bill Catterall, Helen Chapel, Roger Cox, Keith Dorrington, Declan Doyle, Barbara Ehrlich, Mark Gardiner, Andy Harris, Rob Harvey, Steve Hebert, Michael James, Thomas Jentsch, Christof Korbmacher, Johannes Krupp, Nancy Lane, Frank Lehman-Horn, Claire Newland, Denis Noble, Richard Olver, Carl van Os, David Patterson, Mauro Pezzia, Michael Sanguinetti, Jochen Röper, Bernard Rossier, Mark Sansom, Peter Schofield, Paul Smith, Nigel Unwin, Angela Vincent and Gary Yellen. They have helped to ensure the text is as correct and up to date as possible. It goes without saying that the mistakes that remain are entirely my own. I also thank Peter Agre, Andrew Engel, Eric Hoffman, Nancy Lane, and David Julius for providing photographs. My niece, Jenny Rye, painstakingly checked all the references and my secretary, Jenny Griffiths, photocopied them all. In a book like this, it is simply not possible to provide a comprehensive review of all the literature and I apologize to those of my colleagues whose work, for reasons of space, I was unable to refer to. Finally, I am indebted to two key individuals. First, to Fiona Gribble, who drew most of the diagrams, with great skill, whilst still managing to continue her research. And secondly, to my publisher, Graham Lees, who persuaded me to write the book in the first place and without whose constant encouragement and support it would never have been completed. I thank them both.

CHAPTER 1

INTRODUCTION

Cell membranes are composed of two layers of lipid molecules and are therefore relatively impermeable to ions. Consequently, ion transport into and out of the cell across the surface membrane, or between different intracellular compartments, is mediated by membrane proteins known as ion channels, pumps, and transporters. This book focuses on the first group of proteins. Ion channels are considered to be gated pores whose opening and closing may be intrinsic or regulated (for example, by ligand binding or changes in the voltage gradient across the membrane). They are found in the membranes of all animal, plant and bacterial cells and play important roles in such diverse processes as nerve and muscle excitation, hormonal secretion, learning and memory, cell proliferation, sensory transduction, the control of salt and water balance and the regulation of blood pressure. Defects in ion channel function may therefore have profound physiological effects.

Ion channel diseases (*channelopathies*) may arise in a number of different ways. First, mutations in the promoter region of an ion channel gene may cause underexpression or overexpression of the channel protein. Second, mutations in the coding region of ion channel genes may lead to the gain or loss of channel function, either of which may have deleterious consequences. For example, mutations producing enhanced activity of the epithelial Na^+ channel are responsible for Liddle's syndrome, whereas mutations that cause reduced channel activity give rise to pseudohypoaldosteronism type 1 (Chapter 13). Third, some diseases result from defective regulation of channel activity by intracellular or extracellular ligands or modulators, due to mutations in the genes encoding the regulatory molecules themselves, or defects in the pathways leading to their production. This is the case for some forms of diabetes mellitus (see Chapter 8). Fourth, autoantibodies to channel proteins may produce disease either by downregulating or by enhancing channel function (Chapter 18). Fifth, ion channels may act as lethal agents, being secreted by cells and inserting into the membrane of a target cell to form large nonselective pores that cause cell lysis and death. Complement and the haemolytic toxin produced by the bacterium *Staphylococcus aureus* are examples of this type of ion channel (Chapter 23). Finally, ion channels are targets for a large and diverse group of toxins that mediate their effects by

1

enhancing or inhibiting channel function. The high-affinity and specificity of these toxins have led to their use as ligands for the purification of ion channel proteins. They are not considered in detail in this book. The physiological importance of ion channel activity is also exemplified by the fact that many therapeutic drugs mediate their effects by interaction with ion channel proteins.

The majority of the diseases discussed in this book arise from mutations in ion channel genes. These naturally occurring mutations have provided fresh insights into our understanding of the functional roles of ion channels and have been valuable in defining functionally important domains of these proteins. There are also a number of instances where the genetic analysis of a disease has led to the cloning of a novel ion channel. The first K^+ channel to be identified (*Shaker*), for example, came from the cloning of the gene which causes the fruitfly *Drosophila* to shake when exposed to ether. The converse is also true. The plethora of studies on the relationship between the structure and function of the voltage-gated Na^+ channel, for example, has greatly assisted our understanding of how mutations in this protein produce their clinical phenotypes.

Many ion channel diseases are genetically heterogeneous and the same clinical phenotype may be caused by mutations in different genes. For example, mutations in three different genes give rise to LQT syndrome, a relatively rare cardiac disorder that causes sudden death from ventricular arrhythmia in young people (see Chapters 5 and 6). Conversely, different mutations in the same gene may result in very different clinical phenotypes: episodic ataxia type 2, familial hemiplegic migraine, and spinocerebellar ataxia type 6 all result from mutations in the voltage-gated Ca^{2+} channel, *CACNL1A4* (Chapter 9).

It is worth emphasising that, as is the case with all single gene disorders, the frequency of most channelopathies in the general population is very low. However, the insight they have given into the relationship between ion channel structure and function, and into the physiological role of the different ion channels, has been invaluable. As William Harvey said in 1657 "nor is there any better way to advance the proper practice of medicine than to give our minds to the discovery of the usual form of nature, by careful investigation of the rarer forms of disease."

CHAPTER 2

FROM GENE TO PROTEIN

The central tenet of molecular biology is that 'DNA makes RNA makes protein'. In this chapter we consider how this is achieved.

The Structure of the Eukaryotic Gene

The structure of a typical human gene is shown in Fig. 2.1. It contains coding regions which are known as *exons,* separated by non-coding regions called *introns* which are usually of greater length than the exons. Upstream of the gene is a region known as the *promoter* which contains control elements that govern the expression of the gene. Additional *enhancer* elements also exist: their position and orientation with respect to the promoter is variable. Both the promoter and the enhancer bind regulatory proteins known as *transcription factors* that either enhance (usually) or inhibit transcription. The activity of the transcription factors determines whether or not a given gene is expressed in a specific cell type. Transcription factors may be regulated themselves by cytosolic second messengers such as protein kinases, which mediate hormonal effects on gene expression.

DNA is Transcribed into RNA

The genetic information in DNA is contained in a linear sequence of four types of deoxyribonucleotides containing the bases adenine (A), thymine (T), guanine (G) or cytosine (C). Each deoxyribonucleotide is linked to the next one through the phosphate groups attached to the sugar moieties, the 5'-carbon of one attaching to the 3'-carbon of the next (Fig. 2.2). Thus one end of a DNA molecule has a free 5'-carbon at its end and the other has a free 3'-carbon: these are known as the 5'- and 3'-ends. Two DNA strands pair to form an anti-parallel double helix, each nucleotide in one helix being paired with a complementary one in the other helix, adenine with thymine and guanine with cytosine. During DNA replication, the double helix unwinds so that the bases are no longer paired and two new DNA molecules can then

3

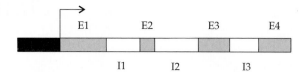

FIGURE 2.1 STRUCTURE OF A EUKARYOTIC GENE

Genes comprise one or more coding regions called exons (E, hatched) separated by non-coding regions called introns (I, white). The promoter (black) sits at the start of the gene and regulates its expression.

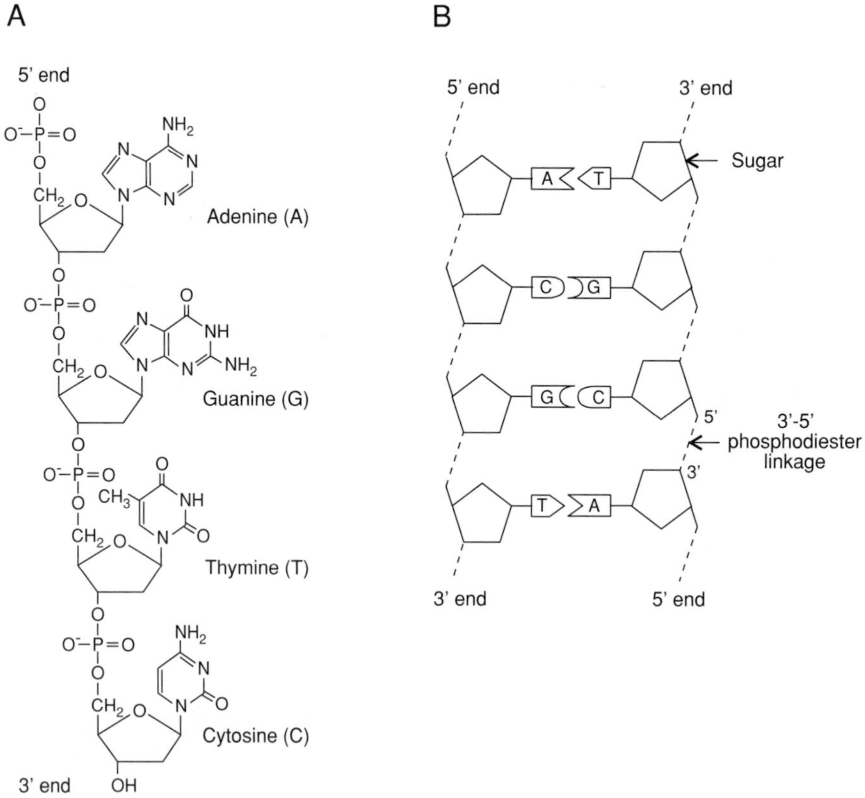

A

B

FIGURE 2.2 STRUCTURE OF DNA

(*A*) Four kinds of deoxyribonucleotides make up DNA. Each consists of a base (A, G, T or C), a ribose moiety and a phosphate group. The sugars are linked together by the phosphate groups, forming a sugar-phosphate backbone with a free 3'-carbon at one end and a free 5'-carbon at the other. (*B*) Two DNA strands are connected in an antiparallel fashion by complementary base pairing: adenine with thymine and cytosine with guanine.

be synthesised using each of the original DNA strands as templates. In a similar way, the DNA molecule serves as the template for synthesis of RNA, a process called *transcription*. The messenger RNA (mRNA) molecule is complementary in sequence to the DNA but contains uracil (U) instead of thymine. Following transcription, the mRNA is exported from the nucleus to the cytoplasm where it acts as a template for protein synthesis. In this way the information encoded in the DNA sequence is *translated* into protein.

Transcription of DNA into RNA usually is unidirectional—the resulting RNA molecule can only be used for protein synthesis. However, the *reverse transcription* of RNA into DNA is not completely forbidden. It occurs, for example, in retroviruses. The genetic information of these viruses is stored as RNA but is converted to DNA within the host cell and inserted into the host genome. The enzyme that catalyses this process (reverse transcriptase) has been of considerable value for molecular biological studies. Translation of RNA into protein is always irreversible.

mRNA Processing

mRNA undergoes several changes before protein synthesis takes place (Fig. 2.3). These modifications take place within the nucleus and include the addition of nucleotides to enhance mRNA stability, the removal of non-coding regions and the editing of the sequence at specific sites. They also mark the mature mRNA for transport to the cytoplasm.

During transcription, an RNA copy of the full-length gene, containing both introns and exons, is first made. This is known as the pre-mRNA. Subsequently, the introns are cut out and the exons spliced together to produce a translationally active mRNA that is exported from the nucleus. The

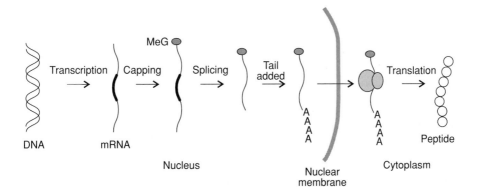

FIGURE 2.3 PROCESSING OF mRNA
The messenger RNA undergoes several modifications before it leaves the nucleus. It is capped with a methylated guanosine (MeG), the non-coding regions are removed by cutting and splicing the mRNA, and a poly-A tail is added. In rare cases, mRNA editing may also occur.

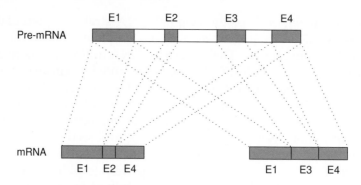

FIGURE 2.4 ALTERNATIVE SPLICING
Different ion channel proteins may be obtained by alternative splicing of
a single gene. E, exon.

exons always appear in the same order as in the gene but the final mRNA
does not necessarily contain them all: some of them may be left out in a
process known as *alternative splicing* (Fig. 2.4). Different mRNAs and thus
different proteins therefore result, contributing to channel diversity. The
Shaker K$^+$ channel, for example, comes in a number of splice variants.

Further modification of the mRNA molecule may occur by editing of
its sequence just prior to translation. This may have dramatic functional
consequences. For example, *mRNA editing* converts the AMPA receptor channel from a Ca^{2+}-permeable to a Ca^{2+}-impermeable one, by substitution of a
single nucleotide (Chapter 16).

During maturation of the mRNA, one end is capped with a methylated
guanosine and a tail of several hundred adenosines is added at the other
end. These end regions govern the efficiency with which the mRNA is translated and the rate at which it is degraded in the cytoplasm. The poly-A tail
is also of practical use to the experimenter, because it may be used to affinity
purify mRNA by using a poly-T column.

The Genetic Code

Each amino acid in a protein sequence is encoded by a set of three nucleotides
known as a *codon*. Since there are only 20 amino acids but 64 possible ways of
arranging four bases in a triplet codon, the genetic code contains considerable
redundancy in the third position of the codon (Table 2.1): for example, both
GAA and GAG code for glutamate. The codon AUG, which specifies methionine, also serves as the signal for initiation of protein synthesis: whether or
not AUG is read as the start signal, or as a methionine within the protein,
depends on a consensus sequence upstream of the initiator methionine. There
are also three stop codons that terminate translation and do not code for an
amino acid (UAA, UAG and UGA).

TABLE 2.1   THE GENETIC CODE

First position (5' end)	Second position				Third position (3' end)
	U	C	A	G	
U	UUU Phe	UCU Ser	UAU Tyr	UGU Cys	U
	UUC Phe	UCC Ser	UAC Tyr	UGC Cys	C
	UUA Leu	UCA Ser	UAA Stop	UGA Stop	A
	UUG Leu	UCG Ser	UAG Stop	UGG Trp	G
C	CUU Leu	CCU Pro	CAU His	CGU Arg	U
	CUC Leu	CCC Pro	CAC His	CGC Arg	C
	CUA Leu	CCA Pro	CAA Gln	CGA Arg	A
	CUG Leu	CCG Pro	CAG Gln	CGG Arg	G
A	AUU Ile	ACU Thr	AAU Asn	AGU Ser	U
	AUC Ile	ACC Thr	AAC Asn	AGC Ser	C
	AUA Ile	ACA Thr	AAA Lys	AGA Arg	A
	AUG Met	ACG Thr	AAG Lys	AGG Arg	G
G	GUU Val	GCU Ala	GAU Asp	GGU Gly	U
	GUC Val	GCC Ala	GAC Asp	GGC Gly	C
	GUA Val	GCA Ala	GAA Glu	GGA Gly	A
	GUG Val	GCG Ala	GAG Glu	GGG Gly	G

Each amino acid is coded for by a triplet of nucleotides (a codon). The genetic code is redundant and some amino acids are coded for by several different triplets. There are three stop signals and AUG forms part of the initiation signal as well as coding for internal methionine residues.

RNA Is Translated into Protein

Translation of the nucleotide sequence of mRNA into the amino acid sequence of the protein it encodes is catalysed by the ribosome. This ribonucleoprotein particle usually attaches to the mRNA cap (at the 5'-end) and then tracks along the molecule until it encounters the start codon (AUG). It then moves along the molecule to the 3'-end, translating each codon in turn so that the protein is assembled one amino acid at a time, starting at the amino-terminus. This is achieved with the help of transfer RNA (tRNA) molecules, which interact both with the nucleotide codon of the mRNA and with the amino acid it specifies. Each tRNA recognises only a single type of amino acid which covalently attaches to its 3'-end. The tRNA also possesses a trinucleotide sequence known as an anticodon, which is complementary to the mRNA codon representing the amino acid and enables the two molecules to interact by base-pairing. Amino acids are thus delivered to the growing polypeptide chain in the order specified by the mRNA.

The translation machinery of eukaryotic cells will accept mRNA from many different tissues and species. This enables alternative cell types, usually known as heterologous expression systems, to be used for the expression of ion channel proteins experimentally. One favoured by electrophysiologists is the *Xenopus* oocyte.

Protein Structure

The conformation of a protein may be described at several levels of complexity. The amino acid sequence is known as its primary structure, the secondary structure refers to the folding of the peptide backbone into α-helices and β-sheets, and the tertiary structure describes the three-dimensional arrangement of the polypeptide. Where a protein is made up of several subunits, its overall conformation, including the subunits, is known as the quaternary structure. The structure of a protein is determined by its amino acid sequence. Theoretically, therefore, it should be possible to predict the three-dimensional structure of a protein simply from its primary sequence. In practice, this goal has not yet been achieved for any protein because the laws governing protein folding have not yet been elucidated.

Primary structure

Individual amino acids are linked together to form a polypeptide chain by peptide bonds, in which the carboxyl group of one amino acid is attached to the amino group of the next (Fig. 2.5). Each polypeptide thus has a free NH_2 group at one end (the amino or N terminus) and a free COOH group at the other (the carboxyl or C terminus). Conventionally protein sequences are written from the N to the C terminus. The side groups of the amino acids protrude from the peptide backbone and vary with the amino acid. Twenty

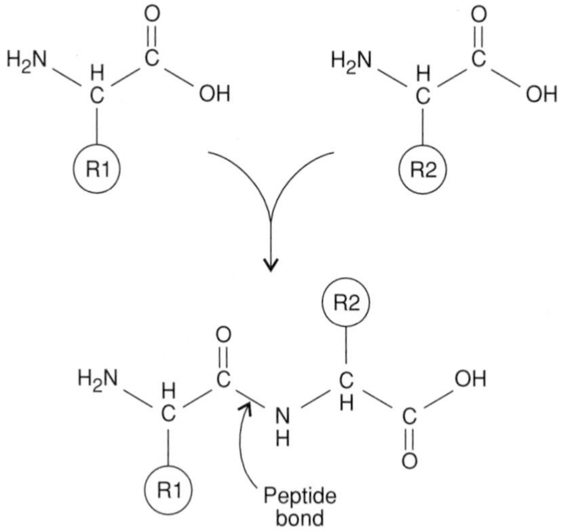

FIGURE 2.5 AMINO ACIDS ARE LINKED BY PEPTIDE BONDS TO FORM A POLYPEPTIDE

Amino acids are linked together through a peptide bond between the $COOH^-$ group of one residue and the NH_2^+ group of the next. R1 and R2 indicate the amino acid side-chains.

different kinds of amino acids are commonly used in proteins (Table 2.2). These may be positively charged (acidic), negatively charged (basic), polar but uncharged or non-polar. Non-polar residues are hydrophobic and stretches of hydrophobic residues therefore tend to be found in the transmembrane domains of ion channel proteins, whereas charged and polar residues are more likely to be located in extracellular or intracellular regions of the protein. Proline has a rigid ring structure and introduces a bend in the polypeptide chain. It tends to be found at the end of an α-helix and in ion channels is sometimes located at the ends of the transmembrane domains. Serine, threonine and tyrosine residues can be modified by the attachment of a negatively-charged phosphate group (*phosphorylation*). This change in the electrostatic properties of the amino acid may alter the local structure, and thereby the function, of the protein. Many types of ion channel are modulated by phosphorylation. *Glycosylation*, the attachment of a sugar residue to the side-chain of an amino acid, may also alter the structure of the protein because the sugar residues tend to be extremely large and bulky. Because glycosylation is only found on extracellular residues, its presence may be helpful in determining the membrane topology of an ion channel.

TABLE 2.2   **PROTEINS ARE MADE UP OF 20 DIFFERENT AMINO ACIDS**

Type	Amino acid	Side chain	Abbreviations		Hydropathy index
Nonpolar	Isoleucine	$-CH(CH_3)CH_2CH_3$	Ile	(I)	4.5
	Valine	$-CH(CH_3)_2$	Val	(V)	4.2
	Leucine	$-CH_2CH(CH_3)_2$	Leu	(L)	3.8
	Phenylalanine	$-CH_2C_6H_5$	Phe	(F)	2.5
	Methionine	$-CH_2CH_2SCH_3$	Met	(M)	1.9
	Alanine	$-CH_3$	Ala	(A)	1.8
	Tryptophan	$-CH_2C(CHNH)C_6H_4$	Trp	(W)	-0.9
	Proline	$-CH_2CH_2CH_2-$	Pro	(P)	-1.6
Uncharged polar	Cysteine	$-CH_2SH$	Cys	(C)	2.5
	Glycine	$-H$	Gly	(G)	-0.4
	Threonine	$-CH(OH)CH_3$	Thr	(T)	-0.7
	Serine	$-CH_2OH$	Ser	(S)	-0.8
	Tyrosine	$-CH_2C_6H_4OH$	Tyr	(Y)	-1.3
	Histidine	$-CH_2C(NHCHNCH)$	His	(H)	-3.2
	Glutamine	$-CH_2CH_2CONH_2$	Gln	(Q)	-3.5
	Asparagine	$-CH_2CONH_2$	Asn	(N)	-3.5
Acidic	Aspartic acid	$-CH_2COO^-$	Asp	(D)	-3.5
	Glutamic acid	$-CH_2CH_2COO^-$	Glu	(E)	-3.5
Basic	Lysine	$-(CH_2)_4NH_3^+$	Lys	(K)	-3.9
	Arginine	$-(CH_2)_3NHC(NH_2)NH_3^+$	Arg	(R)	-4.5

Amino acids are indicated by both the triple and single letter nomenclature. The hydropathy index is that of Kyte and Dolittle (1982). After Aidley and Stanfield (1996)

Secondary structure

Two main types of secondary structure are found in ion channel proteins: the α-helix and the β-sheet. Both result from hydrogen bonding between the NH and C=O groups of the polypeptide backbone. The α-helix is stabilized by hydrogen bonds formed between the C=O group of one peptide bond and the NH group of a peptide bond four residues further along the polypeptide chain (Fig. 2.6). This twists the polypeptide backbone into a right-handed

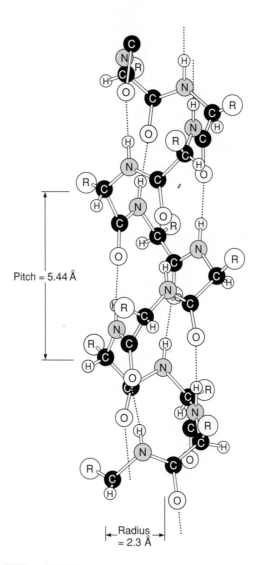

Pitch = 5.44 Å

Radius = 2.3 Å

FIGURE 2.6 THE α-HELIX
The structure of a right-handed α-helix. Carbon, nitrogen and oxygen atoms are labelled. Hydrogen ions are shown as small spheres. The amino acids side-chains are represented by R. The dotted lines indicate hydrogen bonds. After Lewin (1997).

helix which rises one full turn every 3.6 residues. It has a pitch of 5.4 Å, so there is a rise of 1.5 Å per residue. An α-helix of 20 amino acids should therefore be ~30 Å long, which is sufficient to span the membrane. It is believed that many of the transmembrane domains of ion channels are formed from α-helices. Some amino acids (for example, alanine, glutamate, leucine and methionine) are better at forming an α-helix than others. The β-sheet is formed by hydrogen bonding between the peptide bonds of amino acids in different polypeptide chains, or within a single polypeptide chain (Fig. 2.7). Parallel strands run in the same direction, whereas antiparrellel strands run

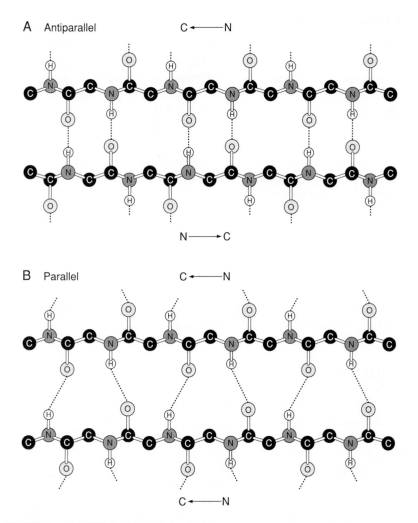

FIGURE 2.7 THE β-SHEET
The structure of the antiparallel (a) and parallel (b) β-sheet. Carbon, nitrogen and oxygen atoms are labelled. Hydrogen ions are shown as small spheres. The amino acids side-chains are represented by R. The dotted lines indicate hydrogen bonds: these may occur between the same, or two different, polypeptides.

in opposite directions. Glycine and alanine residues favour this type of structure. It is used to form the pore of several types of ion channel.

Non-covalent interactions such as ionic bonds, hydrogen bonding, hydrophobic interactions and van der Waals forces help to stabilize the three-dimensional structure of the protein. Covalent bonds may form, however, between the thiol (SH) groups of cysteine residues which link to form a disulphide bridge. Because the cytoplasm contains enzymes that break disulphide bonds, they are not usually found in the intracellular domains of ion channels.

Membrane insertion

The protein is inserted into the membrane of the endoplasmic reticulum at the same time that it is translated. The membrane topology is therefore determined during translation. The N terminus of some ion channels carries a short leader sequence that facilitates insertion of the protein into the membrane. This sequence is cleaved following membrane insertion leaving the N terminus outside the cell. Hydrophobic sequences in ion channel proteins which lack leader sequences are thought to play a similar role, stopping and starting insertion into the bilayer. The protein moves through the endoplasmic reticulum to the Golgi apparatus and from there to the plasma membrane. Glycosylation may take place *en route*.

BASIC GENETICS

The genetic makeup of an organism is known as its *genotype*. The physical appearance of an organism is known as its *phenotype*: it results from the interaction of its genetic makeup and the environment.

Chromosome Structure

Humans have 23 pairs of chromosomes. Twenty-two of these are non-sex chromosomes and are called *autosomes*, whereas the twenty-third pair are the sex chromosomes which may be either XX (in females) or XY in males. One chromosome of each pair comes from the female parent and the other from the male. Each chromosome consists of two identical sister *chromatids*. The chromatids are attached at the *centromere*, which divides the chromatid into a short arm (denoted p) and a long arm (called q). During *mitosis*, or asexual cell division, the chromatids separate, one being assigned to each daughter cell (Fig. 2.8). In this way, both daughter cells end up with a full complement of genetic material from each parent. They are referred to as *diploid* cells because they contain two chromosomes of each type. The chromatids are then duplicated, during the interval between cell divisions. *Meiosis* (the process of sexual division) differs from mitosis in that two successive cell divisions take

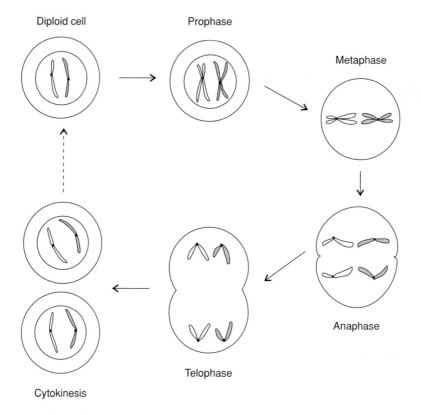

FIGURE 2.8 MITOSIS

Between cell divisions, the DNA exists as an uncoiled mass of chromatin. During prophase, the chromatin condenses to form well-defined chromosomes, each of which consists of two chromatids attached at the centromere. Metaphase begins with the breakdown of the nuclear envelope. The chromosomes are then pulled towards the centre of the cell by specialised spindle microtubules which attach to the centromere: they eventually line up halfway between the poles of the cell. Anaphase is initiated by the separation of the two chromatids of the chromosome and their movement to opposite ends of the cell. During telophase the chromatids arrive at the poles of the cell and the nuclear envelope begins to reform. The two daughter cells then separate, in a process called cytokinesis. Subsequently, the chromatids are replicated.

place so that each of the four daughter cells ends up with only a single chromatid from each pair of chromosomes (Fig. 2.9). These are known as *haploid* cells. The sorting of the chromatids to the four daughter cells is random. Chromosomes only exist as discrete structures during the process of cell division; between divisions the DNA uncoils to form a dense mass known as chromatin.

Within each chromosome, the genes are arranged in a linear sequence. Each gene has two copies, one on each chromosome of a pair. These are

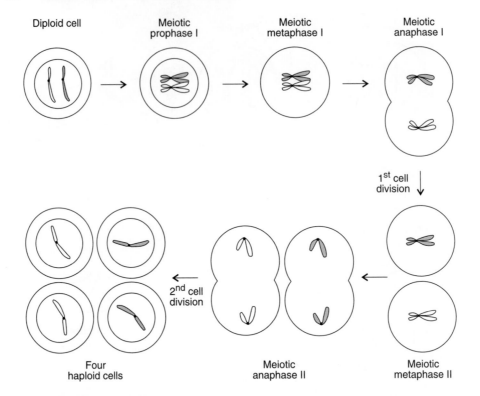

Diploid cell   Meiotic prophase I   Meiotic metaphase I   Meiotic anaphase I

1st cell division

2nd cell division

Four haploid cells   Meiotic anaphase II   Meiotic metaphase II

FIGURE 2.9 MEIOSIS

Meiosis involves two cell divisions. Each chromosome first replicates to produce two identical sister chromatids, as in an ordinary cell division. The process of meiosis and mitosis then diverge. In meiosis, each pair of chromatids joins up with the homologous pair of chromatids derived from the other parent. Recombination, in which a fragment of the maternal chromatid is exchanged for that of the corresponding paternal chromatid, may then take place. In this way, the genetic material is rearranged during meiosis. Subsequently, the chromosomes divide, one of each pair moving to the opposite end of the cell. A second cell division then occurs without further DNA replication, so that four cells are produced, each with half the DNA content of the original cell.

known as *alleles*. Mitosis ensures that each daughter cell contains both alleles. By contrast, only a single allele is passed on to each daughter cell at meiosis. This is derived randomly from either the male or female parent. The definition of an allele is one of two or more versions of the same gene with different sequences. Thus Δ508 and G551D are both alleles of the CFTR gene.

There are at least three different kinds of chromosome map (Fig. 2.10). The first of these is the *banding pattern*, which is obtained by staining the chromosome with a chemical dye. The dye stains the chromosome unevenly, generating a distinct series of bands. In this book, the location of a gene is usually given with reference to the chromosome banding pattern. For example, the sodium channel gene *SCN4A* is located on chromosome 17q23–25. This means that it

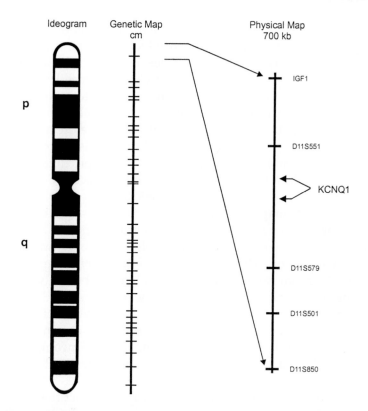

FIGURE 2.10 RELATIONSHIP BETWEEN DIFFERENT KINDS OF
CHROMOSOME MAPS
The banding pattern (ideogram) of chromosome 11 is shown on the left.
The genetic map is given in the centre: the horizontal lines indicate the
location of different polymorphic markers. On the right is shown the
physical map of the section of chromosome corresponding to the 11p15.1
band, with the positions of various markers and of the K⁺ channel gene
KCNQ1 indicated. Adapted from Wang *et al.* (1996).

is found on the long arm of chromosome 17, in the region encompassed by
bands 23–25. A finer *genetic map* is provided by linkage analysis (see later) which
locates the position of a gene with respect to reference markers on the chromosome. These genetic markers now cover the whole of each human chromosome
at intervals of a few centimorgans (one centimorgan corresponds to around 90-
100 million base-pairs). The *physical* map of the chromosome refers to its DNA
sequence and identifies the precise location of a gene: it is measured in base
pairs (bp). The human genome project aims to obtain the complete sequence
of each chromosome by the year 2003.

Chromosomal Modifications

Homologous recombination is a term used to describe the process in which
genes are exchanged between the two non-sister chromatids of a chromosome

pair during meiosis (Fig. 2.11). It results in a double-stranded DNA molecule containing a stretch of hybrid DNA, in which one strand is derived from each parent. This means that the two alleles of a gene, which originate from different parents, can be exchanged between chromosomes. It thus contributes to genetic variability. Recombination is more likely to take place between genes that are far apart on the chromosome than between genes that are close together. The frequency of recombination is therefore a measure of the distance between two genes and can be used to establish a *genetic map* of the chromosome. A recombination frequency of 1% is approximately equivalent to 1 centimorgan, which means the genes recombine once in every 100 meiotic events. The genetic map built up from linkage analysis is not necessarily the same as the physical map of a chromosome (its DNA sequence), as different parts of the chromosomes vary widely in their tendency to recombine.

Females possess two X chromosomes. However, only one is active and the other is switched off. This *X-chromosome inactivation* compensates for the fact that two X chromosomes are present in females but only one in males. The X chromosome which is switched off may vary between different tissues. It is therefore possible for a woman to be heterozygous for a dominant disease-causing mutation in a gene carried on the X chromosome and yet not exhibit symptoms. It is possible this accounts for the incomplete dominant inheritance of X-linked disorders such as Charcot-Marie-Tooth disease (Chapter 20).

Genomic imprinting is a term used to describe the fact that an allele may behave differently according to whether it has been inherited from the male or from the female parent. Imprinting is thought to result from DNA methylation which turns the imprinted gene off in either the oocyte or sperm; this inhibition is maintained during any successive mitosis. Imprinting is reset during gametogenesis in every generation. There is evidence to suggest that the $GABA_A$ receptor β_3 subunit might be an imprinted gene (Chapter 18).

Genetics of Disease

An *monogenic* disease is one in which the disease is produced by a single gene while a *polygenic* disease is one in which multiple genes contribute to the disease phenotype. Many of the ion channel diseases described in this book result from mutation of a single gene. An *autosomal dominant* disease is one in which the phenotype is determined by the possession of a single mutant allele on a non-sex chromosome (an *autosome*). Thus if one parent carries a dominant mutation, the child has a 50% chance of inheriting it. The extent to which an autosomal dominant gene is expressed in individuals who possess this mutation is referred to as *penetrance*. An *autosomal recessive* disease is one in which only *homozygotes* who carry two mutant alleles are symptomatic. *Heterozygotes* who carry one normal allele and one mutant allele are entirely asymptomatic: they are denoted as *carriers* of the disease as they can pass on their mutant allele on to their children. The child of two carriers has

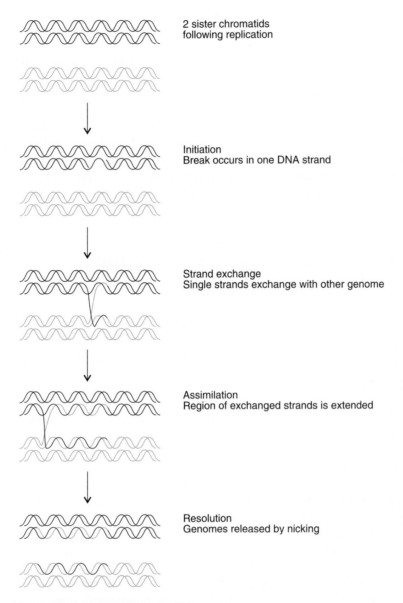

2 sister chromatids
following replication

Initiation
Break occurs in one DNA strand

Strand exchange
Single strands exchange with other genome

Assimilation
Region of exchanged strands is extended

Resolution
Genomes released by nicking

FIGURE 2.11 RECOMBINATION
Recombination is initiated by breakage of the double-stranded DNA molecule and the generation of a free 3'-end of single-stranded DNA. This is
followed by reunion of the free end with the complementary DNA strand
in one of the chromatids of the other chromosome. The region of exchange
is then extended. Finally, the invading DNA strands are cut and the free
ends ligated with those of the recipient chromatids. This results in the
exchange of a short stretch of single-stranded DNA between two non-
sister chromatids of a chromosome pair. After Lewin (1997).

a one in four chance of inheriting a mutant allele from each parent and thus of being affected by the disease.

Mutations may result in either gain of function or loss of function. A dominant disease that results in loss of function, may arise in one of two ways. First, a 50% reduction in the protein may not be sufficient for normal function. Second, the mutant protein may inactivate the wild-type protein and reduce its functional activity. In the case of an ion channel, this *dominant negative* effect may occur if the presence of a single mutant subunit within a multimeric channel complex results in defective channel function. Individuals heterozygous for a dominant disease will express both normal and mutant subunits in the same cell, and heteropolymerization will lead to defective channels.

Mutations May Cause Defective Ion Channels by Different Mechanisms

Conventionally, a nucleotide change that occurs in less than 1% of the population is known as a *mutation*. If it occurs more frequently, it is called a *polymorphism* (from the Greek, meaning multiple forms). A number of human or animal diseases result from mutations in ion channel genes and such natural mutations have provided a large amount of information about the relationship between channel structure and function.

Mutations within the coding region may alter the protein sequence (Fig. 2.12). The genetic code is read as a sequence of non-overlapping triplets. Thus there are three possible ways of translating a DNA sequence, which are determined by the nucleotide with which you start. These are known as *reading frames*. A mutation that substitutes one nucleotide for another may alter a single amino acid within the protein, but it will not affect the reading frame and the subsequent sequence will be translated correctly. If substitutions occur in the first or second nucleotide of a codon, they nearly always result in an amino acid change. Substitutions that occur at the third position of a codon do not necessarily alter the amino acid sequence and may therefore have no functional effect. A mutation that inserts or deletes a single base pair will change the reading frame of the subsequent sequence. Such a *frameshift* means that the amino acid sequence of the protein will be completely different after the site of the mutation. The protein is therefore likely to be nonfunctional. Furthermore, in many cases a novel stop codon may be introduced so that mutant protein is truncated prematurely. Because of the triplet code, the insertion (or deletion) of three nucleotides does not produce a frameshift; nor does the combination of an insertion and a deletion.

Mutations can also occur within introns, or at the exon-intron boundaries. This may prevent the splice site from being recognised and lead to incorrect splicing of the gene. Generally, such splice-site mutations result in the introduction of an intron in the mRNA. The intron is then translated so that the subsequent protein sequence is incorrect. Translation continues until a novel

original sequence
codon   ACC ATC GGT TAT GGC
amino acid   T I G Y G

point mutation
codon   ACC ATC **AGT** TAT GGC
amino acid   T I **S** Y G

Nonsense mutation
codon   ACC ATC AGT **TAG** GGC
amino acid   T I G *

frameshift
codon   ACC **GAT CGG TTA TGG** C
amino acid   T **D** **R** **L** **W**

FIGURE 2.12 *MUTATIONS*
A triplet of nucleotides codes for a single amino acid. Mutation of a single
nucleotide (point mutation) may result in substitution of a novel amino
acid for that in the original sequence. Occasionally, a mutation may result
in the introduction of a stop codon (nonsense mutation) which results in
truncation of the protein. Insertion (or deletion) of a nucleotide produces
a frameshift that results in a change in all subsequent amino acids.

stop codon is encountered and as these occur more frequently within introns,
the protein is often prematurely truncated. Incorrect splicing may lead to
episodic ataxia type II (Chapter 9) and PHHI (Chapter 8). Mutations may
also occur within the promoter or enhancer regions of the gene and cause
aberrant expression. To date, however, no disease-causing mutations have
been reported in the promoter region of ion channel genes.

Transposons are wandering pieces of DNA that can insert randomly into
the chromosome. If this occurs within a coding region, it has deleterious
consequences. Fortunately, because <0.1% of human DNA is exonic and the
rest is either intronic or intervening DNA between genes, this is not a common
occurrence. Mammalian transposons come in two types: short intervening
nucleotide elements (SINES) and long intervening nucleotide elements
(LINES). If a transposon inserts into the chromosome of a germ cell, it
will be inherited along with the normal genome. Examples of channel diseases produced by transposons include murine myotonia (Chapter 10) and
hyperekplexia (Chapter 17).

CHAPTER 3

HOW ION CHANNELS WORK

In order to understand how defective ion channels give rise to disease it is necessary to understand some of their basic properties. In this chapter we first consider the properties of the single channel, then examine how single-channel currents summate to produce macroscopic currents, and finally explore how macroscopic currents produce membrane potential changes. This will provide the basis for a more detailed discussion of the properties of specific ion channels in later chapters. The goal of this chapter is to provide a very basic introduction to ion channel properties rather than a comprehensive review. Those readers with some knowledge of the biophysics of ion channels may therefore prefer to skip this chapter. More detailed treatments can be found in the excellent books by Hille (1992) and Aidley and Stanfield (1996).

PROPERTIES OF SINGLE-CHANNEL CURRENTS

Let us first look at those channel properties that can be observed by recording the activity of only one channel, at a fixed membrane potential. A typical *single-channel current* recording is shown in Fig. 3.1. The quiet part of the current record represents the time the channel is closed. At the points indicated by the arrows, the channel opens for a short interval and then closes; after a variable length of time, it reopens again. This opening and closing of the channel is known as *gating*. The length of time the channel remains open or closed is not constant but instead varies randomly. For this reason, it is important to determine the *mean open time* and *mean closed time* by measuring the duration of a large number of openings and closings. The *open probability* of the channel is defined as the fraction of time the channel spends in the open state and can be obtained by dividing the sum of all the open times by the duration of the recording. Thus an open probability of 0.5 indicates that the channel spends, on average, 50% of its time in the open state.

The recording illustrated in Fig. 3.1 depicts several opening events. It is apparent from these that the amplitude of the current that flows when the

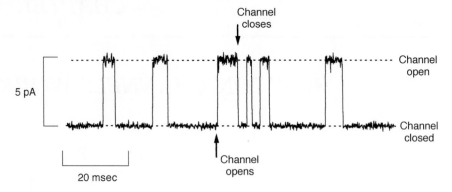

FIGURE 3.1 SINGLE-CHANNEL CURRENTS
Ion channels act as gated pores in the cell membrane. When the channel
opens (first arrow), ions move through it creating a tiny current, shown
as an upward deflection of the trace. At a fixed membrane potential, as
here, the amplitude of the single-channel current is constant. The length
of time the channel remains open, however, is variable. Simulated single-
channel current records supplied by Paul Smith.

channel is open is constant (provided that, as here, the membrane potential
does not change). The channel is therefore said to have a single conductance
state, i.e., it is either open or closed. Although most channels exhibit a single
conductance state, like this one, it is not true of all of them. Glutamate
channels, for example, have multiple conductance states and several different
current amplitudes can be recorded from the same channel under identical
conditions (Chapter 16). The magnitude of the single-channel current is determined by the ion concentrations on each side of the membrane, by the potential across the membrane, and by the ease with which the ion can move
through the channel pore (its *permeability*).

Permeation

When an ion channel opens, permeant ions are able to move through it. The
direction in which they move is governed by the electrochemical gradient
across the membrane. The rate at which they move is influenced not only
by the electrochemical gradient but also by the permeability of the channel
to the permeating ion. We will first consider the role of the electrochemical
gradient. This represents the sum of the chemical gradient across the membrane and the electric field experienced by the ion.

The electrochemical gradient

Imagine a container divided into two equal parts by an impermeable
membrane. One compartment contains a high concentration of an uncharged

molecule and the other contains a very low concentration. If the membrane separating the two compartments is now made permeable to the molecule, the molecules will move down their concentration gradient from the high concentration to the lower one. Now consider what happens if the compartments contain ions. Ions are charged and their movement will therefore be influenced not only by the concentration gradient they experience but also by the electric field across the membrane. For example, positively charged ions will be attracted towards a negative charge and repelled by a positive one. The total force on an ion will therefore be determined by the combined effects of the chemical and electrical gradients. Consider Fig. 3.2, which shows a model cell in which compartment A contains a high concentration of K^+ ions (as is found inside a cell) and compartment B contains a low K^+ concentration (as is found in plasma). If the membrane separating these compartments is made permeable to K^+, as would occur if K^+ channels were to open, the cation will tend to move down its concentration gradient from compartment A to compartment B. This will result in an increase in positive charge in compartment B, which will tend to oppose the further movement of K^+ ions. In other words, the movement of K^+ creates a potential difference between the two compartments and K^+ flux will cease once this electrical gradient exactly balances the chemical one.

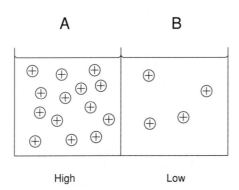

FIGURE 3.2 ION MOVEMENTS ARE DETERMINED BY THE ELECTROCHEMICAL GRADIENT

Compartments A and B contain a high and a low concentration of K^+ ions, respectively. If the barrier (the membrane) separating them is suddenly made permeable to potassium, then K^+ ions will move from A to B down their concentration gradient. Their movement will also be influenced by the voltage difference between the two compartments. For example, if compartment B is positive with respect to compartment A, then K^+ movement will be opposed by the electric field. At the equilibrium potential, the electrical and chemical forces on the ion exactly match each other, and there is no net flux of K^+.

The equilibrium potential

It is evident from this discussion that there will be some potential at which the electrical force on the ion exactly balances the opposing force of the concentration gradient. This potential is known as the *equilibrium potential* of the ion and it is given by the Nernst equation.

For ion X,

$$E_X = \frac{RT}{zF} \ln \frac{[X]_o}{[X]_i} \tag{1}$$

where E_X is the equilibrium potential (in volts), R is the gas constant (8.314 $JK^{-1}mol^{-1}$), T is the absolute temperature (in degrees Kelvin), z is the valency (charge) of the ion (positive for cations and negative for anions), and F is the Faraday constant (96,500 C mol^{-1}). $[X]_o$ and $[X]_i$ are the external and internal concentrations of X, respectively.

Equation (1) is often written in a simplified form, where E_X is given in mV and the temperature is taken as 20 °C (or 293 °K):

$$E_X = \frac{58}{z} \log_{10} \frac{[X]_o}{[X]_i} \tag{2}$$

With physiological concentrations of 140 mM internal K^+ and 5 mM external K^+, the calculated K^+ equilibrium potential will therefore be -84 mV. This is close to the resting membrane potential of most cells (-60 to -100 mV) and indicates that their resting permeability is primarily determined by K^+-selective channels. It is apparent from Eq. (2) that if the external K^+ concentration is plotted on a logarithmic scale against the potassium equilibrium potential, E_K, the relationship will be a straight line with a slope of 58 mV for a 10-fold change in $[K^+]_o$. The intercept will be $58/(\log_{10}/[K^+]_i)$, so that when the intracellular and extracellular K^+ concentrations are equal the membrane potential will be 0 mV. Similarly, for a divalent cation such as Ca^{2+}, the slope of the relationship would be 29 mV per 10-fold change in $[Ca^{2+}]_o$.

Some channels are permeable to more than one ion. The equilibrium potential of the single-channel current is then a function both of the electro-chemical gradients for the individual ions and of their relative permeabilities. For example, in the case of the nicotinic acetylcholine receptor, which is primarily permeable to both Na^+ and K^+, the equilibrium potential is given by

$$E = \frac{RT}{zF} \ln \left\{ \frac{P_{Na} [Na]_o + P_K [K]_o}{P_{Na} [Na]_i + P_K [K]_i} \right\} \tag{3}$$

where P_{Na} is the permeability of the channel to Na^+ and P_K its permeability to K^+. Permeability is discussed in more detail later.

Typical ion concentrations

It is clear from this discussion that the movement of an ion through an open channel depends both on the electric field across the membrane and

TABLE 3.1   ION CONCENTRATIONS AND EQUILIBRIUM POTENTIALS

Ion	Extracellular concentration (mM)	Intracellular concentration (mM)	Equilibrium potential (mV)
Sodium	135–145	12	+66
Potassium	3.5–5	140	−93
Calcium	2.25–2.52	10^{-7} M (free)	+135
Chloride	115	2.5–50	−42
pH (arterial)	7.37–7.42	7.1–7.2	−15

Extracellular ion concentrations refer to the typical range found in human blood. Intracellular ion concentrations are given for a typical mammalian cell. The equilibrium potentials are calculated for 37°C using Eq (1) and the middle of the concentration range where one is indicated.

on the ion concentrations on each side of the membrane. Table 3.1 gives the normal concentration ranges of the ions in human plasma and the intracellular ion concentrations typical of mammalian cells. These are maintained by a battery of pumps and transporters in the plasma membrane: some of these simply exchange external ions for internal ones (as in the case of the Cl/HCO_3 exchanger), whereas others use the energy of ATP to pump ions uphill, against their concentration gradients. The best known of these is the Na/K-ATPase, which in many cells accounts for as much as 60% of the total ATP consumption. It transports three Na^+ out of the cell for every two K^+ that enter, resulting in the net extrusion of one charge per pump cycle. The Na/K-ATPase is therefore said to be electrogenic. It is worth emphasising that the outward current generated by the Na/K-ATPase contributes <5 mV hyperpolarization to the resting potential in most cells. Its chief function is to maintain the ion concentration gradients.

The current–voltage relation

Let us now consider what happens when the membrane potential experienced by an ion channel is changed. Figure 3.3 shows a typical single-channel *current–voltage relationship* obtained by plotting the current amplitude against the applied membrane potential. If the ion concentrations on either side of the membrane are not very different, as in this case, the current flow through the open channel is determined by Ohm's law. In other words

$$i = \frac{V}{R} \tag{4}$$

where i is the single-channel current (in amps), V is the voltage gradient across the membrane (in volts), and R is the resistance to current flow through the open channel (in ohms). Ohm's law predicts a linear relationship between membrane potential and the single-channel current amplitude. The *single-channel conductance* (γ) is given by the slope of this relationship, as conductance is simply the reciprocal of resistance. Conventionally, the single-channel

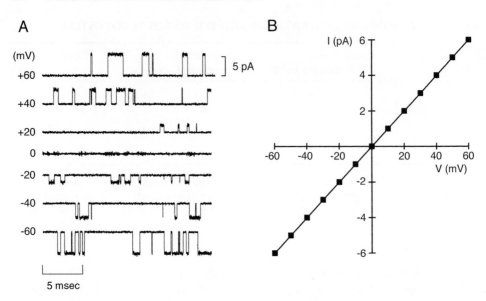

A

B

FIGURE 3.3 SINGLE-CHANNEL CURRENT–VOLTAGE RELATION
The amplitude of the single-channel current varies with membrane potential (*A*). When the solutions on either side of the membrane contain the
same concentration of permeant ions, the currents reverse in sign at a
membrane potential of 0 mV (the equilibrium potential of the ion). If the
membrane potential is made more positive, or more negative, the current
amplitude increases. Inward currents occur at negative membrane potentials and channel openings are shown as downward deflections of the
current trace. Outward currents are recorded at positive membrane potentials and are shown as upward deflections of the current trace. Plotting
the current amplitude against the membrane potential gives the current–
voltage relation (*B*). The slope of the line gives the single-channel conductance, in this case 100 pS. Simulated single-channel current records supplied by Paul Smith.

current is indicated by *i* and the macroscopic current by *I*, whereas the single-
channel conductance is denoted by γ and the macroscopic conductance by *G*.
The potential at which the current reverses in sign is, of course, its equilibrium
potential and, with the symmetrical ion concentrations used in Fig. 3.3, will
be 0 mV. A change in membrane potential to a more positive value is known
as *depolarization; hyperpolarization* is a change to more negative potentials.
    If the ion concentrations on the inside and outside of the membranes are
very different, Ohm's law does not hold. This is the case for Ca^{2+} because
the free intracellular Ca^{2+} concentration is <500 nM compared with a concentration of 2–5 mM in plasma (Table 3.1). Intuitively, it is easy to see that this
huge concentration gradient will make it less easy for Ca^{2+} to move out of
the cell, and thus that the single-channel current–voltage relationship is unlikely to be linear. The constant field equation, which takes ion concentration
gradients into account, gives a better description of the current–voltage rela-

tionship when intracellular and extracellular concentrations are very different. This states

$$I = P\, z^2\, \frac{VF^2}{RT}\, \ln \left\{ \frac{[X]_i - [X]_o\, \exp\,(-zVF/RT)}{1 - \exp\,(-zVF/RT)} \right\} \qquad (5)$$

where V is the membrane potential, P is the permeability of the channel to ion X (in cm^3/sec), $[X]_i$ and $[X]_o$ are the intracellular and extracellular concentrations of ion X, respectively, and z is the valency of ion X.

Block by impermeant ions

In some cases, the single-channel current-voltage relation is not linear even with equal concentrations of permeant ions on either side of the membrane. One reason this occurs is because of the presence of blocking ions. Inwardly rectifying K^+ channels are blocked by internal Mg^{2+} ions, which can enter the pore of the channel from the cytoplasmic side but are unable to pass through it into the extracellular solution (Matsuda et al., 1987; Vandenburg, 1987). They therefore block the outward movement of permeant ions such as K^+. The block is more intense at positive membrane potentials as the larger depolarization facilitates the movement of Mg^{2+} into the pore. Conversely, at hyperpolarized potentials, Mg^{2+} is prevented from entering the channel by the positive voltage field. This accounts for the characteristic inwardly rectifying current–voltage relation that gives these channels their name (see Chapter 8 for more details). Voltage-dependent block is not confined to inward rectifiers and is exhibited by many other types of channel. It may be caused by extracellular as well as intracellular cations (see Chapter 16).

Permeability and selectivity

The movement of an ion through an open channel is not only a function of its electrochemical gradient. It is also dependent on the relative permeability of the channel to the ion. The question of what determines the permeability properties of ion channels will be discussed in detail in the following chapters, but it is worth considering some general principles here. One factor is clearly the relative sizes of the pore and the ion: large ions are physically unable to permeate small pores. Thus as Hille (1992) has noted, ion channels act as molecular sieves. In most cases, however, ion channels do not simply act as pores that allow the free diffusion of all cations (or anions) below a certain size. Rather, they show considerable discrimination in the kinds of ions to which they are permeable. Thus Na^+ channels are highly permeable to Na^+ but not to K^+ ions, whereas K^+ channels are ~100 times more permeable to K^+ than to Na^+ ions. Because Na^+ has a smaller ionic radius than K^+, this high selectivity cannot be achieved simply by physical occlusion. Other factors must also be important. It is usually assumed that ion selectivity takes place where the pore is narrowest, at a region known as the *selectivity filter*, and that it is the amino acids located at the selectivity filter that determine which ions can permeate. Cation-selective channels, for example, often have nega-

tively charged residues at, or near, their selectivity filters, which attract positive ions and repel negative ones. In addition to the size and the charge of
the amino acid side chains, their degree of hydrophobicity also influences
ion permeation. The molecular basis of ion selectivity is just beginning to be
understood for some of the voltage-gated Na^+, Ca^{2+}, and K^+ channels, but
remains unclear for many other types of channel.

Gating

A channel can be either open or closed. The process of transition from the
open to the closed state (and vice versa) is known as *gating*. Some channels,
such as the K^+ channels, which determine the resting potential of the cell,
open and close randomly at all membrane potentials. Their gating is said to
be voltage independent. Other ion channels are normally closed but their
open probability can be greatly enhanced by a change in membrane potential
(*voltage-gated channels*) or by the binding of extracellular or intracellular
ligands (*ligand-gated channels*). In a similar manner, channels that are normally
open at the resting potential can be closed by a change in the membrane
voltage or by ligand binding. The gating of a channel may also be subject to
modulation, a process in which the gating of a channel is modified, most
commonly by one of a number of cytosolic substances.

What happens when a channel opens or closes? The actual transition is
too fast to be resolved and must therefore occur within <10 μsec, which is
about the frequency response of most recording electronics. It must also
involve some conformational change in the protein. Indeed, analysis of electron microscopic images has revealed that the structure of the nicotinic acetylcholine receptor (nAChR) channel is different in the presence and absence of
its ligand, i.e., in the open and closed states (Unwin, 1993, 1995) (Chapter 15).

Ligand-gated channels

Ligand-gated channels are generally named after the ligand that gates
them. These may be extracellular, as in the case of the neurotransmitters
acetylcholine and glycine, or intracellular such as cyclic AMP, Ca^{2+}, and ATP.
Binding of the ligand to one or more specific sites on the channel protein
produces a conformational change that allosterically opens the ion pore (Fig.
3.4). While the ligand remains bound to its receptor, the channel may open
and close several times. This gating behaviour is terminated by dissociation
of the ligand, which causes the channel to enter a permanently closed state
from which it is unable to reopen without ligand binding. Many channels
also undergo *desensitization*, a phenomenon observed at high agonist concentrations in which the channel enters a permanently closed state despite the
presence of bound ligand. Although ligand binding usually causes the channel to open, in some cases it may result instead in channel closure. The
inhibition of the ATP-sensitive K^+ channel by intracellular ATP is one example
(see Chapter 8).

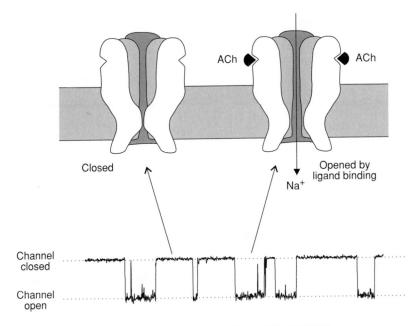

ACh

ACh

Closed

Opened by
ligand binding

Na⁺

Channel
closed

Channel
open

FIGURE 3.4 LIGAND-GATED CHANNEL ACTIVATION

Ligand-gated channels open in response to binding of a ligand. The pore of the acetylcholine receptor channel is closed in the absence of acetylcholine (left). Binding of acetylcholine (ACh) to its receptor site on the channel protein produces a conformational change that results in opening of the channel pore (right). Ions can then move through the pore, resulting in single-channel currents such as those shown in the recording below.

To understand the gating of ligand-gated ion channels we need to identify the location of the ligand-binding site, the mechanism by which ligand binding influences channel gating, and the mechanism of desensitization. To a large extent this has now been achieved for the nicotinic acetylcholine receptor (Chapter 15). Less is known, however, for other types of ligand-gated ion channels.

Voltage-gated channels

At the resting potential of the cell, most voltage-gated channels are closed. When the membrane potential is changed, however, the channel undergoes a series of conformational changes that result in the opening of the channel pore (Fig. 3.5). This voltage-dependent *activation* is sometimes (but not invariably) followed by a further conformational transition (*inactivation*) to an inactivated state. In this state the channel no longer conducts ions. It is also refractory to further changes in voltage and subsequent depolarisation will fail to induce channel opening until recovery has taken place. Recovery from inactivation occurs after a variable period following return to the resting potential. During this period the channel presumably undergoes another

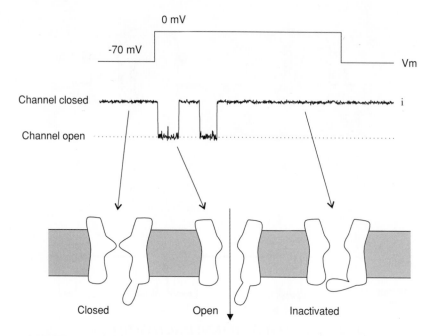

FIGURE 3.5 VOLTAGE-DEPENDENT GATING

Voltage-gated channels open in response to a change in the membrane potential. At negative membrane potentials, most voltage-gated Na^+ channels are closed. Depolarization produces a voltage-dependent conformational change in the channel protein that leads to opening of the channel pore. This is followed by a further conformational change that causes the channel to enter an inactivated state in which it no longer conducts ions. On repolarization the channel returns once more to the closed state (not shown).

conformational change that returns it to the closed state. Both depolarization and hyperpolarization can activate ion channels. Examples of depolarization-activated channels include the voltage-gated Na^+ and K^+ channels (see Chapters 5 and 6). An example of a hyperpolarization-activated channel (I_f) is described in Chapter 11.

Voltage-dependent activation requires that the channel sense a change in the voltage field across the membrane. Although this potential change is measured in millivolts, when the thickness of the membrane is taken into account the field experienced by the channel can be very large: of the order of 100,000 $V.cm^{-1}$. It is therefore perhaps less surprising that a potential change can alter the protein conformation. Likely candidates for the voltage sensor are charged amino acids located within membrane-spanning domains. Their movement is thought to trigger a further conformational change that opens the channel pore.

The key issues that need to be resolved in order to understand voltage-dependent gating include the location of the voltage sensor, the way in which voltage sensing results in channel opening, and the molecular basis of inactivation. Considerable progress in these areas has been made for the voltage-gated Na^+ and K^+ channels (Chapters 5 and 6). Less is known, however, about Cl^- channels (Chapter 10).

Channel modulation

The gating of almost all ion channels is subject to modulation by one or more of a wide range of substances (Kaczmarek and Levitan, 1987). These include monovalent and divalent cations (such as H^+ and Ca^{2+}), metabolites such as ATP and MgADP, fatty acids, phosphorylation, GTP-binding proteins, and even gases (such as oxygen). For example, one type of Ca^{2+}-activated K^+ channel is opened by depolarization but the voltage dependence of activation is shifted to more negative membrane potentials in the presence of intracellular Ca^{2+} (Chapter 7). Calcium therefore serves as a modulator of the channel. Sometimes the distinction between a channel modulator and the principal ligand seems merely semantic, as both are required for normal activity.

Many hormones and neurotransmitters mediate their effects on ion channels indirectly by the activation of some second messenger system that modulates ion channel activity, rather than by directly binding to the channel. The significance of this is that the modulatory response is slow both to turn on and to turn off. This results in a slow synaptic potential that facilitates the integration of signals in excitable cells. Modulation can also produce a sustained change in channel activity, which may outlast the presence of the hormone or transmitter. For example, phosphorylation of an ion channel may persist for some time if phosphatase activity is low. Second messenger regulation also allows for signal amplification and for divergence of action, as a single modulator may act on many different target proteins. Likewise, a single type of channel may be regulated by a number of different neurotransmitters that mediate their effects via a common second messenger.

Defective ion channel modulation may cause disease. Mutations in key metabolic enzymes, for example, cause maturity-onset diabetes of the young, by impairing the normal regulation of ATP-sensitive K^+ channels in pancreatic β-cells (Chapter 8). The enormous complexity of the cell means that many diseases have secondary effects on ion channel activity. It is simply not possible to mention them all in this book, and I have limited myself to a few key examples in which the defective channel regulation is the primary cause of the disease phenotype.

Kinetic analysis of single-channel currents

The kinetics of channel opening and closing provides information about the rates of transition between the closed and open states. A detailed description of single-channel kinetics is beyond the scope of this book but it is worth considering a few basic principles here. For a more detailed discussion you should refer to Aidley and Stanfield (1996) or Hille (1992).

Let us consider the simplest possible model in which a channel can exist in only two states: open (O) or closed (C). The kinetic behaviour of this channel is given by

$$C \underset{k_{-1}}{\overset{k_1}{\rightleftharpoons}} O$$

where k_1 and k_{-1} are the rate constants (in sec^{-1}) for entering and leaving the open state, respectively. The time spent in any one state is given by the reciprocal of the sum of the rate constants for leaving that state. Thus the mean open time is simply $1/k_{-1}$ and the mean closed time is $1/k_1$. The rate constants may therefore easily be obtained by measuring the mean open and closed times (Fig. 3.6). With a single open (or closed) state, as here, the open (or closed) time distribution will follow a single exponential and the mean open (or closed) time will be given by the time constant of the distribution (i.e. τ_o or τ_c). The open probability of the channel is simply the mean open time divided by the total time, i.e., $\tau_o / (\tau_o + \tau_c)$ or $k_1/ (k_1 + k_{-1})$. It therefore depends on both rate constants.

For a voltage-independent channel, the rate constants k_1 and k_{-1} will be unaffected by membrane potential. In the case of a voltage-dependent channel, however, one or both of the rate constants are voltage dependent. Conse-

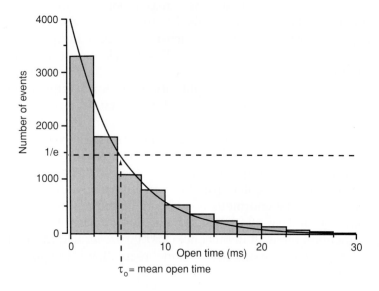

FIGURE 3.6 OPEN-TIME HISTOGRAM
The time a channel spends open is variable. The mean open time (τ_o) can be obtained from the open-time histogram. This is produced by sorting the open times into bins of constant width (here 2.5 ms). In this example, the open times show a single exponential distribution. This means that the channel has only a single open state. The time at which the number of events falls to $1/e$ of its maximal value is the mean open time (τ_o). Figure supplied by Paul Smith.

quently, the mean open and closed times vary with membrane potential, and the open probability is also voltage dependent. The higher the rate constant, the shorter the time spent in the state it exits. Thus for a channel whose open probability increases with depolarization, k_1 will increase with membrane potential and/or k_{-1} will decrease.

In many cases, ion channels have more than one closed state, so that the channel may pass through additional states during its transition from the resting to fully open state. Thus the scheme shown earlier is too simple and additional closed states must be included (for further details see Aidley and Stanfield, 1996). The inactivation of voltage-gated channels, such as the Na^+ channel, also requires a more complex model. In this case, it is known that the inactivated state is accessed from the open state. Although there is considerable evidence that the Na^+ channel has multiple closed states, for simplicity we will assume that there is only a single closed state. Thus

$$C \underset{\beta_m}{\overset{\alpha_m}{\rightleftharpoons}} O \underset{\beta_h}{\overset{\alpha_h}{\rightleftharpoons}} I$$

Conventionally, α_m and β_m are used to describe the rate constants for entering and leaving the open state of the voltage-gated Na^+ channel and α_h and β_h are used for entering and leaving the inactivated state. In the absence of inactivation, the steady-state open probability is simply $\alpha_m/(\alpha_m + \beta_m)$. Steady-state inactivation is given by $\alpha_h / (\alpha_h + \beta_h)$.

FROM SINGLE CHANNELS TO MACROSCOPIC CURRENTS

Thus far we have been concerned with the properties of a single ion channel. Such single-channel currents may be resolved by recording from a very small patch of membrane that contains only one or two channels, using the patch clamp method described in Chapter 4. Unless the patch of membrane is very small, however, it is likely to contain many ion channels and they will not necessarily all be of the same type. This is also the case when we record from the whole of the membrane of a cell. Let us now look at what happens when it is not possible to resolve single-channel currents, but instead we record the simultaneous activity of many ion channels (the *macroscopic current*).

The Macroscopic Current

For simplicity, we first consider a membrane that contains only a single kind of ion channel. The *macroscopic current* (I) represents the summed current through all of the channels in a patch of this membrane. It is related to the single-channel current (i) in the following way:

$$I = N P i \tag{6}$$

where P is the channel open probability and N is the number of channels in the membrane. If the channel open probability does not vary with voltage or time, the macroscopic current will simply appear as a voltage- and time-independent current. However, in many cases the open probability is not constant. In this case, the time course of the current will vary because of time-dependent changes in the channel open probability. We look briefly at one channel where this is the case: the voltage-dependent K^+ channel.

In response to depolarisation, the macroscopic K^+ current does not change instantaneously. Instead, the current increases slowly after a delay and only

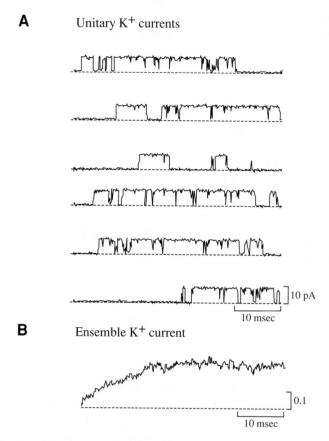

A   Unitary K^+ currents

]10 pA

10 msec

B   Ensemble K^+ current

]0.1

10 msec

FIGURE 3.7 SINGLE-CHANNEL CURRENTS SUMMATE TO GIVE MACROSCOPIC CURRENTS

(A) Single K^+ channel currents (*mSlo*) from channels expressed in *Xenopus* oocyte in response to a depolarization from -80 mV to $+60$ mV. This was recorded in the inside out patch clamp configuration and filtered at 8 kHz, with 10 μM internal calcium. (B) Ensemble current obtained by averaging 200 sweeps. Current is expressed as the probability of the channel being open (Po) vs time. Figure courtesy of Gargi Talukder and Richard Aldrich.

reaches a steady state level several milliseconds after the start of the voltage step (Fig. 3.7). This slow turn on of the current is known as *activation*. How do the single-channel K^+ currents give rise to the characteristic kinetics of the macroscopic K^+ current? This may be easily seen by summing (or averaging in this case) the single K^+ channel currents elicited by a series of depolarizations to the same potential. Figure 3.7 shows that single-channel openings occur after a variable delay and increase in frequency and duration throughout the course of the voltage pulse. In other words, there is a time-dependent increase in the channel open probability. Because each K^+ channel opens at random, and there are many of them, the activation phase of the macroscopic K^+ current is a smooth curve.

The macroscopic K^+ current–voltage relation differs from the single-channel current–voltage relation not only in its magnitude, but also in its voltage dependence. This is because the magnitude of the macroscopic current is determined both by the single-channel current amplitude, and by the channel open probability. Consider Fig. 3.8, which shows the current–voltage relation of the single K^+ channel, the voltage dependence of the K^+ channel open probability, and the macroscopic K^+ current–voltage relation in schematic form. At -140 mV, the single-channel current is large and inward. There is no macroscopic current, however, as the channel open probability is extremely small. As the membrane is depolarised above -40 mV, the channel open probability increases. Because this potential is positive to the K^+ equilibrium potential, the single channel currents are now outward. Thus the macroscopic current at -40 mV is small and outward. The macroscopic current then increases exponentially with membrane potential, as the open probability increases. Finally, at $+40$ mV the maximum open probability is attained. Above this potential, the macroscopic current is simply determined by the single-channel current multiplied by the number of channels in the membrane: it will be therefore be a linear function of membrane potential.

The Membrane Current

Cell membranes usually contain several different types of channel rather than a single kind as we have thus far assumed. The total current that flows across the cell membrane (the *membrane current*) represents the sum of the ion fluxes through all these different kinds of channel. Clearly, if one wishes to measure the macroscopic current through a single type of ion channel in isolation, it is necessary to block the currents flowing through all of the other kinds of ion channel in the membrane. This may be achieved by using drugs which act as selective channel blockers or by the removal of permeating ions (if these differ from those that permeate the channel of interest). There is a huge range of drugs and toxins that block ion channels. Some of these show great specificity: tetrodotoxin, for example, blocks only voltage-gated Na^+ channels. Pharmacological agents with high specificity therefore provide useful tools for the identification of different channel types. If they are also of sufficiently high affinity they may be used for the purification of channel proteins.

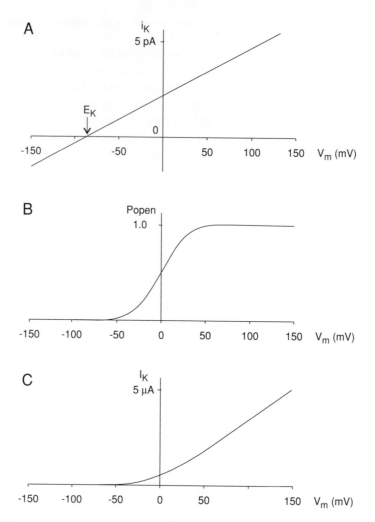

FIGURE 3.8 ORIGIN OF THE MACROSCOPIC I-V RELATION
The shape of the macroscopic current–voltage relation is determined by
both the single-channel current–voltage relation and the voltage dependence of the channel open probability. Schematic illustrating how the
macroscopic current–voltage (I–V) relation (C) depends on the single-
channel I–V relation (A) and the voltage dependence of the channel open
probability (B). The figure depicts the case for a voltage-dependent K$^+$
channel (A). The single-channel i–V relation is linear and the current
reverses at the K$^+$ equilibrium potential (E_K) of -80 mV. (B) At potentials
negative to -60 mV, the channel opens only very rarely but its open
probability increases with depolarization, reaching a maximum value at
potentials above $+50$ mV. (C) Unlike the single-channel currents, no macroscopic currents are observed at potentials negative to -50 mV (as so
few channels are open). Increasing depolarization produces an exponential
rise in the macroscopic current amplitude, as both the single-channel
current amplitude and the number of open channels are increased. The
macroscopic currents increase linearly with potential (like the single-
channel currents) once the channel open probability has attained its maximal value.

FROM WHOLE-CELL CURRENTS TO MEMBRANE POTENTIAL CHANGES

It is axiomatic that an ionic current flowing across the membrane will alter the membrane potential. This is why it is necessary to use a voltage clamp to hold the membrane potential constant when one wishes to record macroscopic or single-channel currents. This technique is described in Chapter 4. The rest of this chapter focuses on how ionic currents give rise to the two main types of potential change observed in cells: the action potential and the synaptic potential.

The Passive Response

If a small amount of positively charged ions flow inward across the cell membrane, a depolarization known as a *passive response* produced (Fig. 3.9B). Its steady-state amplitude is determined by the resistance of the cell membrane and is given by Ohm's law [Eq. (4)]. The higher the membrane resistance, the larger the voltage response. The resistance of the membrane is largely determined by the number, open probability and conductance of the ion channels that are open. At the resting potential of most cells, these are mostly K^+ channels. Their activity serves to keep the resting membrane potential close to the K^+ equilibrium potential.

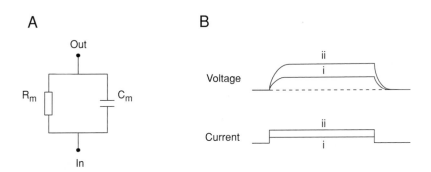

FIGURE 3.9 THE PASSIVE RESPONSE
The cell membrane acts like an *RC* circuit which consists of a resistance (R_m) and a capacitance (C_m) connected in parallel (*A*). If a small amount of current is injected into this circuit (i.e., flows across the membrane), a small voltage response will be produced (*B, i*). The amplitude of this response is determined by the membrane resistance, and its time course is influenced by the membrane capacitance. If a larger amount of current is injected, a greater voltage deflection will occur (*B, ii*). A voltage change like the ones shown here, which does not elicit an action potential, is sometimes refered to as the passive response of the cell membrane.

The time course of the membrane response to a current step is determined by the *capacitance* (C_m) of the membrane. A capacitor is formed by a narrow gap between two conductors. Because the membrane is very thin, is made of insulating material, and separates two conducting solutions, it acts as a capacitor. Biological membranes have a capacitance of around 1 $\mu F/cm^2$. A property of capacitors is that they separate charge, so that a potential difference may be maintained between one side of the capacitor and the other (in the case of a biological membrane, this is the membrane potential). The membrane may be modelled as a resistance and a capacitance connected in parallel (Fig. 3.9A). A current flowing through this circuit has both resistive and capacitative components. The presence of the capacitative component means that the voltage response is not instantaneous (as would be the case it there was only a resistor), but increases only gradually (Fig. 3.9B). The larger the membrane capacitance, the slower the response.

Action Potentials

In an excitable membrane, such as a nerve or muscle fibre, a larger inward current may produce a membrane depolarization that is sufficient to elicit a regenerative potential change known as an *action potential* (Fig. 3.10). The ionic basis of the action potential has been extensively studied using the giant axon of the squid. The squid action potential can be described by the interplay of two principal currents: the voltage-gated Na^+ and K^+ currents, which govern the rising and falling phases of the potential change, respectively. Membrane depolarization causes an initial activation of voltage-gated Na^+ channels followed shortly afterwards by the activation of voltage-gated K^+

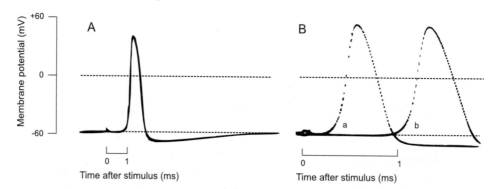

FIGURE 3.10 THE ACTION POTENTIAL OF THE SQUID AXON
(*A*) Action potential recorded intracellularly from a squid giant axon. The lower dashed line indicates the resting potential. From Baker, Hodgkin & Shaw (1961). (*B*) Propagating action potential recorded by two intracellular electrodes spaced 16 mm apart along a squid giant axon. The action potential takes longer to reach the more distant electrode (b). The conduction velocity can be calculated from this time difference and was ~21 m/s. Modified from Del Castillo & Moore (1959).

channels. Because the Na^+ channels activate more rapidly than K^+ channels on depolarization there is an initial net inward current that leads to a further depolarization. Consequently, more Na^+ channels are activated, which increases the inward current and depolarizes the membrane still further. In this way, a regenerative increase in membrane potential (an action potential) is produced. The depolarization does not quite reach the equilibrium potential for Na^+ ions ($\sim$ +50 mV) because two processes limit its amplitude. First, the Na^+ channels begin to inactivate, thus reducing the Na^+ current. Second, K^+ channels open. The resulting net outward current returns the membrane towards E_K. Because the voltage-gated K^+ channels take some time to close, the membrane potential may transiently undershoot the resting potential, producing an *after-hyperpolarization*. The duration of the after-hyperpolarization reflects the time taken for the voltage-gated K^+ channels to close.

The potential at which the inward Na^+ current exactly balances the outward current through resting K^+ channels is known as the *threshold potential*. It is a critical potential, as any increase in Na^+ current elicits an action potential, whereas any reduction in inward current (or increase in outward current) prevents action potential generation. Following an action potential, there is an *absolute refractory period* during which a second action potential cannot be elicited, no matter what the stimulus strength, because the Na^+ channels have not yet recovered from inactivation. This is followed by a *relative refectory period* during which an action potential can only be evoked if the stimulus strength is increased: both the presence of some remaining Na^+ inactivation and the fact that the voltage-gated K^+ channels have not yet closed contribute to the relative refectory period. The regenerative nature of the action potential means that it is able to propagate without decrement along nerve and muscle fibres.

Several ion channel mutations increase nerve or muscle excitability by either enhancing the inward current, or reducing the outward current. Because this produces a larger depolarization, the threshold potential is more likely to be reached and a subsequent action potential initiated. Other mutations produce a depolarizing block of action potential activity. This results from a maintained membrane depolarization of sufficient amplitude to inactivate the voltage-dependent Na^+ channels. These mutations are discussed in more detail in the subsequent chapters.

Many other types of ion channel may contribute to the action potentials of nerve and muscle fibres. For example, the cardiac action potential is mediated by voltage-dependent Na^+, Ca^{2+}, and at least four kinds of K^+ channel; several different kinds of K^+ channel contribute to the repolarization of action potentials in mammalian neurones; and chloride channels play an important role in the electrical activity of skeletal muscle. The functional importance of these different ion channels is exemplified by the fact that mutations in the genes that encode them produce a range of nerve and muscle diseases in man.

The Synaptic Potential

A *synapse* is a specialised junction where transmission takes place between two cells. The presynaptic cell is invariably a nerve cell but the postsynaptic

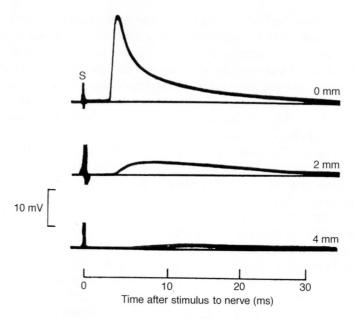

FIGURE 3.11 THE SYNAPTIC POTENTIAL
Excitatory postsynaptic potential (epsp) recorded at different distances
from the postsynaptic region of a frog skeletal muscle fibre. The nerve is
stimulated at the point marked S, resulting in the release of acetylcholine
from the nerve terminal which diffuses across the synaptic cleft and elicits
an epsp. The figures on the right indicate the distance of the recording
microelectrode from the synapse and demonstrate that the amplitude of
the epsp becomes smaller the further the distance away from the synapse.
This is because the epsp is conducted passively. From Fatt and Katz (1951).

cell may be a nerve, muscle, or even a gland cell. Transmission across the
synapse may be accomplished either by chemical or by electrical means. At
chemical synapses, a chemical transmitter released from the presynaptic cell
diffuses across the synaptic cleft and interacts with receptors in the membrane
of the postsynaptic cell. Many of these receptors function as ligand-gated ion
channels; they are described in Chapters 15–18. At electrical synapses, the
pre- and postsynaptic cells are coupled by gap junctions. These are formed
from the gap junction channels discussed in Chapter 20.

The *synaptic potential* describes the potential change in the postsynaptic
membrane which results from synaptic transmission. An *excitatory postsynaptic potential (epsp)* causes excitation of the postsynaptic cell (Fig. 3.11); similarly, an *inhibitory postsynaptic potential (ipsp)* results in inhibition. Epsps produce depolarization of the postsynaptic cell and, if they are large enough,
may elicit an action potential. Ipsps render the cell less excitable; they often
(but not invariably) cause membrane hyperpolarization.

The archetypal chemical synapse is the neuromuscular junction of skeletal
muscle, where binding of acetylcholine (ACh) opens the nicotinic acetylcho-

line receptor channel (nAChR). The operation of this channel is discussed in detail in Chapter 15. Here we will only look at how activation of nAChRs produces an epsp. The nAChR differs from the voltage-gated Na^+ and K^+ channels we have considered in two ways. First, the opening and closing of the channel is governed by binding of a ligand, in this case acetylcholine. Second, nAChR channels are permeable to both Na^+ and K^+ ions. This means that with physiological ion concentration gradients the reversal potential of the nAChR current lies close to 0 mV.

If a small amount of acetylcholine is released from the nerve terminal, only a few nAChR channels will be opened. The inward current that results will be largely shunted by outward currents flowing through K^+ channels that are open at rest in the muscle membrane. Consequently, the potential change will be very small and will not trigger an action potential in the muscle cell. If a large amount of ACh is released, however, many nAChRs are opened and a large inward current is produced. The membrane permeability will now be dominated by current flow through nAChR channels and the membrane will, unlike the action potential, depolarise to around 0 mV, the reversal potential of the AchR (and of the synaptic current). The epsp is not a regenerative response, and thus it does not propagate actively along the muscle fibre (Fig. 3.11). Instead, it passively depolarizes the adjacent extra-synaptic region of the muscle fibre where voltage-dependent Na^+ channels are located and, if the depolarization is sufficiently large, an action potential is elicited.

The time course of the synaptic potential reflects both the kinetics of ligand binding and unbinding and the rate at which the transmitter is removed from the synaptic cleft. The nAChR channels, for example, remain open only as long as acetylcholine is bound, and because binding is transient and acetylcholine in the synaptic cleft is rapidly removed, the synaptic current which flows through the nAChR channels is brief. Consequently, the membrane potential quickly returns to its resting level.

CHAPTER 4

STUDYING ION CHANNELS

In the experimental sciences, the introduction of a new technique often heralds a major advance in understanding and results in rapid expansion of the field. This is certainly the case for the study of ion channels, where the development of both electrophysiological and molecular biological methods has facilitated analysis of the relationship between ion channel structure and function. In order to fully appreciate the work described in the following chapters, it is helpful to have a basic understanding of how these techniques work. This chapter, therefore, looks at the different ways of studying ion channels. It is not intended as a practical guide: those wishing to attempt these techniques themselves are referred to the many excellent manuals that are available (Rudy and Iversen, 1992; Neher and Sakmann, 1995; Sambrook *et al.*, 1987).

INVESTIGATING ION CHANNEL FUNCTION

The most direct way of investigating ion channel function is to record the current which flows through the open channel or to measure the changes in membrane potential that this produces.

Potential Measurements

The potential across the cell membrane can be measured by inserting a glass microelectrode with a very fine tip into the cell and measuring the difference between the potential it records and that registered by a second electrode (the reference electrode) in the bath solution outside the cell (Fig. 4.1A). Usually, cells have resting membrane potentials in the range of -60 to -100 mV (inside negative). A second microelectrode may be inserted into the cell and used to pass current in order to alter the membrane potential. If a small amount of positive current is injected, a depolarization will be produced whose amplitude is determined by the resistance of the cell membrane (Fig.

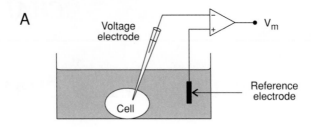

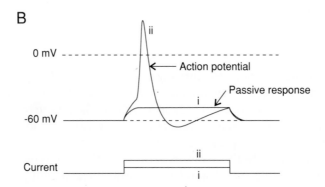

FIGURE 4.1 MEASUREMENT OF MEMBRANE POTENTIAL
(A) Membrane potential (V_m) is measured as the difference between a voltage electrode inserted into the cell and a reference electrode outside the cell in the extracellular solution. The voltage electrode is a fine glass capillary filled with a highly conducting salt solution (usually 3 M KCl). An Ag/AgCl wire is used to form an electrical connection between this solution and the recording electronics. (B) Injection of positive current into an excitable cell elicits a passive depolarizing response or an action potential, depending on the current magnitude.

4.1B). This is known as a passive response. In an excitable cell, such as a nerve cell, injection of a larger amount of positive current may produce a depolarization that is large enough to elicit an action potential (Fig. 4.1B).

Voltage Clamp

As its name suggests, the *voltage clamp* technique allows the membrane potential to be held (clamped) at a constant value. The current that flows through the membrane at any particular potential can then be measured (I_t). This current is the sum of the ionic current (I_i), which represents current flow through open ion channels, and the capacitative current (I_c), which is largely due to the charging of the membrane capacitance (Fig. 4.2). The magnitude of the capacity current is determined by the capacitance of the cell membrane (C_m), which is normally 1 $\mu F/cm^2$, and it flows only while the voltage is changing. Thus,

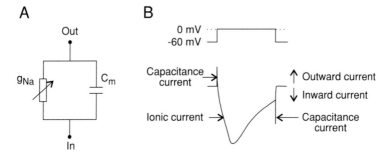

A

Out

g_{Na} C_m

In

B

0 mV
-60 mV

Capacitance current

Ionic current →

↑ Outward current

↓ Inward current

← Capacitance current

FIGURE 4.2 THE MEMBRANE CURRENT CONSISTS OF CAPACITATIVE AND IONIC COMPONENTS
(A) Simplified electrical circuit of a membrane that contains only voltage-gated Na⁺ channels. C_m, membrane capacitance; gNa, Na⁺ conductance. (B) The current elicited by a voltage clamp depolarization consists of a brief initial capacity current followed by an ionic current. In this case, the ionic current only flows through voltage-gated Na⁺ channels. The Na⁺ current flows from the outside to the inside of the membrane.

$$I_t = I_i + C_m\frac{dV}{dt} \tag{1}$$

Clearly, when the potential is held constant there will be no capacity current and the ionic current will be the same as the total membrane current. This is one advantage of the voltage clamp method. A further advantage is that it prevents the regenerative potential response (the action potential) that is triggered by the activation of voltage-gated ion channels. Thus it enables the currents responsible for the action potential to be investigated in a quantitative way.

The voltage clamp was invented by Marmont and Cole, but was more fully developed by Hodgkin and Huxley, who used it to great effect to provide a mathematical description of the action potential of the squid axon (Hodgkin & Huxley, 1952). A simple voltage clamp arrangement is shown in Fig. 4.3. The principle of the technique is to inject a current which is equal in amplitude but opposite in sign to that which flows across the cell membrane. As there is therefore no net current flow across the membrane, the membrane potential remains constant. Furthermore, by measuring the current which has to be injected to clamp the potential, one also measures the current that is flowing across the membrane. Conventionally, cationic currents flowing inward across the membrane (from outside to inside) are defined as negative currents and are shown as downward deflections of the current trace.

Voltage clamps such as the one shown in Fig. 4.3 measure the current that flows through the whole of the cell membrane which is known as the *whole-cell current*. Because this current flows through hundreds or thousands of ion channels simultaneously it is also sometimes referred to as the *macro-*

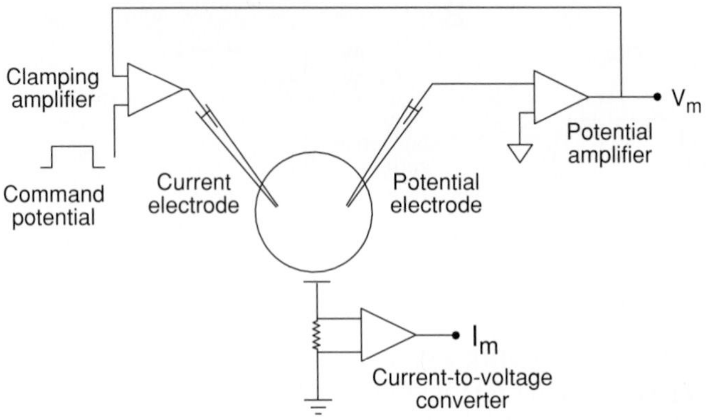

FIGURE 4.3 *A SIMPLE TWO-ELECTRODE VOLTAGE CLAMP*
Two microelectrodes are inserted into the cell, one to record the membrane potential and the other to pass current. The output from the voltage electrode is fed into one input of the clamping amplifier, which injects current into the cell to hold the membrane at the desired potential. This potential (the command potential) is applied to the other input of the clamping amplifier. A feature of operational amplifiers is that they act to maintain the same voltage at their inputs, thus the clamping amplifier injects current into the cell to keep the membrane potential and the command potential at the same value. The current injected is equal in amplitude, but opposite in sign, to that flowing across the plasma membrane. The current (I_m) can be measured using a current-to-voltage converter placed either in the injection circuit or in the ground circuit, as shown.

scopic current (to distinguish it from the *single-channel current*). The whole-cell current is the sum of the currents flowing through several different kinds of ion channel in the cell. Ion substitution experiments and selective channel blockers may be used to separate this current into its constituent components.

A good voltage clamp requires that the membrane potential be changed rapidly so that the capacity current is over before the ionic current starts to flow. The speed of the voltage response is limited by the rate at which current can be injected through the microelectrode. This depends on the electrode resistance: the higher the electrode resistance, the more difficult it is to inject current sufficiently fast. Since microelectrodes with tips tiny enough to be inserted into small cells without causing damage necessarily have a very high resistance, in practice there is a lower size limit on cells that can be successfully voltage-clamped this way. For small cells, the patch clamp method is essential.

Patch Clamp

The *patch clamp* technique was invented by Neher and Sakmann in 1976 and was further refined in 1981 (Hamill *et al.*, 1981). It may be used to record the activity of single ion channels or, in the whole-cell configuration (see later), to measure whole-cell currents from small cells. The ability to record single-channel currents revolutionized the study of ion channels because it allowed the single-channel conductance and channel kinetics to be observed directly. Indeed, the patch clamp technique provides biophysicists with a tool that is the envy of many biochemists: a simple way of measuring the activity of a single protein in real time under (relatively) physiological conditions.

The basis of the patch clamp method is the formation of a high resistance seal ($> 10\,G\Omega$) between the cell membrane and the glass wall of a micropipette placed against the surface of the cell. The high resistance of this seal means that the current flowing through the ion channels in the patch of membrane spanning the tip of the pipette can be recorded with a very low level of background current noise. This is important because the small amplitude of single-channel currents (just a few pA or less) means that they can easily be swamped by background electrical noise. Background noise is caused by spontaneous current fluctuations resulting from small changes in potential difference across resistances in the recording electronics, the seal resistance, and even the membrane itself. A second advantage of the high seal resistance is that the potential across the patch membrane can be altered simply by applying a voltage to the pipette. A low seal resistance acts as a voltage divider, so that the voltage seen by the patch membrane is not the same as that applied to the pipette.

Patch Clamp Configurations

There are four main patch clamp configurations: *cell attached, inside out, outside out,* and *whole cell* (Hamill *et al.*, 1981). The first three of these may be used to record single-channel currents from a small patch of membrane while the last is used to record the whole-cell current flowing through all of the cell membrane. Each configuration has its own particular advantages and disadvantages.

The cell-attached configuration is the starting point for all of the others (Fig. 4.4). In this configuration, the pipette is sealed to the membrane of an intact cell and the internal surface of the channel is therefore exposed to its normal cytosolic environment. Thus the cell-attached configuration is particularly useful for studying the properties of ion channels under physiological conditions. It has also been used to determine whether the effects of neurotransmitters, hormones and nutrients are mediated by a direct interaction with an ion channel or indirectly, via the generation of a soluble cytosolic second messenger. This is because the glass-membrane seal is too tight to permit the passage of substances between the bath and pipette solutions so that any effect of a substance added to the bath solution on the activity of ion channels in patch membrane must be mediated via an intracellular route. The cell-attached configuration has, however, two major disadvantages. First, the composition of the intracellular solution is unknown. Second, the potential across the patch membrane is determined not only by that applied to the

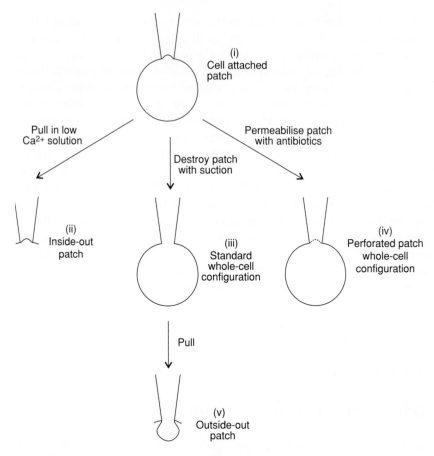

FIGURE 4.4 PATCH CLAMP CONFIGURATIONS
The cell-attached patch clamp configuration (i) is obtained by forming a
seal between the patch pipette and the cell membrane. Withdrawal of the
pipette from the cell surface produces an inside-out patch (ii). This is
facilitated if the patch is excised into an intracellular (low Ca^{2+}) solution.
If the patch membrane is destroyed in the cell-attached configuration, the
standard whole-cell configuration is produced (iii); permeabilisation with
pore-forming antibiotics yields the perforated-patch whole-cell configuration (iv). Withdrawal of the pipette from the standard whole-cell configuration produces an outside-out patch (v).

pipette, but also by the resting potential of the cell, which is not only unknown,
but may not even be constant.
    The two other single-channel recording configurations are possible be-
cause the high mechanical stability of the seal results in the rupture of the
membrane before the seal (Fig. 4.4). If the pipette is withdrawn from the cell
surface after forming a cell-attached patch, an isolated membrane patch is
produced that has its intracellular membrane surface facing the bath solution.

This is known as an inside-out patch. An outside-out patch is produced when the pipette is withdrawn from the whole-cell configuration; as may be inferred from its name, the extracellular membrane surface of the outside-out patch faces the bath solution. The two excised patch configurations allow the properties of the single-channel currents to be studied under controlled ionic conditions and measurements made in different solutions without detachment of the patch. Inside-out patches are used to test the effect of putative cytosolic regulators of channel activity, whereas outside-out patches may be used to study channels gated by extracellular ligands.

The whole-cell configuration allows the summed activity of all of the ion channels in the cell membrane (the whole-cell current) to be measured simultaneously (Fig. 4.4). It comes in two versions. The *standard whole-cell* configuration is obtained by forming a cell-attached patch and then destroying the patch membrane with strong suction. This provides electrical access to the cell interior and so enables measurement of the whole-cell current. It has the advantage that the intracellular solution can be manipulated, but the concomitant disadvantage that soluble cytosolic constituents are lost from the cell because they dialyse with the pipette solution. The *perforated patch whole-cell* configuration avoids this problem. In this configuration, the patch membrane is permeabilised with a pore-forming antibiotic such as amphotericin (Chapter 23) rather than being completely destroyed. The pores formed by the antibiotic are permeable to monovalent cations, such as K^+, but divalent cations, anions and uncharged molecules with a molecular weight greater than ~500 are largely impermeant. The perforated patch configuration therefore allows electrical access to the cell interior while retaining cell metabolism and intracellular messenger systems intact.

The whole-cell configuration is one of the most useful variants of the patch clamp method because it enables cells as small as 5 μm in diameter to be voltage clamped. This is possible because of the high resistance of the seal and the low pipette resistance, properties that prevent cell damage and allow a large amount of current to be injected rapidly in order to control the membrane potential. The standard whole-cell configuration also differs from the conventional microelectrode voltage clamp in a third way: the intracellular solution dialyses with that of the pipette. As we have seen, this is not always an advantage.

A typical patch clamp circuit is shown in Fig. 4.5. A key difference between the patch clamp and the two-electrode voltage clamp method illustrated in Fig. 4.3, is that the patch clamp uses a single electrode both to control the membrane potential and to measure current. Without this, the various patch clamp configurations would not be possible. Another difference is that the patch clamp amplifier is sufficiently sensitive to resolve the tiny currents (pA) flowing through single ion channels.

Noise Analysis

In some cases the single-channel current is simply too small to be measured using the patch clamp method. In this case, *noise analysis* (also called

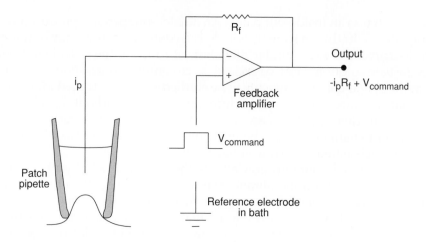

FIGURE 4.5 PATCH CLAMP CIRCUIT
The patch pipette is connected to the inverting input of a feedback amplifier. Because the input resistance of this amplifier is essentially infinite,
all the current recorded by the pipette flows through the feedback resistor
(R_f). This has a very high value, typically 10 GΩ. Its high resistance enables
the tiny single-channel currents (~1 pA) to be measured, since the patch
current (I_p) is given by the voltage drop across R_f, i.e., $V=I_pR_f$. The command potential ($V_{command}$) is applied to the other input of the feedback
amplifier and the amplifier passes current through the feedback resistor
to keep the voltage at the inverting input the same as the command
voltage. This means that the desired potential is also applied to the pipette
and thus to the patch membrane.

fluctuation analysis) can be used to obtain an estimate of the single-channel
current amplitude and the mean channel open time. The method relies
on the fact that the channel fluctuates randomly between open and closed
states. The current flowing through a single channel therefore also fluctuates.
In a large patch of membrane which contains many hundreds of ion
channels these fluctuations will summate to produce a constant current
noise. Embedded in the current noise, however, will be information about
individual current fluctuations. Fourier analysis may be used to obtain a
power density spectrum of the current noise from which the mean single-
channel current amplitude and the mean open time can be derived (Anderson and Stevens, 1973). Noise analysis may be useful when single-channel
currents cannot be resolved. However, it is important to recognise that
the method relies on several assumptions, which may not always be
valid. In particular, it is assumed that the channel population behaves
homogeneously and that there is a single conductance state. The channel
open probability must also be neither very low nor very high, because
then the fluctuations will be few.

Methods for Studying
Intracellular Ion Channels

Ion channels in intracellular membranes have proved less easy to study as they are not easily accessible to the patch clamp method. One way around this problem has been to isolate intracellular membranes, such as those of the sarcoplasmic reticulum, by cell fractionation. Such isolated membranes spontaneously form small vesicles of ~300 nm diameter that can be studied following fusion with an artificial lipid bilayer (Fig. 4.6). The bilayer can be voltage clamped, using a similar technique to that described earlier, in order to record the current. Pure lipid bilayers are impermeable to ions and have a very low level of background current noise so that if only a few vesicles fuse with the bilayer it is possible to record single-channel currents. In addition to intracellular ion channels, artificial bilayers have often been used to study plasmalemma ion channels. Another variant of the approach is to reconstitute the purified channel protein into the bilayer. Indeed, the very first recordings of single-channel currents were obtained by adding the pore-forming antibiotic gramicidin to an artificial bilayer (Hladky and Haydon, 1970; see also Chapter 23).

An advantage of the bilayer approach is that it enables the effect of the lipid environment of the channel to be examined, as bilayers may be formed from different types of lipids (such as charged and uncharged ones). The disadvantage, of course, is that the channel is removed from its native environment. Two additional methods of recording from intracellular membranes are described in Chapter 14.

OBTAINING THE PRIMARY SEQUENCE

Just as the patch clamp method led to an explosion of information about ion channel properties, so the revolution in molecular biological techniques in the late 1970s provided the tools for analysis of the primary structure of ion channels.

A Plethora of Enzymes

Reverse transcriptase is an enzyme that makes a DNA copy of an RNA molecule. RNA is very difficult to work with as it is easily degraded and, unlike DNA, it cannot be readily cut and joined with enzymes. Reverse transcriptase makes it possible to work with DNA instead.

A variety of different enzymes are used to manipulate DNA. Bacterial enzymes, known as *restriction enzymes,* are used to cut DNA at specific sites. In conjunction with DNA ligase, which can rejoin the cut DNA ends together, restriction enzymes can be used to cut and paste DNA sequences at will.

A

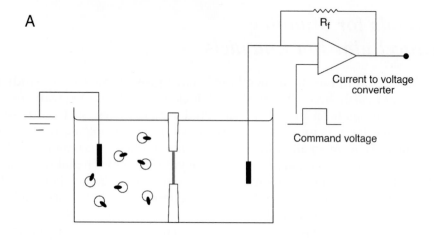

B

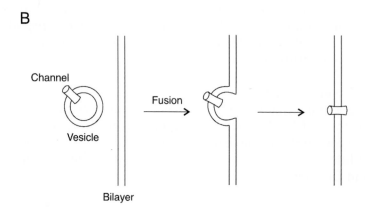

FIGURE 4.6 BILAYER RECORDING METHOD
A bilayer is formed by painting lipid across a small hole that connects two aqueous compartments and allowing it to thin to a bilayer (A). Membrane vesicles are then added to one side of the bilayer, which is known as the *cis* side (the opposite side is known as the *trans* side). When some of these vesicles fuse with the bilayer, the channels they contain are incorporated into the bilayer (B). The bilayer may be voltage clamped in order to record currents flowing through these channels. The orientation of the channels in the membrane depends on whether the vesicles from which they derive were "inside-out" or "outside-out"; because all vesicles do not necessarily share the same orientation, the same may also be true of channels in the bilayer.

Restriction enzymes work by recognising a specific sequence of DNA, usually a palindromic sequence of 4–8 bp. There is a large range of restriction enzymes (>100), most of which recognize a different DNA sequence.

The *polymerase chain reaction* (PCR) is a method for making many copies (> 1 billion) of a selected piece of DNA, provided that at least part of its nucleotide sequence is already known. Figure 4.7 illustrates this method. The technique was developed by Kerry Mullis and colleagues and relies on an enzyme that not only is able to replicate DNA but also is able to withstand high temperatures without denaturation. Such enzymes are found in thermophilic bacteria, such as those that live in the hot waters of Yellowstone National Park in the United States. The PCR technique is so sensitive that it is possible to detect a single DNA molecule in a sample. Indeed, it has been used to amplify the DNA (or mRNA) extracted from single nerve cells and so determine which kinds of ion channels are expressed by individual neurons in the brain (Geiger *et al.*, 1995). However, the extreme sensitivity of the PCR method is not without its problems and care must to be taken to avoid amplifying stray DNA (such as that floating about in the air within the laboratory).

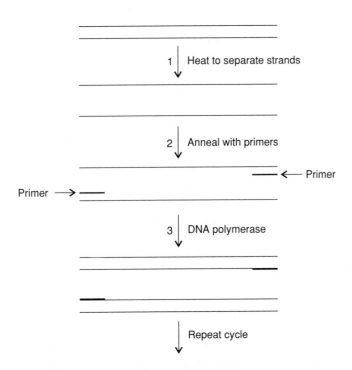

FIGURE 4.7 THE POLYMERASE CHAIN REACTION (PCR)
DNA amplification using PCR involves successive cycles, each of which consists of three main steps. First, double-stranded DNA is heated to denature it into its two constituent strands. Second, it is cooled and annealed with two PCR primers. These are oligonucleotides complementary in sequence to the DNA on opposite strands on either side of the target region. Third, the mixture is heated to a temperature at which the polymerase enzyme synthesises a single strand of DNA, beginning at the 3'-end of each primer. The entire sequence is then repeated many times, which results in an exponential increase in the number of DNA molecules.

Cloning of Ion Channels

How does one go about cloning an ion channel? Historically, there was only one way. That was to purify the protein and determine part of its amino acid sequence. This sequence could then be used to design a range of probes for screening a cDNA library. The very first channels to be cloned—the nicotinic acetylcholine receptor from the *Torpedo* electric organ and the voltage-gated Na^+ channel of the electric eel—were cloned this way (Noda *et al.*, 1982, 1984). The method is facilitated if a high-affinity ligand for the channel is available, but it is still very difficult and laborious for a membrane protein. Furthermore, for many channels a high-affinity ligand simply does not exist. Fortunately, there are now other ways to clone ion channels. These include expression cloning, homology screening of cDNA libraries or computer databases, and genetic approaches such as positional cloning (see page 61). Since several of these methods require the use of a cDNA library, let us first consider what this is.

A complementary DNA (cDNA) library is a very useful tool for cloning ion channels. As its name implies, it is a library of many different cDNA clones. cDNA is the name given to DNA derived by reverse transcription of messenger RNA; it therefore contains the complete coding sequence of the channel but no intronic sequences. To make a cDNA library, messenger RNA (mRNA) is extracted from a tissue which expresses a high level of the channel to be cloned, because this will be enriched in mRNA molecules that code for the channel. Reverse transcriptase is used to obtain the cDNA, which is then cloned into a plasmid or bacteriophage vector. For every original mRNA molecule, we now have a vector containing the equivalent cDNA. This library may be screened using a radioactive DNA probe with a sequence complementary to the channel to be identified. Clearly, this is fairly easy if we know part of the sequence already, as is the case if the protein has been purified and partially sequenced. However, it is much more difficult if, as is usually the case, the protein sequence is not known. In this case, homology screening or expression cloning is used (see later). Once the relevant clone has been isolated and sequenced, it can be tested for channel function in expression studies.

Homology screening is based on the fact that members of the same ion channel family have regions of similar structure. It is therefore possible to construct probes which are based on short stretches of DNA sequence that are strongly conserved between different family members. These may be used to screen cDNA libraries from different sources to identify related proteins, such as further members of the same channel family. The probe consists of a radiolabeled DNA sequence complementary to that selected and can therefore hybridize (interact) with homologous sequences in the cDNA library. Many types of voltage-gated K^+ channels were found this way by using probes with homology to the *Shaker* K^+ channel (see Chapter 6). Homology screening is a powerful method for isolating related ion channels, but it cannot be used to clone channels that have a very different sequence from the parent one.

How can one clone a channel which shows no homology to any known channel without having to purify the protein? One approach is expression cloning. Lily Jan and colleagues used this method to obtain one of the first inwardly rectifying K^+ channels, Kir2.1 (Kubo *et al.*, 1993). They isolated mRNA from cardiac muscle, which strongly expresses Kir2.1, injected it into *Xenopus* oocytes, and looked for expression of a current with properties resembling those of the native channel. The mRNA was then fractionated by size into a number of different pools, each of which was then tested for functional activity using the oocyte system (Fig. 4.8). The pool which produced the largest currents was then further divided and successive rounds of fractionation and expression were carried out until finally only a single RNA—that which coded for Kir2.1—was left. Because this method requires a lot of mRNA, a cDNA library is sometimes made from one of the pools that shows high activity. This has the advantage that unlimited amounts of mRNA can then be obtained for further testing.

Increasingly, ion channels are being cloned by homology screening of computer databases. Not only do these databases contain a number of full-length sequences of proteins of unknown function, they also contain large numbers of short sequences known as expressed sequence tags (ESTs). The ESTs may code for part of an unidentified ion channel. A range of computer programs exist that can search through the database and pick out EST sequences with homology to a chosen "virtual probe" sequence. Real oligonucleotide probes based on the sequence of identified ESTs are then used to screen a cDNA library and isolate a full length clone, which is then expressed and examined for functional channel activity. This method was used to clone the small conductance Ca^{2+}-activated K^+ channel (Kohler *et al.*, 1996).

Functional Expression of Cloned Channels

Once a putative ion channel has been cloned it must be expressed in order to confirm that it does indeed act as a channel and to determine its properties. *Heterologous expression* refers to the expression of a protein from cDNA (or mRNA) that has been introduced into a cell by the experimenter. It may be either *transient* or *stable*. Transient expression lasts only a few days as the cDNA (or RNA) is gradually degraded and diluted out as the cell divides. Stable expression requires that the DNA becomes incorporated into one of the cell's chromosomes, is replicated along with the host DNA at mitosis, and continues to be expressed in the daughter cells. Usually, ion channel proteins are heterologously expressed in one of a number of cell types whose properties have been well characterised. Among these are the *Xenopus* oocyte and a number of different mammalian cell lines. The choice of expression system is governed by several factors. The cell must be capable of expressing large amounts of heterologous protein reliably and the level of endogenous channel activity should be low.

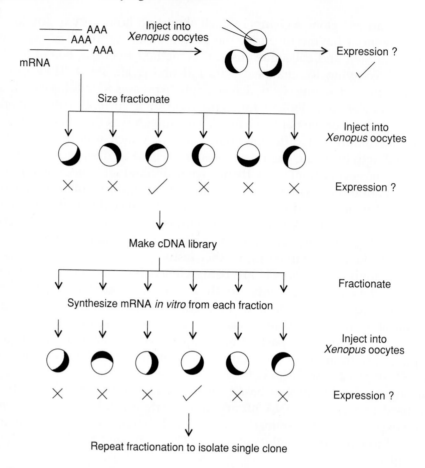

FIGURE 4.8 EXPRESSION CLONING
Expression cloning utilises the functional properties of the channel. Messenger RNA is isolated from a cell expressing a large amount of the channel of interest and injected into *Xenopus* oocyte. If a novel current is detected, the mRNA is size-fractionated, and the different fractions injected into a fresh set of oocytes. A cDNA library is made from the fraction that expresses the largest currents. A large quantity of mRNA is then made, size-fractionated, and injected into oocytes. The pool that expresses the largest current is then further fractionated and reinjected and the cycle repeated until one ends up with a single clone that codes for an ion channel.

   There are two ways to achieve expression: using cDNA or using mRNA. These may be introduced into the cell by a variety of methods which include direct injection, the uptake of liposomes with adherent DNA or RNA (transfection) and infection with recombinant viruses containing the DNA (or RNA) of interest. DNA is also often introduced by transient permeabilization of the surface membrane by electric shock (electroporation).

Tissue Distribution

Determination of the functional role of an ion channel is greatly facilitated if one knows in which tissues it is expressed and whether expression is developmentally regulated. This may be studied at the mRNA level by Northern blotting and at the protein level by Western blotting. In these techniques, mRNA or protein is extracted from the tissue of interest and hybridised with a DNA probe, or an antibody, respectively. To study expression at finer resolution and to identify in which cells, or part of a cell, a particular mRNA (or protein) is expressed, it is necessary to use *in situ* hybridization (or immunocytochemistry). Both methods are carried out on thin sections of tissue prepared for light microscopy. For *in situ* hybridization the sections are incubated with a labeled nucleotide probe of sequence complementary to the mRNA of interest. Traditionally, hybridization is detected using a radioactively labeled probe, although other nonisotopic methods now exist. Immunocytochemistry relies on an antibody specific to the channel of interest to detect the protein. It is worth noting that there may be marked differences in the level of mRNA and protein, depending on the rate of protein turnover and the stability of the mRNA. Furthermore, the mRNA and the protein are not necessarily found in the same part of the cell. For example, channels are normally manufactured in neuronal cell bodies, but many types of channels are only found in dendritic membranes. Thus, in order to be certain of where the protein (as opposed to mRNA) is expressed, immunocytochemistry is essential. Unfortunately, specific antibodies are not always available.

INVESTIGATING ION CHANNEL STRUCTURE

The first step in correlating ion channel structure and function is to examine the amino acid sequence. This can easily be inferred from the DNA sequence using the genetic code. The amino acid sequence gives a number of clues about how the protein might be arranged in the membrane and which regions may be of functional importance.

Topology

The hydropathy profile of the channel gives an indication of the way in which the protein is likely to be folded in the membrane because hydrophobic regions are likely to lie within the membrane, whereas hydrophilic regions are likely to contact the extracellular or intracellular solution. Candidates for membrane-spanning segments are those where a clear peak in hydrophobicity is found over a region of $\sim$20 amino acids, which is the minimum required to form a helix capable of spanning the bilayer (which is ~30 Å thick). Hydropathy plots may be used to construct simple models of the membrane topology of a channel (Fig. 4.9). Interpretation of such plots is not unambiguous, however. For example, a hydrophobic region (which may be less than 22 amino acids) may dip down into the membrane, forming a hairpin loop that enters and leaves the membrane from the same side (the pore region of many ion channels is formed from such hairpin loops). Consequently, the initial topology predicted for many of the ion channels discussed in this book was revised later: in several cases, changes in the putative number of transmembrane domains occurred and/or shifts in the location of the N or C terminus from one side of the membrane to the other. Nevertheless, hydropathy plots provide a useful first guess at a channel's topology.

How can the topology suggested by the hydropathy plot be confirmed? One way is to use antibodies to specific sequences and see whether these label the channel at the outside, or inside of the membrane. Another way is examine the glycosylation of the channel, because glycosylation (attachment of sugars) is only found on residues in extracellular domains. Glycosylation sites can also be introduced into the cDNA sequence and the mutant protein examined to see if glycosylation takes place and thus that the site is located on the outside of the membrane. It is axiomatic that neurotransmitter-binding sites and the binding sites for certain toxins must be accessible to the extracel-

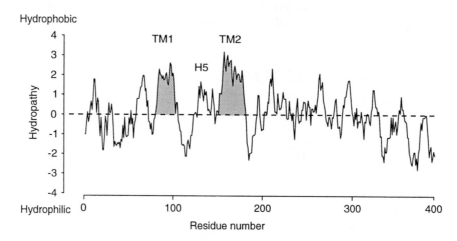

FIGURE 4.9 HYDROPATHY ANALYSIS SUGGESTS THE
TRANSMEMBRANE TOPOLOGY
Hydropathy plot of Kir6.2. Amino acids are numbered from the N to the C terminus. Hydrophobic resides are plotted above the line and hydrophilic ones below the line. A continuous stretch of more than 20 hydrophobic amino acids is considered sufficient to span the membrane. There are two such transmembrane domains (TM1 and TM2) in Kir6.2. The region marked H5, although not long enough to cross the bilayer, forms a hairpin loop that dips into and out of membrane and lines the upper region of the pore.

lular solution. Identification of the residues involved in binding these molecules will therefore also illuminate the channel topology. Studies such as these have shown that initial guesses of membrane topology based on hydrophobicity plots are not always correct.

X-ray crystallography provides a picture of the three-dimensional structure of a protein. Unfortunately, with one or two exceptions (Chapters 8 and 23), it has not been possible to obtain crystal structures of ion channels. Indeed, it has proved exceedingly difficult to obtain crystallographic data on any membrane protein, largely because of the difficulty in obtaining crystals. The crystallographic analysis of an ion-selective channel, described in Chapter 8, is therefore a major advance. The three-dimensional structure of a number of ion channels has been examined with high resolution electron microscopy coupled with computer analysis of the EM images. Detailed structures of the nicotinic ACh channel and the water channel have been obtained his way (see Chapters 15 and 19). Although nuclear magnetic resonance has been instrumental in unravelling the structure of many soluble proteins, this technique has not been widely applied to ion channels.

Mutational Analysis

How does one determine the functionally important regions of a channel: the bits that form the pore, the voltage sensor, or the ligand-binding sites? The method favoured to date has been to alter the amino acid sequence and see how that affects channel function. Interpretation of the results is not always straightforward, however. A mutation that reduces channel sensitivity to a ligand does not necessarily imply that the mutated residue forms part of the ligand-binding site: instead, it may interfere with channel gating or with the mechanism involved in transducing ligand binding into channel opening. Despite this caveat, most of our current knowledge about the relationship between channel structure and function has come from analysis of the properties of mutant channels. The chief difficulty with this approach is choosing which amino acid to alter. The voltage-gated Na^+ channel, for example, is ~2000 amino acids long, so it is clearly impossible to mutate every residue. Some strategy for selecting functionally important residues is therefore required.

Sometimes examination of the channel sequence combined with inspired guesswork has been the key, as in the case of the voltage-gated Na^+ channel where the regular sequence of positively charged residues in the S4 domain suggested that this might act as the voltage sensor (Stühmer et al., 1989). Chimeric channels, made by swapping parts of two related channels with different properties, have been used where nothing can be easily deduced from the amino acid sequence. The name *chimera* derives from the mythical Greek monster, which had the head of a lion, the body of a goat, and the tail of a serpent. Chimeras have been very useful in defining functionally important regions of a channel. For example, the pore of the voltage-gated K^+ channel was identified by making chimeras between two channels of different conductance and investigating whether substitution of a given do-main altered the single-channel conductance (Hartmann *et al.*, 1991).

Once a functionally important region has been identified, individual amino acids within it are mutated, one at a time, to determine which are the key residues. This technique is known as *site-directed mutagenesis*. It has been instrumental in unravelling the relationship between ion channel structure and function. In some cases, natural mutations may serendipitously provide information about important functional domains of an ion channel. This is the case for those channel mutations that result in human or animal disease, which are the subject of this book. For example, analysis of mutant CFTR Cl⁻ channels, which cause a mild form of cystic fibrosis, has identified residues which line the channel pore (Sheppard *et al.*, 1993).

Cysteine scanning mutagenesis is a technique that has provided insight into which residues contribute to the pore of an ion channel. In this method, individual amino acids within the putative pore region are replaced, in turn, with a cysteine residue. The effects of thiol-reactive reagents, which interact specifically with the thiol moiety (−SH) of cysteine residues, are then tested on the mutant channel. The rationale behind these experiments is that if a thiol reagent reacts with a cysteine residue that lines the channel pore, it may thereby block the conduction pathway.

Mutations are usually described by stating first the original amino acid, then its position in the protein sequence, and subsequently the amino acid to which it has been changed. Mutation of an arginine residue at position 50 to glycine is thus called R50G (in the single-letter code).

GENETIC ANALYSIS OF ION CHANNELS AND DISEASE

This section considers some of the methods that have been used to identify the chromosomal location of a gene and to clone ion channels based on their association with an inherited disease.

Chromosomal Location

The chromosomal locations of most of the ion channel genes described in this book have been identified using *fluorescent in situ hybridization* (FISH). In this technique, a probe consisting of a stretch of genomic DNA containing the gene of interest is hybridised to normal human chromosomes in meta-phase. Metaphase is one of the stages that occurs during cell division in which the DNA (which is normally found in the nucleus as an amorphous mass) condenses into distinct chromosomes (Fig. 2.8). The probe is labelled so that its site of hybridisation with the chromosome can be detected visually. The banding pattern of the chromosome is analysed concurrently using a different staining technique, which enables the location of the probe to be defined precisely.

Positional Cloning

Positional cloning (originally called *reverse genetics*) is used to describe the process by which the gene responsible for a disease is cloned by first mapping its position on the chromosome and then narrowing down its position until the correct gene is identified and cloned. Strictly speaking, the term positional cloning is used to refer to the cloning of a gene for which there was no prior functional information. It contrasts with functional cloning in which a knowledge of the function of the gene is used to clone it. The positional cloning approach is only applicable if the disease of interest is caused by a single gene. The first ion channel gene to be identified by positional cloning was not a human one, but the *Drosophila* gene which encodes the *Shaker* K$^+$ channel (Papazian *et al.*, 1987).

The first step in the positional cloning of a human disease gene is to collect large families (*pedigrees*) who suffer from the disease (Fig. 4.10). The

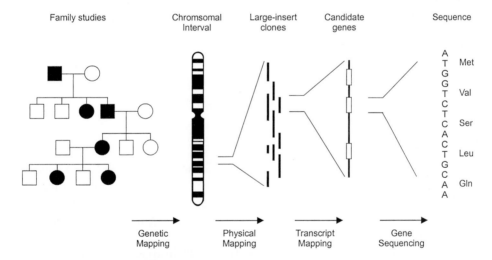

FIGURE 4.10 POSITIONAL CLONING

Outline of the different steps involved in sequencing a disease-associated gene by positional cloning. The gene is first localised to a region of a particular chromosome by genetic mapping of affected families using polymorphic markers. These markers now cover the whole of the human genome, which greatly facilitates gene mapping. The chromosome region implicated by gene mapping is then screened with overlapping DNA clones to produce a physical map of the area. Genes within the target region are then sequenced and scrutinised for the presence of mutations in affected individuals. In some cases, the target region contains genes whose known function may relate to the pathology of the disease: these are known as 'candidate genes'. After Schuler *et al* (1996).

DNA from both normal and affected family members is then screened with multiple markers to identify one or more that are linked to the disease. These markers, and the way in which they are used to establish the position of a disease-causing gene on the genetic map of the chromosome, are described in more detail later (see *polymorphic markers* and *linkage*). Gene mapping permits the chromosomal location of a disease-causing gene to be defined within 2–5 megabases. The target region is then further screened with over-lapping DNA clones to produce a physical map of the area. This is achieved using one of a number of different mapping reagents. One example is the yeast artificial chromosome (YAC) libraries which consist of yeast cells that contain individual fragments of human genomic DNA. An alternative strat-egy once the target area has been defined is to screen a database of gene sequences to identify any possible candidate genes within the region that might theoretically be responsible for the disease. If these are found, the gene is sequenced and the sequence of normal and affected patients is com-pared.

Once a candidate gene for a disease has been identified, *single stranded conformational polymorphism* analysis may be used to detect variations in the sequence of the gene. This method is based on the fact that the conformation of a single-stranded DNA molecule depends on its nucleotide sequence be-cause it forms secondary structures by intramolecular base-pairing. When subjected to an electric field, DNA will move at a rate that is determined not only by its molecular weight but also by its conformation. This property may be used to detect sequence differences, because even a point mutation (a single base-pair change) can alter the DNA conformation enough to cause a detectable shift in its mobility on an electrophoretic gel.

Linkage

Linkage is used to determine whether a disease is associated with a specific gene or region of a chromosome (Fig. 4.11). In this method, one first locates a *polymorphic marker* close to the gene. Linkage studies are facilitated by the fact that such markers have now been identified at regular intervals on each human chromosome.

The basis of the linkage method is that each individual has two copies of the marker, one from each parent. The two parental markers may be distinguished because they have different sizes and thus different mobilities on an electrophoretic gel. As many family members as possible are screened to determine whether a given marker cosegregates with the disease. The larger the family, the more confident one can be as to whether, or not, there is linkage. The distance between the polymorphic marker and the gene can be estimated from the frequency with which recombination occurs between them: the closer the two are located, the less often recombination is likely to happen. Practically, recombination is recognized when an affected family member does not carry the marker that is usually associated with the disease.

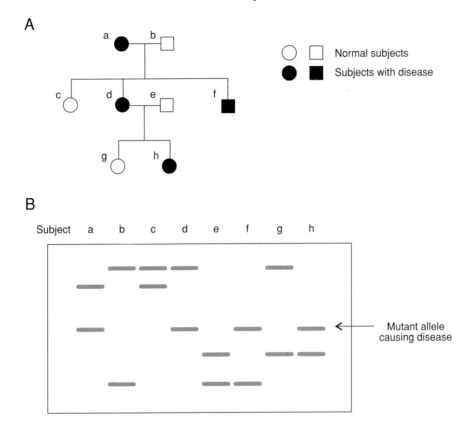

FIGURE 4.11 LINKAGE ANALYSIS

Linkage analysis using a polymorphic DNA marker (VNTR) can be used to locate the position of a gene. (A) Pedigree of a family in which some individuals have an autosomal dominant disease. Circles and squares indicate males and females, respectively. (B) A polymorphic marker whose chromosome position is known is assessed for linkage to the disease-causing gene. The marker consists of a DNA sequence referred to as a VNTR, for *variable number of tandem repeats*. Its is composed of a short block of DNA sequence that is repeated many times. The number of the repeats, and hence the length of the VNTR, varies between different individuals. Because the size of a piece of DNA determines the rate at which it runs on an electrophoretic gel, VNTRs with different numbers of repeat elements will run at different rates. In the example shown, all individuals with the disease possess the same size repeat element. This means that it is likely the VNTR lies very close (within 1 centimorgan) of the gene that causes the disease. Since the chromosomal position of the VNTR is known, this linkage provides an indication of the locus of the gene responsible for the disease.

Polymorphic markers

Polymorphic markers are used to determine the position of a gene on a chromosome: they constitute stretches of DNA that differ between individuals. A complete set of polymorphic markers for the human genome now exists, located at intervals of $\sim$ 2–5 million bp on each chromosome. There are a number of different types of polymorphic marker. Those which are most commonly used today are called *microsatellite markers*. These are hypervariable regions of DNA in which short blocks of sequence (usually a repeat of the bases cytosine and thymine, CT) are repeated many times. The number of these repeats, and thus the length of the hypervariable region, varies between individuals.

Genetically Engineered Animals

Increasingly, the functional role of a protein is being investigated using *knockout* animals, in which the gene of interest has been disrupted so that the protein is no longer made, or *transgenic* animals in which a foreign gene has been introduced. Usually, mice are used for these studies because they reproduce relatively rapidly and their genetics is well understood. It is worth noting, however, that although knockout animals have sometimes provided important and unexpected information about ion channel function, as in the case of the epithelial Na^+ channel (Hummler *et al.*, 1996; see Chapter 13), in other case the results have not been so clear-cut, either because the mutant mice are phenotypically normal, or because they have a different disease phenotype from that of man.

The simplest method of making a *transgenic* animal is inject a fertilized mouse oocyte with DNA encoding the gene of interest and an appropriate promotor. The oocyte is then implanted into a pseudopregnant mouse and, with luck, some of the offspring will express the transgene. The first generation mice will be heterozygotes, but homozygous mice can be obtained by selective breeding. One problem with this method is that it is rather inefficient, since only 10–30% of offspring are likely to carry the transgene and these cannot be identified until after birth. Another is that the foreign DNA integrates randomly into the genome, which may have undesirable consequences. For example, if the transgene inserts into the coding region of another gene it may disrupt its function, which complicates interpretation of the function of the transgene. Whether or not the transgene expresses also depends on its position in the chromosome.

It is therefore preferable to insert the transgene at a specific place in the chromosome. Such gene targeting is also essential to inactivate ('*knockout*') the wild-type gene or replace it with a mutant version. It may be achieved by exploiting the natural process of homologous recombination, in which the chromosomes of a pair exchange genetic material (see Fig. 2.11). A targeting vector is used to replace a short stretch of chromosomal DNA. This consists of the mutant gene (or part of gene) flanked by two arms homologous in sequence to the genomic DNA. Recombination between homologous sequences in the vector and target DNAs results in the substitution of the vector for the genomic DNA. A selection marker, such as a bacterial antibiotic resistance gene, is also normally included in the targeting vector, to enable homologous recombinants to be selected. To knockout a gene it is not necessary to delete the whole gene of interest, only to inactivate its function, which may be achieved by deleting a small section of the gene, such as the promoter and the first part of the coding sequence.

Because homologous recombination is a relatively rare event, it is now common to use embryonic stem cells (ES cells), which are derived from the mouse blastocyst, for the recombination step. These may be rapidly screened *in vitro* to identify recombinant cells in which the gene has been knocked out. Those cells that carry the foreign DNA can then be injected into a blastocyst and implanted into a pseudopregnant mouse. With luck, the embryos will be viable and develop into chimeric mice in which some cells will be derived from the recipient blastocyst and others from the injected ES cells. By using a tissue marker, such as fur colour, it is possible to identify these chimeric animals. Some of the chimeric animals will carry the foreign gene in their germ cells and their offspring will therefore carry the foreign gene in all their cells. As it is not easy to identify those mice in which the germ cells derive from the ES cells, and therefore carry the foreign DNA, this necessitates an intensive breeding program and usually requires much time, patience and money. Any offspring that carry the foreign DNA will heterozygotes and further breeding is needed to obtain homozygous animals.

The Cre/loxP system

A number of technical advances in gene targeting have been made in recent years. One of these is the Cre/loxP system. This enables the production of *conditional knockout* animals, which only express the foreign gene in selected tissues, and of *inducible knockout* animals in which expression of the mutant gene can be turned on at the developmental stage determined by the investigator. It can also be used to introduce point mutations into the target gene and to remove marker genes that are used to select homologous recombinants.

The Cre/loxP system consists of two elements: a bacterial enzyme Cre recombinase and a 34-bp DNA sequence referred to as loxP. Cre recombinase recognises the loxP sequence and mediates recombination between two loxP sites. The loxP system therefore involves two steps. First the loxP sites are inserted into the genome of the ES cell and the recombinant clones are identified and isolated. Secondly, Cre recombinase expression is activated, resulting in Cre-mediated site-specific recombination. Depending of the location and orientation of the loxP sites this may lead to deletion, insertion, or inversion of a target sequence in a single strand of DNA, or recombination between two DNA molecules. One of the most straightforward uses of the Cre/loxP system is to remove a selectable marker, after homologous recombinants have been identified in ES cells; another is to knockout a specific gene. This may be achieved by flanking the marker (or gene) with loxP sites.

The beauty of the loxP system is that recombination occurs only in the presence of Cre recombinase. By expressing Cre recombinase either *in vitro* (in ES cells) or *in vivo* (in the animal), it is possible to achieve great flexibility in gene targeting. If recombination is carried out *in vitro*, all the tissues of the mouse will express the mutant gene throughout life. Alternatively, Cre recombinase expression may be activated *in vivo*, which enables more precise control of gene targeting. For example, if the enzyme is expressed under the control of a tissue-specific promotor, the mutant gene will only be expressed in selected tissues: a conditional knockout is produced if the mutation results in gene inactivation. By using an inducible promotor that is switched on upon administration of the inducer to the adult mouse, it is also possible to activate Cre recombinase expression (and thus recombination) at will. This is particularly useful if it is desired to inactivate a gene once the animal is adult—for example, because it causes embryonic death. Inducible gene knockouts may also be valuable for assessing the functional role of neuronal ion channels because the plasticity of the nervous system means compensatory developmental changes may occur if conventional knockout technology is used.

CHAPTER 5

VOLTAGE-GATED Na⁺ CHANNELS

VOLTAGE-GATED Na⁺ CHANNELS

Voltage-gated Na^+ channels are responsible for the Na^+ current that underlies the rapid upstroke of the action potential in nerve and muscle fibres. Mutations in sodium channel genes may therefore be expected to result in diseases of nerve and muscle, including cardiac muscle. Mutations in the gene encoding the α-subunit of the human skeletal muscle Na^+ channel give rise to a group of diseases known collectively as the periodic paralyses, while mutations in the cardiac muscle Na^+ channel gene result in long QT syndrome. Febrile convulsions and epilepsy are associated with a mutation in the gene encoding the β-subunit of the human neuronal Na^+ channel.

Our understanding of how Na^+ channel mutations produce their different clinical phenotypes has been facilitated by the plethora of studies on the relationship between the channel structure and function that have appeared since the first Na^+ channel was cloned in 1984 (Noda *et al.*, 1984). These studies have focused on understanding three of the most important properties of Na^+ channels: their voltage-dependent activation in response to depolarization, their inactivation during a maintained depolarization and the high selectivity of the channel pore to Na^+ ions. This chapter considers what has been learnt from this work and how the various Na^+ channel mutations cause defects in channel function.

Basic Na⁺ Channel Properties

One of the most distinctive properties of the voltage-gated Na^+ channel is its marked voltage sensitivity. At the resting potential, the open probability of voltage-gated Na^+ channels is extremely low and very few channels ever open. Depolarization dramatically increases the probability of channel opening: a 9 mV depolarization, for example, may increase the open probability by as much as one order of magnitude. The channel open probability is also time dependent. In response to depolarization, the channels open briefly (for <1 msec) after a short latency and then close to an inactivated state which persists until the membrane is hyperpolarized. This produces a macroscopic

Na⁺ current that rises to a peak within about 1 msec and then declines (Fig. 5.1). Voltage-gated Na⁺ channels are highly selective for Na⁺ ($P_K : P_{Na} \sim 0.08$) and have a single-channel conductance of 10–30 pS in 100 mM extracellular Na⁺.

Subunit Composition

Voltage-dependent Na⁺ channels are usually composed of a principal, or α, subunit and one or more smaller β-subunits (Catterall *et al.*, 1986). Functional activity is observed when the α-subunit is expressed in heterologous systems on its own, indicating that it possesses all the necessary structural elements for channel formation. The β-subunit enhances the Na⁺ current amplitude and modifies its properties (Isom *et al.*, 1992; Jia *et al.*, 1994). The α-subunit of the rat skeletal muscle Na⁺ channel, for example, inactivates more slowly than the native channel, whereas channels formed by coexpression of both α- and β-subunits inactivate at a similar rate to the native channel (Fig. 5.1). The skeletal muscle Na⁺ channel has only a single β-subunit, whereas the brain Na⁺ channel has two β-subunits, β_1 and β_2. These associate with the α-subunit in a 1:1 or 1:1:1 stoichiometry.

Voltage-gated Na⁺ channels comprise a major gene family and multiple types of α- and β-subunits have been cloned. The rat, for example, has at

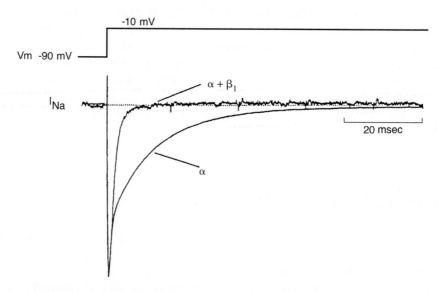

FIGURE 5.1 THE β-SUBUNIT REGULATES Na⁺ CHANNEL INACTIVATION

Na⁺ currents recorded from two different *Xenopus* oocytes expressing either the α-subunit of the rat Na⁺ channel alone, or the α-subunit in combination with the β_1-subunit. The traces have been normalized to the peak current level. From Ji *et al.* (1994).

TABLE 5.1   **MAMMALIAN Na⁺ CHANNEL GENES**

Gene	Protein	Chromosome (human)	Major site of expression	Reference
SCN1A	α_1	2q24	Brain and spinal cord	Malo *et al*, (1994)
SCN2A	α_2	2q23–q24.3	Brain and spinal cord	Litt *et al*, (1989)
SCN3A	α_3	2q24–q31	Brain and spinal cord	Malo *et al*, (1994)
SCN4A	α_4	17q23–q25	Skeletal muscle	George *et al*, (1991)
SCN5A	α_5	3p21	Cardiac muscle	George *et al*, (1995)
SCN6A	α_6	2q21–q23	Heart and uterus	Han *et al*, (1991)
SCN7A	α_7	—	Glia	
SCN8A	α_8	12q13	Brain and spinal cord	Burgess *et al*, (1995)
SCN1B	β_1	19q13.1		Makita *et al*, (1994a,b)

least four kinds of brain Na⁺ channel α-subunit, as well as different types expressed in peripheral neurones, skeletal muscle and cardiac muscle. Their human equivalents are shown in Table 5.1. An additional putative voltage-gated Na⁺ channel α-subunit has also been identified in heart and in uterine smooth muscle; because this channel shows only ~50% sequence identity with the other α-subunits it appears to belong to a separate subfamily. Two kinds of β-subunit have also been identified. The different Na⁺ channel genes are expressed in different tissues and may be differentially expressed during development. The Na⁺ channels they encode may also have different properties: for example, the cardiac Na⁺ channel is less sensitive to the Na⁺ channel blocker tetrodotoxin (TTX) than the brain Na⁺ channel.

In man, the gene encoding the cardiac muscle Na⁺ channel α-subunit (*SCN5A*) is located on chromosome 3p21, and that coding for the adult skeletal muscle Na⁺ channel α-subunit (*SCN4A*) lies on chromosome 17q23-25 (Table 5.1). A cluster of genes on human chromosome 2q21-24 encode several types of brain Na⁺ channel α-subunits (*SCN1A, SCN2A, SCN3A, SCN6A*).

Structure

The α-subunit of voltage-dependent Na⁺ channels is composed of around 2000 amino acid residues (Fig. 5.2). The sequence contains four homologous repeats (I to IV) that probably arose by duplication of a single primordial gene. Hydropathy analysis suggests that each of these repeats contains six transmembrane segments (called S1, S2, etc.), and that the residues that link the repeats, like the N and C termini, lie on the cytoplasmic side of the membrane. Evidence discussed in detail later in this chapter supports the view that the S4 segment, which contains between 6 and 8 charged residues, serves as the primary voltage sensor of the channel; that part of the extracellular loop which links S5 and S6 dips down into the membrane and lines the

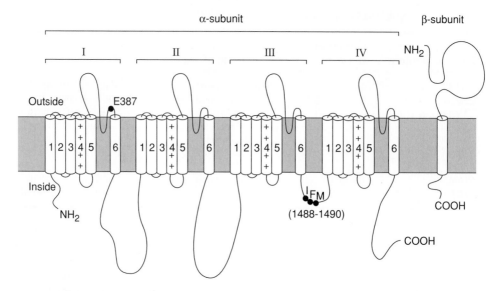

*FIGURE 5.2 PUTATIVE TOPOLOGY OF THE VOLTAGE-GATED Na⁺
CHANNEL*
Topology of α- and β-subunits of the rat brain Na⁺ channel (type IIA)
deduced from hydropathy analysis in conjunction with mutagenesis and
antibody studies. Residues mentioned in the text are indicated: E387 is
involved in TTX binding and IFM (1488–1490) form part of the inactivation gate.

walls of the ion pore; and that the cytosolic linker between repeats III and
IV plays a key role in the inactivation process.

The tertiary structure of the α-subunit remains a matter of speculation.
The fact that there are four homologous repeats argues that these domains
assemble in a four-fold symmetry surrounding a central permeation pathway.
Current models locate the S6 segments in the centre of the complex, adjacent
to the S5 segments, with the S5-S6 linker dipping down into the membrane
to form the pore. The S4 segment is completely surrounded by protein and
is placed within the centre of each repeat. The S1-S3 segments are probably
located adjacent to the lipid bilayer and interact both with residues in the
other protein segments and with membrane lipids.

The β-subunit consists of a single transmembrane domain with an extracellular N terminus and an intracellular C terminus (Fig. 5.2). β-subunits
tend to be heavily glycosylated; the glycosylated form of the rat brain Na⁺
channel β_1-subunit, for instance, has a molecular mass of 36 kDa whereas
the unglycosylated form is 23 kDa (Isom *et al.*, 1992). A single gene encodes
the β_1-subunit expressed in human heart, brain and skeletal muscle. It is
located on chromosome 19 (Makita *et al.*, 1994b).

Correlating Na$^+$ Channel Structure and Function

This section considers some of the experimental evidence that has led to our current view of the relationship between the primary structure of the α-subunit of the voltage-gated Na$^+$ channel and its functional properties. Because this is the first chapter to discuss channel structure and function, the description is more comprehensive than in later chapters and serves an introduction to this type of analysis.

The pore

A combination of site-directed mutagenesis studies and toxin-binding studies have determined that the loop between the fifth and sixth transmembrane domains forms the outer part of the Na$^+$ channel pore. Tetrodotoxin (TTX) has been used to map the outer mouth, and local anaesthetics to define the inner mouth, of the Na$^+$ channel pore. Only the former approach is considered here.

TTX is found in the ovaries and liver of the puffer fish *Fugu*. The firm white flesh of this fish is eaten raw (as sashimi) in Japan and is an expensive delicacy. The testes are also highly prized. Part of the pleasure of eating fugu lies in the tingling sensation in the lips and tongue produced by low concentrations of TTX, coupled with the knowledge that you are dicing with death: despite the fact that all fugu chefs are licensed, poorly prepared fugu causes several fatal cases of TTX poisoning each year. One reason for this is that the puffer fish is hermaphrodite and it is difficult to separate the testes from the ovary, which contains toxic amounts of TTX.

TTX sits in the outer mouth of the Na$^+$ channel and physically plugs the pore. It was realised early on that this meant that identification of the residues involved in binding TTX would help determine the location of the ion channel pore. Noda and colleagues (1989) found that neutralization of the negatively charged glutamate residue at position 387 (to glycine) abolished the ability of TTX to inhibit the rat brain Na$^+$ channel, which implies that this amino acid must be located at the outer mouth of the pore (Fig. 5.3). Glutamate 387 lies within the region of repeat I that links S5 and S6, and is sometimes known as SS1-SS2 (Fig. 5.2). Mutation of each of the negatively charged residues in the equivalent position in the other three repeats also prevented TTX binding (Terlau *et al.*, 1991). This indicated that the SS1-SS2 region of all four repeats contributed to the pore of the Na$^+$ channel and defined the receptor site for the positively charged TTX molecule as a ring of negative charges surrounding the external opening of the pore.

Both cardiac muscle and denervated skeletal muscle express a Na$^+$ channel that is >200-fold less sensitive to TTX than the brain Na$^+$ channel. When this channel was cloned it was found that it possessed the same ring of negative charges in the SS1-SS2 region as the brain Na$^+$ channel, despite its

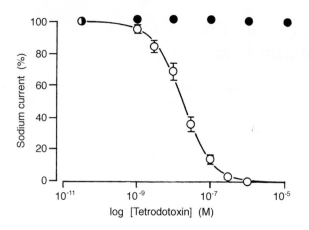

FIGURE 5.3 TTX DEFINES THE OUTER MOUTH OF THE Na⁺ CHANNEL
(A) Effect of TTX on wild-type Na$^+$ channels (○) and mutant Na$^+$ channels in which glutamate 387 has been replaced by glycine (●). From Noda *et al.* (1989).

insensitivity to TTX. Further comparison of the sequences revealed that a neighbouring tyrosine, at position 385 in the brain Na$^+$ channel, was replaced by a cysteine in the cardiac channel. Mutation of this cysteine to tyrosine rendered the cardiac Na$^+$ channel highly sensitive to TTX, demonstrating that this residue is also important for toxin binding and confirming the importance of the SS1-SS2 region in forming the outer part of the pore (Satin *et al.*, 1992).

Perhaps not surprisingly, mutations that alter TTX sensitivity also have effects on the single-channel conductance and ion selectivity. Neutralization of another charge in the SS1-SS2 region (D384N) almost completely abolished the Na$^+$ current but did not affect the gating current (gating currents are caused by the movement of charged groups within the membrane and precede channel activation). This argued that the mutant channel was present in the membrane and had a normal voltage-dependence, but was unable to conduct ions (Pusch *et al.*, 1991). It also suggested that the SS1-SS2 region forms part of the pore of the Na$^+$ channel. Confirmation of this idea was provided by other mutations in the SS1-SS2 region that produced dramatic alterations in ion selectivity (Heinemann *et al.*, 1992). Consider Fig. 5.4, which compares the sequence of this region in the four repeats of the voltage-gated Na$^+$ and Ca^{2+} channels. The arrowed position indicates a negatively charged glutamate residue that is conserved in all four repeats of the Ca^{2+} channel (see also Chapter 9). A negatively charged residue, either aspartate or glutamate, is also present in two of the Na$^+$ channel repeats, but in repeat III it is replaced by a positively charged lysine (K1422) and in repeat IV with a neutral alanine

Na⁺ I	L	M	T	Q	**D**	F	W
Na⁺ II	V	L	C	G	**E**	W	I
Na⁺ III	V	A	T	F	***K***	G	W
Na⁺ IV	I	T	T	S	A	G	W

Ca²⁺ I	C	I	T	M	**E**	G	W
Ca²⁺ II	I	L	T	G	**E**	D	W
Ca²⁺ III	V	S	T	G	**E**	G	W
Ca²⁺ IV	S	A	T	G	**E**	A	W

FIGURE 5.4 RESIDUES IN THE SS1–SS2 DOMAIN DETERMINE Na⁺ CHANNEL SELECTIVITY
Lineup of amino acid residues in the pore (SS1-SS2) region in each of the four repeats of Na⁺ and Ca²⁺ channels. Residues implicated in Na⁺ selectivity are indicated by the arrow: negatively charged residues are shown in bold (**E**, glutamic acid; **D**, aspartic acid), and positively charged residues are shown in bold italics (***K***, lysine). A, alanine. Mutant Na⁺ channels in which either alanine or lysine was replaced with glutamate had Ca²⁺ channel properties.

(A1712). These differences in charge account for the high permeability of Ca^{2+} through the Ca^{2+} channel, because mutation of K1422 or A1712 to glutamate caused the Na⁺ channel to resemble the Ca^{2+} channel in its selectivity properties (Heinemann *et al.*, 1992). Other Ca^{2+} channel properties, such as block by divalent cations and permeability to monovalent cations, are also conferred by these residues. This provides good evidence that K1422 and A1712 constitute part of the selectivity filter of the Na⁺ channel and lie within the permeation pathway.

Activation

Activation of voltage-dependent Na⁺ channels results from a voltage-driven change in the protein conformation that results in the opening of the ion pore. The first step towards understanding the molecular basis of voltage-dependent activation is to identify the residues that serve as the voltage sensor. Activation is associated with non-linear capacititative currents, known as gating currents, which reflect the movement of charge within the membrane. This suggests the voltage sensor is likely to consist of one or more charged amino acids located in the transmembrane voltage field and that these amino acids might move through the membrane voltage field during the activation process. When the primary sequence of the Na⁺ channel was examined, the fourth transmembrane domain in every repeat (S4) stood out as an obvious candidate for a voltage sensor because it contains a positively charged residue, usually arginine or lysine, at every third position, with two nonpolar residues interposed between them. This structure repeats several times to form an amphipathic helix (Fig. 5.5A). The idea that the positive charges in the S4 region act as the voltage sensor is supported by the fact that the S4 segment is found in most voltage-sensitive cation-selective channels so far cloned, and that its primary sequence is highly conserved.

How does one determine whether the S4 segment does indeed act as a voltage sensor? The general approach to this problem has been to mutate individual residues in the S4 domain and then look to see if this has influenced the voltage dependence of activation. Such studies have shown that a reduction in the net positive charge in the S4 segment, produced by replacing positive charges with neutral or negatively-charged ones, leads to a decrease in the steepness of the voltage-dependence of activation (Stühmer *et al.*, 1989; Fig. 5.6). In general, the more charges that are replaced, the shallower the slope of the relationship. The picture is complicated, however, by the fact that individual residues do not always contribute equally to the overall gating charge. The size and shape of the hydrophobic (neutral) residues are also important (Auld *et al.*, 1990) perhaps because they influence the ability of the S4 segment to move through the membrane during the process of activation.

The mechanism by which the S4 segments serve as the voltage sensor is still not completely understood. There is evidence, however, that the S4 segment actually moves physically through the membrane, the external charges becoming exposed on the outer surface and the inner charges becoming buried in the membrane. This evidence comes from experiments in which

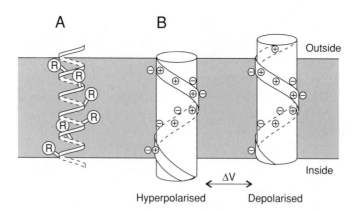

FIGURE 5.5 THE S4 HELIX MOVES DURING VOLTAGE-DEPENDENT ACTIVATION

(A) The S4 segment of repeat I of the rat brain Na⁺ channel portrayed with a ball-and-stick model. Positively charged arginine (R) residues are indicated. (B) Sliding helix model illustrating S4 movement in response to depolarization. From Catterall (1986).

A

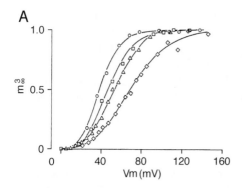

B

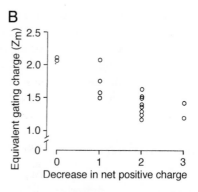

FIGURE 5.6 THE AMOUNT OF CHARGE IN THE S4 SEGMENT
INFLUENCES Na⁺ CHANNEL ACTIVATION
(A) Voltage dependence of steady-state activation of wild-type Na⁺ channels (O) and of mutant Na⁺ channels in which one (□), two (△), or three
(◇) charged residues in the S4 segment of repeat I were neutralized. The
more charges are neutralised, the less steep the activation curve. (B) The
equivalent gating charge (z_m) decreases as the total number of net positive
charges in the S4 segment of repeat I of the α-subunit of the Na⁺ channel is
decreased. Replacement of a positively charged residue with a negatively
charged one is counted as a decrease of two charges. z_m is a measure
of the steepness of the relationship between membrane potential and
activation. Currents were recorded from *Xenopus* oocytes injected with
mRNA encoding the rat brain Na⁺ channel. After Stühmer *et al.* (1989).

each of the charged residues in the S4 domain was replaced in turn by
a cysteine (Yang *et al.*, 1996). Methanethiosulfonate (MTS) reagents react
specifically with cysteines and thereby influence the rate and voltage dependence of inactivation. This will only occur if the cysteine is exposed to the
solution. It is therefore possible to determine the position of a given residue
(relative to the membrane) by testing the effects of external, or internal,
application of MTS reagents on inactivation; and by repeating the experiment
at different membrane potentials to work out if its position is influenced by
voltage. This type of experiment has revealed that the S4 domain moves from
the inside to the outside of the membrane in response to depolarization,
exposing two positive charges that were previously buried in the membrane
as it does so.

One idea that has been put forward to explain how the S4 helix moves
is known as the 'sliding helix' or 'helical screw' model (Fig. 5.5B). This
suggests that the S4 segment adopts an α-helical structure in which the
positively charged arginine residues are stabilized by interaction with negatively charged residues in adjacent transmembrane domains. In response to
a voltage change the S4 helix moves across the membrane in a spiral path,
exchanging ion pairs between the positively charged S4 residues and fixed
negative charges in the surrounding transmembrane segments. Because there
are four S4 segments, the movement of a single positive charge across the

membrane could account for the observed voltage dependence of gating (gating studies predict 4 to 6 charges are transferred). However this model is probably too simple. One reason for thinking so is that the model implies that there will be an equivalent amount of charge movement produced by the different charged residues in the S4 helices, whereas, in practice, neutralisation of individual residues may have very different effects on the voltage dependence of gating.

Inactivation

Na⁺ channels display two modes of inactivation, fast and slow, which are probably mediated by different molecular mechanisms. Fast inactivation describes the rapid decay of Na⁺ currents observed in response to short depolarizations. Slow inactivation occurs when nerve and muscle fibres are depolarized for seconds or minutes. We will first consider the molecular basis of fast inactivation.

Fast inactivation

Even before the advent of genetic manipulation of ion channels, it was known that the fast inactivation site was located on the inside of the membrane, since intracellular perfusion of the squid axon with the proteolytic enzyme pronase removed inactivation (Armstrong et al., 1973). This means that the inactivation gate must be located on the inside of the membrane and that it is accessible to cytoplasmic enzymes. Based on this and other studies, Armstrong and Bezanilla (1977) proposed that inactivation resulted from the movement of part of the protein, "the inactivation particle", into the inner mouth of the pore, where it bound and blocked ion fluxes by plugging the pore. They further suggested that the inactivation particle was tethered to the cytoplasmic side of the channel and thus accessible to proteolysis. Once the sequence of the Na⁺ channel was known, it seemed plausible that one or more of the large cytoplasmic loops that link the various repeats might participate in inactivation. Two types of experiments have provided evidence that the loop between repeats III and IV is important for Na⁺ channel inactivation; those that involved the use of antibodies and those that employed site-directed mutagenesis.

Vassilev and colleagues (1988) found that antibodies directed against the region between repeats III and IV of the Na⁺ channel slowed inactivation whereas antibodies directed at other intracellular regions had no effect. The rate at which the active antibody bound and the extent to which it slowed inactivation were voltage dependent. At negative membrane potentials, where Na⁺ channels do not inactivate, the antibody bound rapidly and inhibited Na⁺ channel inactivation on depolarization. When applied at more positive membrane potentials, however, where Na⁺ channels are partially inactivated, the antibody was much less effective. Based on these results it was proposed that the linker between repeats III and IV serves as the inactivation gate, and becomes inaccessible to antibody binding when the channel is inactivated.

A dramatic (>20-fold) slowing of Na^+ current inactivation was also observed when the mRNA encoding the Na^+ channel was cut between repeats III and IV and the two parts of the molecule were then coexpressed in *Xenopus* oocytes (Stühmer *et al.*, 1989). By contrast, no difference was found if the mRNA was cut between repeats II and III. The voltage-dependence of inactivation was unaffected when the linker between repeats III and IV was cut, despite the marked slowing of inactivation. These effects are very similar to those observed when native sodium channels are treated with pronase. As one might expect from these results, the amino acid sequence of the channel suggests that the cytoplasmic loop linking repeats III and IV of the Na^+ channel contains several cleavage sites for proteolytic enzymes.

Subsequent studies showed that the charged residues within the inactivation loop between repeats III and IV are not required for fast Na^+ channel inactivation (Patton *et al.*, 1992). Instead, a cluster of three hydrophobic residues, isoleucine-phenylalanine-methionine (IFM), were found to be important. The phenylalanine residue at position 1489 is the most critical, as mutation of this amino acid to glutamine completely blocks inactivation (Fig. 5.7A) (West *et al.*, 1992). Mutation of the adjacent two residues, I1488 or M1490, to glutamine also slowed inactivation, but to a lesser extent. On the basis of such studies it has been proposed that the IFM residues serve as a hydrophobic latch that stabilizes the inactivated state of Na^+ channel. This is sometimes known as the 'hinged-lid' model of inactivation because of the resemblance of the inactivation loop to the hinged lids of allosteric enzymes. These enzymes have rigid peptide loops that fold over the active site and control substrate access. Conformational changes induced by the binding of allosteric ligands move the hinged lid away and allow substrate access and catalytic activity. By analogy, it is proposed that the inactivation loop of the Na^+ channel may function as a rigid lid that controls access to the pore and that it is latched in the closed position in the inactivated state by a hydrophobic interaction that involves the IFM residues (Fig. 5.7B).

The site at which the inactivation gate binds to inhibit ion fluxes is likely to be located at the cytoplasmic mouth of the pore. Identification of this acceptor site therefore gives a clue to the identity of those residues that contribute to the intracellular mouth of the permeation pathway. Using an approach similar to that outlined earlier, mutagenesis studies suggest that the cytoplasmic ends of transmembrane domains S5 and S6 and parts of the cytoplasmic loops which link them to adjacent segments, contribute to the acceptor site for the inactivation gate.

Slow inactivation

Slow inactivation is not only kinetically distinct from fast inactivation, it also involves different structural elements. Slow inactivation is preserved even after fast inactivation has been abolished by intracellular application of proteolytic enzymes; and mutations within the III-IV linker that abolish fast inactivation have no effect on slow inactivation. The mechanism of slow inactivation is not fully understood. However, mutations at the cytoplasmic

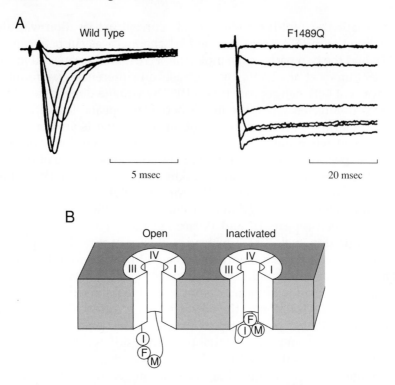

FIGURE 5.7 MECHANISM OF Na⁺ CHANNEL INACTIVATION
Phenylalanine 1489 (F) is critical for inactivation of Na⁺ channels. (A) Currents recorded from *Xenopus* oocytes in response to depolarizing pulses from −90 to +55 mV from a holding potential of -100 mV. Oocytes were coinjected with mRNAs encoding the β_1-subunit and either the wild-type α-subunit of the rat brain Na⁺ channel or a mutant α-subunit in which residue F1488 was replaced with glutamine (Q). Note the different time scales. (B) Model of the hinged-lid mechanism of inactivation. The intracellular loop connecting repeats III and IV is depicted as forming a hinged lid. Critical residues for inactivation, IFM 1488–1490, are indicated. From West *et al.* (1992).

ends of S5 and S6, which are thought to form the inner mouth of the pore, disrupt slow inactivation (Hayward *et al.*, 1997). This suggests that slow inactivation may involve a conformational change at the inner mouth of the channel that results in channel closure.

Summary of structure-function studies

It is now fairly well established that the S4 segment of the voltage-gated Na⁺ channel acts as the primary voltage sensor, that the SS1-SS2 region connecting the S5 and S6 segments forms at least the outer part of the ion pore, and that both the cytoplasmic loop linking repeats III and IV (the inactivation ball) and the internal ends of S5 and S6 (the acceptor site) are involved in inactivation. It is far less clear to what extent other parts of the protein contribute to activation, inactivation or ion selectivity, although there is good reason to believe that other domains will be involved because many mutations have been identified in unexpected places that alter one or more of these important Na^+ channel properties. In general, however, the effects of Na^+ channel mutations that give rise to human disease can be explained on the basis of our current knowledge of the Na^+ channel.

DISEASES OF MUSCLE Na⁺ CHANNELS

Hyperkalaemic Periodic Paralysis, Paramyotonia Congenita, and Potassium-Aggravated Myotonia

Mutations in the α-subunit of the human skeletal muscle Na^+ channel are associated with hyperkalaemic periodic paralysis (HyperPP), paramyotonia congenita (PC), and a diverse group of disorders known collectively as potassium-aggravated myotonias (PAM) (Cannon, 1996; Barchi, 1995). These diseases are characterised by skeletal muscle hyperexcitability or muscle weakness that is exacerbated by an increase in the plasma potassium concentration or by exposure to cold temperatures. Onset of these conditions generally occurs within the first or second decade of life. All of the disorders are inherited as dominant traits and show linkage to the skeletal muscle Na^+ channel gene (SCN4A) on chromosome 17q23-25 (Fontaine et al., 1990; Ebers et al., 1991) (Table 5.1).

Clinical features

Myotonia [from myo (muscle) and tonus (tension)] describes the muscle stiffness experienced by affected individuals. The ability of the muscle to relax is impaired so that patients have difficulty in opening a clenched fist, releasing the grip of a hand shake, or opening their eyes after squinting in bright light. The myotonia results from enhanced muscle excitability, and repetitive activity (known as the myotonic run) is observed in the electromyogram (EMG). Periodic paralysis lies at the other end of the spectrum of muscle excitability and describes the episodic attacks of muscle weakness that can occur. It is associated with a silent EMG. This is because the muscle fibres are depolarised from their normal resting potential of ~-90 mV to about -50 mV, a potential at which Na^+ channels are inactivated and action potentials therefore cannot be generated. Both myotonia and paralysis may be experienced by the same patient, sometimes within a few minutes of each other.

Hyperkalaemic periodic paralysis may occur spontaneously, but attacks are usually precipitated by exercise, stress, fasting or eating K^+-rich foods (such as bananas). Patients are unaffected during exercise but exhibit muscle weakness and paralysis within 10 min of ceasing activity. This may be so severe that the patient is unable to remain standing. Paralysis is often preceded by signs of muscle hyperexcitability such as myotonia or fasciculations (rippling contractions). The duration of paralysis is variable and can last from minutes to hours. HyperPP is associated with a mild elevation of the blood potassium concentration and can be precipitated by the ingestion of potassium-rich foods which raise the plasma K^+ from its normal level of ~4 mM to between 5 and 7 mM. Since plasma potassium rises during strenuous exercise to as much as 8 mM[1] (Medbo and Sejersted, 1990), it seems possible that K^+ may also be a precipitating factor in exercise-induced attacks: if this is the case, however, some additional factor must protect the muscle during the activity itself because the attack is only initiated on cessation of exercise.

Patients with paramyotonia congenita often discover their problem during cold weather. They find that their hands become clamped to the cold metal handlebars of their bicycle, that they are unable to release their grip on the spade they have been using to shovel snow, or that they become stiff and weak after playing football in winter. In these individuals, myotonia is precipitated by cold and, paradoxically, is aggravated by exercise (in contrast to most classical myotonias). In some patients, the myotonia may also be followed by prolonged paralysis. Potassium-aggravated myotonia is, as its name suggests, characterised by myotonia without muscle weakness or paralysis. Its clinical features are similar to those of classical myotonia (Chapter 10), but can be distinguished by the fact that the myotonia of PAM patients is exacerbated by mild elevation of the plasma K^+ concentration.

The periodic paralyses and myotonias are painless and not life-threatening, although they are obviously very inconvenient. Given that the intercostal and diaphragm muscles are both skeletal muscle, it is perhaps surprising that respiratory problems are extremely rare—only a few cases, in which respiratory muscle stiffness was associated with the mutation G1306E, have ever been reported. Over time, individuals may develop a generalised myopathy associated with the appearance of vacuoles in the muscle fibres. The cause of this remains unknown but it may be a result of cumulative Ca^{2+}-induced toxicity as a consequence of the myotonia.

It is possible that the periodic paralyses are more common than is recognised. Because they all show autosomal dominant inheritance, there are often several affected family members. Some individuals may never consult a doctor as their families know of ways in which to prevent or reduce the severity of an attack, such as ingestion of a carbohydrate-rich meal or following a gradual 'warm-down' routine after exercise.

Molecular basis

Studies on muscle biopsies from affected patients first showed that the periodic paralyses were associated with a failure of the Na^+ current to inactivate fully (Lehmann-Horn et al., 1983; 1987). Following the cloning of the gene for the human skeletal muscle Na^+ channel (George et al., 1992), the mutations involved were rapidly identified (for a review see Barchi, 1995).

---

[1] Although 8 mM K would cause cardiac arrest in a resting subject, this does not occur during exercise, apparently due to the elevated catecholamine levels (Paterson, 1996).

Understanding how these mutations give rise to the disease phenotype has been greatly facilitated by the wealth of biophysical information available on the relationship between the structure and function of Na⁺ channel, discussed earlier. In turn, analysis of their properties has provided fresh insights into this relationship.

Around 20 disease-causing mutations have been identified in the human skeletal muscle Na⁺ channel to date, 4 of which produce hyperkalaemic periodic paralysis, 9 of which cause paramyotonia congenita, and 6 of which are associated with potassium-aggravated myotonia (Fig. 5.8 and Table 5.2). Certain mutations, however, occur more frequently than others. The mutations T704M and M1592V are found in more than 90% of families with HyperPP, whereas paramyotonia congenita is most frequently associated with T1313M and R1448H. The potassium-aggravated myotonias may result from a spectrum of mutations at residue 1306, in which the wild-type glycine residue is replaced by alanine (A), valine (V) or glutamate (E). The severity of the clinical symptoms is influenced by the degree to which the substitution differs from glycine, with E > V > A. All the known Na⁺ channel mutations

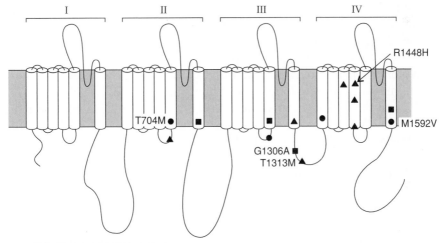

■ Potassium-aggravated myotonia
● Hyperkalaemic periodic paralysis
▲ Paramyotonia congenita

FIGURE 5.8 MUTATIONS ASSOCIATED WITH HyperPP, PARAMYOTONIA CONGENITA AND POTASSIUM-AGGRAVATED MYOTONIA
Schematic representation of the α-subunit of the human skeletal muscle Na⁺ channel SCN4A with mutations indicated. Residues associated with the most frequent mutations are numbered. Note that the human skeletal muscle Na⁺ channel is 1836 residues long and that the numbering does not correspond to that for the rat brain Na⁺ channel in Fig. 5.2. ●, hyperkalaemic periodic paralysis; ▲, paramyotonia congenita; ■, potassium-aggravated myotonia.

TABLE 5.2   DISEASE-CAUSING MUTATIONS IN SNC4A

Phenotype	Codon change	Phenotype	Codon change
HyperPP	T704M	PC	I693T
HyperPP	A1156T	PC	V1293I
HyperPP	M1360V	PC	T1313M
HyperPP	M1592V	PC	L1433R
		PC	R1448H
PAM	S804F	PC	R1448C
PAM	I1160V	PC	R1448P
PAM	G1306A	PC	V1458F
PAM	G1306V	PC	F1473S
PAM	G1306E		
PAM	V1589M		

Note: HyperPP, hyperkalaemic periodic paralysis; PC, paramyotonia congentia; PAM, potassium-aggravated myotonias. For references, see Barchi (1995) and Cannon (1997).

are substitutions that result in a single amino acid change and no frameshifts, deletions, or splicing errors have been reported. For unknown reasons, no mutations have been observed in repeat I of the Na⁺ channel.

The Na⁺ channel mutations may be grouped into three major classes: those which occur in the S4 domain of repeat IV (e.g., R1448H), those which occur in the inactivation loop between repeats III and IV (e.g., T1313M, G1306E), and those that occur at the cytoplasmic end of segments 5 or 6 (e.g., T704M, M1592V), which lie at the inner mouth of the pore and may form part of the receptor for the inactivation ball. Most of these mutations affect inactivation and produce a persistent inward Na⁺ current, although they do so by different mechanisms and to differing extents.

Mutations in the acceptor site for the inactivation gate

Mutations which occur in the acceptor site for the inactivation gate produce macroscopic Na⁺ currents that do not inactivate completely and exhibit a small persistent component of current with an amplitude ~5% of that of the peak current (Fig. 5.9) (Cannon *et al.*, 1991). In contrast, wild-type Na⁺ currents inactivate almost completely. The basis of this difference has been established by analysis of the single-channel currents carried by mutant channels. In response to depolarization, wild-type Na⁺ channels open briefly and then inactivate. Na⁺ channels with the M1592V mutation also exhibit this type of activity. Sometimes, however, the mutant channel enters a different mode in which it continues to open and close throughout the pulse for several successive depolarizations. The openings are also of longer duration. This behaviour gives rise to the sustained component of the macroscopic current. On rare occasions, wild-type Na⁺ channels also show this mode of gating. Thus it appears that the M1592V mutation does not cause a new mode of gating, but rather favours entry into a noninactivating state with a longer

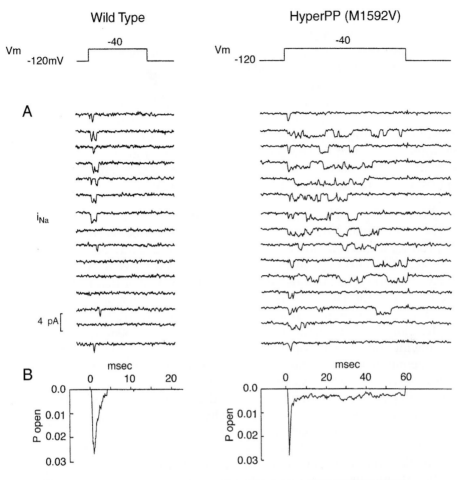

FIGURE 5.9 HyperPP MUTATIONS ALTER Na$^+$ INACTIVATION
Na$^+$ channels bearing the HyperPP mutation M1592V (in the inactivation ball acceptor site) show altered inactivation. (A) Single Na$^+$ channel currents were recorded from cell-attached patches on normal and HyperPP myotubes in response to successive depolarizations. (B) Ensemble (average) currents recorded by averaging unitary currents records such as those shown in *A*. The latency to the first channel opening and the single-channel current amplitudes are unchanged, but HyperPP channels have longer open times and can reopen during the depolarization. From Cannon *et al.* (1991).

open time which occurs very rarely in normal channels. There is no change in the voltage dependence of inactivation and the sustained current is observed over the whole voltage range.

Mutations in the S4 domain

In contrast to mutations in the receptor for the inactivation ball, mutations in the S4 domain cause a dramatic slowing of the rate of fast inactivation and a reduction in its apparent voltage dependence (Ji *et al.*, 1996; Barchi, 1995). In wild-type Na$^+$ channels, the voltage dependence of inactivation reflects that of activation and there is no intrinsic voltage dependence to inactivation itself. Thus the S4 mutations are believed to slow inactivation by reducing the degree of coupling between activation and inactivation. It is of interest that these mutations do not affect the voltage dependence of activation; however, as discussed earlier, not all residues in the S4 domains contribute equally to voltage sensitivity. Analysis of the S4 mutations has provided new insights into the role of the S4 domain of repeat IV and revealed that it plays a greater role in coupling activation to fast inactivation than in serving as a voltage sensor for activation. In this regard it differs from the other three S4 segments.

Mutations in the inactivation loop

Mutations which occur in the inactivation linker between repeats III and IV produce a marked slowing of inactivation and a modest increase in the persistent Na$^+$ current (Mitrovic *et al.*, 1995; Hayward *et al.*, 1996), as shown in Fig. 5.10. These mutations lie close to the IFM triplet that has been postulated to form the latch of the inactivation gate, which suggests they may destabilize the gate and allow it to reopen, so slowing the time course of inactivation. Indeed, single-channel analysis of the G1306E mutation reveals that the mutant channels open more frequently and for longer durations: the first opening may also occur later than is observed for the wild-type channel (Mitrovic *et al.*, 1995). As explained below, the magnitude of the non-inactivating Na$^+$ current determines whether the patient experiences no effect, myotonia, or muscle paralysis. This is clearly illustrated by the fact that the severity of myotonia in patients with mutations in G1306 is well correlated with the degree to which inactivation is slowed: the larger the effect on inactivation, the more severe the myotonia (Mitrovic *et al.*, 1995).

In patients who carry the mutation T1313M, the myotonia is enhanced by cold. Analysis of T1313M mutant channels indicates that fast inactivation is much slower, and recovery from inactivation is much faster, than in wild-type channels (Hayward *et al.*, 1996). Although cooling reduces the rate and extent of inactivation in both wild-type and T1313M Na$^+$ channels, the persistent inward current will be larger in the latter case, so that the threshold for myotonia will be reached more rapidly.

Mutations affecting slow inactivation

In addition to the alterations in fast inactivation described earlier, other effects have also been reported. The two most prevalent HyperPP mutations (T704M and M1592V) disrupt slow inactivation and are associated with Na$^+$ currents that fail to inactivate fully, even during a depolarization of several minutes (Cummins and Sigworth, 1996; Hayward *et al.*, 1997). These mutations are predicted to lie at the cytoplasmic ends of S5 in repeat II and S6 in repeat IV, respectively, regions that are thought to contribute to the inner mouth of the pore. The demonstration that T704M and M1592V impair slow inactivation provided the first evidence that conformational changes in the inner mouth of the Na$^+$ channel are involved in slow inactivation.

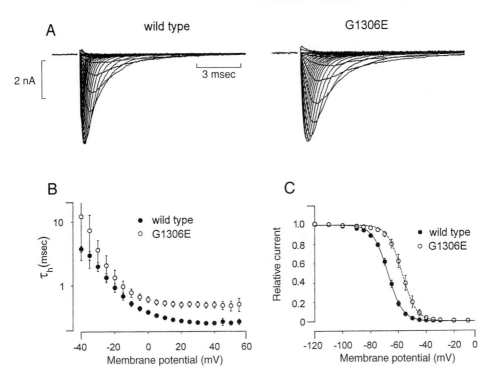

FIGURE 5.10 PAM MUTATIONS SLOW Na$^+$ CURRENT
INACTIVATION

The mutation G1306E in the inactivation loop, which is found in
potassium-aggravated myotonia, causes slower inactivation of Na$^+$ currents. (A) Whole-cell Na$^+$ currents recorded from HEK293 cells cotrans-
fected with cDNA encoding the β-subunit and either the wild type (left)
or the mutant (right) α-subunit of the skeletal muscle Na$^+$ channel. Currents were evoked by depolarizations to between -80 and $+80$ mV in
5 mV increments from a holding potential of -120 mV. (B) Time constants
of inactivation for wild-type and mutant Na$^+$ currents. (C) Voltage dependence of steady-state inactivation for wild-type and mutant Na$^+$ currents.
From Hayward *et al.* (1996).

Impairment of slow inactivation may be expected to accentuate muscle
depolarization and increase the likelihood of paralysis. Indeed, for many
years there was an argument as to whether a defect in slow inactivation was
necessary to produce the sustained paralysis, lasting from minutes to hours,
observed in HyperPP patients. This issue is now resolved. If a mutation
disrupts slow inactivation, then the patient will almost certainly suffer from
episodic paralysis. However, not all mutations that cause muscle weakness
affect slow inactivation: for example, M1360V does not. Thus disruption of
fast inactivation alone is sufficient to cause paralysis. This may be because
slow inactivation is never complete, even in wild-type channels, so that a
defect in fast inactivation is still capable of producing a persistent current
that can cause a maintained depolarization and paralysis.

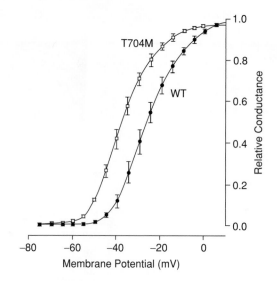

FIGURE 5.11 HyperPP MUTATIONS MAY AFFECT Na⁺ CHANNEL ACTIVATION
Voltage dependence of Na⁺ conductance measured for wild-type Na⁺ channels (●) and for Na⁺ channels bearing the hyperPP mutation T704M (□). Mutant channels activate at more hyperpolarised potentials. From Cannon (1997).

Mutations affecting activation

In contrast to inactivation, Na⁺ channel activation is not altered by most of the mutations that produce the periodic paralyses. One exception is T704M, which causes a 10 mV shift in the voltage dependence of activation to more negative membrane potentials (Fig. 5.11). Two other mutations (V1293I and G1306E) produce a smaller shift in the voltage dependence of activation. This effect would tend to favour membrane depolarization. It is perhaps surprising that none of the three PMC mutations in R1448, which lies with the S4 domain, affect activation.

Summary

To recapitulate, all mutations associated with HyperPP, PC, and PAM have been shown to alter fast inactivation, while two also disrupt slow inactivation and three cause a shift the voltage dependence of activation. Mutations in the S4 domain cause a slowing of the time course of inactivation and a shift in its voltage dependence, those which occur in the linker between repeats III and IV markedly slow inactivation and produce a sustained Na⁺ current, and the principal effect of those in the S5 and S6 segments is to cause a persistent component of Na⁺ current.

A sustained Na⁺ current can produce myotonia or paralysis

This section considers how Na⁺ channel mutations that affect inactivation give rise to enhanced muscle excitability on the one hand and to paralysis on the other; and why both of these conditions are sometimes experienced by the same patient at different times.

It is clear that mutations which cause a persistent component of Na$^+$ current will produce a tonic depolarisation of the muscle membrane. Whether myotonia or paralysis occurs depends on the size of this depolarization and thus on the magnitude of the sustained current. A small depolarization will make the membrane hyperexcitable by lowering the action potential threshold, whereas a more severe depolarization may lead to Na$^+$ channel inactivation and so produce paralysis. In computer simulations of the muscle action potential, myotonic runs were produced when 0.8–2% of Na$^+$ channels did not inactivate, whereas when >2% of Na$^+$ channels failed to inactivate the membrane depolarised to -40 mV and action potentials failed to be elicited (Fig. 5.12) (Cannon *et al.*, 1993). In agreement with these findings, there is a reasonable correlation between the type and the severity of the inactivation defect and the clinical phenotype.

The two most common mutations in HyperPP (M1592V and T704M, located in the acceptor site for the inactivation ball) cause persistent inward currents at all membrane potentials. This current is large enough to cause muscle depolarization and block electrical activity. Consequently, patients experience muscle weakness and paralysis. In some (but not all) attacks, the paralysis is preceded by myotonia. This is expected because there is a variable period of time in which the muscles become hyperexcitable before they fully depolarize and paralysis ensues. Mutations that produce smaller sustained currents (V1589M), and thus may be expected to produce a smaller membrane depolarization, give rise to myotonia without paralysis.

Some mutations do not give rise to a persistent Na$^+$ current over the whole voltage range. These mutations shift the relationship between steady-state inactivation and membrane potential along the voltage axis to more positive potentials (Fig. 5.10) or they reduce the steepness of the relationship. As a result, the activation and inactivation curves overlap. This also produces a persistent Na$^+$ current, known as a 'window' current, but only over a limited range of voltages. This is the case for the G1306E mutation, which resides within the inactivation loop and which causes potassium-aggravated myotonia. The window current produced by this mutation depolarises the membrane sufficiently to generate action potential activity and thereby myotonia. However, the amplitude of the persistent window current decreases with depolarization and above -50 mV it is not sufficient to produce a sustained depolarization. Therefore this mutation does not cause paralysis.

Mutations that cause a slowing of the time constant of inactivation (such as those in the S4 domain) produce action potentials of longer duration. This allows the voltage-gated K$^+$ channels to remain open for a longer time and leads to a greater accumulation of extracellular K$^+$ in the transverse tubules of the muscle. The consequence of this K$^+$ accumulation is a depolarization, which promotes myotonia and may eventually lead to paralysis. K$^+$ accumulation is also expected to occur with the other *SCN4A* mutations and to facilitate the development of myotonia.

The dominant nature of HyperPP is explained by the fact that only a small sustained Na$^+$ current is needed to cause depolarization and thereby

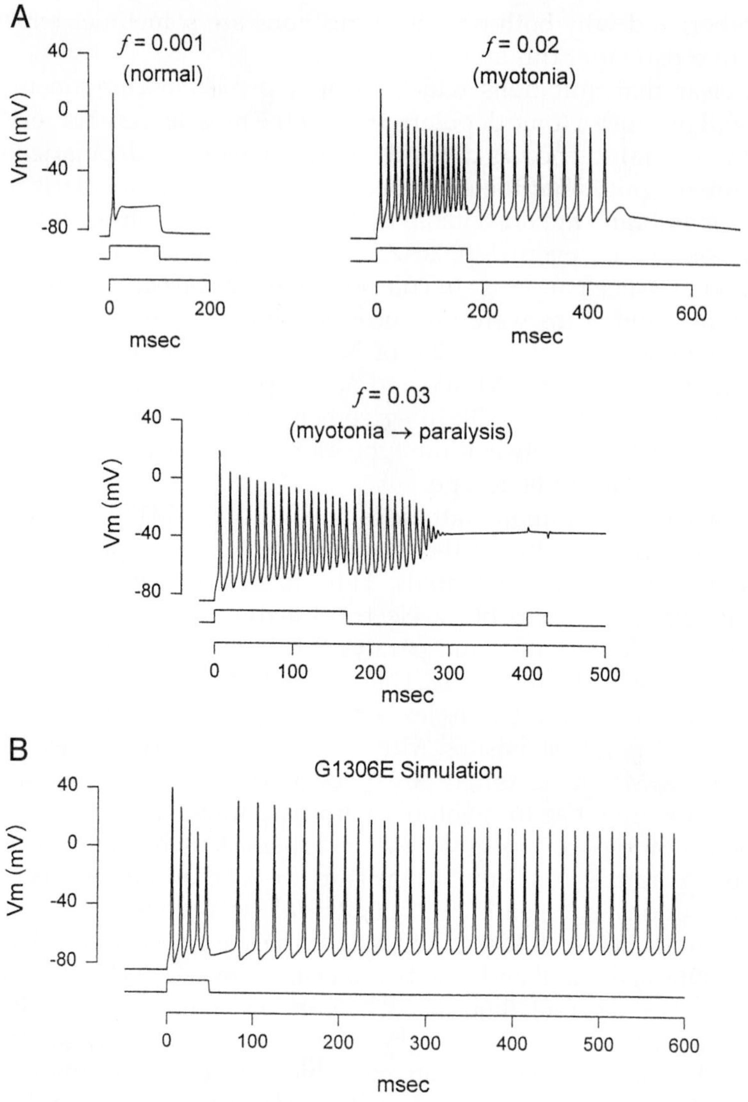

FIGURE 5.12 *COMPUTER SIMULATIONS OF ELECTRICAL ACTIVITY IN A MODEL MUSCLE CELL*

(A) The fraction of the Na^+ current that failed to inactivate (f) was varied to simulate HyperPP. (i) Electrical activity elicited by a depolarization in a normal muscle fibre ($f=0.001$). (ii) A small persistent Na^+ current ($f=0.02$) induced myotonia; note that action potentials continue to be elicited even after the stimulus was removed. (iii) A larger sustained Na^+ current ($f=0.03$) generates a prolonged myotonic discharge that progresses to a sustained depolarization (which would cause paralysis). (B) When inactivation is slowed and its voltage dependence is shifted to positive potentials to simulate the effects of the G1306E mutation, a prolonged myotonic discharge occurs. From Cannon *et al.* (1993).

induce either myotonia or paralysis. In the heterozygote, where roughly half the channels will be of the mutant type, the persistent current is still large enough to produce a depolarization of clinical significance. The reason an increase in plasma potassium precipitates an attack may be because the depolarization it produces increases the open probability of the Na$^+$ channel and unmasks its inactivation defect. The mechanism by which cold enhances muscle stiffness in patients with PMC is not completely clear. However, it forms the basis of a useful diagnostic tool, because these patients will develop a stiff tongue and slurred speech if given ice cream or a cold milkshake to drink.

Equine Periodic Paralysis

Periodic paralysis is not confined to man. Some American thoroughbred quarter horses also have a mutation in their Na$^+$ channels that causes attacks of muscle paralysis when plasma potassium rises (Rudolph et al., 1992). This is known as equine hyperkalaemic periodic paralysis (E-HPP). Affected horses have very well-developed muscles, perhaps because the myotonic contractions they sometimes experience between attacks of paralysis are equivalent to performing continuous exercise. Although quarter horses were originally bred for quarter-mile racing, hence their name, they are now more favoured as show horses. Well-developed musculature is a highly desirable trait in a show horse and thus animals with E-HPP win many prizes. The disease is inherited in an autosomal dominant fashion and as a result of a programme of selective breeding for muscular physiques, as many as 1 in 50 quarter horses are at risk of carrying the E-HPP mutation. All of them can be traced to a single ancestor, a stallion called Impressive (not named, in fact, for the quantity of offspring he sired, but rather for his powerful muscles). Although the affected animals appear healthy, like their human counterparts they experience recurrent attacks of muscle weakness associated with rest after exercise, or ingestion of feed high in K$^+$, such as alfalfa. Periodic paralysis is a particular problem for a horse, because they develop life-threatening pulmonary oedema if they lie down for a long time.

The E-HPP mutation (a phenylalanine-to-leucine substitution) occurs in the S3 domain of repeat IV of the horse skeletal muscle Na$^+$ channel. This substitution produces defective inactivation, and the mutant Na$^+$ channel undergoes long bursts of openings during a maintained depolarization (Cannon et al., 1995). The consequence is a sustained inward current which causes depolarization and thereby paralysis. The E-HPP phenotype is particularly sensitive to plasma K$^+$ levels, and K$^+$ injections (which precipitate paralysis) are used clinically to determine if the animal carries the defective gene (Fig. 5.13).

Long QT Syndrome

Long QT syndrome (LQT3) is a relatively rare inherited cardiac disorder that causes abrupt loss of consciousness or sudden death from ventricular arrhythmia in children or young adults (SADS foundation task force, 1996). Clinically, it can be identified from the electrocardiogram (ECG), which has an abnormally long QT interval (hence the name of the disease). Long QT syndrome presents in young, otherwise healthy people as fainting attacks, which are often precipitated by emotional or physical stress. A typical case is that of a young girl who suffered her first blackout at the age of 3 (Schwartz *et al.*, 1975). An ECG confirmed she had long QT syndrome. As she grew older, the fainting attacks increased in frequency and sometimes exertions such as running for a bus caused her to lose consciousness. When she was 19 years old, she participated in a live television show when the stress and excitement precipitated a fatal arrhythmia.

Electrocardiograms of LQT patients show a QT interval that is about 2–5% longer than normal. The duration of the QT interval is determined by the duration of the ventricular action potential, and a long QT interval indicates that action potential repolarisation occurs later than usual (Fig. 5.14). LQT patients also show a specific type of episodic arrhythmia known as *torsades de pointes*—literally, twisting of the points—so called because the QRS wave changes continuously and appears to twist around the isoelectric line of the electrocardiogram. The rapid contractions associated with *torsades de pointes* result in reduced ventricular refilling. As a consequence, cardiac output is less and the blood flow to the brain decreases, which may cause loss of consciousness. *Torsades de pointes* may degenerate into ventricular fibrillation, which can be fatal.

Four different genetic loci for long QT syndrome have been mapped. They lie on chromosomes 11 (11p15.5: LQT1), 7 (7q35-36: LQT2), 3 (3p21-24: LQT3), and 4 (4q25-27: LQT4). The first three of these have been identified and shown to encode cardiac ion channels; the identity of the gene responsible for LQT4 is unknown. Those forms of the disease linked to chromosomes 7 and 11 result from mutations in voltage-gated K⁺ channels and are considered later (Chapter 6). LQT3 is due to mutations in the cardiac muscle sodium channel gene (*SCN5A*), which is located on chromosome 3 (Wang *et al.*, 1995b,c). Several mutations have been identified (Fig. 5.15). The most severe LQT3 phenotype is produced by a three amino acid deletion in the inactivation loop which links repeats III and IV, ΔKPQ1505–1507. Milder forms of LQT3 are associated with an arginine-to-histidine substitution at position 1644 (R1644H) and an asparagine-to-serine at residue 1325 (N1325S). Expression of mutant Na⁺ channels in *Xenopus* oocytes has shown that all three mutations produce abnormalities in Na⁺ channel inactivation (Bennett *et al.*, 1995; Dumaine *et al.*, 1996). Whereas the wild-type currents were inactivated almost completely during a

---

FIGURE 5.13 *EQUINE PERIODIC PARALYSIS*
Injection of potassium solution induces an attack of paralysis in an American quarter horse carrying the EPP mutation. Photographs supplied by Eric Hoffman.

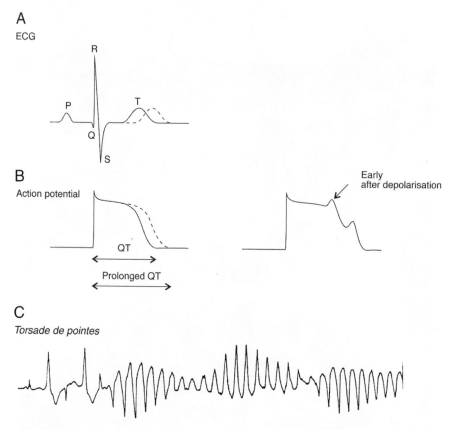

A

ECG

B

Action potential

Early
after depolarisation

C

Torsade de pointes

FIGURE 5.14
Relationship of the ventricular action potential (B) to the electrocardiogram (ECG) (A). ECGs recorded from patients with long QT syndrome
show a prolonged QT interval that reflects the longer duration of the
action potential. (C) Electrocardiogram recorded from a patient showing
torsade de pointes.

100-msec pulse, each of the mutant channels showed a significant component
of non-inactivating current (Fig. 5.16). The relative magnitudes of the sustained
currents were ΔKPQ > N1325S > R1644H. Thus a larger component of non-
inactivating current is associated with the more severe phenotype.

Single-channel recordings revealed two differences in the properties of the
mutant channels that accounted for the sustained current found in macroscopic
recordings (Fig. 5.16A). Wild-type Na⁺ channels generally only open transiently and then inactivate in response to depolarization. By contrast, all of the
mutant channels continued to open and close throughout a depolarizing pulse.
In most cases these openings were very brief, as in wild-type channels, but
occasionally long-lasting bursts of openings occurred. The latter were more
common for ΔKPQ mutant channels, which explains the enhanced current.

The finding that a deletion in the loop between repeats III and IV reduces
inactivation of Na⁺ currents is not unexpected. As explained earlier, this

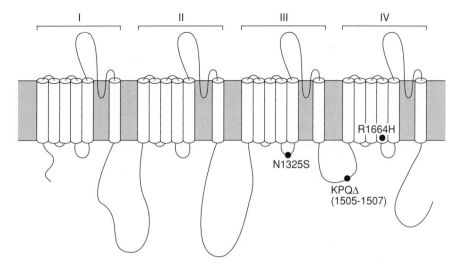

FIGURE 5.15 MUTATIONS ASSOCIATED WITH LQT3
Schematic representation of the α-subunit of the human cardiac muscle
Na⁺ channel *SCN5A* with the location of LQT3 mutations indicated. Note
that the human cardiac muscle Na⁺ channel is 2016 residues long and
that the numbering does not correspond to that for the rat brain Na⁺
channel in Fig. 5.2 or for the human skeletal muscle Na⁺ channel (1836
residues) in Fig. 5.8.

region of the Na⁺ channel serves as the inactivation particle and mutations
in this region disrupt inactivation. The R1644H mutation occurs at the end
of the S4 segment in repeat IV, within the putative voltage sensor. Mutations
in this region slow inactivation by disrupting the coupling between activation
and inactivation, and in *SCN4A* they lead to paramyotonia congenita. The
other LQT3 mutation is located in the S4–S5 loop of domain III and is believed
to interfere with inactivation by destabilizing the structure of the acceptor
site for the inactivation particle.

Why does slowing of Na⁺ current inactivation affect the cardiac action
potential duration? Action potential duration is governed by the balance
between inward currents, which tend to depolarize the membrane, and outward currents, which act to hyperpolarize it. Repolarization occurs when
outward currents prevail. Although the cardiac Na⁺ channel is chiefly responsible for the rapid upstroke of the action potential, it also conducts a small
current during the plateau phase of the action potential. Thus if inactivation
were incomplete, more inward current would be present and cardiac repolarization would be delayed. Pharmacological evidence also supports the idea
that defects in Na⁺ channel inactivation cause long QT syndrome. For example, scorpion toxin, which slows the rate of Na⁺ channel inactivation, can
prolong the cardiac action potential and produce arrhythmias. The way in
which a longer action potential duration precipitates arrhythmia, and why
this occurs more frequently with stress, is discussed more fully in Chapter 6.

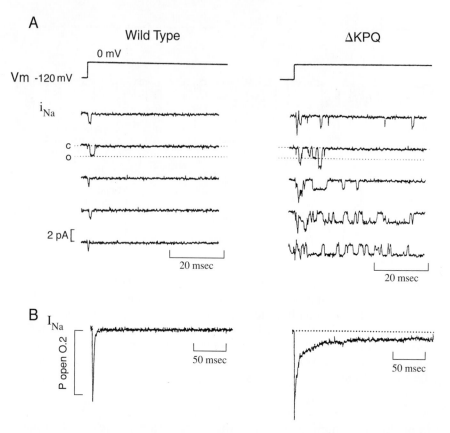

FIGURE 5.16 EFFECT OF LQT3 MUTATIONS ON CARDIAC Na⁺
CHANNEL CURRENTS
Single-channel currents recorded from inside-out patches excised from
Xenopus oocytes injected with mRNA encoding the wild-type human
cardiac Na⁺ channels (left) or Na⁺ channels carrying the 1505–1507
KPQ deletion (KPQ; right). Currents were elicited by depolarization
from −120 mV to 0 mV. (B) Ensemble wild-type and KPQ Na⁺ currents
obtained by averaging many single-channel records. From Bennett *et al.*
(1995).

Agents that inhibit the sustained Na⁺ currents would clearly be of therapeutic value in chromosome 3-linked LQT syndrome. Indeed, class Ib antiarrhythmic agents such as mexiletine shorten the QT interval in LQT3 patients
(Schwartz *et al.*, 1995), and reduce the sustained current through LQT3-mutant
Na⁺ channels (Dumaine *et al.*, 1996). It is important to remember, however,
that LQT syndrome is a heterogeneous disease and that these agents may
not be beneficial in patients with genetically different forms of LQT. Thus it
is vital to genotype individuals before treatment.

DISEASES OF NEURONAL
Na⁺ CHANNELS

Generalised Epilepsy with Febrile Seizures

Febrile seizures—in which convulsions occur in conjunction with fever—affect around 3% of children under six. A small number of these children also develop afebrile generalised epilepsy in later life. This condition has been termed generalised epilepsy with febrile seizures (GEFS) and in one large Australian family it has been linked to chromosome 19q13.1, the locus of the gene (*SCN1B*) that encodes the β_1-subunit of the voltage-gated Na⁺ channel (Wallace *et al.*, 1998).

The β_1-subunit consists of a single transmembrane domain with a short intracellular C terminus and a larger extracellular N terminus (Fig. 5.17A). The extracellular domain contains an immunoglobulin-like fold motif, whose structure is maintained by a disulphide bridge between two highly conserved

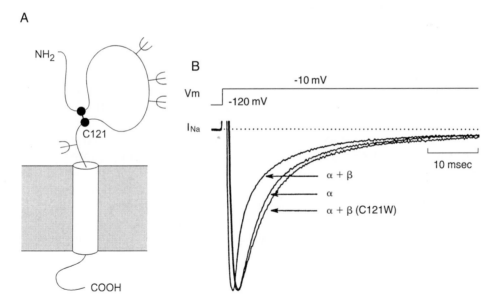

FIGURE 5.17 MUTANT β-SUBUNIT FAILS TO SPEED INACTIVATION

(A) Schematic representation of the human neuronal Na⁺ channel β_1-subunit showing the position of the putative disulphide bridge in the extracellular loop and the mutation that occurs in GEFS. (B) Na⁺ currents recorded from *Xenopus* oocytes expressing the rat brain Na⁺ channel α-subunit alone (α), or the α-subunit together with either the wild-type human β_1-subunit ($\alpha + \beta$) or the human β_1-subunit containing the C121W mutation ($\alpha + \beta$ C121W). Currents were recorded in response to a depolarization from -120mV to -10mV. From Wallace *et al.* (1998).

cysteine residues. A mutation in one of these cysteines (C121W) was identified in the Australian GEFS family, which is predicted to disrupt the disulphide bridge and alter the structure of the extracellular domain (Wallace *et al.*, 1998).

The presence of the β-subunit accelerates both the rate of inactivation and the rate of recovery from inactivation of the voltage-gated Na⁺ channel, as described above. This modulatory effect is abolished if the β_1-subunit carries the C121W mutation (Fig. 5.17B) (Wallace *et al.*, 1998). The loss of β_1-subunit modulation is predicted to cause a persistent inward Na⁺ current that may produce neuronal hyperexcitability and account for the seizures that characterise GEFS.

Motor Endplate Disease of Mice

A null mutation in the mouse Na⁺ channel α-subunit *Scn8a*, which is widely expressed in the brain and spinal cord, gives rise to motor endplate disease (*med*) (Burgess *et al.*, 1995). This is a recessive neurological disorder character-ised by progressive paralysis of the hind limbs, severe muscle wasting, ataxia, and early death. The cerebellar Purkinje cells also degenerate. Muscle atrophy and ataxia presumably result from the observed failure of neuromuscular transmission, because *Scn8a* is not expressed in skeletal or cardiac muscle.

A mild mutation within the *med* locus gives rise to the mouse mutant *jolting*, which displays an ataxia gait and rhythmic tremors of the head and neck in response to attempted movement. The *jolting* mutation results in the substitution of threonine for a conserved alanine residue at position 1071, which lies within the cytoplasmic S4–S5 linker of domain III. When the equivalent mutation was introduced into the rat IIA Na⁺ channel, it caused a shift in the voltage dependence of inactivation to more depolarized potentials (Kohrman *et al.*, 1996). This is expected to increase the firing threshold of cerebellar Purkinje cells, reducing their activity and causing the observed ataxia. Somewhat surprisingly, given that the mutation lies within the ac-ceptor site for the inactivation ball, inactivation of the mutant Na⁺ channels was unaffected.

The human equivalent of *Scn8a* maps to chromosome 12q13, but the gene has not yet been cloned.

CHAPTER 6

VOLTAGE-GATED K⁺ CHANNELS

Potassium channels are found in virtually all cells. They fall into two main structural families: those which possess six transmembrane domains and those which are formed from only two transmembrane domains. The latter include the inwardly rectifying K^+ channels which are discussed in Chapter 8. The six transmembrane K^+ channels may be further subdivided into six conserved gene families (Fig. 6.1). These comprise the voltage-gated K^+ channels (K_V channels), the KCNQ channels, the *eag*-like K^+ channels, and three kinds of Ca^{2+} activated K^+ channels (BK, IK and SK). The different Ca^{2+} activated K^+ channels are considered in Chapter 7. In this Chapter we focus on the K_V, KCNQ and *eag* types of K^+ channel. Mutations in the genes encoding members of these K^+ channel subfamilies lead to a number of human diseases, such as episodic ataxia, long QT syndrome and epilepsy.

The K_V, KCNQ and *eag*-like K^+ channels are typically closed at the resting potential of the cell, but open on membrane depolarization. They are involved in the repolarization of the action potential, and thus in the electrical excitability of nerve and muscle fibres, including cardiac muscle. They also modulate both synaptic transmission and secretion from endocrine cells (such as chromaffin cells and pancreatic β-cells). Most information about the relationship between channel structure and function has been obtained for K_V channels and the next sections therefore focus on the properties of this K^+ channel family. The differences that have been found for KCNQ and *eag*-like K^+ channels are considered later.

K_V CHANNELS

The fruitfly *Drosophila melanogaster* has made a very significant contribution to our understanding of the molecular biology of voltage-gated K^+ channels. The first K^+ channel to be identified came from the cloning of the *Shaker* gene which causes flies to shake when exposed to ether (ether is used as an anaesthetic when sexing, counting or examining fruitflies) (Papazian *et al.*, 1987). Following the identification of *Shaker*, a family of related channels was

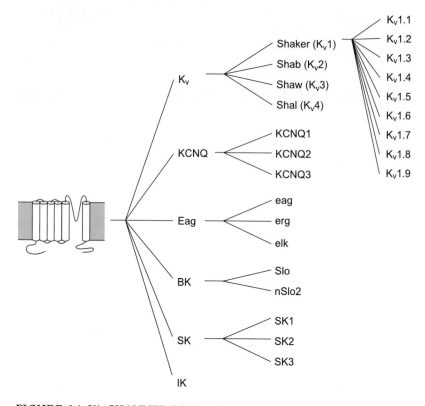

FIGURE 6.1 K⁺ CHANNEL LINEAGES
Hierarchical classification scheme for the different six-transmembrane K⁺ channel genes. After Wei *et al.* (1996).

subsequently cloned from *Drosophila*, known as *Shab, Shaw* and *Shal*. Their mammalian counterparts are referred to as K_V1 (*Shaker*), K_V2 (*Shab*) K_V3 (*Shaw*) and K_V4 (*Shal*) (Dolly and Parcej, 1996) (Fig. 6.1).

K_V *Channel Structure*

Voltage-gated K⁺ channels are made up of pore-forming α-subunits that may associate with one of a number of different types of β-subunits. Nine different subfamilies of K_V channel α-subunits have been described (K_V1–9). At the amino acid level, there is about 70% homology within, and 40% homology between, the different subfamilies.

The α-subunit of the K_V channel corresponds to a single domain of the Na⁺ channel and four such subunits come together to form the K⁺ channel pore. Each α-subunit consists of six transmembrane-spanning segments (S1–S6), which are highly conserved, and intracellular N and C termini of variable length (Fig. 6.2A). The S4 segment shows considerable homology with that of the voltage-gated Na⁺ channel: in particular, it is amphipathic and has a positive charge at every third position. There is evidence that it is involved

A

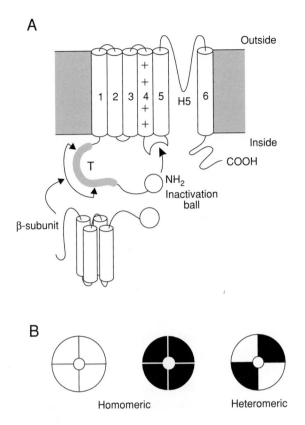

B

Homomeric   Heteromeric

FIGURE 6.2
(A) Putative topology of the *Shaker* K$_V$ channel α-subunit. The distal N terminus possesses a group of residues that act as an inactivation particle by plugging the pore. It also contains a domain required for tetramerization with other α-subunits (T), and a motif involved in interaction with the β-subunit. A receptor site for the inactivation ball lies in the S4-S5 linker. The H5 region lines the pore of the channel. Some K$_V$ channels have β-subunits which lie intracellularly. These have 4 putative α-helical domains and, in some cases, an inactivation ball. (B) K$_V$ channels are made up of 4 α-subunits. Homomeric channels are formed from identical subunits and heteromeric channels from different subunits.

in voltage-dependent activation of K$_V$ channels and that the amino terminus is concerned with voltage-dependent inactivation. The linker between the S5 and S6 segment dips back down into the membrane and participates in formation of the channel pore. This region, known as SS1-SS2 in Na$^+$ channels, is commonly referred to as the H5 region (or the pore loop) in K$^+$ channels.

K$_V$ channel diversity

There turns out to be a huge variety of voltage-gated K$^+$ channels, far more than was ever suspected. Some of this diversity arises because there are different genes which encode different K$_V$ channels. However, several other factors also contribute. First, substantial alternative splicing of K_V channel genes occurs. As described in Chapter 2, all genes contain expressed sequences called exons and non-coding sequences called introns. The introns are spliced out in the nucleus to produce a transcriptionally competent mRNA. Clearly, if the splicing results in the incorporation of different (alternative) exons then different mRNAs will be produced. A second factor generating diversity is that hetero-oligomeric channels formed from different types of K_V channel subunits may occur, and these channels may have hybrid properties. Finally, the properties of the K_V channel α-subunit may be modified by association with different types of β-subunit.

K_V channels are made up of four subunits

Electrophysiological studies first provided support for the idea that the K_V channel is indeed composed of four α-subunits. When mutant *Shaker* subunits with reduced charybdotoxin sensitivity were coexpressed with wild-type subunits, the toxin sensitivity of the expressed channels fit the prediction of a tetrameric model (MacKinnon, 1991). Direct evidence that the voltage-gated K_V channel is made up of 4 subunits subsequently came from studies carried out by Li and colleagues (1994). They took the purified *Shaker* K$^+$ channel protein and imaged it using the electron microscope. As can be seen in Fig. 6.3, the protein is approximately square in outline, with a central hole that probably corresponds to the pore. Each side of the channel is about 80 Å in length and has a thickness of about 50 Å. In some images, the *Shaker* protein appeared to be partially dissociated and 4 globular domains, each of

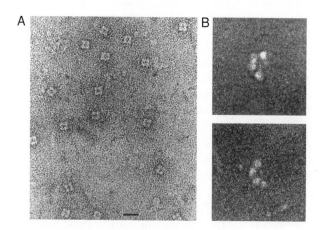

FIGURE 6.3 K_V CHANNELS ARE TETRAMERS
Electron microscope pictures of purified *Shaker* channels. (A) Most channels retain a fourfold symmetry. The bar indicates 150 Å. (B) Some channels were partially dissociated and can be seen to consist of 4 subunits. The bar indicates 50 Å. From Li *et al.* (1994).

which presumably corresponds to a single subunit, could be clearly resolved. Further evidence that the *Shaker* K^+ channel is made up of 4 subunits is that the molecular mass of the purified protein measured under non-denaturing conditions was 350 kDa, about 4 times that measured under denaturing conditions (82 kDa), which splits the subunits apart (Li *et al.*, 1994).

K_V channels can form hetero-oligomers

K_V channels may either be homo-oligomers composed of identical subunits, or hetero-oligomers made up of different types of subunit (Fig. 6.2B). Clearly, when *Xenopus* oocytes are injected with mRNA encoding a single type of subunit the K_V channel expressed must be homomeric. The fact that coinjection of oocytes with mRNAs encoding two different types of subunit may give rise to currents with hybrid properties indicates that heteromeric channels are also viable. K_V channels are not completely promiscuous, however, and only appear to form functional channels with members of the same subfamily. For example, $K_V1.1$ can heteropolymerize with $K_V1.2$ and $K_V1.5$ (Christie *et al.*, 1990), but not with K_V4 channels. Heteromeric K_V channels also exist in native cells. This is demonstrated by the fact that it is possible to immunoprecipitate $K_V1.1$ from the brain neurones using an antibody specific for $K_V1.2$, and *vice-versa* (Sheng *et al.*, 1993). This would only be possible if $K_V1.1$ and $K_V1.2$ physically associate in a tight complex.

The assembly of K_V channel subunits into tetramers requires a stretch of 114 amino acids in a cytosolic N-terminal region of the protein that lies close to the start of the first transmembrane domain (Fig. 6.1; Li *et al.*, 1992). This region has been called the tetramerization domain. It not only confers on the channel the ability to form tetramers but also specifies which tetramers can form. Clearly, other residues, such as those within the transmembrane domains, must also participate in linking the subunits together.

Accessory subunits

A family of voltage-gated K^+ channel β-subunits has also been identified (Rettig *et al.*, 1994; Majumder *et al.*, 1995). The β-subunits have a molecular mass of around 40 kDa and hydrophobicity plots indicate that they are probably cytoplasmic and do not span the membrane. At least 3 different β-subunit genes have been identified at the time of writing, and further diversity is generated by alternative splicing. There is no sequence homology between K_V channel β-subunits and those of Na^+ or Ca^{2+} channels. Dendrotoxin is a snake venom produced by the green mamba *Dendroaspis augusticeps*. It blocks certain K_V channels with high affinity, by binding to the α-subunit, and has been used as a biochemical tool to purify K_V channels. Dendrotoxin-binding studies indicate that the β-subunit tightly associates with the α-subunit in a $1:1$ stoichiometry, suggesting that 4 α-subunits and 4 β-subunits combine to form the whole K_V channel complex (Parcej *et al.*, 1992).

Not all combinations of α- and β-subunits are possible, since a single type of β-subunit does not associate with all types of α-subunit (or *vice-versa*).

For example, the β_1-subunit only associates with α-subunits of the K_V1 family (Sewing *et al.*, 1996). This specificity is conferred by a domain within the N terminus of the α-subunit (residues 112–201) that overlaps with the tetramerization domain (Fig. 6.1). Shamitienko and colleagues (1997) examined the subunit composition of native neuronal K_V1 channels. Their studies have shown that only a few of the large number of theoretically possible combinations actually occur. Most channels contained β_2-subunits, together with $K_V1.2$, either as a homotetramer, or in combination with other subunits ($K_V1.3$, 1.4 and 1.6). There was also a small population of homotetrameric $K_V1.4$ channels, associated with β_1-subunits.

Coexpression of any of the different β-subunits with a compatible pore-forming α-subunit enhances the amplitude of the K⁺ current. This is because the β_1-subunit promotes the surface expression of coexpressed α-subunits, thus increasing K_V channel density (Shi *et al.*, 1996). The presence of the β_1-subunit also modifies the inactivation properties of some members of the K_V1 family. Fig. 6.4 shows that coexpression of the β_1 subunit with the $K_V1.1$ α-subunit produces a very profound increase in the rate of inactivation (Rettig *et al.*, 1994). The β_1-subunit also confers a redox-sensitive inactivation on some types of α-subunit. In contrast, the β_2-subunit does not modify the inactivation properties of the α-subunit but only increases the current amplitude.

As described above, it was the genetic analysis of *Drosophila* mutants that led to the identification of the first K_V channel α-subunit genes. A mutation in the *Drosophila* locus which encodes the equivalent of a K_V channel β-subunit has also been identified: this mutation, *hyperkinetic*, results in very active flies (Choiunard *et al.* 1995).

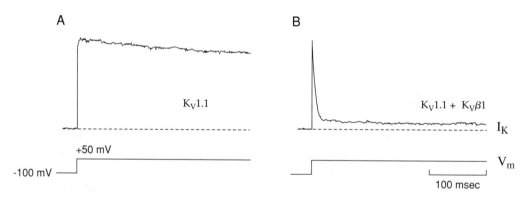

FIGURE 6.4 K_V β-SUBUNITS REGULATE CHANNEL GATING
β-subunits modify the properties of K⁺ channels formed by K_V α-subunits. Outward K⁺ currents were recorded from *Xenopus* oocytes injected with mRNA encoding either $K_V1.1$ (ie: only the α-subunit) or both $K_V1.1$ and $K_V\beta1$ mRNAs. Currents were elicited by a depolarization to +50 mV from a holding potential of -100 mV. From Rettig *et al.* (1994).

Correlating Structure and Function

The K$_V$ channel pore

The outer part of the K$_V$ channel pore is lined by the H5 region, while part of the S6 segment and the loop between S4 and S5 contribute to its inner mouth (Fig. 6.5). The K$^+$ selectivity filter resides entirely within the H5 pore loop since transfer of this region between different K$_V$ channels confers the permeability properties of the donor channel upon the recipient. Hartmann and colleagues (1991) swapped the H5 region of K$_V$2.1 for the equivalent sequence of K$_V$1.2. These two channels differ by only 9 amino acids out of the 21 in this region but they show marked differences in their single-channel conductance, ion selectivity and sensitivity to tetraethylammonium ions (TEA$^+$). Transfer of the H5 region conferred the characteristics of the parent protein on the recipient channel, as is expected if this region lines the channel pore. Changes in individual amino acid residues in this region also altered the permeability properties of the channel (Heginbotham *et al.*, 1994). In addition, mutations in the intracellular loop connecting S4 and S5, and in the intracellular end of the S6 segment, have significant effects on ion selectivity and conductance. Thus these regions may contribute to the inner mouth of the pore.

The pore loop is highly conserved between K$^+$ selective channels. In particular, it contains the sequence (T/S)xxTxGYG which has been termed the K$^+$ selectivity sequence (in some K$^+$ channels the tyrosine (Y) residue is replaced by phenylalanine (F)). Deletion of the tyrosine (Y) and glycine (G) in voltage-gated K$^+$ channels produces remarkable results. *Shaker* channels, for example, are no longer selective for K$^+$ but exhibit permeability to other

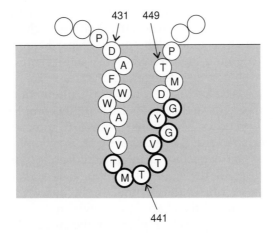

FIGURE 6.5 THE PORE LOOP OF THE Shaker CHANNEL
Amino acids are indicated by their single letter code. The heavily ringed amino acids form the K$^+$ channel signature sequence. Arrows indicate residues that when mutated alter block by external (431,449) or internal (431) TEA$^+$.

monovalent cations (such as Na^+, Li^+, Cs^+) as well as K^+ (Heginbotham *et al.*, 1992a). The single-channel current amplitude is about the same as that for the wild-type K_V channel, which indicates that although the deletion dramatically alters the channel selectivity it does not alter the ion translocation rate. The presence of the YG residues within the pore loop thus seems to confer K^+ selectivity. To identify other amino acids which contribute to K^+ selectivity, Heginbotham and colleagues (1994) carried out scanning mutagenesis of the K^+ channel pore, mutating each residue in the K^+ channel signature sequence in turn. They identified four critical residues. These were located at positions 1,5,6, and 8 in the K^+ channel signature sequence, corresponding to T,V,G and G in the *Shaker* K^+ channel (Fig. 6.5). Non-conservative mutations of these resides invariably resulted in non-selective cation channels. Perhaps somewhat surprisingly, the aromatic tyrosine (Y) residue was not critical.

The outer mouth of the K_V-channel has been mapped using charybdotoxin (CTX), a polypeptide which is found in scorpion venom. The CTX molecule is too large to pass through the pore, but it enters the external mouth of the channel and plugs the pore. It is this ability to block K_V channels that accounts for the neurotoxicity of CTX. Mutagenesis studies have shown that clusters of residues, located on both sides of the H5 region of the *Shaker* channel, interact with CTX (MacKinnon *et al.*, 1990). Residues in the same region are also involved in the binding of another K-channel blocker, the tetraethylammonium (TEA^+) ion (MacKinnon and Yellen, 1990). TEA^+ is smaller than CTX and able to permeate further into the pore before it gets stuck. High affinity TEA^+ block requires the presence of an aromatic residue at position 449 (in *Shaker*, Fig. 6.5): four residues, one contributed by each subunit, participate in the binding site (Heginbotham and MacKinnon, 1992). Residue 449 thus defines the external edge of the pore. It is located in a similar position to those residues in the Na^+ channel that bind tetrodotoxin.

Peptide toxins whose atomic structure is known, such as CTX and agitoxin (from the venom of the funnel web spider), have been used as molecular measuring sticks to define the three-dimensional structure of the outer entrance of the *Shaker* K^+ channel (Ranganathan *et al.*, 1996; Naranjo and Miller, 1996). This approach is based on the idea that the interaction surfaces between the toxin and its target site on the channel are highly complementary—almost mirror images of each other. Mutations in the toxin molecule which decrease block may be expected to result in a less snug fit, and conversely, those which enhance block may be expected to interact more closely. Since the three-dimensional structure of the toxin is known, it is possible to predict the structure of the binding site of the channel. In this way, the toxin can be used to probe the architecture of the channel with which it interacts. Such studies suggest that the pore loops (H5 region) of the *Shaker* K^+ channel do not extend right through the membrane, but instead form a shallow cone with the selectivity filter located about 5 Å from the extracellular solution.

The inner mouth of the pore has been mapped by studying the block by intracellular TEA^+. Threonine 441 in the *Shaker* K^+ channel, interacts with internal, but not external TEA^+, indicating it must be on the inner side of the selectivity filter (Fig. 6.5; Yellen *et al.*, 1991). There is also evidence that both the S4-S5 cytoplasmic linker and the S6 segment contribute to the inner mouth of the pore since mutations in these regions influence the single-channel conductance, alter the ion selectivity and affect block by quaternary ammonium ions (Choi *et al.*, 1993; Slesinger *et al.*, 1993).

The crystal structure of a K$^+$ channel pore

In the summer of 1998, the ion channel community was electrified by the publication of the first crystal structure of an ion-selective channel (Doyle *et al.*, 1998). The structure of a K$^+$ channel (KcsA) from the bacterium *Streptomyces lividans* was solved at 3.2 Å resolution. The choice of a bacterial channel was deliberate, because it made it possible to obtain sufficient amounts of protein for X-ray crystallography. KcsA has only two transmembrane domains, connected by a pore loop. Nevertheless, its amino acid sequence is more closely homologous to that of the voltage-gated K$^+$ channels than the Kir channels (which also have two TMs) described in Chapter 8. Indeed, within the pore region, the sequence of KcsA is nearly identical to that of K$_V$ channels.

Crystallographic analysis reveals that the KcsA channel is a tetramer, with four identical subunits arranged symmetrically around a central pore. Each subunit consists of two transmembrane α-helices (S1 and S2) which are linked by a stretch of ~30 amino acids that form a pore helix and a 'turret' that extends above the membrane. The subunits are arranged in such a way that they form a structure similar to an inverted teepee (Fig. 6.6). The poles of the teepee are formed by the four S2 helices. These lie close together at the inner side of the membrane but widen out towards the extracellular side, forming a funnel-shaped tent that is lined by the S2 helices at its stem and by the pore helices at its outer mouth. The pore helix is connected to the S2 helix by a stretch of amino acids that contains the K-signature sequence (TVGYG) and forms a narrow selectivity filter close to the extracellular side of the membrane (Fig. 6.7). Below the selectivity filter lies a large cavity within the membrane, about 10 Å in diameter. The pore helices are arranged so that their helix dipole is focused on this cavity; put another way, their carboxyl ends point towards the cavity. Finally a water-filled tunnel, 18 Å in length, connects the central cavity to the intracellular solution. The overall length of the KcsA pore is 45 Å.

How does K$^+$ move through the pore?

Let us now consider how a K$^+$ ion moves through the pore, from the intracellular solution to the extracellular solution. Negatively charged amino acids located at the intracellular entrance of the pore serve to attract K$^+$ ions into the vicinity of the channel and raise their local concentration. The walls of the tunnel are lined by hydrophobic residues which enables the hydrated K$^+$ ion to pass easily from the intracellular solution into the central cavity. Its passage is facilitated by the helix dipoles of the pore helices which provide a diffuse negative charge cloud, attracting K$^+$ ions into the large central water-filled cavity.

The K$^+$ ion must next pass through the selectivity filter. This region is so narrow that the K$^+$ ion must shed most of its waters of hydration in order to enter. The side-chains of the amino acids which form the selectivity filter are directed away from the pore. Doyle and his colleagues therefore propose that the carbonyl oxygens of the amino acid backbone may point into the pore, forming a stack of sequential rings, and that the presence of a K$^+$ ion creates a dipole effect that induces the carbonyl oxygens to become negatively charged. These rings of oxygen may substitute for the waters of hydration during the passage of a K$^+$ ion through the selectivity filter, forming 'stepping stones' that facilitate the passage of the ion. The size of the selectivity filter is determined by interaction of the side chains of the valine and tyrosine residues in the VGYG motif with residues in the pore helices. The tyrosine residues, for example, form a massive sheet of aromatic amino acids that are positioned like a 'cuff' around the selectivity filter, and act like molecular springs to hold it open at its proper diameter. This structure suggests an explanation for why Na$^+$ ions cannot permeate the K$^+$ channel, despite their smaller unhydrated diameter. It is argued that the carbonyl groups lining the selectivity filter are held too far apart for them to interact easily with the smaller Na$^+$ ion, making its dehydration energetically unfavourable.

The selectivity filter is 12 Å long and can accommodate two K$^+$ atoms, about 7.5 Å apart. This explains why K$^+$ does not remain bound in the selectivity filter. The presence of a second K$^+$ ion entering the selectivity filter from its inner end creates an electrostatic repulsive force that pushes the outer K$^+$ ion on through the pore and out into the extracellular solution. This design explains how the K$^+$ channel is able to combine high selectivity (K$^+$ over Na$^+$) with a very high ion transport rate.

The elegant work of Doyle and colleagues has provided us with a detailed idea of how K$^+$ actually moves though the pore of the K$^+$ channel. It is interesting to note that some of the basic features of this model had already been suggested by earlier studies combining site-directed mutagenesis with electrophysiological analysis of the properties of the mutant channels, a fact which validates this approach to studying the relationship between channel structure and function.

Activation

Experiments on voltage-gated Na$^+$ channels, described in Chapter 5, have defined the S4 segment as its primary voltage sensor. Similar experiments suggest that the positive charges in the S4 segment also serve as the main component of the voltage sensor in K$_V$ channels (Papazian et al., 1991), and that in response to depolarization the S4 segment is displaced outward across the membrane, giving rise to a gating current and triggering the conformational change that opens the channel pore (Larsson et al., 1996).

Inactivation

K$_V$ channels display two main modes of inactivation: fast and slow. Additional modes of inactivation may also occur but will not be considered here.

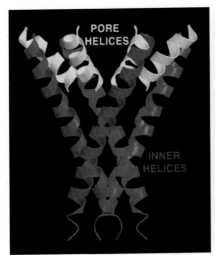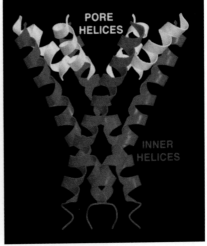

FIGURE 6.6 TEEPEE ARCHITECTURE OF THE KcsA CHANNEL
Ribbon representation illustrating the inverted teepee-like structure formed
by the four S2 α-helices of the bacterial KscA channel (inner helices, shown
in red). The four pore helices (white) are slotted between them and line the
upper part of the structure. The pictures form a stereo pair that can be used
to give a 3-dimensional image of the channel structure. From Doyle *et al.* (1998).

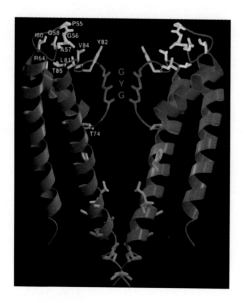

FIGURE 6.7 KcsA STRUCTURE

Two subunits of KscA are shown. The outer helices are the S1 domains, and the inner helices the S2 domains. Mutations in the *Shaker* K_V channel that affect the pore properties have been mapped onto their equivalent positions in the KscA sequence and are shown in colour. Mutation of any of the white side chains alters the affinity of block by charybdotoxin or agitoxin2. Changing the yellow side chains affects the block by extracellular TEA^+: thus, this is the external TEA^+ binding site. The orange side chains form the internal TEA^+ binding site. If the green side chains are mutated to cysteine they are affected by intracellular thiol reagents regardless of whether the channel is open or not, while the pink side chains are modified only when the channel is open. This suggests that the channel gate must lie somewhere between these two sets of residues. The red side chains (GYG) are essential for K^+ selectivity and form part of the selectivity filter. From Doyle *et al.* (1998).

The cytoplasmic amino terminus of the *Shaker* channel is responsible for its fast voltage-dependent inactivation, which is therefore called N-type inactivation. In some mammalian K$_V$ channels fast inactivation is conferred instead by the β-subunit, but since the molecular mechanism resembles that found for the *Shaker* channel, it is often also referred to as N-type inactivation. Fast inactivating K$_V$ currents are also known as A-currents. Some K$_V$ channels, such as those of the squid axon and certain mammalian neurones do not undergo fast inactivation at all. These channels are often referred to as delayed rectifiers. K$_V$ channels also show an additional form of inactivation which is known as C-type inactivation because it involves residues in the C-terminus of the protein.

N-type inactivation

Fast inactivation of K$_V$ channels, like that of voltage-gated Na$^+$ channels, results from occlusion of the inner mouth of the pore by part of the channel protein (Fig. 6.8). In the case of *Shaker* channels, the culprit is the cytosolic N terminus of the channel (Hoshi *et al.*, 1990). The most distal part of this region serves as an inactivation 'ball', which can swing into the channel pore and block ion fluxes, while the more proximal region acts as a 'chain' which tethers the 'ball' to the channel. Although Armstrong and colleagues proposed this ball-and-chain model for inactivation as long ago as 1973, it was not until 1990 that conclusive proof that this idea is correct was provided by experiments on *Shaker* K$_V$ channels. Hoshi and coworkers (1990) made a series of deletions in the N terminus of the channel. They found that deletion of the first 22 amino acids of the protein removed inactivation, while deletions of between 23 and 83 residues increased the rate of inactivation. This is expected if the first 22 amino acids act as the blocking particle and residues 23–83 form part of the chain (reducing the chain length might speed inactiva-

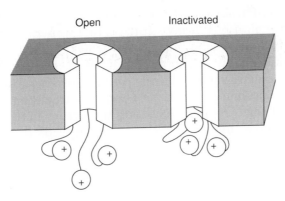

FIGURE 6.8 THE BALL-AND-CHAIN MODEL OF N-TYPE INACTIVATION

Only three of the four K$_V$ α-subunits are shown. Inactivation occurs when one of the 'balls' docks with its receptor site on the channel and plugs the pore.

tion as the ball has less range of movement and less distance to travel). Further support for the ball-and-chain model came from the observation that a synthetic peptide corresponding to the first 20 residues of the *Shaker* B channel (the 'ball' peptide) restored fast inactivation to channels whose N terminus had been deleted and which therefore did not inactivate (Zagotta *et al.*, 1990). Mutations of single positively charged residues in the ball peptide slowed inactivation suggesting that the net charge on the ball is responsible for attracting the ball into the pore. The binding interactions between the ball and the pore appear to be predominantly hydrophobic, however, since mutation of the leucine at position 7 to a hydrophilic amino acid abolished inactivation. Since there are four subunits to each channel, there must also be four balls on chains: however, they function independently and only one of them is needed to inactivate the channel (MacKinnon *et al.*, 1993). A similar ball-and-chain mechanism for fast inactivation is found for other K_V channels. In some channels, such as $K_V1.4$, $K_V3.4$ and members of the K_V4 family, the inactivation ball resides on the N-terminus of the α-subunit, while in other cases it is found on the β_1-subunit (Rettig *et al.*, 1994).

Where is the receptor for the inactivation ball located? Single amino acid substitutions in the cytosolic loop that connects the S4 and S5 transmembrane segments also alter the rate of fast inactivation and first suggested that this region might form the target for the inactivation ball (Isacoff *et al.*, 1991). In a subsequent study, Holmgren and colleagues (1996) mutated individual residues within this region to cysteines and then examined the effect of cysteine-modifying reagents on the binding kinetics of a soluble ball peptide. They found that modifying the charge of residue 391 (in *Shaker*) markedly altered the kinetics of block, which suggests that this residue is located at or near the receptor site for the inactivation ball.

C-type inactivation

C-type inactivation was first described in *Shaker* K⁺ channel mutants which lacked the N-terminal inactivation ball and differed in their C terminal domains (Hoshi *et al.*, 1991). It now appears that it involves residues at the outer mouth of the pore. For example, TEA⁺ slows C-type inactivation when applied to the outside, but not to the inside, of the membrane (the reverse is true for N-type inactivation). Effects of other external cations have also been described, including a reduction of C-type inactivation by extracellular K⁺. It is believed that C-type inactivation involves a local conformational change at the external mouth of the pore which leads to constriction and occlusion of the pore (Liu *et al.*, 1996a).

Modulation

Phosphorylation of many K_V channels modulates their gating (for review, see Jonas and Kaczmarek, 1996). One of the most dramatic examples of this is provided by the $K_V3.4$ channel, where phosphorylation by protein kinase C abolishes N-type inactivation (Levin *et al.*, 1995). This phosphorylation takes place at two serine residues (15 and 21) which are located within the inactivation ball.

Summary of structure-function studies

It is now clear that K$_V$ channels assemble as either homomeric or hetero-meric tetramers. Within each monomer, the S4 segment acts as the voltage sensor for channel activation. Fast inactivation of *Shaker* channels is produced by the N-terminus which acts as a 'ball-on-a-chain', swinging into the inner mouth of the channel and blocking the pore. In other K$_V$ channels, the 'ball' is not located on the α-subunit, but on an accessory β-subunit. Residues within the S4-S5 linker contribute to the acceptor site for the inactivation ball. The outer part of the ion pore is formed by the pore loop (H5) that connects the fifth and sixth TMs, while the cytosolic end of the S6 segment and the S4-S5 linker contribute to its inner mouth.

K$_V$ CHANNELS AND DISEASE

Episodic Ataxia Type-1

Familial periodic cerebellar ataxia is a rare autosomal dominant neurological disorder that affects both peripheral and central nerve functions. It causes attacks of imbalance and uncoordinated movements (ataxia), often accompa-nied by nausea, vertigo and headache. These attacks are usually triggered by physical or emotional stress (for example, the excitement of a football game) or occasionally by alcohol, and they may last from several minutes to many hours. Neurological function is normal, or near normal, between at-tacks. The age of onset ranges from early childhood to no later than adoles-cence. Familial periodic cerebellar ataxia is a clinically heterogeneous disease and two subtypes are distinguished, FPCA/−M and FPCA/+M, based on whether or not the patient experiences myokymia (myokymia refers to spon-taneous contractions of skeletal muscle which make the surface of the muscle seem to ripple). Both subtypes map to different chromosomes and therefore result from mutations in different genes. FPCA/+M is commonly referred to as episodic ataxia type-1 (EA-1). It may be considered the human equivalent of the *Shaker* mutation in *Drosophila*.

Linkage studies have identified the gene responsible for EA-1 as *KCNA1*, which maps to chromosome 12p13 (Browne *et al.*, 1994). This gene encodes the voltage-gated potassium channel K$_V$1.1, which is expressed in the synaptic terminals and dendrites of many brain neurones (Wang *et al.*, 1994). A differ-ent missense mutation was identified in each of the affected six families studied: these occur at positions which are highly conserved between K$_V$ channel families. Two mutations reside in the first TM, two in the intracel-lular loop between TMs 1 and 2, and one is located at the base of the sixth TM on the intracellular side of the membrane (Fig. 6.9). When these muta-tions were introduced into the wild-type K$_V$1.1 channel and expressed in *Xenopus* oocytes, four of them did not produce functional channels but two (V408A and F184C) formed channels with novel properties (Adelman *et al.*, 1995).

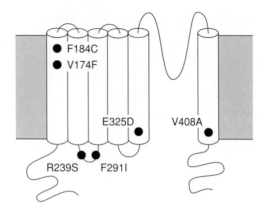

FIGURE 6.9 MUTATIONS ASSOCIATED WITH EPISODIC ATAXIA TYPE-1
Predicted topology of $K_V1.1$ showing the location of mutations associated with episodic ataxia type-1. Note that $K_V1.1$ does not have an inactivation ball.

Valine 408 resides in the cytosolic end of the sixth TM, which is believed to comprise the inner part of the pore. K_V channels expressing the V408A mutation deactivated (closed) much more rapidly than wild-type channels. There were no significant differences in the voltage-dependence of activation. This suggests that valine 408 is important for stabilizing the open state of the channel. The mutant channel also shows accelerated and enhanced C-type inactivation, suggesting an additional role for this residue in C-type inactivation. The other mutation that produced functional K_V channels was a phenylalanine for cysteine substitution (F184C) within the C-terminal part of the first TM, close to the extracellular edge of the membrane. Like V408, this residue is also highly conserved. The voltage-dependence of activation of F184C channels was shifted by ~25 mV to more positive potentials and the currents also activated more slowly. In contrast to the voltage-gated Na⁺ channel, where our current knowledge of the protein can account for the functional effects of most disease-causing mutations, the effects the K_V channel mutations V408A and F184C are not readily explained. Instead, these mutations provide fresh information about the relationship between K_V channel structure and function and suggest that current models may be too simple. Whatever the precise molecular mechanism, however, both mutations are expected to reduce the K⁺ conductance and prolong the action potential.

All individuals with episodic ataxia type-1 so far examined are heterozygotes, so presumably neurones expressing the $K_V1.1$ gene will contain both mutant and wild-type channels. The dominant inheritance of the disease may therefore be explained by a 'dominant negative' effect in which the mutant subunits combine with, and thereby modify or inactivate, the wild-type subunits. The amplitude of the outward K⁺ conductance will therefore be reduced, thus prolonging the action potential and inducing the repetitive firing observed in patients. This would be expected to cause excessive and unregulated transmitter release and produce the clinical symptoms of ataxia and myokymia. Several issues, however, remain to be resolved, such as why attacks are triggered by stress, and why carbonic anhydrase inhibitors (such as acetazolamide) are able to relieve an attack.

KCNQ CHANNELS

The first member of the KCNQ family to be cloned, *KCNQ1*, was isolated by virtue of the fact that mutations in this gene give rise to the most common form of long QT syndrome, LQT1. A partial sequence of the gene was first identified by positional cloning (Wang *et al.*, 1996) and the full length sequence was obtained subsequently, making this yet another example of where genetic analysis of a human disease has led to the molecular identification of an ion channel. *KCNQ1* was originally named *KVLQT1*, and is referred to in this way in most of the earlier references cited: in this chapter, however, I use the new nomenclature. Two other members of the KCNQ family, *KCNQ2* and *KCNQ3*, were subsequently cloned by exploiting their homology to *KCNQ1*. Mutations in these two genes give rise to a form of idiopathic epilepsy known as benign familial neonatal convulsions.

Structure

The primary sequence of the KCNQ subunits predicts a membrane topology similar to that of the K_V channels, comprising six transmembrane-spanning segments (S1-S6) with a typical S4 domain, a pore loop linking S5 and S6, and intracellular N and C termini (Fig. 6.10). Both KCNQ2 and KCNQ3 have longer C-termini than KCNQ1. KCNQ channels differ from K_V channels most notably in the lack of the tetramerization domain that mediates subunit associations between specific members of the K_V channel family. KCNQ channels form functional channels by interaction with accessory subunits, as in the case of KCNQ1, or by heteromeric association with each other (KCNQ2 and KCNQ3).

KCNQ1 encodes the major subunit of the cardiac I_{Ks} channel, which is involved in repolarization of the ventricular action potential. When KCNQ1 was heterologously expressed it produced a voltage-dependent outward K^+ current that had very different properties from those of I_{Ks}, suggesting that the native channel might possess an additional regulatory subunit. K_VLQT1 mRNA is most strongly expressed in the heart, but lower levels are found in pancreas, kidney, lung and placenta. A similar tissue distribution is found for minK (I_sK), a small 130 amino acid peptide with a single transmembrane domain whose function was the subject of much controversy for many years. Coexpression of minK with K_VLQT1, modifies the properties of K_VLQT1 so that they match those of I_{Ks} (Sanguinetti *et al.*, 1996; Barhanin *et al.*, 1996). In

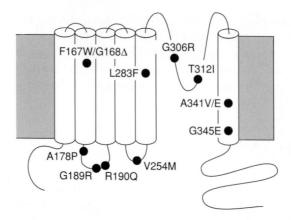

FIGURE 6.10 MUTATIONS ASSOCIATED WITH LQT1
Predicted topology of KCNQ1 with the location of LQT1-associated mutations indicated. The mutations are numbered according to their positions
in the full-length protein and therefore differ from those given in the
original paper.

particular, the currents are much larger and they activate more slowly (Fig.
6.11). minK may therefore be considered to serve as the β-subunit of the I_{Ks}
channel. It is encoded by the gene *KCNE1*.

Heterologous expression studies have shown that both KCNQ2 and
KCNQ3 are able to form homomeric channels that resemble KCNQ1 in their
permeability, voltage dependence and kinetics. In both cases, however, the
amplitude of the expressed currents is extremely low. Coexpression of
KCNQ2 and KCNQ3 gives rise to much larger currents than is observed when
either KCNQ2 or KCNQ3 are expressed alone (Yang *et al.*, 1998; Schroeder *et
al.*, 1998). This suggests that the native channel is a heteromer of KCNQ2
and KCNQ3, an idea which is consistent with their overlapping tissue distribution and the fact that antibodies directed against KCNQ2 are able to coimmunoprecipitate KCNQ3, and *vice-versa*. As described later, the heteromeric
KCNQ2/KCNQ3 channel corresponds to the M-channel of neurones (Wang
et al., 1998).

There is evidence that KCNQ2/KCNQ3 currents are modulated by phosphorylation. First, agents that elevate cytosolic cAMP levels stimulate
KCNQ2/KCNQ3 currents in intact cells. Second, direct application of the
catalytic subunit of protein kinase A, together with MgATP, to the intracellular membrane surface of an excised patch markedly enhances the current
(Schoeder *et al.*, 1998). Mutagenesis studies have localised the critical residue
to the N terminus of KCNQ2.

Tissue distribution

KCNQ1 mRNA is expressed strongly in the heart, with lower levels in
pancreas, kidney, lung, placenta and ear. KCNQ2 and KCNQ3 show expression patterns that largely overlap. They are widely distributed throughout

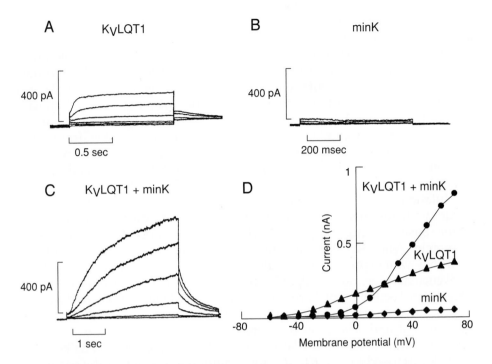

FIGURE 6.11 COEXPRESSION OF KCNQ1 AND minK PRODUCES I_{Ks}
Expression of KCNQ1 (A), minK (B) or KCNQ1 plus minK (C) in transfected COS cells. Currents were elicited by a series of depolarizations
from −60 mV to +20 mV, from a holding potential of −80 mV. Expression
of KCNQ1, but not minK, produces voltage-gated outward currents. Coexpression of minK with KCNQ1 enhances the current magnitude and slows
the kinetics. (D) Current-voltage relations for KCNQ1, minK and the
current formed by KCNQ1 + minK.

the brain, being expressed at high levels in hippocampus, chordate nucleus,
and amygdala.

KCNQ CHANNELS AND DISEASE

Long QT Syndrome

Long QT (LQT) syndrome is a cardiac disorder that causes arrhythmias,
syncope and sudden death (see Chapter 5 for further details). It is genetically
heterogeneous, being caused by mutations in both voltage-gated Na^+ and
K^+ channel genes. Long QT syndrome is characterised by a prolonged QT
interval in the electrocardiogram, which reflects the delayed repolarization
of the ventricular action potential. A markedly prolonged action potential
predisposes the heart to a specific type of life-threatening arrhythmia known
as *torsade de pointes*, that can precipitate ventricular fibrillation and sudden
death. In most families, *torsade de pointes* is triggered by physical or emotional

stress and it may be responsible for a number of cases in which children are said to have died of fright or anger.

The most common form of long QT syndrome, LQT1, which accounts for over 50% of inherited cases of the disorder, is linked to chromosome 11p15.5. It results from mutations in the K+ channel gene *KCNQ1* (Wang *et al.*, 1996). Two forms of LQT1 have been described: an autosomal dominant form of the disorder, known as Romano-Ward (RW) syndrome, and a much rarer recessive form known as Jervall-Lange-Nielsen (JLN) syndrome. The major clinical difference between these syndromes is that patients with JLN syndrome suffer from profound congenital bilateral deafness, in addition to the cardiac abnormalities characteristic of all LQT disorders.

Mutations in KCNQ1 have been identified in affected members of 16 families with dominant LQT1, and include 1 intragenic deletion and 10 different missense mutations (Fig. 6.10). These mutations occur in all regions of the protein. When KCNQ1 containing the mutations causing Romano-Ward syndrome were expressed, either alone or in the presence of minK, no functional currents were observed, indicating that LQT-1 results from a reduction in K+ channel activity. Furthermore, the RW mutant subunits exerted a dominant negative effect, leading to downregulation of the current amplitude, when mutant and wild-type subunits were coexpressed. This explains the dominant nature of Romano-Ward LQT1. It also demonstrates that, like the K_V channels discussed earlier, KCNQ1 forms multimeric channels (by analogy, it is expected to be a tetramer).

Mutations in minK also give rise to LQT1 (Splawski *et al.*, 1997). Patients who are heterozygous for the mutations D76N and S74L present with LQT syndrome, whereas homozygotes or compound heterozygotes exhibit both LQT syndrome and deafness. When coexpressed with KCNQ1, these minK mutations decrease the K+ current by shifting activation to more depolarized levels.

KCNQ1 and minK mutations prolong the cardiac action potential

The duration of the cardiac action potential is determined by a delicate balance between the inward and outward currents that flow during the plateau phase. Prolongation of the action potential duration can be caused either by a persistent inward current (for example, through Na+ channels, as discussed in Chapter 5) or by a reduction in outward K+ currents. Three potassium currents contribute to repolarization of the cardiac action potential from the plateau phase: an inwardly rectifying K+ current (I_{K1}; Chapter 8), a rapidly activating K+ current (I_{Kr}) and a slowly activating K+ current (I_{Ks}). Dysfunction of either of the latter two K+ currents causes long QT syndrome. The I_{Kr} current is encoded by the gene *HERG*, which maps to the LQT2 locus and is described below. The I_{Ks} channel is a complex of KCNQ1 and minK. All the KCNQ1 and minK mutations analysed to date either abolish, or markedly decrease, the K+ current and are therefore expected to prolong the cardiac action potential, and cause a increase in the QT interval. The way in which this predisposes to cardiac arrhythmia and *torsade de points* is considered in detail below (page 121).

Mutations in KCNQ1 and minK may cause deafness

Expression of KCNQ1 and minK is not limited to the heart. Both are also expressed in the stria vascularis of the inner ear. This may explain why patients with recessive mutations in *KCNQ1* (JLN syndrome) suffer from profound congenital deafness. Abnormalities in the C terminus of KCNQ1 have been found in some of these patients (Fig. 6.10). In two families, for example, a homozygous deletion of 7-bp and an 8-bp insertion at the same site was identified (Neyroud *et al.,* 1997). This produced a frameshift at residue 415 and truncation of the protein close to the end of the normal coding region. A number of missense mutations have also been described (eg., W305S). The dominant negative effect of the JLN mutations is much less than that observed for the RW mutations, and correlates with the recessive mode of inheritance of JLN syndrome. Individuals homozygous for mutations in minK also suffer from deafness.

Why do recessive KCNQ1 and minK mutations lead to deafness? KCNQ1 is expressed in the marginal cells of the stria vascularis, that secrete the endolymph of the inner ear (Fig. 6.12). This is a K^+ rich fluid that bathes the stereocilia of the sensory hair cells. A high endolymph K^+ concentration is required for hair cell function. Lack of endolymph leads to death of the hair cells and degeneration of the ganglion cells in the auditory pathway, as is observed in mice in which the gene encoding minK (which modulates KCNQ1) has been knocked out. These animals also suffer hearing and balance

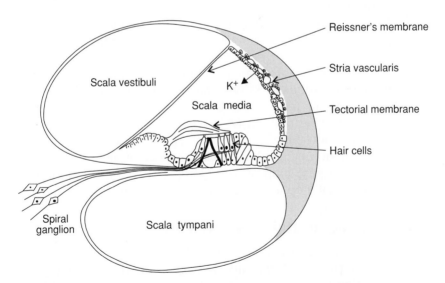

FIGURE 6.12 KCNQ1 IS EXPRESSED IN THE INNER EAR
Diagram of the inner ear showing the location of the marginal cells of the stria vascularis which express KCNQ1 and minK, and secrete the K^+ rich endolymph. A reduction in the fluid that fills the endotic space results in the collapse of Reissner's membrane onto the sensory hair cells, and their subsequent degeneration.

defects, such as a defective righting reflex and circling behaviour (Vetter *et al.*, 1996). The lack of KCNQ1 and minK may be expected to lead to a similar loss of endolymph K⁺ secretion and hair cell degeneration in man. Why some mutations should cause RW syndrome whereas others cause JLN syndrome has not been fully established.

Benign Familial Neonatal Epilepsy

When my niece was born, she stopped breathing during her first day of life and suffered from recurrent brief seizures until she was ∼3 months old, after which they ceased. This pattern is typical of benign familial neonatal convulsions (BFNC), a disorder associated with neonatal convulsions that usually present within the first 3 days after birth and show spontaneous remission by the third month of life. Paroxysmal fetal convulsions during the last two months of pregnancy have also been reported. Neurocognitive development is usually normal, but there is an increased risk of subsequent epilepsy in 10–15% of individuals in later life. In most families the gene responsible for BFNC maps to chromosome 20q13.3 and is inherited in an autosomal dominant manner. A second locus for BFNC has also been identified, which lies on chromosome 8q24.

Mutations in both *KCNQ2* and *KCNQ3* are associated with BFNC: the former gene lies on on chromosome 20q13.3 and the latter on chromosome 8q24. In one large Australian pedigree BFNC was linked with a 5-bp insertion at codon 534 of *KCNQ2* (Biervert *et al.*, 1998). This produces a frameshift that results in the truncation of >300 amino acids and produces a non-functional protein. No dominant negative effect was found when wild-type and mutant KCNQ2 mRNAs were coexpressed, suggesting that a 50% loss of channel activity is sufficient to explain the clinical symptoms observed in heterozygotes. Additional mutations in *KCNQ2* have also been identified, including two missense mutations (one within the pore (Y284C) and another within the sixth transmembrane domain (A306T)), two frameshift mutations and one splice-site mutation (Singh *et al.*, 1998). All of these mutations result in reduced expression of functional channels (Shroeder *et al.*, 1998). A missense mutation within the pore loop of *KCNQ3* (G263V) is also associated with benign familial neonatal epilepsy (Charlier *et al.*, 1998).

How do KCNQ mutations cause epilepsy?

In 1998, Wang and colleagues demonstrated that the M-channel comprises a heteromeric complex of KCNQ2 and KCNQ3. This channel plays a critical role in determining the electrical excitability of many neurones. It is slowly activated when the membrane is depolarized to around the threshold level for action potential firing, thereby hyperpolarizing the membrane back towards its resting level and reducing neuronal excitability. The M-current does not inactivate and there is a substantial steadystate current at a membrane potential of −30 mV. It also deactivates slowly on repolarization. These properties mean that the M-current has a major impact on neuronal excitability, limiting the spiking frequency and reducing the responsiveness of the neurone to synaptic inputs.

The M-current derives its name from the fact that it is inhibited by activation of muscarinic acetylcholine receptors (Brown and Adams, 1980). When the M-current is turned off, the ability of the cell to respond to synaptic inputs is enhanced and excitation may produce a burst of action potentials rather than a single spike. In addition to sympathetic ganglion neurones, where it was first identified, the M-current is found in hippocampal neurones where it is inhibited by activation of a variety of neurotransmitter receptors, including those for muscarine, serotonin and opiates. The way in which receptor activation leads to suppression of the M-current remains a subject of intensive investigation (for review see Marrion, 1997).

All of the BNFC mutations that have been studied to date result in reduced expression of the mutant protein. This may be expected to lead to neuronal hyperexcitability, accounting for the epileptic seizures. Because the M-channel is a heteromer of KCNQ2 or KCNQ3 it is not unexpected that loss-of-function mutations in either type of subunit may cause BNFC. In contrast to the LQT1 mutations in KCNQ1 described earlier, none of the BNFC mutations examined to date exerts a dominant negative effect on the activity of the heteromeric channel. Instead, it appears that a reduction in the M-current of as little as 25% is sufficient to cause the neuronal hyperexcitability that characterises BNFC. This implies that the expression level of KCNQ2 (or KCNQ3) is critical during the first few months after birth—the time when epilepsy is observed. Because the symptoms of BNFC are transient, other currents must be able to compensate for the reduction in the M-current in later life.

Eag-LIKE K+ CHANNELS

In the late 1960's, go-go dancing was very popular. Consequently when mutant fruitflies were discovered which shook their legs when anaesthetised with ether, the gene responsible was called ether-à-go-go or *eag*. It turned out to be a gene coding for a voltage-gated K+ channel. A family of *eag*-like genes was subsequently isolated from *Drosophila*, known *eag*, *erg* and *elk* (Fig. 6.1). A related gene, *HERG*, was cloned at the same time by homology screening of a human hippocampal cDNA library (Warmke and Ganetsky, 1994). *HERG* shares only 49% amino acid identity with *eag* and therefore is not the human homologue of *eag* but a member of a different subfamily of *eag*-related genes. It was subsequently discovered that *HERG* was strongly expressed in the heart, suggesting that it may play a role in repolarization of the cardiac action potential. This hypothesis was confirmed when it was found that mutations in *HERG* cause a form of long QT syndrome linked to chromosome 7 (LQT2; Curran *et al.*, 1995).

Structure and Functional Properties

The molecular architecture of HERG resembles that of K_V channels in having six transmembrane domains, intracellular N and C termini and a pore loop linking the S5 and S6 domains (Fig. 6.13). Like other voltage-gated K⁺ channels, the S4 segment is highly charged and thought to act as the voltage sensor. The primary structure of HERG channels only differs from that of K_V channels in two ways. First, the K⁺ channel signature sequence that forms the pore contains a GFG motif, rather than the GYG motif characteristic of most K⁺ channels. Second, the proximal part of the C terminus contains a highly conserved stretch of amino acid residues that is homologous to the cyclic nucleotide-binding domains of cAMP-gated ion channels. However, there is no effect of cyclic AMP or cyclic GMP on HERG currents expressed in *Xenopus* oocytes.

When HERG is heterologously expressed in *Xenopus* oocytes, outward currents are elicited by depolarization to potentials greater than −50mV (Sanguinetti *et al.*, 1995; Smith *et al.*, 1996c). The amplitude of these currents increases with depolarizations up to +10 mV and then declines, so that the current-voltage relation is bell-shaped (Fig. 6.14). This unusual voltage dependence results from that fact that HERG shows slow activation and very fast inactivation kinetics. Thus, unlike most voltage-gated K⁺ channels, inactivation is almost complete before activation has barely begun. Both activation and inactivation increase with depolarization so that the current amplitude first increases (because of activation) but then decreases (due to inactivation) as the membrane potential is made progressively more positive. Although the currents elicited by depolarization are small, those which flow when the membrane is repolarised to negative membrane potentials can be very large (Fig. 6.14). This is because the rate at which HERG channels recover from inactivation on repolarization is very fast in relation to that of deactivation (channel closing), so that the tail current amplitude is unaffected by inactivation and appears large. A significant amount of outward current will therefore flow through HERG channels during action potential repolarization.

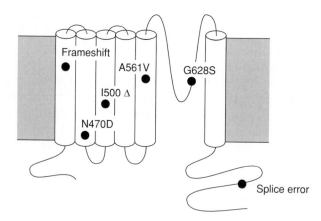

FIGURE 6.13 MUTATIONS ASSOCIATED WITH LQT2
Predicted topology of HERG showing the location of some of the LQT2-associated mutations.

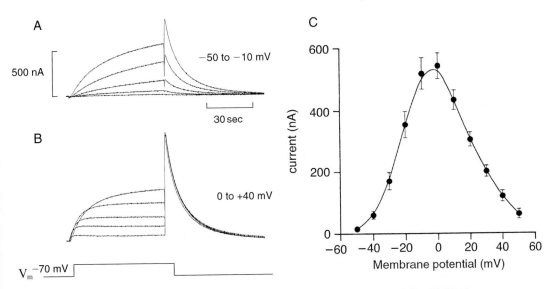

FIGURE 6.14 HERG CURRENTS EXHIBIT UNUSUAL PROPERTIES
A,B. Currents elicited by depolarizing voltage steps from a holding potential of −70 mV, recorded from *Xenopus* oocytes injected with HERG
mRNA. (A) the currents progressively increase with voltage. (B) the currents progressively decrease with voltage. (C) Current-voltage relationship
for HERG currents measured in oocytes injected with HERG mRNA. From
Sanguinetti *et al.* (1995).

The very fast inactivation exhibited by HERG does not result from N-type
(ball-and-chain) inactivation, since it is unaffected by deletion of the N terminus
of the protein (Spector *et al.*, 1996b). Instead, inactivation of HERG currents
resembles the C-type inactivation described for *Shaker* K⁺ channels: thus, it is
slowed by external, but not internal, TEA⁺ and eliminated by mutations at the
outer mouth of the channel pore, in a region known to be important for C-type
inactivation (Smith *et al.*, 1996c). This argues that inactivation of HERG currents
involves a mechanism similar to C-type inactivation, but that it occurs very
much more rapidly than is observed for *Shaker* K⁺ currents.

HERG Constitutes Part of the Cardiac I_{Kr} Channel

Many of the properties of HERG (when expressed in *Xenopus* oocytes) are
the same as those of the delayed rectifier current, I_{Kr}, of cardiac muscle
(Sanguinetti *et al.*, 1995). Most importantly, the voltage dependence of activation and the unusual rectification properties are nearly identical. Like I_{Kr},
HERG currents are also paradoxically enhanced by increasing extracellular
potassium between 1 and 20 mM. This is the opposite of what is expected,
since the driving force for the outward movement of K⁺ is actually reduced
by elevating $[K]_o$. The mechanism by which $[K^+]_o$ enhances HERG currents

is unknown but a similar phenomenon has been observed for another K_V-channel, $K_V1.4$ (Pardo *et al.*, 1992). HERG also shares other properties of I_{Kr} including a high K⁺ selectivity and inhibition by La^{3+} and Co^{2+} ions. It is also sensitive to class III antiarrthymic drugs (Spector *et al.*, 1996a). On the basis of this evidence, it is now generally accepted that HERG constitutes the cardiac I_{Kr} channel.

Eag-LIKE K⁺ CHANNELS AND DISEASE

HERG and LQT2

LQT2, one of the less common forms of inherited LQT syndrome, was mapped by linkage analysis to chromosome 7q35–36. It was later shown that *HERG* maps to the same location and that mutations in *HERG* produce LQT2. The disease shows dominant inheritance and only heterozygous mutations have been reported: possibly, homozygous mutations would be lethal. At least twenty different mutations have been found to date. These include deletions, missense mutations in conserved areas, and a splice-donor mutation at an intron-exon border (Fig. 6.13; Curran *et al.*, 1995). Two of the deletion mutations result in truncated proteins that when expressed in *Xenopus* oocytes neither produce functional channels nor suppress the activity of wild-type channels (Sanguinetti *et al.*, 1996). As all individuals with these mutations are reported to be heterozygotes, they must possess half the usual complement of HERG channels, which suggests that a 50% reduction in HERG is not sufficient for normal cardiac function.

Three of the missense mutations result in single amino acid substitutions; these occur in the S2 transmembrane segment (N470D), the S5 segment (A561V) and the K⁺ signature sequence of the pore loop (G628S). Expression of the A561V and G628S mutant channels did not produce currents in oocytes, but caused a dominant-negative suppression of wild-type HERG currents. This confirms that the HERG channel is formed from several monomers (by analogy with other K_V channels it is expected to be tetrameric). It is somewhat surprising that the G628S mutation did not produce functional channels since the analogous mutation in the *Shaker* K⁺ channel does express (but gives rise to non-selective cation currents, see earlier). In contrast to the other *HERG* mutations, substitution of aspartic acid for asparagine in the second TM (N470D) did result in functional channels. However, the currents were of smaller amplitude, activated at more negative potentials and deactivated more slowly. This mutant also caused a dominant-negative suppression of wild-type HERG currents. Thus, as predicted from the prolongation of the action potential in LQT, all the LQT2 mutations have the common effect of reducing HERG currents. It seems likely that the degree of reduction will be found to correlate with the extent of action potential prolongation and the severity of the long QT syndrome in patients.

Although *HERG* is most strongly expressed in the heart it was originally cloned from hippocampus. However, effects of *HERG* mutations on neuronal function have not been reported, perhaps because other K⁺ currents are able to compensate for the loss of HERG. It is not often appreciated that the total membrane conductance during the plateau phase of the cardiac action potential is very small, smaller even than at the resting potential. This means that a small change in current will produce a large effect on the ventricular plateau potential. By contrast, in neurons, the conductance during the action potential is very much larger and a change in current of similar magnitude will have much less effect on the membrane potential.

Drosophila Mutations

The cloning of the *HERG* gene and identification of its role in human pathophysiology was made possible by the fact that the related gene, *eag*, is also associated with a mutant phenotype in *Drosophila* (Warmke *et al.*, 1991). Analysis of *eag* mutant flies reveals that the motor neurones fire repetitively, so that transmitter release is enhanced. This accounts for the myotonic phenotype. Thus the *eag* gene is involved in suppressing neuronal excitability. Mammalian homologues of *eag* have also been identified, but disease-causing mutations in these genes have not been reported. The *Drosophila* homologue of *HERG* has also been cloned (Wang *et al.*, 1997). It is encoded by the *seizure* locus. Mutations in this gene cause temperature-induced hyperactivity and paralysis.

How Do Mutations in KCNQ1 or HERG Cause Torsade de Pointes?

By itself, a small increase in the duration of the cardiac action potential, such as that observed in LQT patients, is not life-threatening and does not elicit *torsade de pointes*. Indeed, it may offer some protection against other types of arrythmia and constitutes the mechanism of action of the Class III anti-arrythmic drugs. In combination with physical or emotional stress, however, a prolonged cardiac action potential duration may precipitate a fatal cardiac arrythmia. Why is this the case? One answer is that an increase in sympathetic nerve stimulation, or in circulating catecholamines, increases the magnitude of the cardiac inward Ca^{2+} current and thereby enhances the heart rate. In normal patients, the magnitude of I_{Ks} is also increased by sympathetic stimulation, which causes a rate-dependent shortening of the cardiac action potential and protects the heart from premature excitation caused by the larger Ca^{2+} current. In patients with LQT syndrome, however, the longer action potential duration means that the increased Ca^{2+} current produces early after-depolarizations, which occur during the repolarization phase of the action potential (Fig. 6.12). One idea is that these early after-depolarizations may trigger additional action potentials at multiple loci, and so initiate or help maintain *torsade de points*. Another view postulates that *torsade de pointes* results from re-entrant excitation. The precise mechanism that underlies *torsade de pointes*, however, is still unresolved and remains a matter of controversy.

Currently, most patients with LQT syndrome are treated with inhibitors of β-adrenergic receptors. This markedly reduces the mortality of symptomatic LQT patients, which in untreated individuals approaches 50% within ten years of diagnosis. Now that the molecular basis of the major types of LQT are known, it may be possible to devise even more effective therapies by targeting them to different genotypes. As discussed elsewhere, elevation of extracellular K^+ may be beneficial in LQT2 (see below; Compton *et al.*, 1996), and Class 1b antiarrythmic agents may be useful in LQT3 (see Chapter 5).

I_{Kr} and I_{Ks} Are Involved in Acquired LQT Syndrome

Although a number of different genetic defects can give rise to LQT syndrome, in most cases the disorder is not inherited but acquired. As described above, repolarization of the cardiac action potential results from activation of I_{Kr} and I_{Ks} and mutations in the genes that encode these channels result in long QT syndrome. It is therefore not unexpected that drugs which block I_{Kr} and I_{Ks} currents prolong the cardiac action potential and induce long QT syndrome. Among these are class III antiarrthymic agents such as sotalol, dofetilide and quinidine, which selectively block I_{Kr}. Although it appears ironic that drugs used to treat cardiac arrythmias should themselves be arrythmogenic and cause significant mortality, it is worth noting that there are multiple types of cardiac arrthymia and that these may have different causes and require different therapy. The class III antiarrthymic drugs, for example, are effective against re-entrant arrythmias. Hypokalaemia and bradycardia are additional precipitating factors in drug-induced LQT syndrome.

The K^+ sensitivity of HERG (I_{Kr}) may be of importance in acquired long QT syndrome. External K^+ ions accumulate in the space between cardiac muscle fibres when action potential activity is high, as at rapid heart rates. This enhances HERG currents, shortening the action potential and enabling more rapid heart rates to be achieved. Modest hypokalaemia is a common clinical problem and would be expected to reduce HERG currents, thus lengthening the cardiac action potential. Class III anti-arrhythmic agents also tend to lengthen the cardiac action potential. By themselves, neither hypokalaemia, nor anti-arrhythmic agents, prolong the action potential sufficiently to trigger *torsade de pointes*, but in combination they can be lethal. It has been suggested that mild elevation of blood K^+ levels may reduce the incidence of side-effects in anti-arrhythmic therapy (Yang and Roden, 1996). Indeed, in patients with mutations in HERG (LQT2), increasing plasma K^+ concentration by ~1.5 mM caused a marked improvement of ventricular action potential repolarization, shortening the QT interval by ~25% (Compton *et al.*, 1996).

Sudden cardiac death is also associated with the antihistamine H_1-receptor antagonists terfenidine and astemizole. These drugs are very potent blockers of I_{Kr} (HERG) and cause a prolonged action cardiac potential and QT interval, early after-depolarizations and *torsade de pointes* (Salata *et al.*, 1995). In most people, terfenidine does not produce cardiac problems as it is rapidly broken down in the liver and its metabolite, terfenidine carboxylate, does not block IK$_r$: as the drug is taken orally, it will encounter the liver before it ever reaches the heart. Those individuals with liver disease, or who are deficient in the oxidative P-450 enzymes that break down terfenidine, or who are coadministered drugs which inhibit these enzymes (such as ketoconazole and macrolide antibiotics), or who take an overdose of terfenidine, are at risk of developing *torsade de pointes*. The drug is now only available on prescription in the United Kingdom.

CHAPTER 7

Ca²⁺-ACTIVATED K⁺ CHANNELS

The Ca^{2+}-activated potassium (K_{Ca}) channels are a structurally diverse group of K^+ channels which share the common property that they are activated by an increase in the intracellular Ca^{2+} concentration ($[Ca^{2+}]_i$). They are found in almost all nerve cells where they play a key role in controlling the action potential waveform and in regulating cell excitability. In secretory epithelia they are also involved in K^+ secretion. To date, no human disease has been firmly associated with any K_{Ca} channel, although there is some evidence to suggest that the small conductance K_{Ca} channels may contribute to the pathology of myotonic dystrophy.

K_{Ca} channels come in three major types, which can be distinguished electrophysiologically by their very different single-channel conductances. They also exhibit different voltage and pharmacological sensitivities and have different functional roles. The maxi K_{Ca} (or BK) channel is so called because of its large (big) single-channel conductance: 100–250 pS in symmetrical 100 mM K^+ solutions. It is activated both by depolarization and by micromolar $[Ca^{2+}]_i$. The small (SK; 5–20 pS) and intermediate (IK; 20–80 pS) conductance calcium-activated K^+ channels are not voltage sensitive and are activated by Ca^{2+} within the submicromolar range. BK channels are blocked by low concentrations of tetraethylammonium (TEA^+) and charybdotoxin (CTX), but are unaffected by apamin, a peptide isolated from honeybee venom. By contrast, SK channels are insensitive to TEA^+, but some—although not all— are blocked by apamin. IK channels are insensitive to apamin but blocked by CTX and clotrimazole.

K_{Ca} channels contribute to the repolarization and after-hyperpolarization (AHP) of vertebrate neurones. The duration of the AHP is variable and in many neurones at least two components can be detected: an initial fast component that lasts for 1–2 msec and a later component (or components) that activates more slowly and can persist for several seconds (Fig. 7.1A). The fast AHP regulates the interval between adjacent spikes. It is mediated by K^+ channels which are activated in response to depolarization and close again soon after repolarization. The slow AHP components are mediated by

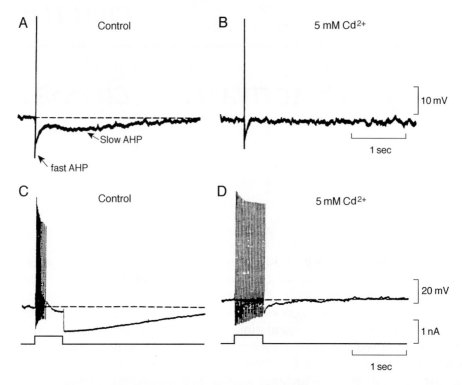

FIGURE 7.1 AFTER-HYPERPOLARIZATIONS
(A) The action potential recorded from a vagal neuron is followed by a fast and then a slow after-hyperpolarization (AHP). (B) Addition of 5 mM Cd^{2+} to the bathing solution, which blocks Ca^{2+} influx, abolished the slow AHP but leaves the fast AHP intact. (C) In response to a maintained stimulus, the firing rate of the vagal neurone rapidly decreases and then ceases. A large slowly decaying hyperpolarization is also observed after the stimulus pulse. Both these effects result from activation of SK channels triggered by Ca^{2+} influx during the train of action potentials, as they are abolished by 5 mM Cd^{2+} (D). From Yarom *et al.* (1985).

SK and IK channels which open in response to the elevation of submembrane $[Ca^{2+}]_i$ that results from Ca^{2+} entry during each action potential. They remain active for as long as Ca^{2+} remains elevated. The slow AHP plays an important role in controlling the firing pattern of the cell. It underlies the decrease in action potential frequency, known as spike-frequency adaptation, which occurs during a maintained stimulus (Fig. 7.1C). This is because Ca^{2+} accumulates during repetitive electrical activity, producing increasing activation of SK currents and a larger AHP. Eventually, the increase in SK current may be sufficient to prevent the cell from reaching its firing threshold. Blocking Ca^{2+} influx with Cd^{2+} prevents SK channel activation and abolishes both the slow AHP and spike frequency adaptation (Fig. 7.1B, D).

MAXI K$_{Ca}$ (BK) CHANNELS

BK channels are found in virtually all cell types, with the exception of the heart. They cause membrane hyperpolarization and contribute to action potential repolarization. They thereby play important roles in neuronal excitability, transmitter release, and the regulation of smooth muscle excitability.

BK channels consist of two subunits, α and β (Knauss et al., 1994b). The α-subunit was the first to be cloned, by identifying the gene that encodes a mutant form of *Drosophila* known as *Slopoke*, which exhibits characteristic jerky movements (Atkinson et al., 1991). The mammalian counterpart of *slopoke* is known as *mslo* (Butler et al., 1993). Expression of either *slopoke* or *mslo* gives rise to K$^+$ currents that are gated both by voltage and by the binding of Ca^{2+} ions to the intracellular side of the channel (Fig. 7.2). Thus the α-subunit serves as the pore-forming subunit and is able to express independently. In contrast to the K$_V$ channels, the different types of BK channels do not arise from a family of homologous α-subunit genes. Rather, alternative splicing of the *mslo* gene produces numerous structurally and functionally distinct α-subunits (Adelman et al., 1992).

The α-subunit of the BK channel was initially thought to have a similar transmembrane organisation to the voltage-gated K$^+$ channels (K$_V$ channels), but it is now believed that there are seven transmembrane domains (S0–S6) and that the N terminus is extracellular (Fig. 7.3; Wallner et al., 1996). The carboxyl tail contains four domains, which also show some hydrophobicity (S7–S10), but these appear to lie intracellularly. As is the case for K$_V$ channels, the S4 domain of BK channels contains a number of positively charged residues and is believed to serve as the voltage sensor. Likewise, the S5—S6 loop forms the pore and contains the K$^+$ selectivity filter. The extracellular N terminus and the first TM (S0) of the α-subunit are required for β-subunit modulation (Wallner et al., 1996). The location of the Ca^{2+}-binding site is still not known, but there is some evidence that residues in the C terminus (between segments S9 and S10) are involved.

BK channels are blocked with high affinity by a number of toxins isolated from scorpion venom, including CTX and iberiotoxin, which bind to the outer mouth of the pore. CTX was used to biochemically purify the BK channel from smooth muscle and show that the native channel consists of two subunits, α and β, arranged in a 1:1 stoichiometry (Knaus et al., 1994a,b). Each channel is thus a complex of four α- and four β-subunits. The β-subunit (31 kDa) has no significant sequence homology to K$_V$ α- or β-subunits and consists of two transmembrane-spanning domains linked by a large extracellular domain, with cytosolic N and C termini (Fig. 7.3). When expressed alone, the β-subunit does not form functional channels. BK channels comprising both α- and β-subunits, however, are more sensitive to activation by Ca^{2+} or voltage than are channels made up of α-subunits alone (McManus et al., 1995). Thus the β-subunit contributes to the Ca^{2+} and voltage sensitivity of the BK channel. It also endows the channel with sensitivity to certain drugs, such as the

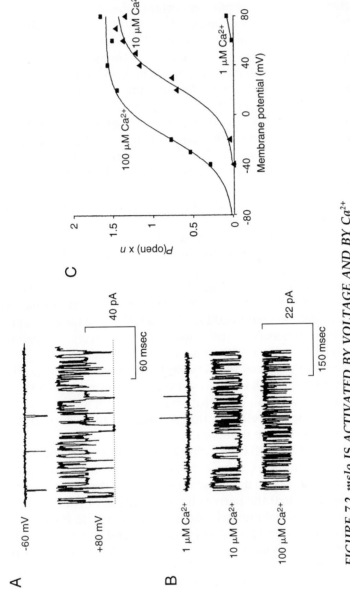

FIGURE 7.2 mslo IS ACTIVATED BY VOLTAGE AND BY Ca^{2+}
(A) Single *mslo* channel currents recorded from an inside-out patch exposed to 100 μM Ca^{2+}. Channel activity was observed at +80 mV but not −60 mV. (B) Increasing the cytoplasmic Ca^{2+} concentration activates *mslo* currents. Membrane potential: +40 mV. (C) Voltage-dependence of *mslo* open probability is influenced by Ca^{2+}. Open probability is defined as P$_{(open)}$ × n, where n = 2, because there were two channels in the patch. From Butler *et al.* (1993).

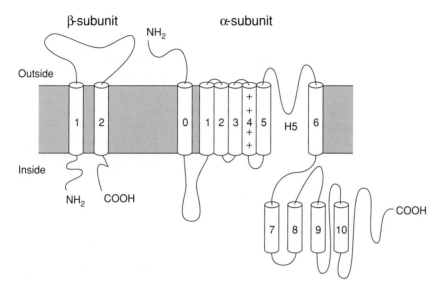

FIGURE 7.3 PUTATIVE MEMBRANE TOPOLOGY OF α- AND β-SUBUNITS OF BK CHANNELS

Proposed membrane topology of BK channel subunits deduced from hydropathy analysis and antibody-binding studies. The S4 transmembrane domain of the α-subunit serves as the voltage sensor, the loop between S4 and S5 forms the pore, and sites involved in Ca^{2+} activation are thought to be located within the cytosolic C terminus of the molecule. The β-subunit plays a regulatory role.

channel activator dehydrosoyasaponin (Fig. 7.4), a natural product isolated from a Ghanese medicinal herb.

Both the α- and β-subunits of the BK channel contribute to the high-affinity-binding site for CTX (Hanner *et al.*, 1997). When expressed alone, the α-subunit is blocked by CTX in a manner consistent with the toxin physically plugging the pore (as is the case for the K_V channels, Chapter 6). Coexpression with the β-subunit increases the sensitivity to CTX ~50-fold (Knaus *et al.*, 1994a). Cross-linking studies also demonstrate that two lysine residues, one in the α-subunit and the other in the β-subunit, lie within 11 Å of each other. These studies therefore indicate that part of the large extracellular loop of the β-subunit lies in close proximity to the toxin-binding site and thus to the extracellular mouth of the pore formed by the α-subunit.

A key question has been whether BK channels require both depolarization and Ca^{2+} to open or whether only one of these is sufficient for channel activation. It is now believed that voltage is the primary activator because strong depolarisation is able to open the BK channel even at nanomolar Ca^{2+} concentrations (Cui *et al.*, 1997). Measurements of gating current have also shown that the maximum amount of gating charge moved is independent of Ca^{2+} (Stefani *et al.*, 1997). The role of Ca^{2+} is to alter the conformation of the channel such that less voltage is needed to move the same amount of charge. In the

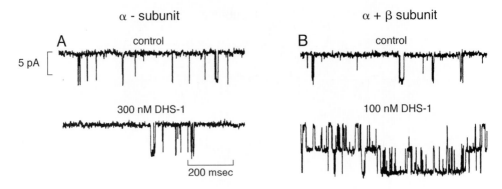

FIGURE 7.4 THE β-SUBUNIT IS REQUIRED FOR DHS-1
SENSITIVITY OF BK CHANNELS
Single-channel currents recorded from channels comprising α-subunits
only (A) or both α- and β-subunits (B). Only channels containing β-
subunits were activated by dehydrosoyasaponin (DHS-1). From McManus
et al. (1995).

presence of Ca^{2+}, therefore, the voltage dependence of channel activation is
shifted to lower membrane potentials. In general, BK channels are closed at
the normal resting potential of the cell, but can be activated by a rise in
$[Ca^{2+}]_i$, such as that which might occur in response to an action potential or
the action of hormones and neurotransmitters.

SMALL K_{Ca} CHANNELS

The small conductance Ca^{2+}-activated K^+ channels (SK channels) were cloned
by computer. Köhler and colleagues (1996) searched the database of expressed
sequence tags (ESTs) with a virtual probe based on the pore sequence of
cloned K^+ channels. They identified a novel EST sequence, homologous but
not identical to known K^+ channels, which was then used for screening of
cDNA libraries. In this way, three types of SK channel were cloned, which
show overlapping but distinct distributions in the brain. In addition to the
brain, SK1 channels are expressed in heart, SK2 channels in the adrenal gland,
and SK3 channels in skeletal muscle.

Although they show no significant amino acid homology except within
the putative pore region, hydrophobicity analysis suggests SK channels have
a similar topology to that of voltage-gated K^+ (K_V) channels (Chapter 6).
There are six putative transmembrane domains (TMs), a hairpin loop linking
TMs 5 and 6 (which is thought to form part of the pore) and intracellular N
and C termini. The fourth TM resembles the S4 voltage sensor of K_V channels
in that it contains three positively charged residues (compared with seven
in K_V channels). Despite the presence of these charged residues, however,
SK channels are not voltage sensitive. Heteromeric SK channels, with novel

properties, can be generated by coassembly of SK1 and SK2 subunits (Ishii et al., 1997a).

When expressed in *Xenopus* oocytes, SK currents can be activated by addition of Ca^{2+} to the intracellular face of the membrane (Fig. 7.5A). Half-maximal activation of all three types of SK channels is produced by ~0.3 μM Ca^{2+}, making them at least ten-fold more Ca^{2+} sensitive than BK channels. The steepness of the relationship between $[Ca^{2+}]_i$ and channel activity suggests that four Ca^{2+} ions are involved in gating the channel. However, Ca^{2+} does not bind directly to the SK channel to effect opening, but instead to the regulatory protein calmodulin, that is constitutively bound to the channel (Xia et al., 1998). The possibility of such an interaction was raised by the fact that the sequence of the SK channel does not possess any obvious consensus motifs for Ca^{2+}-binding, despite the fact that channel gating is regulated by Ca^{2+} with very high affinity. This stimulated Xia and colleagues (1998) to look for an associated Ca^{2+}-binding protein. Using a variety of biochemical methods, they demonstrated that calmodulin interacts with a domain in the C terminus of the SK channel in a Ca^{2+}-independent fashion. Thus, calmodulin is expected to be constitutively bound to the channel *in vivo*. Indeed, calmodulin is so tightly bound to the channel that it is not possible to disrupt the association readily. In order to demonstrate its func-

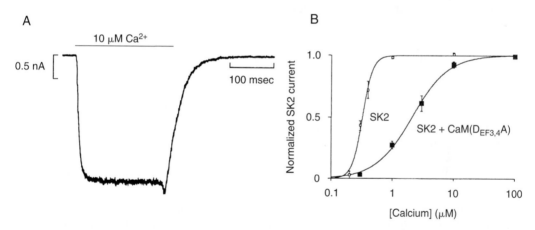

FIGURE 7.5 CALMODULIN CONFERS Ca^{2+} SENSITIVITY ON SK CHANNELS

(A) SK2 current activated by application of 10 μM Ca^{2+} to the intracellular side of an inside-out patch excised from a *Xenopus* oocyte heterologously expressing SK2 channels. The holding potential was -80 mV. (B) Calmodulin mediates the gating of SK channels. Relationship between the Ca^{2+} in the intracellular solution and SK current at -100 mV. SK2 channels were expressed alone or with a mutant calmodulin (CaM($D_{EF3,4}$A) in *Xenopus* oocytes. When expressed alone, the SK2 channel associated with wild-type calmodulin endogenously expressed by the oocyte. The Ca^{2+} concentration that produced maximal activation was 0.3 μM for SK2, and 2 μM for SK2 plus CaM($D_{EF3,4}$A). The Hill coefficients were 5 and 1, respectively. From Xia et al. (1998).

tional role, therefore, Xia and colleagues coexpressed SK2 channels together with either wild-type calmodulin, or a mutant calmodulin with reduced Ca^{2+} affinity, in *Xenopus* oocytes. They observed that the Ca^{2+} sensitivity of SK2 channel was substantially lower when it was coexpressed with the mutant calmodulin (Fig. 7.5B). Thus calmodulin may be considered as the β-subunit of the small K_{Ca} channel, that, when it binds Ca^{2+}, induces a conformational change in the SK α-subunit that opens the channel pore.

The sensitivity to apamin varies between SK channels, with SK2 and SK3, but not SK1, channels being blocked by picomolar concentrations of apamin. Two amino acids, located within the outer pore region of the SK channel, constitute the primary determinants of apamin sensitivity (Ishii *et al.*, 1997a).

Myotonic Muscular Dystrophy

Myotonic muscular dystrophy (MMD) is one of the most common inherited human myopathies, affecting 1 in 8,000 to 18,000 people. It shows autosomal dominant inheritance with high penetrance. The hallmark of the disease is muscle stiffness (myotonia), which results from repetitive bursts of electrical activity in the muscle, and can be observed in the electromyogram. Muscle wasting and weakness also occur. In addition, patients may experience oph-thalamic (cataracts, ptosis), endocrine, cardiac, and mental problems. When the gene is inherited from the female parent, the condition may present in childhood, but more usually it becomes clinically overt between the ages of 20 and 50. Patients usually present with weakness of the hands and difficulty in walking.

Genetic mapping localised the MMD gene to chromosome 19q13.3. Subse-quently, it was shown that the disease is produced by an expanded CTG repeat in the 3'-untranslated region of a gene that encodes a protein with homology to protein kinases (Fu *et al.*, 1992). The length of the CTG expansions increases in successive generations and is associated with increasing disease severity. The role of the DM protein kinase in MMD remains unclear. Its function has not been established, and mice in which the gene encoding this protein has been 'knocked out' do not develop the disease (Jansen *et al.*, 1996). The presence of the CTG repeat expansion also affects the expression of a flanking gene known as DMAHP, which contains a homeodomain sequence. Because homeodomains are implicated in the binding of proteins to DNA, the DMAHP protein may be involved in regulating the expression of other genes, one or more of which may be involved in the pathology of MMD.

There is accumulating evidence that SK3 channels may play a role in MMD. As described earlier, this channel is apamin sensitive. Apamin binding is not detectable in normal human skeletal muscle, but is markedly increased in muscle biopsies from MMD patients (Renaud *et al.*, 1986), suggesting that the amount of SK3 protein may be increased. Apamin decreases myotonic activity when injected into the muscle of MMD patients (Behrens *et al.*, 1994), consistent with the idea that overexpression of SK3 may contribute to the MMD phenotype. Why an increase in SK3 channel density should cause myotonia is still far from clear. One possible explanation, however, is that it results in an increased K^+ current and thus in an enhanced accumulation of K^+ ions in the T-tubules of skeletal muscle. This would tend to depolarize the muscle fibre and lead to the repetitive muscle activity that characterises the disease. Indeed, muscle biopsies have shown that the resting potential of myotonic muscle is more depolarized than normal muscle. Muscle atrophy in MMD might be a consequence of the enhanced Ca^{2+} entry produced by myotonia and depolarization; several of the ion channel diseases described in this book which cause myotonia (or depolarizing muscle block) are associated with increased Ca^{2+} influx and muscle damage.

To date, it remains unclear whether the increase in SK3 expression is the cause or consequence of the muscle dystrophy. Because apamin binding is induced following denervation (Schmid-Antomarchi et al., 1985), the high level of apamin binding in MD muscle may simply reflect a secondary increase in SK3 resulting from muscle degeneration.

INTERMEDIATE K_{Ca} CHANNELS

Ca^{2+}-activated K^+ channels of intermediate conductance (IK channels) were also cloned using a database search strategy (Ishii et al., 1997b). Structurally, IK channels resemble SK channels, but they have a shorter N terminus. They are strongly expressed in smooth muscle but not in brain. The Gardos channel is also a member of the IK family. This channel has the distinction of being the first Ca^{2+}-activated K^+ channel to be identified, in red blood cells in 1958, and is named after its discoverer. Half-maximal activation of IK channels occurs at approximately the same Ca^{2+} concentration as SK channels, but the relationship between Ca^{2+} and channel activity is less steep. This means that IK channels will be open at physiological concentrations of $[Ca^{2+}]_i$ and also suggests that fewer Ca^{2+} ions need to bind to open the channel. IK channels are not blocked by IBX or apamin, but are inhibited by CTX and clotrimazole.

Clotrimazole is currently in clinical trials for the treatment of sickle cell disease (Brugnara et al., 1996). This disease is caused by a single point mutation in the haemoglobin molecule, which renders it less soluble in the deoxygenated form. Consequently, it precipitates out of solution within the cell more readily. When this happens, the erythrocyte is no longer sufficiently distensible to pass through the small capillaries so that the sickled cells block the microvasculature and cause local tissue hypoxia. Such a sickle cell crisis is extremely painful and, because the sickled cells are more fragile and have a shorter life, may lead to chronic anaemia. Kidney failure may result if sickled erythrocytes block the renal vasculature. The sickle attacks are triggered by hypoxia, by protons, and by elevated temperature, all of which cause dehydration of the red blood cell and thus an increase in the haemoglobin concentration, which precipitates sickling. IK channels contribute to cell sickling as their activation results in an efflux of K^+, which leads to Cl^- and water loss. By blocking IK channels, and thus K^+ fluxes, clotrimazole reduces dehydration of the erythrocyte and thus the chance of cell sickling.

CHAPTER 8

INWARDLY RECTIFYING K⁺ CHANNELS

Inwardly rectifying potassium channels (Kir channels or inward rectifiers) constitute a large family of voltage-independent K^+ channels, that are structurally and functionally distinct from the K_V channels. All members of this family show the property of *inward rectification,* a term which describes the fact that the inward current evoked by a given hyperpolarization from the potassium equilibrium potential (E_K) is greater than that elicited by a depolarization of the same amplitude (Fig. 8.1).

Inward rectifiers have two major physiological roles: they stabilize the resting potential near the K^+ equilibrium potential and they are involved in K^+ transport across membranes. Kir channels involved in regulating cell excitability tend to show strong inward rectification. The stronger the degree of rectification, the sharper the threshold for excitation, because depolarization will rapidly decrease the K^+ conductance and so switch off its effect on membrane potential. Strong inward rectifiers are found in neurones, cardiac muscle and skeletal muscle. Kir channels that show weak rectification are able to pass more outward current. They are involved in K^+ fluxes across epithelia, but in excitable cells their activation also acts to suppress excitability. Some types of Kir channels mediate the effects of hormones and neurotransmitters on cell excitability, such as the G-protein-regulated inward rectifiers of heart and brain that couple the activation of certain ligand-gated receptors (which activate G-proteins) to electrical activity. The ATP-sensitive K^+ channels (K_{ATP} channels) are regulated by cytosolic nucleotides and link cell metabolism to electrical activity and K^+ fluxes. They are important in the regulation of insulin secretion from pancreatic β-cells, the response to cardiac and cerebral ischaemia and the control of vascular smooth muscle tone.

To date, mutations in Kir channels, or defects in their regulation, have been shown to result in neuronal degeneration, failure of renal salt absorption and defective insulin secretion (both increased and decreased secretion have been described).

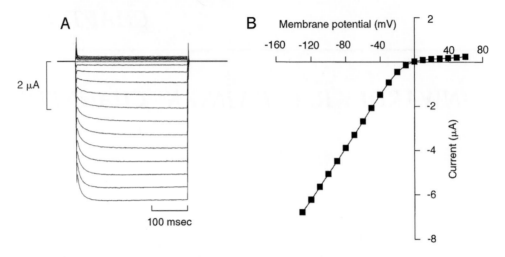

FIGURE 8.1 INWARD RECTIFICATION
Inward rectifiers pass more current response to a hyperpolarizing voltage
step than to a depolarizing voltage step of the same amplitude. Currents
(left) and associated current-voltage relation (right) were elicited by a
series of voltage steps from +60 to −120 mV and recorded from a *Xenopus*
oocyte injected with mRNA encoding the strong inward rectifier Kir2.1.
Figure supplied by Carina Ämmälä.

Structure of Kir Channels

Kir channels comprise a large superfamily

The first inward rectifiers to be cloned, in 1993, were ROMK1, IRK1 and
GIRK1 (Ho *et al.*, 1993; Kubo *et al.*, 1993a,b). Originally, the names of these
channels were based on their functional roles and the tissue from which they
were cloned. Subsequently, Doupnik, Davidson and Lester (1995) suggested
a more rational nomenclature based on the extent of homology in the amino
acid sequence, and ROMK1, IRK1 and GIRK1 became Kir1.1a, Kir2.1 and
Kir3.1, respectively. However, both nomenclatures are still to be found in
the literature. In this chapter, I refer to Kir channels by their new names. The
relationship between the different Kir genes is shown in Fig. 8.2. They fall
into 6 main subfamilies. Differential splicing of Kir1.1 results in a number of
different isoforms but splice variants have not been reported for most Kir
channel genes. Indeed, in some cases (such as Kir6.2) the gene consists of a
single exon.

Primary structure of Kir channels

Kir channels are much smaller than voltage-gated K⁺ (K$_V$) channels, containing only 390–500 amino acids and having molecular masses of ~40 kDa.

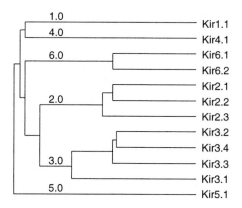

FIGURE 8.2 KIR CHANNEL LINEAGES
Relationship between different members of the Kir channel family. The
length of the lines indicates the percentage of amino acid identity. Six
major sub-families are evident.

They possess only two putative transmembrane domains, TM1 and TM2,
linked by a loop which dips back down into the membrane to line the outer
part of the pore (Fig. 8.3). This loop is referred to as the pore loop (or H5
region) and it shows significant sequence homology to the pore loops of
other K^+ channels. The inward rectifiers therefore correspond to the S5 and
S6 domains of the voltage-gated K^+ channels. There is no equivalent to the
S4 domain that is involved in the voltage-dependent activation of K_V channels,

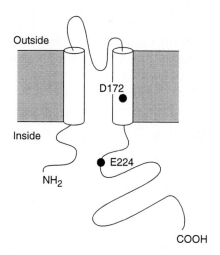

FIGURE 8.3 PUTATIVE TOPOLOGY OF KIR CHANNELS
Putative membrane topology of Kir channels, deduced from hydropathy
analysis. Residues mentioned in the text are indicated, and numbered
according to their position in Kir2.1. Both are involved in inward rectification.

which may explain why the open probability of inward rectifier channels shows little voltage dependence. Both the N and C termini of Kir channels are located in the cytoplasm.

In members of the Kir2.0 and Kir4.0 families, the extreme C-terminus contains a motif that interacts with a 95 kD protein known as the postsynaptic density-95 protein (PSD-95). As its name suggests, this protein is found in the post-synaptic region of nerve cells and may provide a means of anchoring Kir channels at the synapse (Cohen *et al.*, 1996). The C terminus of Kir6.2 appears to be uniquely important for functional channel activity, since this subunit does not express independently of its regulatory subunit unless the last 26–36 residues of the C terminus are deleted (Tucker *et al.*, 1997; see later).

Heteropolymerization occurs between Kir subfamily members

As is the case for the voltage-gated K$^+$ channels, the pore of the Kir channel is formed from four subunits. There is evidence that the ability of Kir channel subunits to coassemble resides in the second transmembrane domain and the proximal part of the C terminus: unless these regions are compatible, tetramerization will not occur (Tinker *et al.*, 1996). This may be contrasted with K$_V$ channels (Chapter 6), where subunit coupling involves the N terminal part of the molecule.

Both homomeric Kir channels, formed from the same type of subunit, and heteromeric Kir channels, formed from different types of subunit, have been identified. When heteropolymerization occurs it generally takes place between members of the same subfamily rather than between members of different subfamilies. Heteropolymerization may either enhance (Fig. 8.4) or reduce expression of functional channel activity. It is also of physiological significance. In particular, the G-protein coupled Kir channels of heart and brain appear to be heteromultimers, the brain channel being composed of Kir3.1 and Kir3.2 subunits (Kofuji *et al.*, 1995) and the cardiac channel comprising Kir3.1 and Kir3.4 subunits (Krapivinsky *et al.*, 1995).

Role of other subunits

Unlike most other inward rectifiers, heterologous expression of either Kir6.1 or Kir6.2 alone does not result in functional channel activity. Each of these proteins requires an additional subunit, a sulphonylurea receptor (SUR), which coassembles with the Kir subunit in a 4:4 stoichiometry to form an octameric channel complex (Fig. 8.5; Clement *et al.*, 1997). This complex forms the ATP-sensitive K$^+$ (K$_{ATP}$) channel (Sakura *et al.*, 1995). Although it is tempting to regard Kir6.2 as the primary α-subunit and SUR as an accessory β-subunit of the K$_{ATP}$ channel, similar to the α- and β-subunits of voltage-gated Na$^+$ and K$^+$ channels, this analogy is not strictly correct. The two K$_{ATP}$ channel subunits are more intimately linked because both must be coexpressed to obtain functional channel activity.

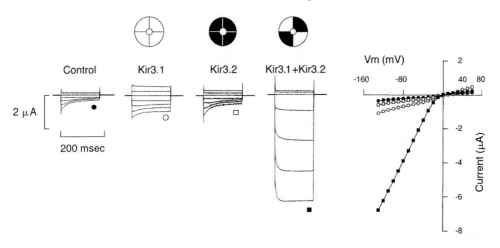

FIGURE 8.4 HETEROMERIZATION MAY ENHANCE THE CURRENT MAGNITUDE

Kir channels are made up of four subunits, which may be either identical (homomeric) or different (heteromeric). Currents recorded from oocytes injected with either Kir3.1 (○) or Kir3.2 (□) mRNA are very small and similar to those recorded from water-injected control oocytes (●). Coinjection of Kir3.1 and Kir3.2 mRNA (■), however, results in large currents.

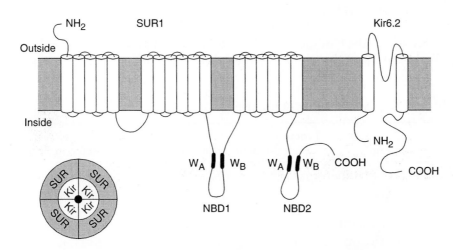

FIGURE 8.5 K_{ATP} CHANNELS ARE FORMED FROM TWO DIFFERENT TYPES OF SUBUNITS

The K_{ATP} channel is an octameric complex of 4 pore-forming Kir6.2 subunits and 4 regulatory sulphonylurea receptor subunits. Hydropathy analysis suggests that Kir6.2 conforms to the usual transmembrane topology of Kir channels. SUR1 contains multiple transmembrane domains and two cytosolic nucleotide binding domains (NBDs), each of which contains a Walker A (W_A) and a Walker B (W_B) motif. These motifs are involved in the activation of the K_{ATP} channel by MgADP.

The high affinity with which sulphonylureas bind to their receptor was exploited by Aguilar-Bryan and her colleagues (1995) to purify and subsequently clone a high-affinity sulphonylurea-binding protein from pancreatic β-cell membranes. The sulphonylurea receptor (SUR1) they isolated turned out to be a member of the ABC-transporter superfamily, which also includes CFTR, the cystic fibrosis gene product (see Chapter 12). SUR1 is 1581 amino acids long and has a molecular mass of ~140 kDa. Hydropathy analysis suggests that there are multiple transmembrane domains: the precise number is unclear, but the most recent model suggest there may be 17 (Fig. 8.5). There are also two large intracellular loops that contain consensus sequences for nucleotide binding and that play an important role in regulating channel K_{ATP} activity.

There are at least two different sulphonylurea receptor genes (SUR1 and SUR2) and further diversity is created by alternative splicing of SUR2 (for review see Ashcroft & Gribble, 1998). SUR1 is expressed primarily in pancreatic β-cells and in the brain, whereas SUR2A is expressed in cardiac and skeletal muscle and SUR2B is found in smooth muscle. When coexpressed with Kir6.2, the various sulphonylurea receptors form K_{ATP} channels with different properties (Inagaki et al., 1996). The interaction of Kir6.x subunits with SUR raises the question of whether other ABC transporters can couple to Kir subunits. This question remains open, although there is some evidence that the related ABC transporter CFTR may couple to Kir1.1 (McNicholas et al., 1996a).

Chromosomal Location

Table 8.1 shows the chromosomal location of the major Kir channel subunits. The two subunits of the β-cell K_{ATP} channel, Kir6.2 and SUR1 map to adjacent

TABLE 8.1   MAMMALIAN KIR CHANNEL GENES

Gene (human)	Protein	Chromosome location (human)	Major site of expression	Reference
KCNJ1	Kir1.1 (ROMK)	11q24	Kidney	Yano et al. (1994)
KCNJ2	Kir2.1 (IRK)	17	Heart	Raab-Graham et al. (1994)
KCNJ3	Kir3.1 (GIRK1)	2q24.1	Heart, brain	Stoffel et al. (1994)
KCNJ6	Kir3.2 (GIRK2)	21q22.1–22.2	Brain	Sakura et al. (1995)
KCNJ9	Kir3.3 (GIRK3)	1q21–q23	Brain	Lessage et al. (1995)
KCNJ5	Kir3.4 (GIRK4)	11q24	Heart	Tucker et al. (1995)
KCNJ8	Kir6.1	12p11.23	Ubiquitous	Inagaki et al. (1995a)
KCNJ11	Kir6.2 (K_{ATP})	11p15.1	Pancreatic β-cells, brain, heart, skeletal muscle	Inagaki et al. (1995b)
SUR1	SUR1	11p15.1	Pancreatic β-cells, brain	Thomas et al. (1995)
SUR2	SUR2	12p12.12	Heart, brain, skeletal muscle	Chutcow et al. (1996)

regions of chromosome 11p15.1 and both reading frames are aligned in the same direction (Inagaki *et al.*, 1995), suggesting they may be subject to similar transcriptional regulation.

Correlating Structure and Function

Inward rectification

All Kir channels show inward rectification but some, like Kir2.1 and 3.1, are strong inward rectifiers while others, such as Kir1.1 and Kir6.2, are weak inward rectifiers. Those currents that exhibit strong inward rectification generally also show an apparent time-dependent activation on hyperpolarization, whereas weak inward rectifier currents turn on instantaneously. It is now clear that inward rectification results from a voltage-dependent block by intracellular cations which move into the inner mouth of the pore under the influence of the voltage field and block the outward flux of K^+ ions. The higher the binding affinity for blocking cations, the stronger the rectification. The apparent activation kinetics of strong inward rectifiers result from a time-dependent unblock of the channel on hyperpolarization.

Clay Armstrong (1969) was the first to postulate that inward rectification could arise from a voltage-dependent block by an internal blocking particle. This stimulated a search for such a blocking ion. Sodium ions were soon ruled out, but Mg^{2+} ions were found to produce inward rectification (Matsuda *et al.*, 1987; Vandenburg, 1987). Their effect is illustrated in Fig 8.6. In the absence of Mg^{2+}, Kir1.1 channels have a linear current-voltage relation in symmetrical K^+ solutions. Intracellular Mg^{2+} ions convert this to an inwardly-rectifying current-voltage relation by producing a voltage-dependent block of outward currents. Surprisingly, some types of Kir channel were found to show rectification even when all internal Mg^{2+} ions had been removed. This mystery was solved when it was discovered that polyamines such as spermine also produce inward rectification (Lopatin *et al.*, 1994). Spermine block plays an important role under physiological conditions because the polyamine is present at high concentrations within cells. The affinity of the block is also very high, partly because spermine has 4 positive charges. The order of potency of blocking ions in producing inward rectification is spermine^{4+} > spermidine^{3+} > putresceine^{2+} ~Mg^{2+}. Those channels that show the greatest degree of inward rectification under physiological conditions, such as Kir2.1, are also those which are most sensitive to spermine.

Two residues appear to be important for Mg^{2+} and polyamine block. One of these is located in the second TM and the other in the C terminus of the protein (Fig. 8.3). The first of these amino acids (position 171 in Kir1.1), is neutral (asparagine) in weak inward rectifiers but negatively charged (aspartate) in strong inward rectifiers (Fig. 8.7a). Substitution of aspartate (D) for asparagine (N) at this position converted the weak inwardly rectifying channel Kir1.1 into a strong inward rectifier (Lu & McKinnon, 1994) and increased the affinity for magnesium and polyamine block (Fig. 8.7B,C). The activation

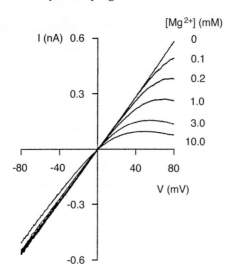

FIGURE 8.6 INTRACELLULAR Mg²⁺ CAUSES INWARD RECTIFICATION

Magnesium causes a dose-dependent block of outward currents through the weak inward rectifier Kir1.1, but does not the affect inward current. Current-voltage relations were recorded at different intracellular Mg^{2+} concentrations with 100 mM KCl on both side of the membrane. From Lu & MacKinnon (1994).

kinetics of the mutant channel were also altered: the current showed a time-dependent increase on hyperpolarization, reflecting the fact that the Mg^{2+} ions that bound in the pore at the holding potential were driven out on hyperpolarization.

In some inward rectifiers an additional residue in the C terminus contributes to inward rectification (Yang *et al.*, 1995). In Kir2.1, mutation of the aspartate in TM2 to asparagine (the reverse mutation to that described in the previous paragraph) alters the activation kinetics to those of Kir1.1, but does not modify the strong inward rectification. However, neutralization of a residue in the C terminus of the channel (E224) produced weak rectification. Thus, two residues appear to be involved in the rectification properties of Kir2.1.

The pore

The pore of Kir channels is formed, in part, from the H5 pore loop, as is the case for K_V channels. Mutations in this region alter ion selectivity and, as described later (page 149), can give rise to genetic disease. Other regions of the protein also contribute to the channel pore. In particular, there is evidence that the cytosolic end of the second transmembrane domain and the proximal part of the C terminus form part of the inner mouth of the pore. As described above, residues located in TM2 and in the C terminus interact with intracellular Mg^{2+} ions to produce inward rectification. Since Mg^{2+} blocks outward currents by entering the conduction pathway and impeding the movement of K^+, these residues must lie within the pore.

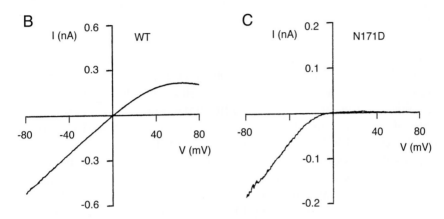

FIGURE 8.7 EFFECTS OF CHANGING RESIDUE 171 ON INWARD RECTIFICATION
(A) Line-up of amino acid residues in the second transmembrane domain of different Kir channels. The N/D site involved in inward rectification is indicated by the arrow. (B) Mutation of N171 to aspartate converts Kir1.1 from a weak to a strong inward rectifier. Current-voltage relations were recorded from cell-attached patches on *Xenopus* oocytes expressing either wild-type or mutant (N171D) Kir1.1. Currents were recorded in symmetrical 100 mM K^+ solutions and 4 mM Mg^{2+} was present in the intracellular solution. From Lu & MacKinnon (1994).

Modulation

A characteristic of inward rectifiers is that they are subject to modulation by a variety of cytoplasmic agents. Of particular physiological importance is the regulation produced by protons, phosphorylation, GTP-binding proteins and adenine nucleotides, which is now considered briefly.

Protons

A number of Kir channels are inhibited by a decrease in intracellular pH, as first shown for the inward rectifier of the starfish egg by Moody and Hagiwara (1982). Both the open probability and the single-channel conductance may be reduced. Proton regulation is of particular importance in the kidney, as changes in internal pH influence both Kir1.1 and K_{ATP} channel activity, and thereby transepithelial K^+ fluxes. Site-directed mutagenesis has shown that the pH sensitivity of Kir1.1 channel activity resides in a single amino acid (lysine 80) that is located in the cytosolic N-terminal region, close to where it enters the membrane (Fakler *et al.*, 1996).

External pH may also influence Kir channel activity, as is the case for Kir2.3 where proton sensitivity is conferred by the presence of a histidine residue at position 117, in the extracellular loop linking TM1 and the pore (Coulter *et al.*, 1995). This modulation may be of functional importance, because Kir2.3 is strongly expressed in hippocampal neurones and extracellular pH falls markedly during cerebral ischaemia. This would be expected to activate Kir2.3, hyperpolarize the neurone, and conserve cellular ATP levels by preventing neuronal excitability.

Phosphorylation

Phosphorylation mediates the effects of a number of hormones and transmitters on Kir channels, and thereby on cell excitability and membrane K⁺ fluxes. For example, patch-clamp studies have established that the increased activity of vascular smooth muscle K_{ATP} channels produced by calcitonin-gene-related peptide (CGRP) is mediated by protein kinase A (PKA) dependent phosphorylation (Quayle *et al.*, 1994). Because the K_{ATP} channel determines the level of vascular smooth muscle tone, an increase in channel activity leads to vasorelaxation. This accounts for the potent effects of CGRP on blood pressure. The precise residues that PKA phosphorylates in the smooth muscle K_{ATP} channel (which probably comprises Kir6.2 and SUR2B subunits) have not yet been identified. In the case of other Kir channels, however, the residues which PKA phosphorylates to regulate channel activity are known (Wishmeyer & Karschin, 1996; Xu *et al.*, 1996)

GTP

Many hormones mediate their effects on cell excitability by activation of GTP-binding proteins, which interact directly with members of the Kir3.0 family to enhance their activity and so stabilise the membrane at a potential close to E_K. In brain neurones, this interaction plays a crucial role in regulating neuronal excitability as it is involved in the generation of slow synaptic potentials: slow depolarizations result from a decrease, and slow hyperpolarizations from an increase, in Kir currents. G-protein regulation is also important in cardiac muscle as it mediates the slowing of the heart rate in response to acetylcholine (ACh). ACh binds to muscarinc receptors which (unlike nicotinic AchRs) do not contain an intrinsic ion channel and mediate their effects by activation of G-proteins. Subsequently, the activated G-protein interacts with effector molecules, which may cause stimulation of second messenger pathways. The action of ACh on the heart, however, occurs far more rapidly than is expected if an intracellular second messenger is involved. This was something of a puzzle until it was discovered that ACh increased the activity of Kir channels in excised membrane patches, thus effectively excluding the possibility that a soluble cytoplasmic second messenger mediates the interaction. It was later found that ACh activates an inhibitory G-protein called G_o which diffuses within the membrane to bind directly to the Kir channel and inhibit its activity. This type of interaction has been termed *membrane-delimited* by Hille (1992) because the components remain associated with the membrane and do not diffuse away into the cytoplasm.

Heterotrimeric G-proteins, like G_o, consist of α-, β- and γ-subunits (Fig. 8.8). The β- and γ-subunits are tightly associated and they are generally referred to as $\beta\gamma$ in recognition of this fact. The association of the α-subunit with the $\beta\gamma$-subunits is less tight, however, and is regulated by GTP. In the inactive state, the G-protein consists of a complex of all three subunits, with GDP bound to the α-subunit. Activation of G-proteins, for example by interaction of ACh with its receptor, results in the exchange of bound GDP for GTP. This causes the α-subunit to dissociate from the $\beta\gamma$-subunits, leaving both α- and $\beta\gamma$-subunits free to interact with other proteins. Subsequently, hydrolysis of bound GTP to GDP occurs, and when the α- and $\beta\gamma$-subunits encounter each other they reassociate and are no longer available to activate their target proteins. After considerable controversy, it is now firmly established that it is the $\beta\gamma$-subunits which interact with the cardiac G-protein-gated K^+ channels to enhance their activity (Fig. 8.9) (Wickman & Clapham, 1995; Krapivinsky *et al.*, 1995). It remains possible that the α-subunit interacts with other types of Kir channel, however, since it is a well-known activator of other effector molecules.

The cardiac G-protein-gated Kir-channel (sometimes called I_{KACh}) is a heteromer composed of Kir3.1 and Kir3.4 subunits. Both types of subunit are able to interact directly with $\beta\gamma$-subunits. A key question is where the binding site for the $\beta\gamma$-subunits is located on the Kir channel. This issue is not yet fully resolved, but there is evidence that both the N and C termini are involved (Huang *et al.*, 1995).

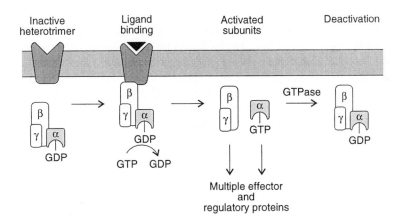

FIGURE 8.8 THE G-PROTEIN CYCLE
Ligand-binding to a G-protein linked receptor allows it to catalyse the exchange of bound GTP for GDP on the α-subunit of the G$\alpha\beta\gamma$ trimer. This causes the $\beta\gamma$ and α-subunits to dissociate and enables them to interact with effector proteins, such as enzymes and ion channels. The α-subunit has intrinsic GTP-ase activity and GTP is slowly hydrolysed to GDP, after which the α-subunit is able to reassociate with $\beta\gamma$ subunits.

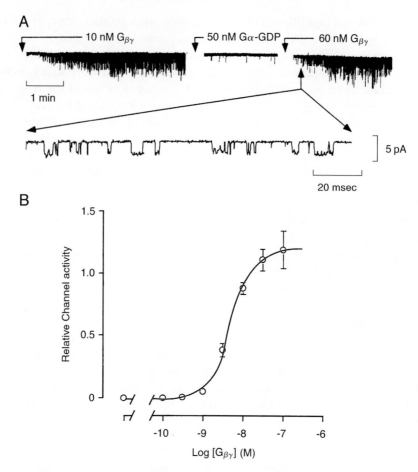

FIGURE 8.9 *G$_{\beta\gamma}$ STIMULATES CARDIAC I$_{KACh}$ CURRENTS*
(A) Addition of purified G$_{\beta\gamma}$ to the intracellular solution activates single
I$_{KACh}$ channels. Channel activity is abolished by the subsequent addition
of G$_\alpha$-GDP (which complexes with G$_{\beta\gamma}$ to form the inactive heterotrimer),
but can be restored by addition of excess G$_{\beta\gamma}$. Single-channel currents
were recorded from an inside-out patch excised from a cardiac ventricular
myocyte. (B) Relationship between channel activity (NPo) and G$_{\beta\gamma}$ concentration. Half-maximal activation occurred at ~5 nM G$_{\beta\gamma}$. From Krapivinsky
et al. (1995b).

Adenine nucleotides

Potassium channels which are inhibited by elevation of intracellular ATP
are generically known as ATP-sensitive K$^+$ channels or K$_{ATP}$ channels. They
were first described by Noma (1984) in cardiac muscle, but they have subsequently been found in a wide variety of tissues including pancreatic β-cells,
smooth and skeletal muscle, brain nerve cells, peripheral axons, and epithelial
cells (for review see Ashcroft & Ashcroft, 1990). Since they are sensitive to
intracellular adenine nucleotide levels, K$_{ATP}$ channels serve to couple cell

metabolism to cell excitability. In some tissues, they also mediate the effects of hormones and transmitters on membrane excitability. Most types of K_{ATP} channel are blocked by sulphonylurea drugs, such as glibenclamide and tolbutamide, which are used in the treatment of non-insulin-dependent diabetes, and they are activated by K-channel openers, such as diazoxide and cromakalim. Native K_{ATP} channels differ in their ATP-sensitivity, single-channel properties and pharmacology, and the variation in tissue sensitivity to different drugs is of major therapeutic importance. Recently, the basis of these differences has become clear, as it turns out that channels sensitive to ATP are genetically heterogenous and include members of several Kir subfamilies (reviewed by Ashcroft & Gribble, 1998).

The archetypal K_{ATP} channel, such as that of the pancreatic β-cell and cardiac muscle, is highly ATP-sensitive, being half-blocked by 10–30 μM ATP (Fig. 8.10). Since the intracellular ATP concentration in most cells lies between 3 and 7 mM, it was initially argued that the K_{ATP} channel would have no physiological role, as it should never be open. Contrary to this belief, however, significant K_{ATP} channel activity was observed in cell-attached patches on β-cells exposed to glucose-free solutions (Ashcroft et al., 1984). The explanation for this paradox is that ATP is not the only regulator of channel activity. The membrane phospholipid phosphatidylinositol-4,5-bisphosphate (PIP$_2$) markedly reduces the inhibitory effect of ATP (Baukrowitz et al., 1998). The higher ATP-sensitivity observed in inside-out patches may therefore be due to gradual washout of PIP$_2$ from the patch membrane following excision. Low concentrations of MgADP also activate K_{ATP} channels and relieve the inhibitory effects of ATP (Kakei et al., 1986). In the intact

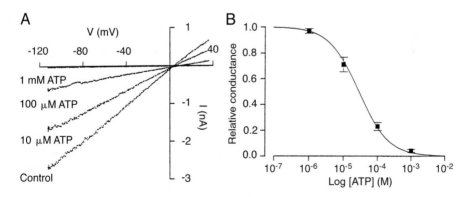

FIGURE 8.10 ATP BLOCK OF K_{ATP} CHANNELS
Addition of ATP to the intracellular solution causes a dose-dependent inhibition of the K_{ATP} channel. (A) Current-voltage relations recorded at different ATP concentrations from a giant inside-out patch containing many hundreds of wild-type (Kir6.2/SUR1) K_{ATP} channels. (B) Dose-dependence of ATP inhibition: half-maximal block (K_i) occurred at ~30 μM ATP. From Gribble et al. (1997).

cell, therefore, K_{ATP} channel activity will be determined by the relative concentrations of ATP, MgADP and PIP_2. The dual regulation by ATP and MgADP provides a means of coupling K_{ATP} channel activity to cell metabolism.

The archetypal K_{ATP} channel, which shows high ATP-sensitivity, is formed from two subunits: Kir6.2 and a sulphonylurea receptor (SUR) (Fig. 8.5). Expression of Kir6.2 by itself does not result in measurable currents, so it was initially difficult to determine which K_{ATP} channel properties were intrinsic to the Kir subunit and which were conferred by association with SUR. This issue has now been resolved, most directly by using a truncated form of Kir6.2, lacking the last 26 or 36 residues of the C-terminus, which is capable of independent expression (Tucker *et al.*, 1997). It appears that Kir6.2 serves as the pore-forming subunit and has intrinsic ATP sensitivity, whereas SUR acts as a regulatory subunit that endows the K_{ATP} channel with sensitivity to sulphonylureas, K^+ channel openers and the stimulatory effects of MgADP. The location of the inhibitory ATP-binding site on Kir6.2 and of the drug-binding sites on SUR have yet to be discovered. It is known, however, that MgADP mediates its potentiatory effect by interaction with the nucleotide-binding domains (NBDs) of SUR (Nichols *et al.*, 1996). This is perhaps not surprising since ATP and ADP have been also shown to bind to the NBDs of the related ABC transporter CFTR (see Chapter 12).

K_{ATP} channels found in epithelia are usually far less sensitive to ATP than those of β-cells or heart, with millimolar concentrations being needed to produce significant block. They also have a smaller single-channel conductance and different kinetics. These channels are probably encoded by members of the Kir1.0 subfamily. It is not known whether ATP binds directly to Kir1.1 or whether an additional subunit confers ATP sensitivity on the channel. However, the C terminus of Kir1.1 contains a motif that has been implicated in ATP binding and hydrolysis in other proteins and there is some evidence that mutations in this motif can alter the channel ATP-sensitivity (McNicholas *et al.*, 1996b).

Summary

Kir channels form a large family of voltage-independent K^+ channels with structural homology but diverse functions. They assemble as either homomeric or heteromeric tetramers. In contrast to the K_V channel subunits, each Kir monomer possesses only two transmembrane domains. The outer part of the pore is formed by the H5 pore loop while the second TM and the C-terminus contribute to its inner mouth. A hallmark of Kir channels is that they are modulated by a wide variety of cytosolic agonists and antagonists, of which the most important are GTP-binding proteins (Kir3.0 subfamily) and ATP (Kir6.0 subfamily). Regions within the cytosolic N and C termini are important for these interactions.

DISEASES ASSOCIATED WITH KIR CHANNELS

Three inherited diseases have been associated with mutations in Kir channels to date, two of which are found in man and one in mouse. Bartter's syndrome is a renal tubular disorder characterised by salt-wasting, hypokalaemia and metabolic acidosis. Some Bartter's kindreds have mutations in the gene for Kir1.1. The *weaver* mouse results from a mutation in the pore region of Kir3.2 which makes the channel permeable to Na^+ ions. The mutation leads to the selective death of brain neurones and thus to the ataxic gait that characterises the *weaver* phenotype. Familial persistent hyperinsulinaemic hypoglycaemia of infancy (PHHI) is associated with mutations in the SUR1 or Kir6.2 subunits of the β-cell K_{ATP} channel. These result in the lack of functional K_{ATP} channel activity and lead to unregulated insulin secretion and, as a consequence, very low blood glucose levels.

Bartter's Syndrome (Kir1.1)

Clinical features

In 1962, Bartter described two patients who had very low plasma K^+ levels and metabolic acidosis, yet elevated plasma renin and aldosterone levels. Subsequently, it has become clear that the syndrome he described is both phenotypically and genetically heterogeneous, and at least three sub-types have been distinguished. The first of these is the classical Bartter's syndrome for which the underlying genetic defect remains unknown. The second, known as the Gitelman variant, is characterised by late age of onset and very low urinary Ca^{2+} and Mg^{2+} concentrations, and results from muta-tions in the gene encoding the NaCl cotransporter (*NCCT*). The third variant is known as antenatal Bartter's syndrome or hyper-prostaglandin E syndrome. It is a life-threatening disorder that presents *in utero* with a marked fetal polyuria and it can precipitate premature birth. Newborns show severe salt-wasting, moderate hypokalaemia and metabolic acidosis, and elevated urinary excre-tion of prostaglandins. In addition, there is a marked loss of Ca^{2+} in the urine and as a consequence, osteopenia (bone loss) and nephrocalcinosis (kidney stones). Unsurprisingly, affected infants show failure to thrive. Recent studies have shown that antenatal Bartter's syndrome is genetically heterogeneous and may result from mutations in the genes encoding the inwardly rectifying K^+ channel Kir1.1 (*KCNJ1*; Bartter's syndrome type II), the NaK2Cl cotrans-porter (*SCL12A1*, Bartter's syndrome type I) or the voltage-gated Cl^- channel CLC-Kb (*CLCNKB*, Bartter's syndrome type III; see Chapter 10). These vari-ants may be distinguished clinically because hypokalaemia is less pronounced (3.0–3.5 mM) in patients with mutations in *KCNJ1*, and the course of the disease is less severe.

Analysis of mutations

At least 11 mutations in *KCNJ1* have been described to date, which are distributed over the entire coding region of the gene (Fig. 8.11; Derst *et al.*, 1997; Int Collab Study Group on Bartter's syndrome, 1997). Two mutations produce truncated proteins that are expected to lead to a non-functional channel. Nine missense mutations that result in amino acid substitutions have also been identified. The functional effects of some of these mutations were examined by Derst and colleagues (1997), who found that when they were introduced into Kir1.1 either no measureable current (D108H or P110L), or only a very small current (V72E, A198T and V315G) was produced. It therefore appears that the loss of Kir1.1 channel activity is the primary cause of Bartter's syndrome. The mechanism by which the Bartter's mutations result in a loss or reduction of channel activity is still unclear, but it is noteworthy that those mutations that occur in the N and C termini are associated with reduced currents, whereas those that occur within the highly conserved trans-membrane regions of the protein result in the total loss of channel activity.

Kir1.1 is alternatively spliced to give several different transcripts that are expressed in different parts of the nephron. To date, however, all mutations associated with Bartter's syndrome have been found within exon 5, which forms the central core of the protein and is common to all Kir1.1 isoforms. Thus, Kir1.1 channels all along the nephron will be affected by the Bartter's mutations. Kir 1.1 is also expressed in several other tissues, including the spleen, lung and eye but specific defects in these organs in Bartter's syndrome have not been reported.

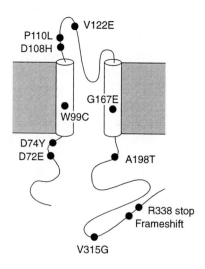

FIGURE 8.11 MUTATIONS IN Kir1.1 CAUSE BARTTER'S SYNDROME

Putative membrane topology of Kir1.1 with the mutations associated with Bartter's syndrome marked.

Why does loss of Kir1.1 cause Bartter's syndrome?

Kir1.1 is expressed in the apical membrane of distal kidney tubule cells, principally those of the thick ascending loop of Henle, the distal convoluted tubule and the collecting ducts (Lee & Hebert, 1995). It plays a key role in K^+ recycling in the loop of Henle, a process which is important for salt uptake. This is illustrated in Fig. 8.12, which shows the ion transport pathways across a cell of the thick ascending loop of Henle. The activity of the Na/K-ATPase in the basolateral membrane generates an electrochemical gradient which facilitates the uptake of Na^+ ions from the tubule lumen via the NaK2Cl cotransporter in the apical membrane. The accompanying Cl^- ions leave the cell via channels in the basolateral membrane; consequently there is a net transport of NaCl from the tubule lumen into the blood. Most of the K^+ ions that enter the cell via the NaK2Cl cotransporter recycle back into the tubule lumen through Kir1.1 channels in the apical membrane. This K^+ recycling ensures a constant supply of K^+ ions that enables the continuous operation of the NaK2Cl cotransporter, and thus NaCl uptake, despite the fact that K^+ ions in the luminal fluid are ~20 times lower than those of Na^+ or Cl^-. In the absence of Kir1.1, as in Bartter's syndrome, K^+ recycling is prevented and NaCl uptake is impeded. This leads to a high salt concentration in the urine which induces an osmotic diuresis and accounts for the salt-wasting, polyuria and low plasma volume characteristic of Bartter's syndrome. A similar phenotype is observed with loop diuretics, such as frusemide, which inhibit the NaK2Cl cotransporter.

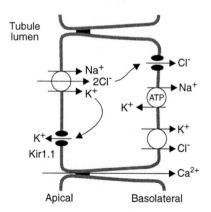

FIGURE 8.12 SALT TRANSPORT IN THE THICK ASCENDING LOOP OF HENLE

The apical membrane of the TAL contains a NaK2Cl cotransporter which mediates the uptake of one Na^+, one K^+ and two Cl^- ions. Energy for this process is provided by the Na^+ gradient generated by the activity of the Na/K-ATPase in the basolateral membrane. Some of the K^+ ions translocated by the NaK2Cl cotransporter are recycled across the apical membrane via Kir1.1 channels. This recycling is required for salt uptake and in its absence there will be Na^+, Cl^- and K^+ loss.

Because much of the K^+ taken up by the NaK2Cl cotransporter recycles across the apical membrane, whereas all of the transported Cl^- leaves via the basolateral membrane, the potential in the interstitial fluid is normally ~6 mV more negative than that in the tubule lumen. This transepithelial potential facilitates the uptake of positive ions, such as Ca^{2+}, via paracellular pathways. If Kir1.1 is non-functional, the transepithelial potential is markedly reduced, hindering Ca^+ uptake and contributing to the hypercalciuria seen in Bartter's syndrome.

Hypokalaemia is also found in Bartter's syndrome. This occurs because a small amount of the Na^+ that passes down the nephron is absorbed in more distal parts of the tubule by the epithelial Na^+ channel (ENaC). As explained in Chapter 13 (page 240), increased activity of ENaC is accompanied by enhanced K^+ secretion, thereby reducing the plasma K^+ concentration. Clearly, K^+ channels other than Kir1.1a must be involved in this K^+ secretion.

Weaver Mouse (Kir3.2)

Phenotype

Weaver mice (*wv/wv*) are so called because they have an ataxic gait and thus 'weave' around when they move. They also exhibit hyperactivity and tremor. These behaviours result from a selective loss of neurones in two regions of the brain, the granule layer of the cerebellum and the substantia nigra, during development. Although the precursors of the cerebellar granule cells develop normally, they fail to differentiate and migrate into the granule layer and they die during the first two weeks of post-natal life. The dopaminergic cells of the substantia nigra also die and the mice are sterile. Heterozygous animals (*wv/+*) possess a significantly smaller cerebellum than wild-type animals but do not exhibit ataxia.

Mutation analysis

The *weaver* phenotype is caused by a point mutation in the gene encoding Kir3.2, which results in substitution of a serine residue for glycine 156 (G156S), which lies within the pore loop of the channel (Fig. 8.13A; Patil *et al.*, 1995). Mutagenesis studies, described in Chapter 6, have shown that the equivalent mutation eliminates the K^+ selectivity of *Shaker* channels (Heginbotham *et al.*, 1994). As was expected from these studies, when the *weaver* mutation was engineered into Kir3.2 the mutant channel no longer discriminated between monovalent cations (Fig.8.13B; Kofuji *et al.*, 1996; Navarro *et al.*, 1996; Slessinger *et al.*, 1996). More surprisingly, it was also constitutively open and insensitive to G-protein activation. These novel properties resulted in a large inward current through Kir3.2 channels at rest, and thus to a marked depolarization of the resting potential. Moreover, *Xenopus* oocytes injected with *wv*Kir3.2 mRNA died within a few days. Oocyte death was prevented by elimination of Ca^{2+} from the extracellular solution, suggesting it might result

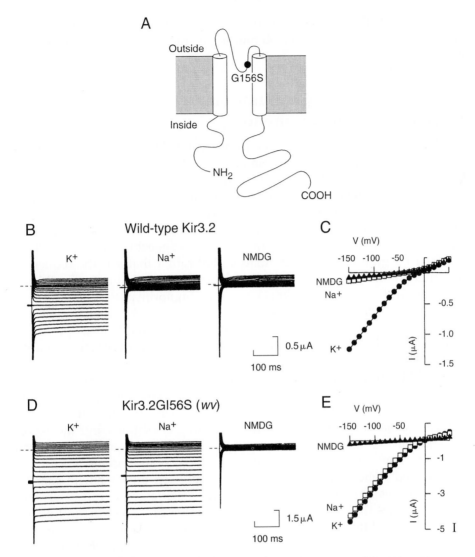

FIGURE 8.13 THE WEAVER KIR3.2 CHANNEL IS Na⁺ PERMEABLE
The *weaver* Kir3.2 channel contains a glycine to serine mutation at position
156, which lies within the K⁺ selectivity filter (A) This mutation renders
the channel permeable to Na⁺ (B–E). Currents (B,C) and current-voltage
relations (B,D) were recorded from *Xenopus* oocytes expressing wild-type
(B,C) or weaver (D,E) Kir3.2. The major cation present in the extracellular
solution is indicated above each trace. From Slesinger *et al.* (1996).

because the membrane depolarization resulting from the inward wvKir3.2
current triggers Ca^{2+} entry, cell swelling and cell death.

    The reason why the G156S mutation should affect the ability of G-proteins
to promote channel activation remains obscure, because the G-protein binding
site is thought to involve residues located in the N and C terminal domains

of the channel. The simplest explanation is that binding of $G_{\beta\gamma}$ is unaffected, but that the mechanism by which binding is translated into channel opening is impaired.

Why does the G156S mutation produce the weaver phenotype?

Studies of cerebellar granule neurones isolated from *wv/wv* mice revealed that they did not possess G-protein activated K⁺ currents, unlike their wild-type littermates (Kofuji *et al.*, 1996). Instead, they exhibited an enhanced resting Na⁺ conductance and a depolarised membrane potential. These differences are explained by the altered selectivity and constitutive activation of the *wv*Kir3.2 channel. Drugs which block *wv*Kir3.2 currents markedly enhanced the viability of *wv/wv* granule cells, suggesting that Kir3.2 is not required for granule cell differentiation, but rather that it is the reduced K⁺ selectivity of the channel that causes cell death. Further support for this idea comes from studies of Kir3.2 knockout mice (Signorini *et al.*, 1997). Although homozygous knockout animals did not express Kir3.2, their brains were morphologically normal and they did not exhibit ataxia. Thus it appears that increased Na⁺ influx, rather than loss of Kir3.2 *per se*, is the cause of neuronal death in *weaver* mice.

In wild-type animals, Kir3.2 coassembles with Kir3.1 to form heteromeric channels. What happens then, when *wv*Kir3.2 is coexpressed with Kir3.1 in heterologous systems? It turns out that heteromeric channels show a much reduced whole-cell current, both under resting conditions and in response to G-protein stimulation (Slessinger *et al.*, 1996). This suggests that neuronal death in *wv/wv* animals will only occur in those brain regions where Kir3.2 is expressed alone, and not where it is coexpressed together with Kir3.1. In support of this idea, Kir 3.2, but not Kir3.1, is expressed in those substantia nigra neurones that die in *wv/wv* mice. Furthermore, no structural or functional abnormalities have been reported in *wv/wv* animals in some of the brain regions where Kir 3.2 and Kir3.1 are coexpressed, such as cerebellar Purkinje cells, pontine nucleii, olfactory bulb, cerebral cortex, septum and amygdala. Although both Kir 3.2 and Kir3.1 are expressed in cerebellar granule cells, which die in *wv/wv* mice, it is possible that the Kir3.1 protein is expressed at too low a level to enable heteromerization to ameliorate the potentially toxic effects of the *weaver* allele.

In summary, the *weaver* mouse results from a gain-of-function mutation in Kir3.2. This causes a constitutive inward current that produces depolarization and cell death. The development of the weaver phenotype correlates with the neuronal cell death that follows age-dependent expression of Kir3.2. Loss of Kir3.2, as in the Kir3.2⁻/⁻ mice, does not result in the *wv/wv* phenotype (Signorini *et al.*, 1997). However, these mice show sporadic stress-induced seizures and enhanced susceptibility to convulsive agents, consistent with a role for the G-protein-activated K⁺ current in reducing cell excitability. No human disease with symptoms matching those of *weaver*, or Kir3.2 knockout mice, has been reported and no disease has been mapped to the human Kir3.2 gene (which is located on chromosome 21q22.1–22.2).

Diseases of Insulin Secretion

The ATP-sensitive K^+ (K_{ATP}) channel plays a key role in glucose-stimulated insulin secretion. It is therefore not surprising that mutations in the genes that encode K_{ATP} channel subunits, or that encode proteins involved in the regulation of K_{ATP} channel activity, affect insulin secretion. Fig. 8.14 illustrates our current model of insulin secretion. When plasma glucose levels rise, glucose uptake and metabolism by the pancreatic β-cell is enhanced, producing an increase in intracellular ATP and a concomitant fall in intracellular MgADP. These changes act synergistically to close K_{ATP} channels in the β-cell membrane because ATP inhibits, whereas MgADP activates, channel activity. Since K_{ATP} channel activity determines the β-cell resting potential, its closure causes a membrane depolarisation that activates voltage-gated Ca^{2+} channels, increases Ca^{2+} influx and so stimulates insulin release. Two classes of therapeutic drugs modulate insulin secretion by interacting with the SUR1 subunit of the K_{ATP} channel. Sulphonylureas, such as glibenclamide,

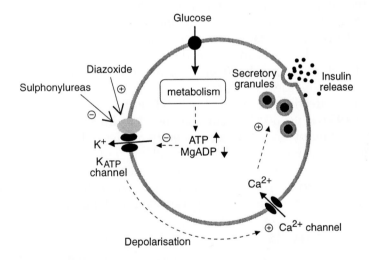

FIGURE 8.14 K_{ATP} CHANNELS PLAY A KEY ROLE IN INSULIN SECRETION
At low extracellular glucose levels (<3mM), K_{ATP} channels are open and their activity sets the resting potential of the pancreatic β-cell at a hyperpolarised level. When blood glucose rises, the uptake and metabolism of the sugar is increased. Some product of this metabolism, possibly changes in the intracellular concentration of adenine nucleotides, causes the K_{ATP} channels to close. This results in membrane depolarization, activation of voltage-gated Ca^{2+} channels and an increase in cytosolic Ca^{2+} that triggers the exocytosis of insulin-containing secretory granules.

inhibit channel activity and are used to enhance insulin secretion in patients with type-2 diabetes mellitus. By contrast, K-channel openers (e.g., diazoxide) activate K_{ATP} channels, hyperpolarizing the β-cell and preventing insulin release. It is clear from this model, that a reduction in K_{ATP} channel activity will enhance insulin secretion and cause hypoglycaemia (low blood glucose), whereas an increase in channel activity may be expected to decrease insulin release and produce hyperglycaemia. Both may occur in man.

Familial Persistent Hyperinsulinaemic Hypoglycaemia of Infancy (PHHI)

Clinical features

Persistent hyperinsulinaemic hypoglycaemia or PHHI is a disorder of glucose homeostasis that is characterised by unregulated insulin secretion and profound hypoglycaemia. Both familial and sporadic forms have been identified. Familial PHHI is an autosomal recessive disease which occurs at low frequency in Europeans (~1 in 40,000) but reaches incidences as high as 1 in 2,700 live births in Arabic families or Ashkenazi Jews, due to the much higher frequency of consanguineous marriages. It usually manifests at birth or within the first year of life and by the time the disease is recognised and treated, the infant may have suffered severe hypoglycaemia and consequent brain damage. The disease is probably under-diagnosed and may contribute to the incidence of post-natal deaths from unknown causes.

Diagnosis of PHHI is made on the basis of persistent hypoglycaemia associated with raised insulin levels. Glucose infusions are required to maintain the plasma glucose level during diagnosis and so prevent brain damage, but the most effective treatment for PHHI is to remove all or part of the pancreas (>95% is usual). As a consequence, insulin and exocrine pancreatic replacement therapy are often subsequently needed. The K⁺ channel opener diazoxide is occasionally used to treat PHHI, but many patients do not respond to this drug. In mild cases of the disease, some mothers have even successfully treated the disease by giving their child chocolate bars throughout the day: a practice that certainly appealed to the patient!

Mutation analysis

Linkage analysis of families with PHHI indicated that the gene responsible for the condition was located on chromosome 11, in the region 11p14–15.1 (Thomas et al., 1995a). Subsequently it was found both K_{ATP} channel subunits map to the same region, and at least 11 different mutations in SUR1, and 2 in Kir6.2, associated with PHHI have now been reported (Thomas et al., 1995b; 1996; Nestorowicz et al., 1996; Shyng et al., 1998). Mutations in SUR1 are the most common cause of PHHI. These mutations occur throughout the protein but are concentrated in the cytosolic loops and the intracellular ends of the transmembrane domains (Fig. 8.15). They may be grouped into two major classes on the basis of their functional effects. Class I mutations result

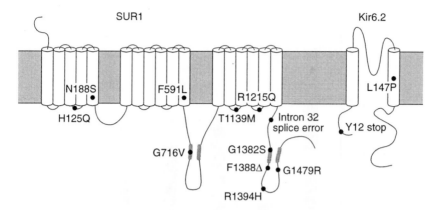

FIGURE 8.15 MUTATIONS IN KIR6.2 OR SUR1 CAUSE PHHI
Putative membrane topology of Kir6.2 and SUR1 with some of the mutations associated with PHHI marked.

in the total loss of functional K_{ATP} channel activity, even in excised membrane
patches. These include the deletion of a phenylalanine at position 1388
(ΔF1388), and a mutation within intron 32 that is postulated to activate cryptic
splice sites and cause premature truncation of the protein. Class II mutations
impair the ability of MgADP to enhance K_{ATP} channel activity, but do not
alter the inhibitory effect of ATP (Fig. 8.16). Importantly, these mutations
also prevent activation of the K_{ATP} channel in response to metabolic inhibition
(Nichols et al., 1996; Shyng et al., 1998). This has led to the recognition that
metabolically-induced changes in MgADP are more important in coupling
metabolism to K_{ATP} channel activity than are changes in $[ATP]_i$. Some Class
II mutations lie within the nucleotide-binding domains (NBDs) of SUR1, and
lend support to the idea that MgADP mediates K_{ATP} channel activation by
interacting with the NBDs. Others, such as F591L, do not and the way in
which they produce their functional effects is unclear.

Two PHHI mutations in Kir6.2 have also been identified. One of these
is a proline-for-leucine substitution at position 147, which lies within the
second TM. It is predicted to disrupt the α-helical structure of the protein,
although no functional data have been reported (Thomas et al., 1996). The
other is a nonsense mutation that truncates the protein within the first transmembrane domain and is expected to produce a non-functional channel.

Functional consequences of PHHI mutations

Most of the mutations associated with PHHI identified to date appear to
fall into two groups: those that result in the total loss of K_{ATP} channel activity,
even in excised patches, and those in which the regulation of the K_{ATP} channel
by MgADP is abolished. In the latter case, channel activity can be observed on
patch excision, but in the intact cell the K_{ATP} channel is always closed because
MgADP is unable to relieve the inhibitory effect of $[ATP]_i$. As a consequence,
the β-cells of these PHHI patients lack K_{ATP} channel activity, even at low blood

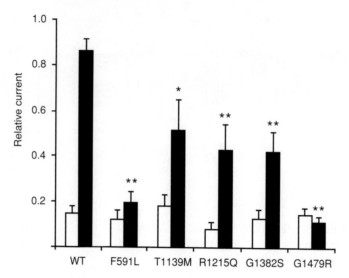

FIGURE 8.16 LACK OF MgADP STIMULATION OF PHHI MUTANT
K_{ATP} CHANNELS
Addition of ATP to the intracellular surface of both wild-type (WT) and
PHHI mutant K_{ATP} channels produces a similar degree of inhibition, as
shown by the white bars. However, MgADP was much less effective at
reversing the inhibitory effect of ATP when the SUR1 subunit of the
channel contained a mutation associated with PHHI, as shown by the
black bars. K_{ATP} channels were formed by coexpression of Kir6.2 and either
wild-type SUR1, or SUR1 containing the mutation indicated. Currents are
plotted relative to their amplitude in the absence of ATP: white bars were
recorded in the presence of 100 μM ATP, black bars in 100 μM ATP +
0.5 mM ADP (Mg²⁺, ~1 mM). From Shyng *et al.* (1998).

glucose levels. This results in a continuous depolarisation of the β-cell and
thereby a high resting intracellular Ca²⁺ concentration (Kane *et al.*, 1996), which
explains the constitutive insulin secretion characteristic of PHHI patients.

Diabetes Mellitus

There are two main types of diabetes mellitus: type-1 (formely known as
insulin-dependent diabetes) and type-2 (formerly known as non-insulin-
dependent diabetes). Type 1 diabetes results from an autoimmune destruction
of the β-cells and will not be considered here. The aetiology of type-2 diabetes
is still obscure but there is evidence that the disease is associated with defective β-cell secretion because although glucose fails to release sufficient insulin,
secretion can be stimulated by drugs, such as the sulphonylureas, and by
hormones. This suggests that metabolic regulation of K_{ATP} channels may be
altered in the β-cells of type-2 diabetics. Despite considerable effort, mutations
in Kir6.2 or in SUR1 associated with type-2 diabetes have not been detected.
However, defective metabolic regulation of K_{ATP} channels is observed in
several rare types of diabetes.

One form of maturity-onset diabetes of the young (MODY2) results from mutations in glucokinase, the enzyme that catalyses the first step in glucose metabolism (the conversion of glucose to glucose-6-phosphate) in liver and β-cells (Randle, 1993). In French families, around 50% of all MODY patients suffer from MODY2 and carry mutations in the glucokinase gene. The disease shows autosomal dominant inheritance and all individuals identified to date are heterozygotes. They show mild hyperglycaemia with fasting blood glucose levels of ~7 mM. Many can be managed on diet alone, but some individuals require sulphonylurea therapy. Mice in which the glucokinase gene has been disrupted ('knocked-out') specifically in the β-cell, develop severe hyperglycaemia and die within 3 days of birth of acute diabetes, while heterozygous animals show symptoms resembling those of MODY2 patients (Terauchi et al., 1995). Because expression of glucokinase in the liver was not altered in these knockout mice, it appears that the β-cell defect is the primary cause of MODY2. Sakura and his colleagues (1998) have shown that there is an almost total loss of glucose-sensitivity of the β-cell K_{ATP} channels in homozygous animals, and a reduction in heterozygotes, which results from impaired glycolytic metabolism. Defective regulation of K_{ATP} channels probably also accounts for the reduction in glucose-stimulated insulin release in MODY2 patients.

Impaired mitochondrial metabolism can also give rise to diabetes, presumably as a consequence of a failure to regulate β-cell K_{ATP} channel activity. Maternally-inherited diabetes with deafness (MIDD) results from a mutation at position 3243 of the mitochondrial DNA, which encodes a leucine transfer RNA (Maasen & Kadowaki, 1996). The disease shows maternal inheritance because all mitochondria derive from the oocyte. Confusingly, MELAS syndrome (mitochondrial myopathy, encephalopathy, lactic acidosis and stroke-like episodes) is associated with the same mutation as MIDD and it is still unclear why some carriers of the 3243 mutation develop MIDD whereas others develop MELAS syndrome. The cause of deafness in MIDD is also unknown.

CHAPTER 9

VOLTAGE-GATED Ca^{2+} CHANNELS

Ca^{2+} ions play crucially important roles in regulating a variety of cellular functions. They initiate muscle contraction, trigger the release of neurotransmitters from nerve terminals and of hormones from secretory cells, regulate gene expression and the cell cycle, and mediate cell death. The intracellular Ca^{2+} concentration ($[Ca^{2+}]_i$) is very much lower than that outside the cell (10^{-7} M as compared with 1–2 mM), and a transient rise in internal Ca^{2+} acts as a second messenger coupling receptor activation to many cellular processes. This increase in $[Ca^{2+}]$ is mediated by voltage- or ligand-gated Ca^{2+} channels that regulate Ca^{2+} influx across the plasma membrane and/ or by ligand-gated Ca^{2+} channels which control the release of Ca^{2+} from intracellular stores. The ligand-gated Ca^{2+} channels are considered in Chapter 14. In this chapter we look at the properties of the voltage-gated Ca^{2+} channels and see how mutations in the genes that encode these proteins lead to defective channel function and so result in a number of diseases of nerve and muscle, including hypokalaemic periodic paralysis, familial hemiplegic migraine, episodic ataxia, and spinocerebellar ataxia.

Basic Ca^{2+} Channel Properties

Voltage-gated Ca^{2+} channels come in a surprising number of different types. These are classified as T, L, N, P, Q, and R and are distinguished by their sensitivity to pharmacological blockers, single-channel conductance, kinetics and voltage dependence. Like the voltage-gated Na^+ channels (Chapter 5), the opening of voltage gated Ca^{2+} channels is strongly voltage-dependent: they are closed at the resting potential of the cell and open only on depolarization. On the basis of their voltage dependence of activation, the voltage-gated Ca^{2+} channels fall into two broad groups: the low and the high threshold-activated channels. T-type Ca^{2+} channels, or low threshold-activated Ca^{2+} channels, are opened by small depolarizations from the resting potential. In contrast, L, N, P, Q, and R channels require stronger depolarizations to open them. The high threshold Ca^{2+} channels can be further separated by their pharmacology. Thus, L-type channels are blocked by the dihydropyridines and N-type channels by two toxins that come from the fish-hunting cone shell *Conus geographus,* called ω-conotoxin GVIA and ω-conotoxin MVIIA. A toxin isolated from the venom of the funnel web spider *Agelenopsis aperta,* ω-agatoxin IVA, blocks P-type channels and, with lower affinity, Q-type channels.

All types of Ca²⁺ current undergo inactivation, but the rate and extent vary with the channel type. For example, T-type channels inactivate rapidly and completely at positive membrane potentials, whereas L-type channels inactivate more slowly (Fig. 9.1). The single-channel conductance also varies. Commonly, single Ca²⁺ channel currents are studied using Ba²⁺ as a charge

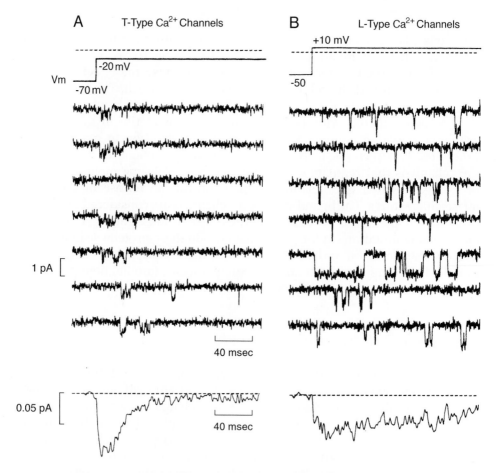

FIGURE 9.1 L-TYPE AND T-TYPE Ca²⁺ CURRENTS
Single-channel currents flowing through L-type and T-type Ca²⁺ channels recorded from a cell-attached patch on a guinea pig cardiac ventricular myocyte. Barium has been used as the permeant ion. Each of the seven upper current traces was elicited by a single depolarization, whereas the bottom trace respresents the average of ~280 such traces. (From Nilius et al. (1985).

carrier, as this ion is not only permeant but also blocks K^+ channels and reduces the inactivation of some types of Ca^{2+} channel. This makes the single-channel currents easier to record. T-type channels have a single-channel conductance of ~8 pS with 100 mM Ba^{2+} as the charge carrier, lower than that of L-type channels, which is ~25 pS. Indeed, T-type channels were so called because they are *tiny* and *transient*, whereas L-type channels are *large* and *long-lasting*; as you might guess, N-type channels are *neither* (they are also *neuron-specific*). L-type Ca^{2+} channels have a wide distribution, being found in skeletal, cardiac and smooth muscle, endocrine cells, and other tissues. In contrast, the P, N, and Q channels appear to be confined to the nervous system and some types of endocrine cells. T-type channels are found in cardiac muscle and some types of nerve cell.

There is an interesting difference between the inactivation of L-type Ca^{2+} currents and that of voltage-gated Na^+ channels, in that the former is enhanced by the presence of Ca^{2+} ions at the inner side of the membrane. Furthermore, because permeating Ca^{2+} ions are able to induce inactivation, the voltage dependence of inactivation mirrors that of the Ca^{2+} current itself (Fig. 9.2). This was first recognised in the protozoan *Paramecium* by the fact that Ba^{2+} currents through L-type Ca^{2+} channels showed little inactivation

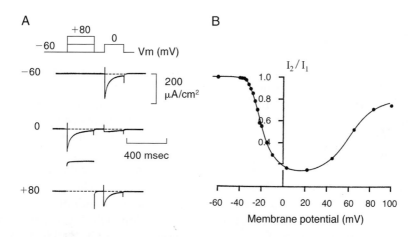

FIGURE 9.2 *Ca²⁺-INDUCED INACTIVATION OF Ca²⁺ CURRENTS*
Ca^{2+}-induced Ca^{2+} inactivation exhibits a U-shaped voltage dependence, reflecting that of the Ca^{2+} current itself. (A) L-type Ca^{2+} currents recorded from insect skeletal muscle in response to paired voltage depolarizations. The purpose of the first pulse, which is of variable amplitude, is to inactivate the Ca^{2+} current. The second pulse is a test pulse to a constant amplitude to measure the extent of inactivation produced by the first pulse. (B) Voltage dependence of inactivation, obtained by plotting the peak Ca^{2+} current during the test pulse (I_2), expressed as a fraction of its value in the absence of a test pulse (I_1), against the membrane potential during the first pulse. From Ashcroft and Stanfield (1982).

(because Ba^{2+} is unable to substitute for Ca^{2+}) (Brehm and Eckert, 1978). The functional importance of Ca^{2+}-induced inactivation is that it provides a negative feedback mechanism that limits Ca^{2+} entry into the cell. In contrast to L-type Ca^{2+} channels, the other types of Ca^{2+} channel exhibit primarily voltage-dependent inactivation.

Modulation

Ca^{2+} channels are subject to complex regulation by a wide range of cytosolic modulators. Among the most important of these are the GTP-binding proteins and the protein kinases. This modulation may be of physiological significance. GTP-binding proteins, for example, mediate the inhibitory effects of neurotransmitters on N-type Ca^{2+} channels (Dolphin, 1998). Phosphorylation of L-type Ca^{2+} channels contributes to the increase in heart rate and contractile force produced by adrenaline. Stimulation of adenylate cyclase in response to β-adrenergic agonists leads to the elevation of cytosolic cyclic AMP levels and activation of protein kinase A. This kinase phosphorylates L-type Ca^{2+} channels and thereby increases the channel open probability. The resulting larger Ca^{2+} current lowers the threshold for action potential initiation and thereby increases the heart rate, whereas the accompanying rise in Ca^{2+} entry enhances Ca^{2+} release from intracellular stores and potentiates contractile force.

Functional Roles

N- and P-type Ca^{2+} channels mediate neurotransmitter release in the mammalian peripheral and central nervous systems. The former is inhibited by a range of neurotransmitters, including serotonin and dopamine, which acts presynaptically to reduce synaptic transmission. L-type Ca^{2+} channels are not involved in neurotransmitter release, although they do provide the Ca^{2+} influx required for the secretion of peptide hormones, such as insulin. They are also important in skeletal muscle, where they function as the voltage sensor for excitation–contraction coupling, and in cardiac muscle, where they provide the pathway for the Ca^{2+} influx that triggers Ca^{2+}-dependent Ca^{2+} release from intracellular stores. Ca^{2+} currents also modulate action potential waveform: for example, L-type Ca^{2+} channels contribute to the plateau phase of the ventricular action potential, while T-type channels contribute to the pacemaker potential of cardiac sinoatrial node cells and spontaneously spiking neurones.

Ca²⁺ Channel Structure

The voltage-gated L-type Ca^{2+} channel of skeletal muscle was the first to be purified and cloned (Curtis and Catterall, 1984; Tanabe et al., 1987). Skeletal muscle was chosen for the initial purification studies because it contains a high density of Ca^{2+} channels and purification was achieved by utilising the ability of the dihydropyridine drugs to bind to the channel with high affinity and specificity (Curtis and Catterall, 1984).

We now know that the skeletal muscle L-type Ca^{2+} channel is a hetero-ligomeric complex of five subunits, α_1, α_2, β, γ and δ, arranged in a $1:1:1:1:1$ stoichiometry (Fig. 9.3) (Catterall, 1985). The α_2 and δ chains are derived by proteolytic cleavage of the same gene product and remain associated by disulphide interactions. The other subunits are encoded by separate genes. They have molecular masses of 190 (α_1), 160 ($\alpha_2\delta$), 52 (β) and 32 (γ) kDa. Hydrophobicity analysis and the location of glycosylation sites indicate that the α_1-, α_2-, δ-, and γ-subunits span the membrane, whereas the β-subunit is cytoplasmic. The δ-subunit appears to act as a membrane anchor for the α_2-subunit, which is mostly extracellular. The α_1-subunit is functionally the most important as it acts as the channel pore, the voltage sensor, and the receptor for many drugs. The other subunits have auxiliary roles: when coexpressed with the α_1 subunit they enhance the current magnitude, alter its kinetic properties, and, in some cases, confer sensitivity to channel modulators.

At least seven different genes that code for α_1-subunits have been identi-fied (Table 9.1). Molecular biologists have used a different nomenclature for these subtypes than that used by electrophysiologists for the channels they compose. L-type Ca^{2+} channels are formed from α_{1S}-, α_{1C}-, or α_{1D}-subunits, N-type channels from α_{1B}-subunits, P- and Q-type channels from α_{1A}-subunits, and R-type Ca^{2+} channels are probably composed of α_{1E}-subunits. T-type

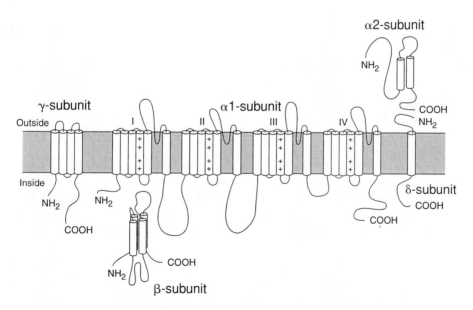

FIGURE 9.3 THE SKELETAL MUSCLE Ca²⁺ CHANNEL CONSISTS OF FIVE SUBUNITS

Putative membrane topology of the α_1-, α_2/δ-, β-, and γ-subunits of the voltage-gated Ca^{2+} channel. After Hockerman et al. (1997).

TABLE 9.1   HUMAN CALCIUM CHANNEL GENES

Gene	Protein	Alternative name	Chromosome location	Tissue distribution	Reference
CACLN1A3	α_{1S}	L-type (skeletal	1q31–32	Skeletal muscle	Gregg et al. (1993a)
CACLN1A4	α_{1A}	P/Q-type	19p13.1	Cerebellum, cortex, hypothalamus	Diriong et al. (1995)
CACLN1A5	α_{1B}	N-type	9q34	Brain, peripheral neurones	Diriong et al. (1995)
CACLN1A1	α_{1C}	L-type (cardiac)	12p13.3	Heart, smooth muscle, lung	Schultz et al. (1993)
CACLN1A2	α_{1D}	L-type (neuroendocrine)	3p14.3	Brain, endocrine tissues	Seino et al. (1992)
CACLN1A6	α_{1E}	R-type?	1q25–31	Brain, heart	Diriong et al. (1995)
CACNA1G	α_{1G}	T-type	17q22	Brain, heart	Perez-Reyes et al. (1998)
CACLN1L21	$\alpha_2\delta$		7q21–22		Powers et al. (1994)
CACLNB1	$\beta1$		17q11.2–22		Gregg et al. (1993b)
CACLNG1	γ		17q23		Powers et al. (1993)

Ca²⁺ channels are made up of α_{1G} subunits. Ca²⁺ channel diversity is further enhanced by the presence of multiple isoforms of each type of α_1-subunit, generated by alternative splicing (Snutch and Reiner, 1992; Zhuchenko et al., 1997). Multiple types of $\alpha_2\delta$-, β- and γ-subunits also exist (at least four types of β-subunits have been cloned). Potentially this diversity could yield a remarkable degree of Ca²⁺ channel heterogeneity, but the exact combinations of subunits that are actually expressed in mammalian tissues remain to be determined.

Like the α-subunit of the voltage-gated Na⁺ channel (Chapter 5), that of the Ca²⁺ channel has four homologous repeats (Fig. 9.4). Each repeat consists of six putative transmembrane domains (TMs) and a loop linking TMs 5 and 6 that dips back down into the membrane and contributes to the pore. The fourth transmembrane domain possesses a number of highly conserved positive charges and, by analogy with what is found for Na⁺ channels (Chapter 5), is thought to serve as the voltage sensor for Ca²⁺ channel opening. The α_1-subunit is a substrate for protein kinase A, protein kinase C, and Ca²⁺-calmodulin–dependent protein kinase. The β-subunit also has consensus sequences for phosphorylation by a number of different protein kinases. Both these subunits may therefore be involved in the modulation of Ca²⁺ currents by phosphorylation.

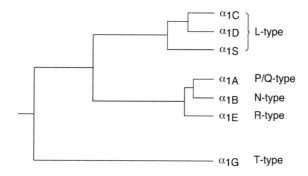

FIGURE 9.4 *Ca²⁺ CHANNEL LINEAGES*
Relationship between different members of the voltage-gated Ca²⁺ channel family. The length of the lines indicates the percentage of amino acid identity. After Tsien (1998).

Correlating Structure and Function

The pore

The hydrophobic loops linking TMs 5 and 6 in each of the four repeats come together to form the pore of the voltage-gated Ca²⁺ channel, as is the case for voltage-gated Na⁺ channels. A ring of four negatively charged glutamate residues, one in each pore loop, form a high-affinity binding site for Ca²⁺ and provide a structural basis for the high Ca²⁺ selectivity of the channel (Yang *et al.*, 1993). The importance of these residues was first suggested by work on Na⁺ channels, in which only two of the repeats possess negatively charged residues at the equivalent position. As described in Chapter 5, mutation of the other two residues to negatively charged ones endows the Na⁺ channel with selectivity properties resembling those of Ca²⁺ channels (Heinemann *et al.*, 1992). Yang and colleagues (1993) explored the effect of making the reverse mutation in Ca²⁺ channels. In the absence of divalent cations, Ca²⁺ channels are highly permeable to monovalent cations, such as Na⁺ and Li⁺. When the Ca²⁺ concentration is increased the current is first reduced, because at micromolar concentrations Ca²⁺ binds to sites within the pore and blocks Na⁺ permeation, and then increases again at millimolar concentrations where Ca²⁺ itself carries significant current. The relationship between Ca²⁺ concentration and current is therefore U shaped. When each of the critical glutamates within the pore loop was mutated, in turn, to the uncharged amino acid glutamine, the ability of Ca²⁺ to block Li⁺ fluxes was markedly diminished (Fig. 9.6; Yang *et al.*, 1993). This provides confirmation that the glutamate residues are responsible for Ca²⁺ selectivity. All four glutamates are required but they contribute unequally to the Ca²⁺-binding site, with that in repeat III having the strongest effect.

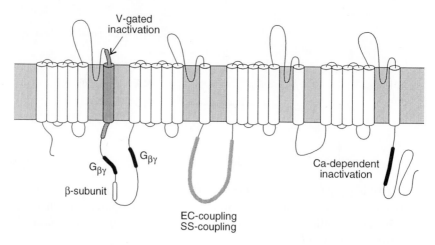

FIGURE 9.5 MEMBRANE TOPOLOGY OF THE α_1-SUBUNIT
Putative membrane topology of an archetypal α_1-subunit with the important functional domains indicated. The β-subunit and the $G_{\beta\gamma}$ subunits of GTP-binding proteins interact with the I–II loop. The II–III loop is involved in EC coupling in α_{1S}-subunits and interacts with synaptic proteins such as syntaxin binding in α_{1A}- and α_{1B}-subunits. The C terminus participates in Ca²⁺-dependent Ca²⁺ inactivation, while in some α_1-subunits, the sixth transmembrane domain of repeat I is involved in voltage-dependent inactivation. EC, excitation–contraction; SS, stimulus–secretion.

Inactivation

Inactivation of voltage-gated Na⁺ channels is caused by the cytosolic loop linking repeats III and IV, which acts as a hinged lid and occludes the inner mouth of the pore (Chapter 5). It was therefore expected that a similar mechanism might account for the voltage-dependent inactivation of Ca²⁺ channels. Surprisingly, this turned out not to be the case, for changes in the III–IV linker of the Ca²⁺ channel α-subunit had little or no effect on inactivation (Zhang *et al.*, 1994). Instead, residues within, and immediately adjacent to, the sixth transmembrane domain of repeat I are important because transplantation of this region between α_{1E}- and α_{1A}-subunits also transferred the inactivation properties characteristic of the donor channel. An additional residue, which lies within the $G_{\beta\gamma}$-binding site in the cytoplasmic loop linking repeats I–II, is also important (Herlitze *et al.*, 1997). Inactivation of α_{1A} channels is slower, and the voltage dependence of steady-state inactivation is shifted to more positive membrane potentials, when this residue is negatively charged (glutamate rather than arginine). How these structural elements contribute to voltage-dependent inactivation of Ca²⁺ channels is still unclear, but there is no evidence that a blocking particle is involved. Instead, inactivation more closely resembles the C-type inactivation of voltage-gated K⁺ channels (see Chapter 6).

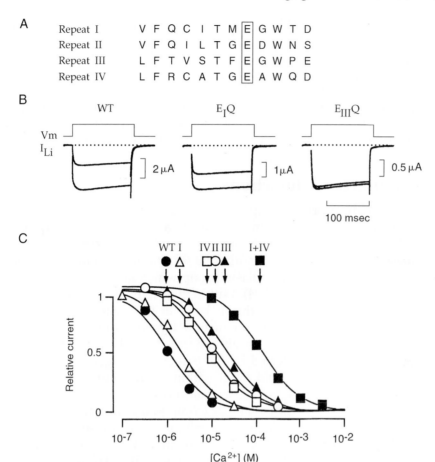

FIGURE 9.6 A RING OF GLUTAMATES ACTS AS A Ca²⁺-BINDING SITE WITHIN THE PORE

(A) Alignment of amino acid sequence in the pore loops of each of the four repeats of the α_{1C}-subunit of the voltage-gated Ca²⁺ channel with the critical glutamates (E) marked. (B) Currents recorded from wild-type (WT) and mutated channels (E_{IQ}, E_{IIIQ}) with Ba²⁺ or Li⁺ as the charge carrier in the absence (large currents) or presence (smaller currents) of 3 mM external Ca²⁺. Glutamate was mutated to glutamine (Q) in repeat I (middle) or repeat III (right). (C) Ca²⁺ block of Li⁺ currents for wild-type (●) and mutated channels. The arrows indicate the concentration at which the block is half-maximal and show that this increased when the glutamate in each of the four repeats (I–IV) is replaced by glutamine. From Yang *et al.*, 1993.

As discussed earlier, L-type Ca²⁺ channels, composed of α_{1S}-, α_{1C}-, or α_{1D}-subunits, undergo Ca²⁺-dependent inactivation. There are two theories which attempt to account for the phenomenon. The first suggests that Ca²⁺ activates a Ca²⁺-dependent phosphatase (calcineurin) that dephosphorylates the channel and turns it off (Chad and Eckert, 1986), whereas the second proposes that Ca^{2+} gates the channel by binding directly to a cytoplasmic site on the channel itself. Most evidence favours the latter hypothesis. A Ca^{2+}-binding motif known as an 'EF hand' is located within the C terminus of the α_{1C}-subunit and appears to act as the Ca^{2+}-binding site. Transfer of this motif into an α_{1E}-subunit, which does not normally show Ca^{2+}-induced inactivation, causes inactivation to become Ca^{2+} dependent (De Leon et al., 1995). How Ca^{2+} binding to the EF hand initiates sustained channel closure is still not known.

Protein–protein interactions

The β-subunit exerts an important modulatory influence on the Ca^{2+} channel. A substantial increase in Ca^{2+} current amplitude and/or changes in the current kinetics are observed when each of the six different α_1-subunits is coexpressed with any of the four different β-subunits that have been cloned. The β-subunit binds to the cytoplasmic loop which links repeats I and II of the α_1-subunit (Pragnell et al., 1994). Within this region lies a conserved motif that is positioned just downstream of the sixth transmembrane domain of repeat I. Mutations within this α_1-interaction domain reduced binding of the β-subunit to the α_1-subunit and largely prevented the changes in Ca^{2+} current amplitude and kinetics conferred by the β-subunit (Pragnell et al., 1994). A highly conserved 30 amino acid motif within the β-subunit is required for its interaction with the α_1-subunit (De Waard et al., 1995).

N-type and P/Q-type Ca^{2+}-channels (composed of α_{1A}- and α_{1B}-subunits) are modulated by a variety of neurotransmitters, which cause a positive shift in the voltage dependence of channel activation. Consequently, the Ca^{2+} current is reduced. This inhibitory modulation is mediated by the $\beta\gamma$-subunits of pertussis toxin-sensitive receptor-activated G-proteins (Dolphin, 1998), which have been shown to bind directly to the intracellular loop connecting repeats I and II of the α_1-subunit (De Waard et al., 1997). The $G_{\beta\gamma}$-subunits bind to two distinct sites within this linker: a motif (QQxxRxLxGY) located in the α_1-interaction domain and a second sequence further downstream (Fig. 9.7). The former contains a consensus sequence–QxxER–implicated in $G_{\beta\gamma}$ binding to other proteins: this motif is also critical for $G_{\beta\gamma}$ modulation of the Ca^{2+} channel as regulation is abolished when it is mutated (De Waard et al., 1997; Herlitze et al., 1997). The $G_{\beta\gamma}$-binding site lies adjacent to that of the β-subunit, which suggests that the binding of one protein may modify the functional effects of the other. Several pieces of evidence suggest that $G_{\beta\gamma}$ modulation of α_{1A} and α_{1B} Ca^{2+} channels also involves residues that lie outside the I–II loop (Dolphin, 1988).

The different types of α_1-subunit differ most strikingly in two principal regions: the cytoplasmic loop linking repeats II and III and the carboxy-terminal region. The former mediates the interaction of the α_{1S}-subunit with effector molecules such as the ryanodine-sensitive Ca^{2+}-release channel in the sarcoplasmic reticulum of skeletal muscle (see Chapter 14). In α_{1A}- and α_{1B}-subunits the same region also interacts with synaptic proteins, such as

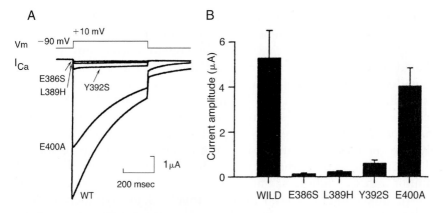

FIGURE 9.7 β-SUBUNIT REGULATION
The β-subunit enhances the Ca^{2+} current amplitude. Mutations in the α_1-subunit interaction domain that prevent modulation by the β-subunit therefore cause a reduced Ca^{2+} current. (A) Barium currents recorded from *Xenopus* oocytes injected with mRNAs encoding either wild-type (WT) or mutated α_{1A}-subunits, together with α_2 and β_1-subunits. (B) Mean current amplitude recorded for wild-type and mutated α_{1A}-subunits. From Pragnell *et al.* (1994).

syntaxin and SNAP-25, which are involved in the docking and release of neurotransmitter vesicles in neurones (Sheng *et al.*, 1994b; Rettig *et al.*, 1996). This may be a mechanism for ensuring that the Ca^{2+} influx required for exocytosis of synaptic vesicles occurs close to the vesicle release sites.

Pharmacology

Voltage-gated Ca^{2+} channels are major targets for therapeutic drugs (Hockerman *et al.*, 1997). There are three major types of Ca^{2+} channel blocker– the phenylalkylamines, the benz(othi)azipines and the dihydropyridines– typified by the drugs verapamil, diltiazem, and nifedipine. These drugs are believed to bind to three separate sites on the α_1-subunit, which are allosterically linked. Although their binding sites have not been fully characterised, mutagenesis studies have implicated residues within TMs 5 and 6 in repeat II, and TM6 in repeat IV, as forming part of the binding site for all three types of blocker. It is worth noting that local anaesthetic block of voltage-gated Na^{+} channels involves an equivalent region of the α-subunit of that channel. The location of the binding sites for conotoxin and agatoxin on N-type and P/Q-type Ca^{2+} channels, respectively, is unknown.

Summary of structure–function studies

It is now well established that the region connecting the S5 and S6 segments forms the outer part of the Ca^{2+} channel pore and acts as the selectivity filter. By analogy with voltage-gated Na^+ channels, it is also believed the S4 segment acts as the voltage sensor. The mechanisms underlying inactivation of Ca^{2+} channels are far less clear. Ca^{2+}-dependent inactivation appears to be mediated by Ca^{2+} binding to a site located within the C terminus, whereas voltage-dependent inactivation involves the S6 segment of repeat I and residues within the cytoplasmic loop linking repeats I and II. How these residues are coupled to closure of the pore is not known. The Ca^{2+} channel interacts with many other proteins, which may either regulate its activity or, alternatively, are themselves influenced by the activity of the Ca^{2+} channel. The binding sites for some of these proteins have been identified. Thus both the β-subunit and $G_{\beta\gamma}$ subunits bind to the cytosolic I–II linker region, while the II–III linker interacts with proteins involved in excitation–contraction coupling and in neurotransmitter release. Important differences in the functional properties of the various α-subunits result from amino acid differences within these regions.

DISEASES OF SKELETAL MUSCLE Ca²⁺ CHANNELS

Muscular Dysgenesis in Mice

Muscular dysgenesis is an autosomal recessive disease of mice in which there is a failure of excitation-contraction coupling. The skeletal muscles of homozygous *mdg* mice are completely paralysed and, consequently, the animals die at birth from respiratory failure. The heterozygous mice, however, appear normal. Dysgenic mice lack functional skeletal muscle Ca^{2+} channel α_1-subunits (α_{1S}-subunits). This is the result of a single nucleotide deletion (*mdg*) which produces a frameshift that leads to truncation of the protein within the fourth repeat and thus results in deletion of the latter part of the pore loop, the sixth transmembrane domain and the C terminus (Chaudhari, 1992). A novel 53 amino acid carboxy terminus is also inserted due to continued translation after the frameshift before a stop codon is encountered. Northern blot analysis revealed that expression of α_{1S}-subunit mRNA was substantially lower in dysgenic mice than in normal mice, explaining the reduced protein level.

Beam and colleagues (1986) realised that the lack of functional α_{1S}-subunits makes myotubes isolated from dysgenic mice an excellent system in which to study the role of the voltage-gated Ca^{2+} channel in excitation-contraction (EC) coupling because all other components of the system are present. In a series of very elegant studies on dysgenic myotubes they showed that the α_{1S}-subunit not only acts as the primary subunit of the voltage-gated Ca^{2+} channel, but also serves as the voltage sensor for EC coupling in skeletal muscle. Contraction of vertebrate skeletal muscle is triggered by depolarization of the plasma membrane. This depolarization is conducted deep into the interior of the fibre along infoldings of the surface membrane known as transverse (T-) tubules (see Fig. 14.4). At the triads, the T-tubular membrane comes into close apposition with membranes of the sarcoplasmic reticulum (SR), which serves as an intracellular calcium store. The two membranes remain separate, however, so for many years it was a puzzle as to how depolarization of the T-tubule membrane triggered Ca²⁺ release from the sarcoplasmic reticulum. Blocking the skeletal muscle Ca²⁺ current did not block EC coupling, so clearly Ca²⁺ ions could not serve as a second messenger (as they do in cardiac muscle).

The first clue to what was going on came with the discovery that myotubes cultured from dysgenic mice lacked both L-type Ca²⁺ currents and EC coupling (Beam *et al.*, 1986). Intracellular calcium stores and the contractile machinery were, however, quite normal. Subsequent progress was rapid. Western blotting revealed that whereas $\alpha_2\delta$-, β- and γ-subunits were present in dysgenic myotubes, no α_{1S}-subunit protein was detectable. Injection of cDNA encoding the normal α_{1S}-subunit into dysgenic myotubes restored both the L-type Ca²⁺ currents (Fig. 9.8) and EC coupling (Tanabe *et al.*, 1988). Furthermore, the nonlinear capacitative currents (charge movements) associated with EC coupling were also restored (Adams *et al.*, 1990). These results clearly demonstrated that the α_{1S}-subunit acted as the voltage sensor for EC coupling and somehow caused the opening of Ca²⁺ release channels in the SR (these channels are described more fully in Chapter 14). They also showed, for the first time, that the α_{1S}-subunit formed the Ca²⁺ channel pore.

In skeletal muscle, EC coupling is independent of extracellular calcium. By contrast, in cardiac muscle, EC coupling requires Ca²⁺ entry through

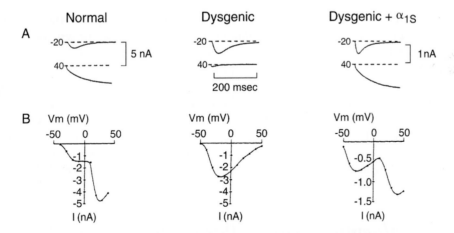

FIGURE 9.8 DYSGENIC MYOTUBES LACK THE α_{1S}-SUBUNIT
Ca²⁺ currents (top) and current–voltage relations (bottom) recorded from normal myotubes (A) exhibit both a low threshold- and a high threshold (L-type)-activated Ca²⁺ current. The high threshold Ca²⁺ current is absent in dysgenic myotubes (B) but can be restored by transfection with the α_{1S}-subunit (C). From Tanabe *et al.* (1988).

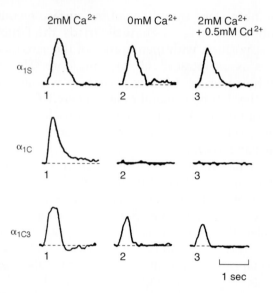

FIGURE 9.9 CARDIAC TYPE E–C COUPLING IS CONFERRED BY
THE II–II LINKER REGION OF THE α_1 SUBUNIT
Electrically evoked contractions in dysgenic myotubes expressing the α_{1S}-subunit (skeletal muscle type), α_{1C}-subunit (cardiac muscle type), or a
chimeric subunit in which the II–III linker of the α_{1C}-subunit has been
replaced with that of the α_{1S}-subunit (labelled α_{1C3}). Contractions were
recorded in the presence of 2 mM Ca^{2+} (1), 0 mM Ca^{2+} (2), or 2 mM Ca^{2+}
plus 0.5 mM Cd^{2+} to block Ca^{2+} currents (3). From Tanabe *et al.* (1990b).

voltage-gated Ca^{2+} channels. The Ca^{2+} which enters then triggers the release
of additional Ca^{2+} from intracellular stores in a process known as Ca^{2+}-dependent Ca^{2+} release. The type of voltage-gated Ca^{2+} channel involved is
an important determinant of EC coupling. When dysgenic myotubes were
transfected with the skeletal muscle α_{1S}-subunits, the resulting EC coupling
was of the skeletal muscle type, whereas when they were injected with cardiac
α_{1C}-subunits, EC coupling was dependent on external Ca^{2+}, as is the case in
cardiac muscle (Tanabe *et al.*, 1990b). Chimeric channels containing parts of
both α_{1S}- and α_{1C}-subunits localised the region critical for determining the
type of EC coupling to the cytoplasmic domain linking repeats II and III (Fig.
9.9; Tanabe *et al.*, 1990a).

Hypokalaemic Periodic Paralysis

Clinical features

Hypokalaemic periodic paralysis (hypoPP), the most common form of
periodic paralysis, is so called because patients have episodic attacks of
skeletal muscle weakness that are associated with low serum potassium

concentrations. Paralysis can be triggered by a carbohydrate-rich meal, by insulin, or by strenuous exercise and may be quite severe, lasting from a few hours to several days. Attacks may also occur during sleep, so that the patient is weak or paralysed on awaking. The disease is inherited as a dominant trait but symptoms do not usually develop until puberty and the penetrance is reduced in females (more than three times as many males as females are affected). For reasons that are not understood, the muscles controlling speech, swallowing and respiration are usually unaffected. During attacks, the plasma potassium may fall from its normal value of 3–5 mM to as low as 1.8 mM. Such severe hypokalaemia may have additional consequences, including reduced gut motility and cardiac dysrhythmia, particularly in subjects with established cardiac disease. In some patients, however, paralysis is observed at plasma K$^+$ levels within the normal range (3 mM).

Although many patients improve spontaneously with age, around 30% develop a progressive myopathy. The disease is treated prophylactically with acetazolamide. Ingestion of potassium at the start of an attack may also be helpful. HypoPP is not confined to man. It also occurs in Burmese cats.

Analysis of mutations

HypoPP is genetically heterogeneous. In some families, the disease maps to chromosome 1q31–32, coincident with the gene (CACNL1A3) encoding the α_1-subunit of the voltage-gated L-type calcium channel in skeletal muscle (α_{1S}; Jurkat-Rott et al., 1994). Three point mutations in this gene have been identified: arginine-to-histidine substitutions at residues 528 and 1239 (R528H and R1239H), which occur with equal frequency, and an arginine-to-glycine substitution (R1239G), which is very rare. These mutations occur within the S4 domain of the second and fourth repeats (Fig. 9.10). The age of onset and the serum K$^+$ level during an attack are lower in patients with the R1239H mutation, while the R528H mutation shows incomplete penetrance.

No functional effects on the Ca^{2+} current were detected when the HypoPP mutations were introduced into the α_1-subunit of the skeletal muscle Ca^{2+} channel and studied following expression in a dysgenic muscle line line, that lacks an endogenous α_1-subunit (Jurkat-Rott et al., 1998). The cause of the HypoPP phenotype thus remains a mystery.

How do HypoPP mutations give rise to paralysis?

Muscle biopsies reveal that the muscle fibres of HypoPP patients are slightly (5–15 mV) more depolarized than normal fibres between attacks and have a reduced excitability. They are further depolarized by lowering the external K$^+$ concentration to 1 mM (Rudel et al., 1984). This agrees with reports that HypoPP muscles are depolarized in vivo during an attack of weakness (Riecker and Bolte, 1996). Thus the muscle weakness and paralysis characteristic of HypoPP appear to result from a membrane depolarization that inactivates the Na$^+$ channel. Unlike HyperPP (Chapter 5), myotonia is never observed in HypoPP. This suggests that the membrane depolarization

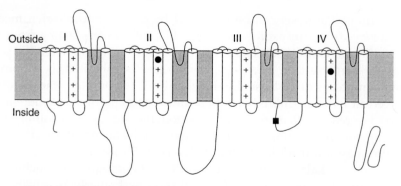

● Hypokalaemic Periodic Paralysis
■ Malignant hyperthermia

FIGURE 9.10 MUTATION OF CACNL1A3 CAUSES TWO TYPES OF HUMAN DISEASE
Putative membrane topology of the α_{1S}-subunit of the voltage-gated Ca^{2+} channel with the mutations associated with hypokalaemic periodic paralysis (●) and malignant hyperthermia (■) indicated.

occurs so slowly that the Na^+ current inactivates without ever being substantially activated. The real puzzle is why a reduction in extracellular K^+ should cause depolarization of HypoPP skeletal muscle, because healthy fibres are *hyperpolarised* when external K^+ is reduced (Rudel *et al.*, 1984).

It seems likely that the low plasma K^+ level is primarily responsible for precipitating an attack of paralysis in HypoPP patients. The ability of insulin or a carbohydrate-rich meal (which elevates insulin levels) to induce an attack may therefore be related to the fact that insulin stimulates the Na/K-ATPase, enhancing K^+ uptake into muscle and lowering the plasma K^+ concentration. Likewise adrenaline, which also stimulates the Na/K-ATPase, can provoke paralysis in HypoPP patients, which may explain why attacks of weakness can be triggered by stress.

To summarize, the skeletal muscle of HypoPP patients undergoes a paradoxical depolarization in response to a reduction in plasma K^+. How this is related to the observed mutations in *CACNL1A3* is still unexplained. A further mystery is why HypoPP is inherited in a dominant fashion. Heterozygous patients with 50% of mutant channels exhibit the disease, yet heterozygous dysgenic mice mice, which also only have half their full complement of α_{1S} subunits, are clinically normal.

Malignant Hyperthermia

Malignant hyperthermia is one of the main causes of death due to anaesthesia. In susceptible individuals, volatile anaesthetics trigger an increase in $[Ca^{2+}]_i$ in skeletal muscle, and the resulting contraction precipitates a potentially fatal rise in body temperature which gives the disease its name. Susceptibility to malignant hyperthermia (MHS) is genetically heterogeneous and in most families has been found to result from mutations in the gene encoding the Ca^{2+} release channel of the sarcoplasmic reticulum—the ryanodine receptor RYR1. The clinical symptoms of the disease are therefore described in conjunction with this channel in Chapter 14. In one large French family, however, a mutation in the gene (*CACLN1A3*) that encodes the α_{1S}-subunit of the skeletal muscle Ca^{2+} channel has been linked to MHS (Monnier *et al.*, 1997). This results in substitution of a histidine residue for the arginine at position 1086 (R1086H), which lies within the cytosolic loop linking repeats III and IV (Fig. 9.10). The functional effects of this mutation have yet to be described, but because patients with MHS have no clinical symptoms unless exposed to inhalation anaesthetics or depolarising neuromuscular blockers, it seems likely that mutation of R1086 does not result in any gross abnormality of the Ca^{2+} channel. Rather, the mutation is expected to lead to an enhanced intracellular Ca^{2+} concentration under anaesthesia, either by interfering with the coupling of the α_{1S}-subunit to RYR1 (which seems most likely) or by directly increasing Ca^{2+} influx through the α_{1S}-subunit. Although mutations in *CACLN1A3* cause hypokalaemic periodic paralysis, as described earlier, individuals carrying the R1086H mutation do not show any clinical symptoms of this disease.

Summary

Mutations in the *CACNL1A3*, which encodes the α_{1S}-subunit of the skeletal muscle Ca^{2+} channel, cause two very different human diseases: hypokalaemic periodic paralysis and malignant hyperthermia. There is currently too little analysis of the effects of the biophysical properties of these mutations to be able to draw detailed conclusions about how they give rise to the different phenotypes.

DISEASES OF NEURONAL CALCIUM CHANNELS

Three human diseases with very different phenotypes are associated with mutations in the same Ca^{2+} channel gene, *CACNL1A4*. These are episodic ataxia type-2 (EA-2), familial hemiplegic migraine (FHM), and spinocerebellar ataxia type-6 (SCA6). Mutations in the equivalent murine gene produce absence epilepsy.

The human *CACNL1A4* gene contains 47 exons and maps to chromosome 19p13.1 (Diriong *et al.*, 1995). It is strongly expressed in the brain, particularly in the Purkinje and granule cells of the cerebellum, the cerebral cortex, the thalamus and the hypothalamus. It is thought to encode the α_{1A} subunit of P- and/or Q-type Ca^{2+} channels; however, when expressed in oocytes the properties of the cloned channel are not identical to either P or Q channels, so the exact relationship is not clear. P-type Ca^{2+} channels were originally identified in Purkinje cells and Q-type channels in cerebellar granule neurones. They may be distinguished by their electrophysiological properties and their sensitivity to pharmacological agents.

Familial Hemiplegic Migraine

Migraine is a debilitating neurological disorder with which many people, including myself, are all too familiar. Indeed, it is one of the most common of chronic disorders with a prevalence of up to 24% in women and 12% in men (Russell et al., 1995) and it costs the United Kingdom half a billion pounds each year in lost productivity. The attacks last from 4 to 72 hr and the symptoms include headache (often confined to one side of the head), nausea, vomiting, and sensitivity to light and noise. Around 20% of patients experience an aura prior to an attack that may manifest as visual disturbances, weakness of the limbs, and difficulty in speaking. For all its debilitating effects, migraine is not entirely without its positive aspects: the remarkable colours and distorted visual perception experienced during the classical aura have inspired many artists and poets.

FHM is a rare autosomal dominant type of migraine with aura. It is associated with paralysis of one half of the body during an attack and in some families with progressive cerebellar atrophy. The disease maps to chromosome 19p13.1 and results from mutations in *CACNL1A4* (Joutel et al., 1993). Four different missense mutations have been identified (Fig. 9.11; Ophoff et al., 1996). One of these results in substitution of an arginine for a glutamine (R192Q) in the fourth transmembrane domain of repeat I. The location of this mutation within the voltage sensor suggests that it may affect the gating of the channel. The second mutation lies within the pore loop and is a substitution of threonine at position 666 with a methionine (T666M). This threonine residue is highly conserved between Ca^{2+} channels and its mutation may alter the ion selectivity or conductance of the channel. The third and fourth mutations are located at the cytoplasmic ends of the sixth transmembrane domain of repeats II and IV, respectively, and consist of a valine-to-alanine (V714A) and an isoleucine-for-leucine (I1811L) substitution, respectively. These residues may form part of the inner mouth of the pore.

Although the region of chromosome 19 containing the FHM locus has also been shown to be involved in the more common forms of migraine (May et al., 1995), no mutations in *CACNL1A4* were identified in six sporadic migraine patients (Ophoff et al., 1996). This is perhaps not unexpected because several studies have suggested that a predisposition to migraine is likely to be genetically heterogeneous.

Episodic Ataxia Type-2

Episodic ataxia type-2 (EA-2) is an autosomal dominant disorder that is characterised by intermittent attacks of cerebellar ataxia, migraine-like symp-

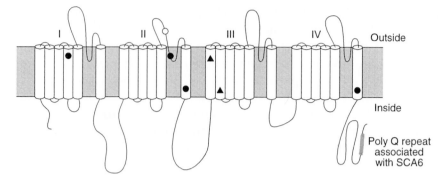

● Familial hemiplegic migraine
▲ Episodic ataxia type-2
○ *tg* mouse

FIGURE 9.11 MUTATION OF CACNL1A4 CAUSES THREE
DIFFERENT HUMAN DISEASES
Putative membrane topology of the α_{1A}-subunit of the voltage-gated Ca^{2+}
channel with the mutations associated with familial hemiplegic migraine
(●), episodic ataxia type 2 (▲), and the *tottering* mouse (○) indicated. The
position of the polyglutamine repeat, which is expanded in spinocerebellar
ataxia type 6, is also shown.

toms and cerebellar atrophy. Typically, these are precipitated by emotional
stress, alcohol or exercise and they may last for several hours. Uncontrolled
eye movements (nystagmus) are also observed between the ataxic attacks.
Treatment with acetazolamide is usually very effective in preventing attacks.

    Like FHM, episodic ataxia type-2 maps to chromosome 19p13.1 and is
associated with mutations in *CACNL1A4* (von Brederlow *et al.*, 1995). Two
different disease-causing mutations have been identified to date (Fig. 9.11;
Ophoff *et al.*, 1996). One of these is a frame-shift mutation, that introduces a
premature stop codon and truncates the protein at the start of the third repeat.
The other is a basepair substitution within intron 28 which is predicted to
cause aberrant splicing of the protein and truncate the protein within the third
repeat. Both mutations are expected to give rise to nonfunctional channels.

Spinocerebellar Ataxia Type-6

The spinocerebellar ataxias are a group of clinically and genetically heterogenous neurological disorders which are characterised by cerebellar dysfunction. At least seven different loci have been mapped, although the genes have
not yet all been identified. SCA6 maps to chromosome 19p13 and results
from mutations in *CACNL1A4* (Zhuchenko *et al.*, 1997). SCA6 is characterised
by a slowly progressive cerebellar ataxia, nystagmus, and proprioceptive
sensory loss that develops over a period of 20–30 years and eventually results
in the patient being confined to a wheelchair. Postmortem examination reveals

a severe loss of cerebellar Purkinje cells and a moderate loss of cerebellar granule and dentate nucleus neurones.

In normal individuals, the C terminus of the α_{1A}-subunit contains a stretch of between 4 and 16 consecutive glutamine residues (Fig. 9.11). This reflects the repetition of a CAG triplet in *CACNL1A4*. SCA6 is associated with expansion of this trinucleotide repeat which results in a polyglutamine stretch of increased length (typically >20). Trinucleotide repeat expansions are the basis of a number of progressive neurodegenerative disorders and an increased number of CAG repeats is also found in Huntington's disease and in several other types of spinocerebellar ataxia. Like these disorders, the length of the trinucleotide expansion in *CACNL1A4* appears to be correlated with disease severity, because the onset of SCA6 occurs earlier in individuals with a larger number of repeats, being ~30 years for 27 repeats as compared with 40–50 years for 22–23 repeats. However, *CACNL1A4* is unusual in that the size of the repeat expansion is relatively small. Its significance is therefore not yet certain.

In other neurodegenerative disorders, a CAG repeat expansion results in a gain-of-function of the encoded protein. It is not yet known whether this is also the case in SCA6, but the observed cerebellar atrophy would be consistent with an enhanced Ca²⁺ influx.

The Tottering Mouse

Epilepsy affects approximately 1% of the human population. It is characterised by recurrent bursts of synchronised electrical discharges that interfere with normal neuronal function. Both generalised and partial seizures are observed. Generalised seizures are commonly known as *grand mal* (convulsive) or *petit mal* (absence). Absence epilepsy is recognised by the fact that the person appears to 'freeze', stopping whatever they are doing, adopting a fixed posture, and staring into space. This behaviour is accompanied by generalised bilateral discharges in the electroencephalogram. An epileptic mouse mutant, the tottering (*tg*) mouse, exhibits absence seizures closely resembling those of man; it also suffers from motor seizures and ataxia. Another mutation at the same locus (*tg^{la}*, the *leaner* mouse) produces a slow and selective degeneration of cerebellar neurones, as a consequence of apoptotic cell death. *Leaner* mice suffer absence seizures and ataxia, but unlike *tottering* mice they do not have motor seizures. Mice possessing the *tg^{rol}* mutation display an intermediate phenotype.

The tottering (*tg*) gene has been cloned and turns out to be the α_1-subunit of the murine type-A voltage-gated Ca²⁺ channel (Fletcher *et al.*, 1996). The *tg* phenotype is associated with the substitution of a leucine residue for a proline at position 601 (Fig. 9.11). This position is similar to that of one of the human FHM mutations (T666M), being located at the start of the pore loop in repeat II. Its location suggests that the *tg* Ca²⁺ channel may have altered pore function. The more severe tg^{la} phenotype is associated with two splicing variants, one of which fails to splice out an intron whereas the other skips an exon. Both introduce a frameshift and therefore are expected to produce truncated proteins.

Conclusions on CACNL1A4 Mutations

To summarise, different types of mutations in the α_{1A}-subunit of the voltage-gated Ca^{2+} channel produce three human diseases: episodic ataxia type-2, familial hemiplegic migraine and spinocerebellar ataxia type-6. EA-2 is caused by truncation of the protein within the third repeat, FHM is associated with missense mutations, and SCA6 is produced by expansion of a polyglutamine repeat in the C-terminal coding region of the protein. All three diseases result in cerebellar atrophy, but they differ in the extent and rate of progression of neuronal degeneration. The relative severity of other symptoms also varies; for example, migraine is most severe in FHM patients. The key question is how the different mutations in *CACNL1A4* relate to the different phenotypes. The answer to this question, however, will have to await analysis of the properties of the mutant Ca^{2+} channels.

There are obvious parallels between the mice and human diseases associated with the α_{1A}-subunit of the voltage-gated Ca^{2+} channel. In particular, missense mutations produce a less severe phenotype, whereas mutations that result in truncated proteins cause severe ataxia and cerebellar degeneration. Absence epilepsy, however, has not been reported for the different human mutations.

Drosophila Mutants

As in mice and men, mutations in the Ca^{2+} channel α_1-subunit genes of the fruitfly *Drosophila* also produce disease (Smith *et al.*, 1996b). Mutations in the *DmCa1D* gene cause embryonic death. *DmCa1A* maps to a chromosomal region which has been implicated in various behavioural and visual defects, including *nightblind-A, lethal(1)L13* (which causes late embryonic death) and courtship-song *cacophony* mutations. It is therefore a strong candidate for these diseases.

Mutations in Voltage-Gated Ca^{2+} Channel β- and γ-Subunits also Cause Disease

The *lethargic* mouse exhibits ataxia and absence seizures resembling those of the *tottering* mouse. Unlike the *tg* mouse, however, it does not show obvious cerebellar atrophy. The *lethargic* phenotype results from a splice site mutation in the gene encoding the β_4-subunit of the voltage-gated Ca^{2+} channel (*Cchb4*), which leads to a translational frameshift and the predicted loss of >60% of the C terminus of the protein (Burgess *et al.*, 1997). The truncated protein lacks the site of interaction with the α_1-subunit and is therefore expected to be nonfunctional. It is suggested that the phenotype results from a reduction in neuronal Ca^{2+} currents, due to the absence of β_4 regulation. *Cchb4* is strongly expressed in cerebellar Purkinje and granule cells and is expressed at lower levels in hippocampus, olfactory bulb, cerebral cortex, and thalamic nuclei, a pattern which is most similar to that of the α_{1A}-subunit. Moreover, the α_{1A}-subunit binds β_4-subunits with much higher affinity than other β-subunits (De Waard *et al.*, 1995). This suggests that it is the loss of α_{1A}-subunit regulation that is primarily responsible for the *lethargic* phenotype, an hypothesis that is consistent with the observation that the phenotype mirrors that of the *tg* mouse, in which the α_{1A}-subunit is lacking.

Mutations in the gene encoding a γ-subunit of the voltage-gated Ca^{2+} channel (*CACNG2*) also cause epilepsy. The *stargazer* mouse is characterised by its ataxic gait and the distinctive way in which it tosses its head back and scans the sky (hence its name). It also exhibits spontaneous episodic seizures, characteristic of absence epilepsy, that typically last for ~6 sec and may occur more than a hundred times an hour. Positional cloning of the gene associated with the *stargazer* phenotype revealed that it encodes a γ-subunit of the voltage-gated Ca^{2+} channel, which was termed stargazin (Lette *et al.*, 1998). Expression of *CACNG2* appears to be restricted to the brain, with the highest levels being found in the cerebellum, olfactory bulb, thalamus and hippocampus. The stargazer phenotype is associated with the insertion of an *ETn* transposon within intron 2 of *CACNG2*, which interrupts the gene and leads to the termination of transcription.

When coexpressed with other neuronal voltage-gated Ca^{2+} channel subunits (α_{1A}, α_2, β_{1A} and δ) stargazin modifies the properties of the Ca^{2+} current. In particular, it shifts the voltage dependence of inactivation to more negative potentials, which reduces the channel availability at the resting potential and is predicted to decrease Ca^{2+} entry in presynaptic nerve terminals. The decrease in stargazin expression found in the *stargazer* mouse may be expected to give rise to enhanced Ca^{2+} entry in the presynaptic terminals, and contribute to the epileptic phenotype.

Polycystic Kidney Disease: A Disease of Ca²⁺ Channels?

Autosomal dominant polycystic kidney disease (ADPKD) is the most common inherited kidney disorder, occurring in ~1 in 1000 people and accounting for ~10% of end-stage renal disease worldwide. It is characterised by the presence of multiple fluid-filled cysts that occur throughout the tubules of both kidneys. These cysts are already present in the fetus and develop progressively through life, eventually resulting in renal failure in ~45% of affected individuals between 40 and 60 years of age. Hypertension is common and many patients also have cysts in other organs, most notably the liver.

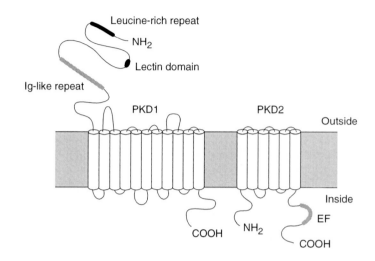

Leucine-rich repeat

NH$_2$

Lectin domain

Ig-like repeat

PKD1

PKD2

Outside

Inside

EF

COOH   NH$_2$

COOH

*FIGURE 9.12 PUTATIVE MEMBRANE TOPOLOGY OF PKD1
AND PKD2*
PKD1 has several features suggestive of a molecule involved in cell–cell
or cell–matrix interactions including leucine-rich repeats, a C-type lectin
domain, and a number of immunoglobulin-like domains. PKD2 contains
an EF-hand domain (EF) that is a putative Ca^{2+}-binding domain.

ADPKD is a genetically heterogeneous disease and has been mapped to
chromosome 16p13.3 (*PKD1*) and chromosome 4q21–23 (*PKD2*); a third gene
is also known to exist. Mutations in PKD1 are the most frequent cause of the
disease, but ~15% of cases are associated with mutations in PKD2. The
latter encodes a 968 amino acid protein that has six putative transmembrane
domains with cytosolic N and C termini (Fig. 9.12) (Mochizuki *et al.*, 1996).
PKD2 shows significant sequence similarity within both its transmembrane
domains and C terminus with the corresponding regions of the α_{1E}-subunit
of the voltage-gated Ca^{2+} channel. Three different nonsense mutations, that
are predicted to result in a truncated protein, have been identified in
ADPKD2-affected individuals. PKD1 possesses 11 putative transmembrane-
spanning domains and a cytosolic C terminus, but the N terminus is extracellular, very large (>2500 amino acids), and contains multiple regions that
have been implicated in cell–cell and cell–matrix interactions (Fig. 9.12)
(Hughes *et al.*, 1995). Homology to PKD2 is confined to the first four transmembrane domains. PKD1 and PKD2 appear to interact within their C-
terminal regions (Tsiokas *et al.*, 1997), suggesting that they are involved in a
common signalling pathway that, when disrupted, leads to renal cyst formation. Whether either protein has any channel activity has yet to be established,
but the similarity to the Ca^{2+} channel α_{1E}-subunit suggests that this might be
a useful avenue of investigation. Interestingly, those *lethargic* mice that survive
to adulthood show evidence of renal cysts.

CHAPTER 10

VOLTAGE-GATED CL⁻ CHANNELS

Chloride channels are important for the control of membrane excitability, transepithelial transport, and the regulation of cell volume and intracellular pH. They are also important in intracellular organelles where they provide an electrical shunt pathway which facilitates acidification of the organelle interior. Many different kinds of chloride channel exist (Franciolini and Petris, 1990), and not all of those that have been characterised electrophysiologically have yet been identified at the molecular level. Three major Cl⁻ channel families, with very different structures, have been cloned to date (Jentsch and Günther, 1997). These are the voltage-gated chloride channels, the cystic fibrosis transmembrane conductance regulator (CFTR) and related channels, and the ligand-gated Cl⁻ channels opened by GABA and glycine. Mutations in the genes encoding members of these Cl⁻ channel families result in cystic fibrosis (CFTR), certain forms of myotonia and inherited kidney stone disease (voltage-gated Cl⁻ channels) and startle disease (the glycine receptor). The ligand-gated Cl⁻ channels are discussed in Chapters 17 and 18, and CFTR is considered in Chapter 12: here, we focus on the voltage-gated Cl⁻ channels.

Structure

The first member of the voltage-gated Cl⁻ channel family was cloned from the electric organ of the ray *Torpedo mamorata* by Jentsch and colleagues in 1990. This channel is designated CLC-0. Subsequently, nine mammalian Cl⁻ channels have been identified (see Jentsch and Günther, 1997). The human genes are named *CLCN1* to *CLCN7*, *CLCNKa* and *CLCNKb* and they encode the Cl⁻ channel proteins CLC-1 to CLC-7, CLC-Ka and CLC-Kb, respectively (Table 10.1). They may be grouped into three main subfamilies, as shown in Fig. 10.1. CLC-1 is expressed primarily in mammalian skeletal muscle where it plays an important role in regulating muscle excitability. CLC-2, CLC-6 and CLC-7 are expressed ubiquitously. CLC-3 and CLC-4 also have a wide tissue distribution, but the former is expressed in brain, heart, lung and kidney, while CLC-4 is found in skeletal muscle, brain and heart. CLC-5, CLC-Ka and CLC-Kb are found predominantly in the kidney and are involved

TABLE 10.1   CHLORIDE CHANNEL PROPERTIES

Gene	Protein	Chromosome (human)	Tissue distribution	Reference
CLCN1	CLC-1	7q35	Skeletal muscle	Koch *et al.* (1992)
CLCN2	CLC-2	3q26-qter	Heart, brain, lung, pancreas, liver	Cid *et al.* (1995)
CLCN3	CLC-3	4	Brain, lung and kidney	Borsani *et al.* (1995)
CLCN4	CLC-4	Xp22.3	Skeletal muscle, brain and heart	Van Slegtenhorst *et al.* (1994)
CLCN5	CLC-5	Xp11.22	Kidney	Fisher *et al.* (1994)
CLCN6	CLC-6	1p36	Ubiquitous	Brandt and Jentsch (1995)
CLCN7	CLC-7	16p13	Ubiquitous	Brandt and Jentsch (1995)
CLCNKa	CLC-Ka	1p36	Kidney	Saito-Ohara *et al.* (1996)
CLCNKb	CLC-Kb	1p36	Kidney	Saito-Ohara *et al.* (1996)
CLC0	CLC-0		*Torpedo* electroplaque	Jentsch *et al.* (1990)

in transepithelial Cl⁻ transport. As CLC-2 is activated by cell swelling, it may play a role in the control of cell volume; it is not, however, the ubiquitous swelling activated Cl⁻ conductance that is found in almost all cells (which has not yet been cloned). The physiological functions of CLC-3, CLC-4, CLC-6 and CLC-7 have not yet been elucidated.

The different voltage-gated Cl⁻ channels have between 650 and 1000 amino acids and molecular masses ranging from ~75 to ~130 kDa. Hydropathy analysis suggests that there are between 8 and 12 transmembrane domains (D1–3 and D5–12), but the precise number is still not resolved (Fig. 10.2). There are two additional domains: D4, which appears to be extracellular, and D13 which lies intracellularly and is of significance because its deletion

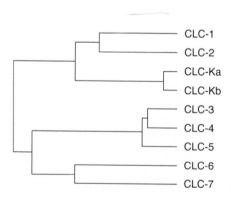

FIGURE 10.1 RELATIONSHIP BETWEEN THE DIFFERENT MAMMALIAN CLC CHANNELS
The length of the lines indicates the degree of similarity between the different proteins. There are three major families: the first comprises CLC-1, CLC-2 and the CLC-K channels; the second, CLC-3 to CLC-5; and the third CLC-6 and CLC-7. After Jentsch and Günther (1997).

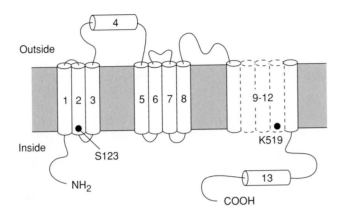

FIGURE 10.2 PREDICTED MEMBRANE TOPOLOGY OF VOLTAGE-GATED Cl⁻ CHANNELS
Putative membrane topology, based on hydropathy analysis. Two regions of intermediate hydrophobicity (D4 and D13) were originally supposed to span the membrane but are now thought not to do so. The region D9 through D12 is a hydrophobic region which is hypothesised to span the membrane but the precise number of transmembrane spanning domains is not established (between 3 and 5 is the favoured number). The positions of residues mentioned in the text are indicated.

results in non-functional channels. Both the N and C termini are intracellular. There is no obvious equivalent of the S4 domain of voltage-gated Na^+ channels. The structure of the archetypal voltage-gated Cl^- channel is therefore very different from that of the voltage-gated Na^+, Ca^{2+} and K^+ channels.

Correlating Structure and Function

The best studied of all the voltage-gated chloride channels is CLC-0. It is strongly anion selective, with a permeability sequence of $Cl^- > Br^- > I^-$, and has a single-channel conductance of ~8 pS in symmetrical 140 mM Cl^-. A curious property of this channel is that it contains two identical, but physically separate, pores, a feature which has led to it being christened a 'double-barrelled' channel. This was first suggested from single-channel recordings which showed that the channel opens in bursts, and that within each burst the channel fluctuates between the closed state and two identical conductance levels in a binomial fashion (Fig. 10.3). Because its peculiar gating behaviour was reminiscent of a double-barrelled shotgun, Chris Miller christened it the 'double-barrelled' channel (Miller, 1982). Further support for the two-pore model came from the study of mutant channels. Mutation of serine 123 in CLC-0, which lies within a region that is highly conserved across all known CLC channels, produced a reduced anion selectivity and a smaller single channel conductance (~1.5 pS) (Ludewig et al., 1996). When concatamers

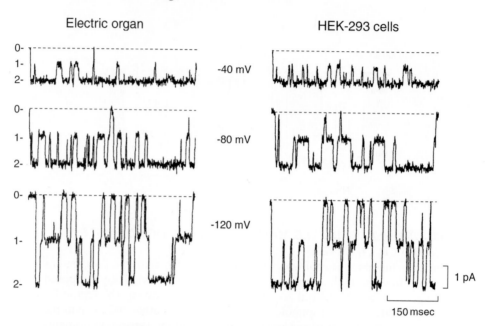

FIGURE 10.3 SINGLE CLC-0 CHANNEL CURRENTS
Single-channel currents recorded after reconstitution in a planar lipid bilayer. Two distinct open conductance levels of equal amplitude are clearly present. Left, native CLC-0 purified from *Torpedo* electric organ. Right, cloned CLC-0 channels expressed in HEK293 cells. The dashed lines show the zero current level. From Middleton *et al.* (1996).

were formed between S123T mutant channels and wild-type channels, by linking the two proteins end to end, both 1.5 pS (mutant) and 8 pS (wild-type) conductances were observed (Fig. 10.4). Furthermore, the 1.5 pS conductance channel had the reduced anion selectivity characteristic of the parent S123T channel. This suggests the channel has two independent conduction pathways. In support of this idea, coinjection of wild-type and mutant monomers produced only 8 pS/8 pS, 1.5 pS/1.5 pS or 8 pS/1.5 pS channels: intermediate conductances were never observed. Similar effects were observed when lysine 519 was mutated (Ludewig *et al.*, 1996; Middleton *et al.*, 1996). These results are explained most simply by assuming each channel is formed by dimerization of two CLC-0 subunits and that it contains two pores which function independently. It is notable that this structure is quite distinct from that of the voltage-gated cation channels discussed in Chapters 5 to 9.

The location of the pore within the CLC-0 protein is still not resolved. However, because mutation of either S123 or K519 alters both the single-channel conductance and halide selectivity, these amino acids are likely to contribute to the permeation pathway. Both residues are located at the cytoplasmic end of a putative transmembrane domain, the second (D2) and eleventh (D12), respectively (Fig. 10.2), and are therefore expected to lie at the inner mouth of the pore. It may not be a simple matter to define the CLC-0

WT   S123T-WT   S123T

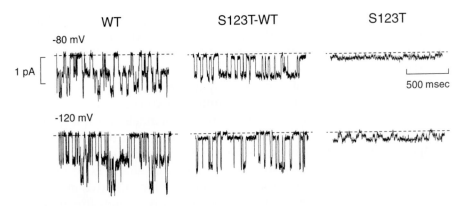

FIGURE 10.4 CLC-0 CHANNELS EXIST AS DIMERS
Single channel currents recorded for a channel formed from a concatamer
of two wild-type CLC-0 subunits (WT) a concatamer of an S123T and a
wild-type subunit (S123T-WT) and the S123T mutant alone (S123T). Note
the equal conductance levels with the WT-WT concatamer (8 pS) and
S123T channel (1.5 pS) and the presence of two conductance levels (8 and
1.5 pS) with the S123T-WT concatamer. From Ludewig *et al.* (1996).

pore, because the homodimeric nature of the channel suggests that many
parts of the protein may be involved, in contrast to the highly localised pore
loops found for voltage-gated cation channels.

The voltage-dependent gating of the CLC-0 channel is also unusual. There
are two independent modes of gating with opposite voltage dependence:
hyperpolarization increases the probability of bursts of channel openings
(slow gating) whereas depolarization increases channel opening within the
burst (fast gating). Slow gating operates simultaneously on both pores of the
double-barrelled channel, whereas fast gating acts independently on a single
pore. The latter is strongly dependent on the concentration and species of
the permeating anion, and increasing extracellular Cl⁻ shifts the voltage
dependence of fast gating to more negative membrane potentials (Pusch *et al.*,
1995a). Mutation of the positively charged lysine at position 519 to glutamate
changes both the gating properties of CLC-0 (kinetics, concentration dependence, halide sensitivity) and also those of the pore (single-channel conductance, selectivity, rectification) (Pusch *et al.*, 1995a). This argues that the voltage dependence of fast gating is conferred by the permeating ion itself acting
as the gating charge. The data are consistent with the idea that binding of a
chloride ion to a site within the inner mouth of the pore is required to open
the channel; although K519 lies at the inner side of the membrane it is not
clear if it contributes to the anion binding site. The fast gating mechanism
of CLC-0 is in marked contrast to the gating of voltage-dependent Na^+ and
K^+ channels (see Chapters 5 and 6).

Far less is known about the relationship between channel structure and
function for the other voltage-gated chloride channels. There is evidence,
however, that like CLC-0, the muscle chloride channel, CLC-1, is a homodimer

(Fahlke *et al.*, 1997b). This was obtained by constructing dimers in which one subunit possessed a mutation (D136G) that produced a marked difference in the voltage sensitivity of the channel. When these heterodimeric channels were expressed in a mammalian cell line, the currents exhibited novel gating properties, which were best explained by assuming a single population of channels with an even number of subunits. Coexpression of homodimeric wild-type channels with homodimeric mutant channels did not lead to the formation of heteromultimeric channels, indicating that the CLC-1 channel is composed of only two subunits. Biochemical analysis of the molecular mass of the native channel also supports the idea that CLC-0 exists as a dimer. In contrast to the dimeric CLC-0 channel, which has two independent pores, the the CLC-1 channel appears to possess only a single functional pore (Fahlke *et al.*, 1998). Like CLC-0, however, gating of CLC-1 does not involve an intrinsic voltage sensor but is instead facilitated by external chloride ions (Ludewig *et al.*, 1997).

It is of interest that CLC-1 and CLC-2 can form heteromeric channels with novel properties (Lorenz *et al.*, 1996), because if such heteromerization also occurs *in vivo* it would increase the functional diversity of Cl⁻ channels.

Cl⁻ CHANNEL DISEASES

Myotonia Congenita and Generalised Myotonia

Mutations in the gene encoding the skeletal muscle Cl⁻ channel *CLCN1* are involved in two forms of myotonia in man: the autosomal dominant disease *myotonia congenita* and the autosomal recessive disease *generalised myotonia*. Myotonia congenita is also known as Thomsen's disease after the Danish physician who first described it, in himself, in 1876. Generalised myotonia was distinguished as a recessive form of myotonia in 1957 by Becker and is consequently sometimes called Becker's type myotonia (or Becker's disease). Both diseases are characterised by muscle stiffness (myotonia), which results from the continued firing of action potentials in the muscles after the cessation of voluntary effort or stimulation. This results in a characteristic repetitive discharge, known as the myotonic run, in the electromyogram. Myotonia congenita is present at birth, but generalised myotonia usually develops gradually during the first decade of life, beginning in the legs and progressing to the arms, neck and facial muscles. Generalised myotonia is three times more common in men than women, suggesting that in females the penetrance is reduced, or the disease has a milder phenotype. The symptoms are also more severe than those of myotonia congenita. In both forms of the disease, the myotonia is accentuated by rest and gradually relieved by exercise. This means that although the patient may be able to walk normally, they may fall over if they suddenly try to run following a period of rest. A gradual 'warm-up' to exercise alleviates some of the symptoms. Patients with significant myotonia often have muscle hypertrophy (overdevelopment) as a consequence of the continuous muscle activity.

Goats show the way

Shirley Bryant was the first to discover the cause of myotonia congenita, by studying a hereditary myotonia in goats. While reading a veterinary journal, he came across an article about a herd of goats in Texas that fell over every time the train went past their field. Apparently, what happened was that the goats were startled by the train, but just like humans afflicted with myotonia, when they tried to run after a period of rest their muscles seized up causing them to fall down. Microelectrode recordings from skeletal muscle fibres isolated from affected goats revealed that they had an extremely high input resistance, which resulted from a marked reduction in their chloride conductance (Bryant, 1969). Inhibition of the chloride conductance in normal muscles by agents such as 9-anthracene carboxylic acid also produced myotonia (Bryant and Morales-Aguilera 1971). This pinpointed a defect in the skeletal muscle chloride channel as a possible cause of myotonia, an idea which received further support when it was shown that muscle biopsies from patients with myotonia congenita also had a reduced Cl⁻ conductance (Lipicky et al., 1971).

Mice mutants identify myotonia as a CLC-1 channel defect

Identification of the mutations which cause myotonia had to await the cloning of the gene encoding the skeletal muscle chloride channel (CLCN1). Shortly after this was achieved, it was shown that disruption of CLCN1 causes an inherited myotonic disorder of mice known as the arrested development of righting response, adr, because affected mice return to the upright position more slowly after being turned on their side (Steinmeyer et al., 1991). In these animals, a transposon of the ETn family has inserted into the gene after the sequence which encodes the ninth putative transmembrane domain. This disrupts the protein and functional channels are not produced. Transposons are wandering pieces of DNA, thought to be originally of viral origin, that may insert randomly into the genome. Murine myotonia can also result from point mutations in CLCN1, which occur independently of ETn insertion (Gronemeier et al., 1994).

Human myotonia results from mutations in CLCN1

The subsequent cloning of the human gene encoding the skeletal muscle chloride channel enabled identification of the mutations that cause myotonia in man. Tight linkage of both dominant and recessive forms of myotonia to the human CLCN1 gene on chromosome 7q35 was found (Koch et al., 1992). Twenty missense mutations, four nonsense mutations, three deletions, one insertion and two splice mutations in the gene have been identified to date (Fig. 10.5; Table 10.2). The original mutation described by Thomsen, which causes domi-

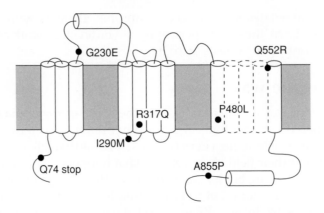

FIGURE 10.5 Cl⁻ CHANNEL MUTATIONS ASSOCIATED WITH MYOTONIA

Predicted topology of CLC-1 with some of the mutations causing myotonia congenita and generalised myotonia indicated. The A855P mutation is found in the goat: all the other mutations occur in man.

nant myotonia, was identified in his descendants as a leucine-for-proline substitution at position 480 (Steinmeyer *et al.*, 1994). At least four other point mutations also cause myotonia congenita (G230E, I290M, R317Q and Q552R). The remaining mutations and the deletions cause generalised (recessive) myotonia: the majority of these patients are heterozygous for the mutation. Mutations have not yet been identified in all affected patients and thus further mutations within the human skeletal muscle Cl⁻ channel gene may be expected.

CLC-1 forms functional chloride channels when expressed in *Xenopus* oocytes. These channels have a very low single channel conductance (~1 pS) so that the large macroscopic conductance found in skeletal muscle must result from a very high channel density in the membrane. Expression of CLC-1 channels bearing myotonia-causing mutations results in either a marked reduction, or the complete loss, of the whole-cell Cl⁻ current.

Analysis of four missense mutations producing dominant myotonia (I290M, R317Q, P480L, Q552R) showed that they cause a marked shift in the voltage-dependence of steady-state activation to more positive potentials (Pusch *et al.*, 1995b). Similar results were found for the A855P mutation which causes myotonia in the goat (Beck *et al.*, 1996). The resulting shift in voltage dependence is sufficient to prevent the channel from contributing to repolarization of the action potential and so predisposes to myotonia. Although these dominant mutations cause a common phenotype, they occur at diverse locations throughout the protein and the fact that the structure of the Cl⁻ channel is far from resolved makes it difficult to postulate precisely how they affect channel gating.

A fifth mutation causing dominant myotonia, G230E, affects the pore properties of CLC-1 (Fahlke *et al.*, 1997a). This mutation markedly enhances the permeability of the channel to cations. For example, the permeability to Na^+ relative to Cl⁻ (P_{Na}/P_{Cl}) is increased at least 7-fold. The relative selectivity to different anions is also altered: thus, the mutant channel has an anion selectivity sequence of $NO_3^- > I^- > Br^- > Cl^-$ compared to that of $Cl^- > Br^- > NO_3^- > I^-$

TABLE 10.2 SOME MUTATIONS IN *CLCN1* ASSOCIATED
WITH MYOTONIA

Mode of inheritance	Mutation	Reference
Recessive	Q74stop	Mailänder *et al.* (1996)
Recessive	R105C	Meyer-Kleine *et al.* (1995)
Recessive	D136G	Heine *et al.* (1994)
Recessive	Y150C	Mailänder *et al.* (1996)
Recessive	V165G	Meyer-Kleine *et al.* (1995)
Recessive	F167L	Meyer-Kleine *et al.* (1995)
Dominant	G200R	Mailänder *et al.* (1996)
Dominant	G230E	George *et al.* (1993)
Recessive	Y261C	Mailänder *et al.* (1996)
Dominant	I290M	Lehman Horn *et al.* (1995)
Recessive	E291K	Meyer-Kleine *et al.* (1995)
Recessive	R300stop	George *et al.* (1994)
Dominant	R317Q	Meyer-Kleine *et al.* (1995)
Recessive	Splice mutation	Lozenz *et al.* (1994)
Recessive	I329T	Meyer-Kleine *et al.* (1995)
Recessive	R338Q	George *et al.* (1994)
Recessive	fs387stop	Meyer-Kleine *et al.* (1995)
Recessive	F413C	Meyer-Kleine *et al.* (1995)
Recessive	A415V	
Recessive	fs429stop	Meyer-Kleine *et al.* (1995)
Recessive	fs433stop	Heine *et al.* (1994)
Recessive	fs503stop	Meyer-Kleine *et al.* (1995)
Dominant	P480L	Steinmeyer *et al.* (1994)
Recessive	G482R	Meyer-Kleine *et al.* (1995)
Recessive	M485V	Meyer-Kleine *et al.* (1995)
Recessive	Splice mutation	Meyer-Kleine *et al.* (1995)
Recessive	R496S	Lorenz *et al.* (1994)
Dominant	Q552R	Lehman-Horn *et al.* (1995)
Dominant recessive	R894stop	Meyer-Kleine *et al.* (1995)

Mutations in *CLCN1* associated with dominant and recessive forms of myotonia. fs, frameshift.

for the wild-type channel. These findings suggest that the glycine residue at position 230 lies within the channel pore and provide a pointer for future studies on the location of the CLC-1 channel pore. The increased cation permeability found with the G230E mutation may arise because the introduction of a nega-tively charged glutamate within the pore favours cation flux and hinders anion flux. Both the greater Na^+ permeability and the smaller currents observed when G230 is mutated are predicted to lead to an enhanced membrane depolarization and are likely to be involved in producing the myotonic phenotype.

The functional basis of the mutations that cause recessive myotonia has not been analysed extensively. Some are nonsense mutations or deletions which result in truncation of the protein and the loss of functional channels (eg., Q74-Stop). Individuals heterozygous for these mutations may be ex-pected to possess half the normal number of Cl⁻ channels, consistent with the idea that a reduction of more than 50% of the Cl⁻ current is required to cause the myotonic phenotype.

One puzzle is how mutations in the same gene give rise to both dominant and recessive forms of myotonia. It has been suggested that this occurs because the CLC-1 channel is formed from more than one subunit, and that a wild-type subunit can combine with a mutant subunit to form a heteromeric channel (Steinmeyer *et al.*, 1994). The mutant subunit then reduces or abolishes the function of channel (the dominant negative effect). Thus the severity of myotonia will depend on the extent to which the wild-type subunits are inactivated. Chloride channel blockers only induce myotonia when they reduce the muscle chloride conductance to below 30% of its normal value. If a single mutant monomer in the dimeric complex is enough to cause complete loss of function, then only one out of four channels will be functional in heterozygous individuals and the chloride conductance will be reduced to 25%. This would produce dominant myotonia. If loss of channel function only occurs when both subunits are defective, this would leave 75% of the channels still functional in heterozygous individuals and result in no overt clinical symptoms. Such mutations would cause recessive myotonia. Thus whether a given mutation is dominant or recessive may depend on how it alters the function of the channel complex. In general, recessive myotonia is clinically more severe than dominant myotonia. It is possible that this is due to the presence of some normal channels in heterozygous patients with dominant myotonia, which results in a residual chloride current.

How does a reduction in the muscle chloride conductance give rise to myotonia?

A single nerve stimulus elicits a single action potential in a normal muscle fibre, but may produce a repetitive train of action potentials in myotonic muscle fibres (Fig. 10.6). How does this happen? In normal skeletal muscle, the chloride conductance accounts for 70–80% of the resting membrane conductance. The absence of this chloride conductance, either due to a lack of functional Cl⁻ channels or as a result of pharmacological inhibition, will produce an increase in the input resistance of the muscle fibre. Consequently, a smaller Na⁺ current will be sufficient to trigger an action potential and muscle excitability will therefore be enhanced. The elevated input resistance also causes a reduced rate of action potential repolarization. A further, and most important, role of the Cl⁻ conductance in muscle is to counteract the depolarizing effect of K⁺ accumulation in the T-tubules that accompanies excitation. T-tubules are invaginations of the surface muscle membrane whose role is to conduct the action potential deep into the interior of the fibre. A single action potential increases the K⁺ concentration in the T-tubules by ~0.3 mM. Normally, this increase in K⁺ has little effect on the membrane potential because the high tubular Cl⁻ conductance clamps the membrane at the chloride equilibrium potential. Myotonic muscle, however, has a much lower Cl⁻ conductance. Consequently, the K⁺ equilibrium potential dominates the membrane potential and a small rise in tubular K⁺ produces a significant depolarisation following the action potential. If several action potentials occur

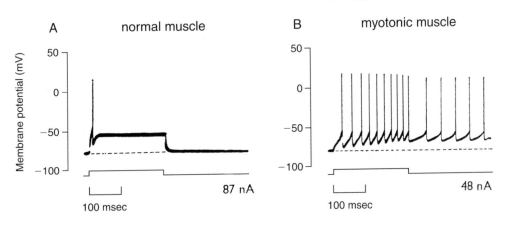

FIGURE 10.6 *MYOTONIA RESULTS FROM MULTIPLE ACTION POTENTIALS*
A depolarizing current pulse evokes a single action potential in a normal goat muscle fibre (A), whereas a weaker stimulus elicits a train of action potentials in the muscle fibre of a myotonic goat (B). Action potentials were recorded from intercostal muscle fibres in vitro at 37°C. From Adrian and Bryant (1974).

in rapid succession, summation of these after-depolarizations may be sufficient to trigger spontaneous action potentials and so produce myotonia.

A characteristic symptom of myotonia congenita is that repetitive muscle activity (causing stiffness) is worse after a period of rest and is alleviated by exercise. It is possible that this amelioration results from the enhanced activity of the muscle Na/K-ATPase induced by exercise, which facilitates the clearance of K^+ ions from the T-tubules.

Nephrolithiasis, a Disease of CLC-5

Kidney stones (nephrolithiasis) affect 12% of men and 5% of women and account for up to 1% of all hospital admissions. They often cause excruciating pain and infection of the kidney tract, which may lead to loss of kidney function. The stones are formed from an excess of salts (usually calcium) which precipitate out of the supersaturated urine. In 45% of patients this hypercalciuria is familial and is inherited on the X chromosome. Three forms of hypercalciuria (Dent's disease, X-linked recessive nephrolithiasis [XRN] and X-linked recessive hypophosphataemic rickets [XLRH]) were mapped to chromosome Xp11.22 (Fisher *et al.*, 1995). This led to the subsequent identification of the renal chloride channel gene *CLCN5* and the demonstration that mutations in this channel are responsible for all three syndromes (Lloyd *et al.*, 1996). Subsequently, an additional renal proximal tubule disorder in Japanese children was also shown to result from mutations in *CLCN5* (Lloyd *et al.*, 1997b). Around twenty disease-causing mutations have been identified to date (Fig. 10.7). Five are nonsense mutations that produce truncated pro-

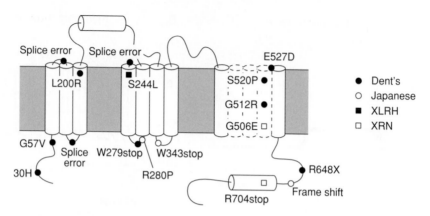

FIGURE 10.7 *Cl⁻ CHANNEL MUTATIONS ASSOCIATED WITH NEPHROLITHIASIS*
Predicted topology of CLC-5 with the mutations causing Dent's disease, X-linked nephrolithiasis (XRN), X-linked recessive hypophosphatemic rickets (XLRN) and the tubular defect in Japanese children (○) indicated.

teins, which are expected to be nonfunctional. Indeed, all disease-causing mutations in *CLCN5* that have been examined to date resulted in a reduction of the Cl⁻ current when CLC-5 was heterologously expressed in *Xenopus* oocytes (Lloyd *et al.*, 1997a).

There are a number of phenotypic differences between Dent's disease, XRN and XLRH and the tubular defect of Japanese children. For instance, Dent's disease is characterised by low-molecular weight proteinuria, nephrocalcinosis, rickets and, eventually, renal failure; while rickets is absent in XRN; nephrocalcinosis and renal failure are more marked in XLRH; and the renal proximal tubulopathy of Japanese is not associated with either rickets or renal failure. There is no obvious correlation between these different phenotypes and the different mutations in CLC-5. For example, mild mutations which reduce, rather than abolish, Cl⁻ currents are found in both Dent's disease (S520P) and in XLRH (S244L).

The fact that hypercalciuria resulted from mutations in a chloride channel came as a surprise, because it was widely expected that it was caused by a defect in calcium handling by the kidney. This discovery tells us that CLC-5 is important in the handling of calcium by the kidney, but leaves us still with the question of how it is involved and precisely how the loss of channel function leads to kidney stone formation. The rickets that accompanies nephrolithiasis is the consequence of Ca²⁺ mobilization from bone to replace that lost in the urine.

Bartter's Syndrome Type III: A Disease of CLC-Kb

Bartter's syndrome is an autosomal recessive disorder characterised by severe salt-wasting, that may lead to a marked depletion of the extracellular volume, low blood pressure, hypokalaemia, hypercalciuria and alkalosis. The clinical features are described more fully in Chapter 8. Mutations in three different genes are known to produce Bartter's syndrome: the NaK2Cl cotransporter (*SCL12A1*, Bartter's syndrome type I), the inwardly-rectifying K^+ channel Kir1.1 (*KCNJ1*; Bartter's syndrome type II) and the voltage-gated Cl⁻ channel CLC-Kb (*CLCNKB*; Bartter's syndrome type III). In contrast to patients with Bartter's syndrome types I and II, patients with mutations in *CLCNKB* do not suffer from nephrocalcinosis, despite elevation of the urinary calcium concentration.

Mutations in *CLCNKB* associated with Bartter's syndrome type III have been identified in seventeen kindreds of diverse ethnic background (Simon *et al.*, 1997b). They include five missense mutations, one nonsense mutation and one splice-site mutation (Fig. 10.8). In the other ten kindreds either the entire *CLCNKB* gene, or a substantial portion of it, was deleted. Some of these deletions appear to have arisen by unequal crossing over between *CLCNKB* and the related Cl⁻ channel gene *CLCNKA*. These genes lie side by side on chromosome 1, in the same transcriptional orientation and are separated by 11 kb. They have an identical genomic organization (both contain 19 introns) and share 94% DNA sequence identity within the exons, suggesting that they were produced by duplication of a single ancestral gene. Crossing-over between the exons of *CLCNKA* and *CLCNKB* in some kindreds has resulted in chimeric genes consisting of some parts of each gene, with the *CLCNKA* portion 5′ to the *CLCNKB* sequence. All the deletions and mutations identified in Bartter's syndrome type III kindreds are presumed to result in a decreased Cl⁻ conductance.

Figure 10.9 explains why a reduction in CLC-Kb results in Bartter's syndrome. The renal tubules reabsorb more than 99% of the fluid filtered by the glomeruli, which amounts to around 180 litres each day. Most salt reabsorp-

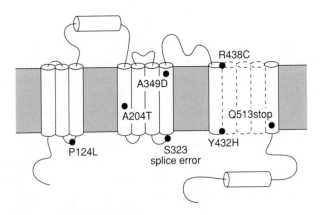

FIGURE 10.8 *Cl⁻ CHANNEL MUTATIONS ASSOCIATED WITH BARTTER'S SYNDROME TYPE III*
Predicted topology of CLC-Kb with the mutations causing Bartter's syndrome type III indicated.

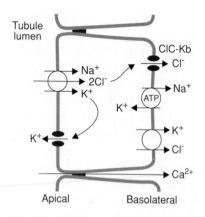

FIGURE 10.9 SALT TRANSPORT IN THE THICK ASCENDING LOOP
OF HENLE

The apical membrane of the cells of the thick ascending loop of Henle
contains a NaK2Cl cotransporter which mediates the uptake of one Na⁺,
one K⁺ and two Cl⁻ ions. Energy for this process is provided by the Na⁺
gradient generated by the activity of the Na/K-ATPase in the basolateral
membrane. The Cl⁻ ions translocated by the NaK2Cl cotransporter leave
the cell via CLC-Kb channels in the basolateral membrane and in their
absence the activity of the NaK2Cl cotransporter will be inhibited, leading
to loss of Na⁺, Cl⁻ and K⁺ in the urine.

tion occurs constitutively within the proximal tubule, but significant reabsorption also takes place across the cells of the thick ascending limb of the loop
of Henle. The latter is essential for the formation of a concentrated urine.
Na^+, K^+ and Cl^- enter the cell via the NaK2Cl cotransporter in the apical
membrane as a consequence of the electrochemical gradient for Na^+ that is
produced by the activity of the basolateral Na/K-ATPase. K^+ ions are recycled
across the apical membrane via the inwardly rectifying K^+ channel, Kir1.1
(see Chapter 8), while Cl^- exits to the plasma through Cl^- channels in the
basolateral membrane. The facts that CLC-Kb is expressed in the loop of
Henle and that mutations in this protein produce Bartter's syndrome suggest
that CLC-Kb comprises the basolateral Cl^- channel. Loss of this channel may
be expected to lead to an intracellular accumulation of Cl^-, which will reduce
the activity of the NaK2Cl cotransporter and lead to the loss of Na^+, Cl^- and
K^+ in the urine. The blood concentrations of these ions will therefore be
lowered. There will also be an associated diuresis, which accounts for the
hypovolaemia and hypotension characteristic of Bartter's syndrome. Patients
with Bartter's syndrome type III exhibit marked variability in the severity of
the clinical features. It is not yet known whether this is linked to variation
in the level of Cl^- conductance produced by the different CLCNKB mutations.

CHAPTER 11

CYCLIC NUCLEOTIDE-GATED CHANNELS

Ionic currents sensitive to cyclic nucleotides have been described in a number of tissues, including photoreceptors, olfactory sensory neurones, neurones, cardiac cells and the kidney (Kaupp, 1995; Zimmerman, 1995). They share the common property that their activation is mediated by the direct binding of cyclic nucleotides. Since many of these channels have not yet been identified at the molecular level, it is not known whether they all share a common structural motif. Two families of channels regulated by cyclic nucleotides have, however, been cloned. These are the cyclic nucleotide-gated channels (CNG channels) and the hyperpolarization-activated cyclic nucleotide-gated (HCN) channels. Both belong to the superfamily of voltage-gated channels. In addition to the CNG and HCN channels, two K^+ channels that are activated by cyclic AMP have been cloned from the plant *Arapidopsis thalia*. These channels, called AKT1 and KAT1, resemble the K_V channels in their molecular structure. In this chapter, we focus on the structure and function of CNG and HCN channels and their role in health and disease.

CYCLIC NUCLEOTIDE-GATED CHANNELS

The importance of the CNG channels was first recognised in vertebrate photo-receptors (rods and cones) and in the olfactory epithelium, where they play important roles in sensory transduction. Subsequently, they have been found in a variety of other tissues including spermatozoa, neurones, kidney, liver, lymphocytes and heart. Mutations in the rod CNG channel give rise to the degenerative disease of rods known as retinitis pigmentosa.

Structure

Despite the fact that their gating is only slightly voltage dependent, CNG channels are members of the superfamily of voltage-gated ion channels. The native channel is believed to be a tetramer composed of two homologous subunits, α and β (Kaupp *et al.*, 1989; Chen *et al.*, 1993; Kaupp, 1995; Liu *et al.*, 1996b). Three main types of α-subunit and one type of β-subunit have been identified in mammals to date, with molecular masses of between 63 and 240 kDa (Fig. 11.1; Table 11.1). Related channels have been cloned from *Drosophila* and the worm *Caenorhabditis elegans*.

The α-subunit has six putative transmembrane (TM) domains and a hydrophobic loop (H5) that dips down into the membrane between the fifth and sixth TMs to form part of the lining of the pore (Fig. 11.2). The fourth TM shows significant homology to the S4 region of voltage-gated channels, although the gating of the channel is only slightly voltage dependent. Both N and C termini are located in the cytoplasm. The latter contains a domain of 80–100 amino acids, which is homologous to the cGMP-binding domains of cGMP kinases and is involved in the binding of cyclic nucleotides. In the olfactory CNG channel there is also a calmodulin-binding site in the N terminus.

The β-subunit has a similar topology to that of the α-subunit, but is much larger (240 kDa as compared with 63 kDa; Körschen *et al.*, 1995). This is because although both subunits are highly homologous in the C terminal and transmembrane regions, the amino-terminal region of the β-subunit contains an additional large domain which is nearly identical to a glutamic acid-rich protein (GARP). The function of the GARP region is completely unknown.

Both the α- and β-subunits are able to form functional homomeric ion channels, when expressed in heterologous systems (Kaupp *et al.*, 1989; Broillet and Firestein, 1997). Homomeric α-channels probably do not occur *in vivo*, since their properties differ from those of native CNG channels. Although it was initially thought that β-subunits do not form functional channels when expressed independently, this turns out not to be the case, for homomeric β-channel activity can be observed in response to stimulation with nitric oxide (Broillet and Firestein, 1997). Homomeric β-channels are expressed in

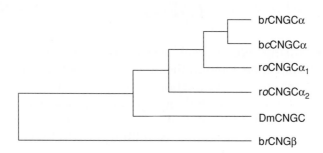

FIGURE 11.1 RELATIONSHIP BETWEEN THE DIFFERENT MEMBERS OF THE CNG FAMILY
The horizontal branch lengths are proportional to the degree of similarity in the amino acid sequences. br, Bovine rod; bc, bovine cone; ro, rat olfactory; Dm, *Drosophila* muscle. After Kaupp (1995).

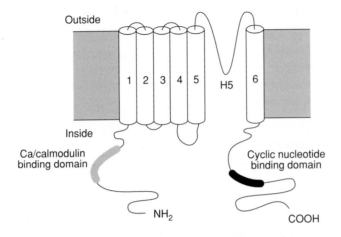

FIGURE 11.2 *PREDICTED TOPOLOGY OF THE α-SUBUNIT*
Postulated membrane topology of the α-subunit of the CNG channel
deduced from hydropathy analysis and mutagenesis studies. The positions
of the Ca^{2+}/calmodulin and cyclic nucleotide-binding sites are indicated.
The pore is formed by the loop between transmembrane domains 5 and 6.

neurons of the vomeronasal organ. Most native CNG channels, however,
appear to be heteromers of α- and β-subunits, because coexpression of both
types of subunits produces channels with properties characteristic of the
native CNG channel. Like the voltage-gated K^+ channels, which they structurally resemble, CNG channels are tetramers (Liu *et al.*, 1996b). Interestingly,
the position (order) of the subunit in the tetramer affects the single-
channel properties.

   The homomeric β-subunit channel exhibits a number of differences from
its α-homomeric counterpart (Broillet and Firestein, 1997). In particular, it is
not activated by cyclic nucleotides—despite the fact that it contains a CNG
motif—but is gated instead by nitric oxide. Also of importance is the fact
that the β-homomeric channel is not blocked by Ca^{2+} and thus exhibits a
much higher Ca^{2+} permeability than α-homomeric channels. When coexpressed with the rod CNG α-subunit, the β-subunit confers sensitivity to the
drug L-*cis*-diltiazem, enhances the sensitivity to cyclic nucleotides and alters
the kinetics of the single-channel currents (Chen *et al.*, 1993).

Chromosomal Location and Tissue Distribution

The human gene for the rod α-subunit (*CNGA1*) is located on chromosome
4 close to the centromere (Table 11.1). It encodes a 686 amino acid protein
with a molecular mass of ~63 kDa. *CNGA3* (cone) maps to chromosome
2q11, but the location of *CNGA2* (olfactory) has not been determined. The
gene for the β-subunit is located on chromosome 16q13 (Ardell *et al.*, 1996)
and is alternatively spliced to produce two proteins of 623 and 909 amino

TABLE 11.1   HUMAN CYCLIC NUCLEOTIDE CHANNEL GENES

Gene	Protein	M_r	Chromosome location (man)	Major sites of expression	Reference
CNGA1	αCNG-1	63 kDa	4p12-cen	Rods, bipolar cells, ganglion cells, kidney	Griffin *et al.* (1993)
CNGA2	αCNG-2			Olfactory neurones, aorta	
CNGA3	αCNG-3		2q11	Cone, spermatozoa, kidney, heart	Wissinger *et al.* (1998)
CNGB1	βCNG-1	140 kDa	16q13	Photoreceptors	Ardell *et al.* (1996)

acids (Chen *et al.*, 1993). Table 11.1 also gives the tissue distribution of the four different CNG α-subunits.

Correlating Structure and Function

Ion permeation and selectivity

Native cyclic nucleotide-gated channels are permeable to monovalent cations, such as Na^+ and K^+, but do not discriminate much between them. Ca^{2+} is also permeable but at the same time acts as a voltage-dependent blocker of monovalent cation permeability (as does Mg^{2+}). This reduces the single-channel conductance from between 20 and 45 pS in the absence of divalent cations to less than 1 pS under physiological conditions, which explains why single-channel currents cannot be resolved. The low single-channel conductance is of physiological significance because the gating of a large number of low conductance channels is associated with less current noise than that of a few high conductance channels. This facilitates the reliable detection of small changes in current and thus enhances sensitivity.

The sequence of the pore (H5) region of both the α- and β-subunits of the CNG channel shares significant homology with that of voltage-gated K^+ channels. The difference in ion selectivity resides in two amino acid residues, tyrosine and glycine, which are present in K_V channels but absent in CNG channels. When these two residues are deleted, the ionic selectivity of the K_V channel resembles that of the CNG channel (Heginbotham *et al.*, 1992; see also Chapter 6). Subsequent studies have confirmed that the H5 region determines the conductance and ion selectivity of the CNG channel and suggest that the pore has a cross-sectional area of 3.8 × 5.0 Å (Picco and Menini, 1993).

Gating

The physiological regulator of the rod channel is cyclic GMP (cGMP) whereas that of the olfactory channel is cyclic AMP (cAMP). Although each type of CNG channel is actually capable of activation by both cAMP and cGMP, the rod channel is far less sensitive to cAMP than cGMP whereas the olfactory channel is equally sensitive to both ligands (Fig. 11.3). To determine the molecular basis of this difference, Goulding et al. (1994) made chimeras between the rod and olfactory α-subunits. As expected, they found that the putative nucleotide-binding domain in the C terminus conferred the appropriate ligand sensitivity on the channel. They narrowed the critical region down to 24 residues lying within the third α-helix of this domain. When this region was transferred from the rod channel to the olfactory channel, the olfactory channel exhibited the ligand-sensitivity characteristic of the rod channel. The reverse chimera produced the opposite effect (Fig. 11.3). Subsequent studies have have shown that residues in the N terminus are also involved in channel activation: these residues do not participate in nucleotide binding but instead influence the conformational change that transduces agonist binding into opening of the pore, by allosterically promoting channel opening (Gordon and Zagotta, 1995; Varnum and Zagotta, 1997).

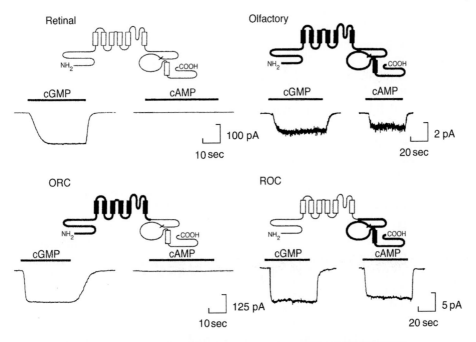

FIGURE 11.3 CHIMERIC MOLECULES IDENTIFY THE CYCLIC NUCLEOTIDE-BINDING SITE
Retinal CNG channels are far less sensitive to cAMP than olfactory channels. To identify the basis of this difference, chimeric molecules were constructed by swapping the C terminal domains of the α-subunits. ORC indicates that the C terminus of the olfactory α-subunit has been replaced with that of the rod, while the rod α-subunit with an olfactory C terminus is denoted ROC. The response of the chimeric channel to cAMP was determined by the origin of its C terminus, arguing that the cyclic nucleotide-binding site lies within this region. From Goulding et al. (1994).

Because the tetrameric CNG channel carries four cGMP binding sites, a key question is how many sites must be occupied before the channel opens. Ruiz and Karpen (1997) studied this problem by recording single-channel currents through α-homomeric channels heterologously expressed in *Xenopus* oocytes. They used a photoaffinity analgue of cGMP, that covalently binds to the binding site, to lock the channel into a state in which exactly one, two, three or four cGMP molecules were bound. Channels in which all four binding sites were occupied stayed open almost all of the time whereas doubly and triply occupied channels opened ~1% and 30% of the time, respectively, and channels in which only a single ligand was bound opened very rarely indeed. These striking findings indicate that maximal channel activation requires all four cGMP-binding sites to be occupied.

The affinity of both the rod and the olfactory CNG channels for their respective ligands can be modulated in a variety of ways. For example, the sensitivity of both types of channel to their respective ligands is decreased by Ca^{2+}/calmodulin. A calmodulin-binding site is present in the amino-terminal region of the olfactory CNG channel α-subunit (Fig. 11.2) but is absent in the α-subunit of the rod channel, where calmodulin sensitivity is conferred by the β-subunit. Binding of Ca^{2+}/calmodulin to the N-terminus of the olfactory α-subunit reduces cGMP sensitivity by disrupting the protein-protein interaction between the N and C terminal domains that promotes channel opening (Varnum and Zagotta, 1997). Modulation of channel activity by Ca^{2+}/calmodulin may be of physiological importance, since intracellular Ca^{2+} will rise following channel activation. This may provide a feedback mechanism that helps to inactivate the channel and restore the resting state.

Ca^{2+} also interacts directly with CNG channels at both external and internal sites (Kaupp, 1995). Extracellular Ca^{2+} blocks homomeric α-channels by binding with high affinity to a ring of four glutamate residues (E363) which lie within the pore (Root and MacKinnon, 1993). This causes a voltage-dependent block of monovalent currents. In this regard, the CNG channel resembles the voltage-gated Ca^{2+} channel, where a ring of glutamate residues within the pore plays a similar role (Chapter 9). In contrast, homomeric β-channels are not blocked by Ca^{2+} (Broillet and Firestein, 1997). This is probably because the β-subunit possesses an aspartate, rather than a glutamate, residue at the critical position. Although the negative charge is conserved, the side chain of the aspartate residue is shorter and extends less far into the pore, and may therefore cause less hindrance to ion flow. While heteromeric $\alpha\beta$ channels (and native channels) are inhibited by external Ca^{2+}, their sensitivity is less than that of α-homomeric channels, presumably because of the presence of the β-subunit (Körschen et al., 1995).

The sensitivity of CNG channels to both the direct and indirect effects of Ca^{2+} varies widely. In olfactory channels, for example, Ca^{2+} is highly permeant and under physiological conditions much of the current is likely to be carried by Ca^{2+} ions. The modulation of channel activity by Ca^{2+}/calmodulin is also more pronounced in olfactory CNG channels, where a 20-fold reduction in cyclic nucleotide sensitivity occurs, compared to a two-fold decrease in rod CNG channels.

Physiological Roles of CNG Channels

The cGMP-gated ion channel plays a central role in visual transduction in rods and cones (Fig. 11.4). In the dark, cGMP levels are high and hold the channel in the open state. This results in a resting influx of Ca^{2+} that stimulates transmitter release. Light activates a G-protein-mediated signalling cascade which leads to activation of a cGMP-dependent phosphodiesterase, hydrolysis of cGMP and a fall in cytosolic cGMP levels. As a consequence, the cGMP-gated channel closes, hyperpolarizing the cell and turning off transmitter release. This signal transduction cascade produces an enormous amplification of the original signal and enables the rod to detect a single photon. The system is unique in that signal transduction leads to inhibition of transmitter release rather than to its activation.

In contrast to the rod CNG channel, that of the olfactory neurone is closed at rest (Fig. 11.5). Activation of odorant receptors in the cilia of the olfactory epithelial cells leads to elevation of cAMP, and thus to opening of the olfactory cAMP-gated ion channel (Nakamura and Gold, 1987). The ensuing Ca^{2+} and Na^+ influx depolarises the cell and stimulates electrical activity. The impor-

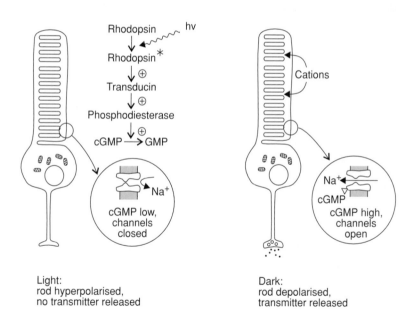

Light:
rod hyperpolarised,
no transmitter released

Dark:
rod depolarised,
transmitter released

FIGURE 11.4 PHYSIOLOGICAL ROLE OF THE ROD CNG CHANNEL

In the dark, CNG channels are open, creating an inward current that depolarises the rod and stimulates transmitter release. Light activates rhodopsin (rhodopsin*) which stimulates the GTP-binding protein transducin. Transducin activates a cGMP phosphodiesterase which lowers cGMP, leading to closing of GNG channels, hyperpolarization and inhibition of transmitter release.

A

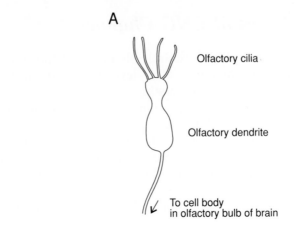

B

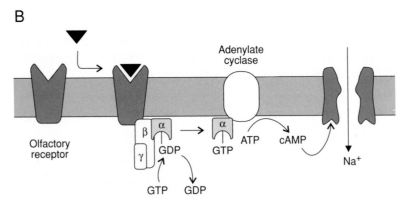

FIGURE 11.5 PHYSIOLOGICAL ROLE OF THE OLFACTORY CNG
CHANNEL
(A) The cell bodies of olfactory neurones are located in the brain but they
send processes to the olfactory epithelium in the nose. These terminate
in olfactory cilia, ~2 μm thick, which bear odorant receptors. (B) Binding
of odorants to specialised receptors causes activation of a GTP-binding
protein. This leads to exchange of bound GDP for GTP, which results in
dissociation of the α-subunit and its subsequent interaction with adenylate
cyclase. Activation of adenylate cyclase promotes the conversion of ATP
to cyclic AMP. Cyclic AMP opens the CNG channels causing an inward
current that depolarises the cell and stimulates electrical activity.

tance of the CNG channel in olfactory transduction is demonstrated by the
fact that mice in which the gene encoding the α-subunit of this channel has
been 'knocked out' do not respond to odorants (Brunet et al., 1996). They
also die within 1–2 days of birth because they fail to suckle: presumably,
suckling is stimulated in normal animals by 'nipple odours'.

CNG CHANNEL MUTATIONS ASSOCIATED WITH DISEASE

Retinitis Pigmentosa

Retinitis pigmentosa (RP) is a term used to describe a genetically heterogenous group of inherited diseases in which patients suffer from a progressive degeneration of both rod and cone photoreceptors. It is relatively common, affecting approximately 1 in 3000 people. Typically, patients experience night blindness by the age of 20. This is followed by a progressive loss, first of the peripheral, and subsequently of the central, visual field, which leads to blindness by middle age.

More than 20 different RP loci have been mapped of which 6 have been identified to date. Four of these encode members of the rod photoreceptor visual transduction cascade, one of which is the α-subunit of the rod cGMP-gated cation channel. About 1% of RP patients carry mutations in the cGMP α-subunit, so it is not a common cause of the disease. Eight mutations have been identified (Fig. 11.6; Dryja et al., 1995). Four of these are nonsense mutations and one is a deletion encompassing most, if not all, of the transcriptional unit: none of these mutations would be expected to encode a functional channel. The remaining mutations are two missense mutations (S316F; S344R) and a frameshift mutation (R654[1-bp del]) which results in truncation of the last 32 amino acids in the C terminus. Two mutations (S316F, R654[1-bp del]) were expressed in vitro and found to encode proteins that were predominantly

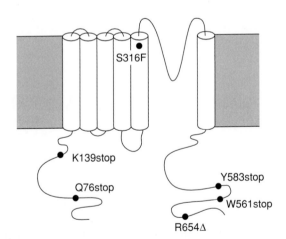

FIGURE 11.6 MUTATIONS IN THE CNG CHANNEL CAUSE RETINITIS PIGMENTOSA
Postulated membrane topology of the α-subunit of the CNG channel with the mutations associated with retinitis pigmentosa indicated.

retained inside the cell instead of being targeted to the plasma membrane. However, those channels which reached the plasma membrane had essentially normal properties. Why the lack of the cGMP-gated cation channel should cause photoreceptor degeneration is not known. The fact that synapses and neurones which are not used normally degenerate, however, suggests that cell death may be a secondary consequence of the failure of photoreceptor transmittor release that is expected to result from the reduction in CNG channel activity.

Mutations which cause RP have also been described in the rhodopsin gene and in the α- and β-subunits of the rod phosphodiesterase, all of which encode proteins of the rod phototransduction cascade. Their mutation may also be expected to impair the function of the rod cGMP-gated cation channel indirectly, suggesting that this constitutes a common path for photoreceptor degeneration.

Caenorhabditis Elegans Mutants

The genes encoding the C. *elegans* equivalents of the α- and β-subunits of the CNG channel (*tax-4* and *tax-2*, respectively) have been cloned by analysis of mutant animals with chemosensory deficits (Coburn and Bargmann, 1996; Komatsu *et al.*, 1996). Unexpectedly, these studies showed that the CNG channels are also expressed in thermosensory neurones and are mutated in animals which fail to respond to changes in temperature. This implies that CNG channels are involved in thermosensation in C. *elegans*. More surprising still was the finding that sensory neurones of *tax-2* mutants showed growth defects: their axons had extra processes and invaded regions from which they were normally excluded. This suggests that the CNG channel may be involved in sensory axon guidance in C. *elegans*. It will be interesting to see if this is also the case in man.

HYPERPOLARIZATION-ACTIVATED CYCLIC-NUCLEOTIDE-GATED CHANNELS

A cation current that is slowly activated by membrane hyperpolarization is found in a variety of cell types, including neurones, cardiac myocytes and photoreceptors. This current was first identified in the cardiac pacemaker cells of the sino-atrial node (SA node cells) where its curious properties led to it being christened I_f- or the 'funny' current (Brown *et al.*, 1979). It was subsequently found in neurones, where it was named I_h (Attwell and Wilson, 1980). Although it belongs to the superfamily of voltage-gated ion channels, the I_f/I_h channel is unusual in that it is open at negative membrane potentials and closes on depolarization. This property underlies the ability of I_f/I_h to support pacemaker activity in the heart and spontaneously spiking neurones.

At potentials around -30 mV, for example, I_f channels are closed, but on membrane hyperpolarization they open, eliciting a slowly-activating inward current that contributes to the slow pacemaker depolarization of neurones and SA node cells. It is important to recognise, however, that I_f/I_h is not the only current that participates in the pacemaker depolarization–at least four different currents are known to be involved in SA node cells.

As is the case for the CNG channels discussed earlier, agents that elevate intracellular cyclic AMP enhance I_f (Brown *et al.*, 1979). This effect is mimicked by direct application of cAMP to the intracellular membrane surface and does not involve protein phosphorylation (DiFrancesco and Tortora, 1991). I_f also resembles the CNG channels in having a very small single-channel conductance (<1 pS; DiFrancesco, 1986) and in being permeable to both K^+ and Na^+ ($P_{Na}/P_K \sim 0.25$), but differs in having no significant permeability to anions or divalent cations. The hyperpolarization-activated current of the heart is insensitive to extracellular application of TEA^+ (20 mM) or Ba^{2+} (1 mM), but is highly sensitive to Cs^+. Similar properties are found for I_h in neurones.

Three different groups simultaneously cloned the genes that encode I_f/I_h and related channels (Gauss *et al.*, 1998; Ludwig *et al.*, 1998; Santoro *et al.*, 1998). These investigators each used a different nomenclature for the ion channels that they isolated. For clarity, in this chapter, I shall refer to the mammalian *h*yperpolarization-activated *c*yclic-*n*ucleotide gated channels as HCN channels. HCN1 corresponds to HAC1 (Ludwig *et al.*, 1998) and BCNG-2 (Santoro *et al.*, 1998), HCN2 to corresponds to HAC2 (Ludwig *et al.*, 1998) and BCNG-1 (Santoro *et al.*, 1998) and HCN3 corresponds to HAC3 (Ludwig *et al.*, 1998). A related channel, named SPIH, was found in sea urchin sperm (Gauss *et al.*, 1998).

The different HCN channels show $\sim 60\%$ sequence identity with each other but only about $\sim 30\%$ sequence identity to the CNG channels and the *eag* family of K^+ channels (Chapter 6). They therefore constitute a separate family of voltage-gated ion channels. Sequence analysis suggests that the HCN channels possess six transmembrane-spanning segments, including an S4 domain and a pore loop, with cytosolic N and C termini. A cyclic nucleotide-binding domain is present in the C terminus of the protein. The pore loop contains the GYG motif that is characteristic of K^+ selective channels (Chapter 6), but the negatively charged aspartate residue that follows this motif in K^+ channels is replaced in HCN channels by a positively charged or neutral residue (arginine in HCN1; alanine in HCN2). In addition, a cluster of conserved threonine residues N-terminal to the GYG motif are lacking in HCN channels. Presumably these differences in the pore sequence account for the enhanced Na^+ permeability of the HCN channel.

HCN1 is expressed in the heart and throughout the brain, while HCN2 and HCN3 have been detected in brain but not in heart. When heterologously expressed in a mammalian cell line, the properties of HCN1 closely matched those of I_f and I_h (Ludwig *et al.*, 1998). We must now await analysis of the relationship between channel structure and function to ascertain why the

HCN channel is activated by hyperpolarization, rather than depolarization. To date, no disease has been associated with mutations in the genes that encode human HCN channels. It would not be surprising, however, if such mutations were to be found in the future, given the role of HCN channels in regulating neuronal and cardiac pacemaking activity.

CHAPTER 12

CYSTIC FIBROSIS TRANSMEMBRANE CONDUCTANCE REGULATOR

The cystic fibrosis transmembrane conductance regulator (CFTR) is a voltage-independent Cl^- channel found in the epithelial cells of many tissues, including those of the intestine, lung, reproductive tracts, pancreatic ducts and sweat glands. It plays a major role in regulating Cl^- fluxes in all these tissues, as emphasized by the fact that mutations in CFTR give rise to cystic fibrosis. Cystic fibrosis (CF) is one of the most common serious human diseases caused by mutations in a single gene, and affects 1 in 2000–2500 people in northern Europe and the United States. Remarkably, around 5% of people are heterozygotes and carry one mutant allele. Fortunately, however, the disease is recessive and heterozygotes are asymptomatic.

CF is primarily a disease of the lung and exocrine glands and is characterised by thick mucous secretions that plug the smaller airways and secretory ducts allowing infection, inflammation and progressive destruction of the tissue. It is also associated with a high salt concentration in the sweat. The latter symptom is documented in many maxims from north European folklore, such as that which states: *Woe to that child which when kissed on the forehead tastes salty. He is bewitched and soon must die.* Despite these early references, CF was not recognised as a distinct disease until 1938, when Dorothy Andersen published the first comprehensive description of its symptoms. Several years later, during a heat wave in New York, the paediatrician Paul di Sant' Agnese noticed that many children admitted to hospital with heat prostration also suffered from cystic fibrosis. Realizing the significance of this observation, he analysed their sweat and found that it contained an abnormally high level of NaCl. His observation forms the basis of the 'sweat' test, which is still used in the diagnosis of CF (the disease is confirmed if the sweat Na^+, or Cl^-, concentration is >60 mM). It took another 30 years, however, before it was shown that the high salt concentration in the sweat of CF patients results from defective Cl^- absorption (Quinton, 1983). It was then recognised that many other symptoms of CF were also a consequence of impaired Cl^- uptake.

Structure of CFTR

The CF gene was mapped to chromosome 7 (7q31) by linkage analysis in affected individuals, and isolated 4 years later using positional cloning methods by a large group of collaborators led by Francis Collins, John Riordan and Lap-Chee Tsui (Riordan *et al.*, 1989). It contains 27 exons and encodes a 1480 amino acid protein with a predicted molecular mass of 168 kDa. It is expressed in epithelial cells, principally those of the pancreas, sweat glands, salivary glands, lung, intestine and reproductive tract. It is also found in the renal tubules. Immunocytochemistry has shown that the protein is localised to the apical membranes of these cells, which face the lumen of the ducts or airways. Much is also found in intracellular membranes. CFTR is not confined to epithelial tissues and is also expressed in neurones and in cardiac muscle but its functional role in these tissues is not fully established.

The product of the CF gene is known as the cystic fibrosis transmembrane conductance regulator (CFTR) because the primary sequence shows consider-able homology to a family of transporters known as ATP-binding cassette (ABC) transporters (Fig. 12.1). This family includes a large number of bacterial permeases and, in eukaryotic cells, the multidrug resistance protein (P-glycoprotein) and the sulphonylurea receptor (Chapter 8; Higgins, 1992). Sequence analysis indicates that CFTR consists of two homologous repeats (Fig. 12.2). Each of these repeats is made up of a set of six transmembrane domains followed by a large cytosolic loop that contains a highly conserved nucleotide-binding domain (NBD). Unlike other ABC transporters, in CFTR the two repeats are linked by a large cytosolic regulatory (R) domain, which contains multiple potential phosphorylation sites. It is currently not clear whether a single CFTR molecule is sufficient to form a Cl$^-$ channel or if the channel is multimeric.

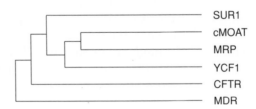

FIGURE 12.1 EXTENT OF RELATEDNESS BETWEEN DIFFERENT ABC TRANSPORTERS
The length of the lines indicates the percentage of amino acid identity. SUR, sulphonylurea receptor. cMOAT, liver canalicular multispecific organic anion transporter. MRP, multidrug resistance-related protein. YCF, yeast cadmium factor. CFTR, cystic fibrosis transmembrane conductance regula-tor. MDR, multidrug resistance protein (P-glycoprotein).

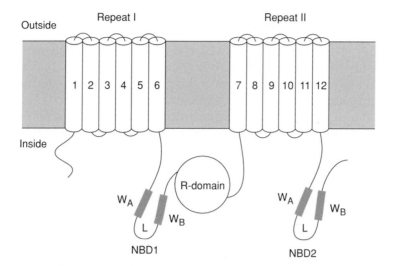

FIGURE 12.2 PUTATIVE TOPOLOGY OF CFTR
Putative membrane topology deduced from hydropathy analysis of CFTR.
R, regulatory domain; NBD, nucleotide binding domains; W_A, Walker A
motif; W_B, Walker B motif; L, linker motif.

Functions of CFTR

CFTR serves as an ion channel

Because of its homology to other ABC transporters, it was first thought
that CFTR might act as a transporter and that this transport function in some
way regulated membrane Cl^- permeability. However, there is now conclusive
evidence that CFTR is itself a Cl^- channel. This evidence derives from three
principal observations. First of all, expression of CFTR in a number of different
systems produces Cl^- currents with properties resembling those of a class of
small conductance Cl^- channels found in native epithelia (Gray et al., 1989;
Kartner et al., 1991; Tabcharani et al., 1991). The expressed channels are voltage
independent, have a single-channel conductance of ~10 pS, and are anion
selective with a Cl^- permeability 10–20 times that of Na^+. The anion selectivity
follows the sequence $Br^- \geq Cl^- > I^- > F^-$. Channel activation is dependent
both on phosphorylation by cyclic AMP-dependent protein kinase (PKA) and
on the additional presence of intracellular MgATP. Importantly, a channel
with these properties is absent from CF epithelia. Second, mutations in CFTR
have been shown to alter the anion selectivity and conductance of the expressed channel, consistent with the idea that CFTR serves as the channel
itself rather than acting as a regulator of an endogenous channel (Anderson
et al., 1991). Finally, confirmation that CFTR is indeed a channel came when
the purified protein was reconstituted into artificial lipid bilayers and single
Cl^- channel currents were recorded following activation with PKA (Bear et
al., 1992).

Regulation of other ion channels

In addition to its intrinsic Cl$^-$ channel activity, CFTR may also act as a regulator of both an outwardly-rectifying Cl$^-$ channel (ORCC) and an epithelial Na$^+$ channel. This regulation is lacking in cystic fibrosis and leads to a further reduction in epithelial Cl$^-$ permeability and an enhanced Na$^+$ permeability. The properties of ORCC differ from those of CFTR in several ways: it has a larger single-channel conductance (30–70 pS), a different halide selectivity (I$^-$ > Cl$^-$ > Br$^-$), is activated by depolarization and has an outwardly rectifying current–voltage relation. Like CFTR, however, ORCC is activated by PKA. This regulatory action is absent in CF, further enhancing the defective Cl$^-$ secretion; however, transfection of CF bronchial epithelial cells with CFTR restores PKA activation of ORCC and corrects the impaired Cl$^-$ secretion (Egan *et al.*, 1992). The mechanism by which CFTR enhances the activity of ORCC is not fully resolved. One controversial hypothesis suggests that CFTR transports ATP from the inside to the outside of the cell, where the nucleotide activates purinergic receptors (probably of the P$_{2U}$ variety) and thereby ORCC (Fig. 12.3) (Schwiebert *et al.*, 1995). While there is no doubt that external ATP can activate ORCC, the ability of CFTR to transport ATP is disputed and others have failed to observe an increase in ATP release associated with CFTR (Reddy *et al.*, 1996). The outwardly-rectifying Cl$^-$ channel can also be activated by extracellular UTP in both normal and CF tissues and has led to the suggestion that aereolised UTP might be beneficial in cystic fibrosis.

CFTR also appears to act as a regulator of the epithelial sodium channel (ENaC), mediating the inhibitory effects of cyclic AMP on ENaC activity, as described in Chapter 13. Loss of this regulation leads to enhanced ENaC activity in CF airway epithelia and accounts for their abnormally high Na$^+$ permeability (Boucher *et al.*, 1986). How CFTR regulates ENaC is currently the subject of investigation. Interestingly, CFTR expressing the Δ508 mutation is unable to regulate ENaC (Mall *et al.*, 1996).

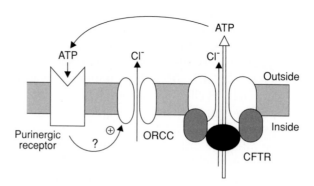

FIGURE 12.3 MODEL FOR CFTR REGULATION OF ORCC
CFTR is proposed to promote the transport of ATP across the cell membrane. ATP then acts on puringeric receptors to enhance the activity of the outwardly-rectifying Cl$^-$ channel (ORCC).

Correlating Structure and Function

The pore

Disease-causing mutations which altered the single-channel conductance first suggested that the sixth transmembrane domain (TM6) might form part of the pore of CFTR (Anderson *et al.*, 1991). Subsequently, cysteine-scanning mutagenesis was used to map the residues in TM6 which actually line the pore and might therefore contribute to the anion selectivity filter (Cheung and Akabas, 1997). In this technique, successive amino acids are replaced with cysteine and the channel is then probed with thiol reagents, which interact specifically with cysteine residues. The results of these experiments indicated that residues at the cytoplasmic end of TM6 are major determinants of anion selectivity (Fig. 12.4). Thus the selectivity filter of CFTR may be located at the cytoplasmic end of the pore, in contrast to that of K$^+$ channels, which appears to reside towards the outer part of the permeation pathway (see Chapter 6).

Further evidence that TM6 contributes to the pore of CFTR was provided by Linsdell and colleagues (1997), who showed that simultaneous mutation of two residues near the middle of TM6 (T338A, T339A) increased the single-channel conductance. This result argues that these residues are located close to the narrowest part of the pore (Linsdell *et al.*, 1997). By comparing the relative permeability to anions of different sizes, the diameter of the pore at its narrowest region has been estimated as ~5.3 Å, larger than that of the unhydrated Cl$^-$ ion (3.6 Å). Mutations outside TM6 also affect the single-channel conductance, but it is not known whether these reflect an allosteric interaction or if additional residues contribute to the channel pore.

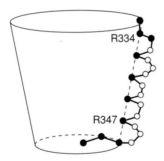

FIGURE 12.4 TM6 FORMS THE PORE OF CFTR
A cartoon of the CFTR channel pore illustrating the TM6 segment residues lining the pore wall. The residues at the end of the segment loop back into the channel, narrowing the lumen and forming the selectivity filter. Residues accessible (●) or inaccessible (○) to the lumen are indicated. Residues which are mutated in some forms of mild CF and produce a reduced single-channel conductance are indicated. From Cheung and Akabas (1997).

Gating

Role of the regulatory domain

The CFTR Cl$^-$ channel is normally closed, but can be activated by protein kinase A (PKA). The R domain contains a large number of consensus sites for PKA-dependent phosphorylation and deletion of this domain (amino acids 708–835) leads to channels which are constitutively active and conduct Cl$^-$ in the absence of cyclic AMP (Rich *et al.*, 1991). The R domain thus appears to be the major site of action of PKA. The mechanism by which it modulates channel activity is not fully established, but a simple model in which the R domain serves as an inhibitory blocking particle and plugs the internal mouth of the pore appears to be incorrect. Instead, the evidence argues that phosphorylation of the R domain stimulates channel activity by facilitating the interaction of ATP with the nucleotide-binding domains of CFTR (Winter and Welsh, 1997).

Role of nucleotide-binding domains

The gating of CFTR channels is coupled to ATP hydrolysis at the nucleotide-binding domains (NBDs). One of the most distinctive features of CFTR is that even when it has been phosphorylated by PKA, it will open only when the intracellular membrane surface is exposed to hydrolysable ATP. Conversely, vanadate prevents the channel from closing by stabilising the ADP-bound state (Baukrowitz *et al.*, 1994). An elegant study by Hwang and colleagues (1994) has shown that both NBDs participate in CFTR channel gating, but that they are not functionally equivalent.

Each NBD contains three highly conserved motifs that have been implicated in ATP binding and hydrolysis (Logan *et al.*, 1994; Ko and Pedersen, 1995). These comprise a Walker A motif (GX$_4$GKT/S)[1], in which a conserved lysine (**K**) is believed to be important in coordinating the negative charge of the phosphate tail of ATP; a Walker B motif (R/KX$_7$h$_4$**D**)[1] which contains an aspartate residue (**D**) that is thought to coordinate the Mg^{2+} ion of MgATP; and a short linker sequence (LSGGQ) lying between the Walker A and Walker B motifs which is also implicated in nucleotide binding (Fig. 12.2).

Carson and colleagues (1995) found that mutation of either of the Walker A lysines, which is expected to reduce ATP-hydrolysis, decreased the channel open probability, but by different mechanisms (Fig. 12.5). Mutations in NBD1 decreased the frequency of the bursts of channel openings, whereas mutations in NBD2 simultaneously prolonged the duration of the bursts and reduced their frequency. They interpreted their results to indicate that hydrolysis at NBD1 initiates a burst of openings, while hydrolysis at NBD2 terminates the burst. In support of this idea, mutation of the Walker B aspartate in NBD2, which is predicted to abolish Mg^{2+} binding and hydrolysis, markedly increased the channel open time (Gunderson and Kopito, 1995). There is thus general agreement with the hypothesis that ATP hydrolysis at NBD2 is required in order for the channel to close. The role of NBD1 is less clear because

[1] X = any residue; h = hydrophobic residue.

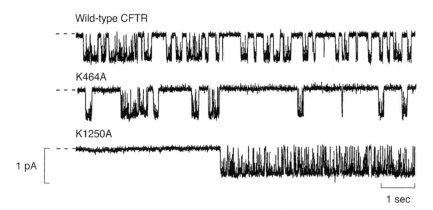

Wild-type CFTR

K464A

K1250A

1 pA

1 sec

FIGURE 12.5 SINGLE-CHANNEL CURRENTS THROUGH
WILD-TYPE AND MUTANT CFTR
Single-channel currents recorded from inside-out patches excised from
HeLa cells transfected with wild-type or mutant (K464A, K1250A) CFTR.
The intracellular solution contained 1 mM ATP and 75 nM protein kinase
A. From Carson et al. (1995).

mutation of the Walker A lysine in NBD1 does not completely abolish channel
activity, as might be expected if ATP hydrolysis is required for channel
opening. There are currently two schools of thought regarding the function
of NBD1. First, ATP hydrolysis at NBD1 may result in channel opening
(Baukrowitz et al., 1994; Carson et al., 1995). Second, ATP hydrolysis at NBD1
may serve to prime CFTR into an active state which can then be opened by
ATP binding to NBD2 (Gunderson and Kopito, 1995). These models cannot
be distinguished simply from the effects of mutating the Walker A lysine in
NBD1, as this may not wipe out ATP hydrolysis altogether. Thus the precise
role of NBD1 remains unresolved.

Summary of structure–function studies

To recapitulate, it is now fairly well established that CFTR not only
functions as a Cl^- channel, but also regulates the activity of at least two other
types of ion channel: the outwardly rectifying Cl^- channel and the epithelial
Na^+ channel. The possibility that CFTR also serves as a transporter (e.g., of
ATP) is hotly debated. There is evidence that the channel pore is formed
from residues within TM6, but whether additional residues also contribute
to the permeation pathway is not known. Nor is it clear whether a single
CFTR molecule is sufficient to form the pore or whether the channel is
multimeric. Gating is regulated by at least two regions of the protein: the R
domain (which must be phosphorylated by PKA for the channel to be active)
and the nucleotide-binding domains. How phosphorylation, or nucleotide
interaction, with these domains is transduced into opening of the channel
pore remains a matter for speculation.

CYSTIC FIBROSIS

Clinical Features

Cystic fibrosis results from the complete absence of, or dramatic reduction in, the epithelial Cl^- conductance, as a consequence of mutations in CFTR. The most important pathological abnormalities occur in the lungs: although these appear normal at birth, subsequent mucous accumulation leads to chronic infection and a progressive destruction of the alveolar tissue. Air-trapping, due to occlusion of the distal airways, and fibrosis also occur. Increasing lung damage is associated with enhanced resistance of the pulmonary vasculature, which sometimes leads to right ventricular hypertrophy (*cor pulmonale*). Cystic fibrosis is also associated with pancreatic insufficiency (i.e., a lack of digestive enzyme secretion) and malabsorption in the gut. In the past this was a major cause of malnutrition but today it can be ameliorated by administration of pancreatic enzymes. As with the lung, pancreatic disease is often progressive and increases with age, as accumulating obstructions in the pancreatic ducts lead to increasing destruction of the acinar tissue. In some individuals, pancreatic degeneration may also be associated with the loss of endocrine tissue and consequent diabetes mellitus. A number of CF infants (~10%) are born with an intestinal obstruction called meconium ileus, which is caused by abnormally solid contents of the small bowel. If the meconium is not excreted, abdominal distension occurs and may lead to perforation of the gut. Similar distal intestinal obstructions may also occur in adults, a condition known as 'meconium ileus equivalent'. Cystic fibrosis is also associated with male infertility because of the obstruction, or congenital absence, of the vas deferens. The altered electrolyte composition of the sweat, which is diagnostic of the disease, is normally not of great threat to the patient, except at high ambient temperatures when children may be prone to heat stroke because of excessive salt loss. Although such temperatures are only reached during heat waves in the United Kingdom, they occur commonly in other regions of the world: children visiting Disneyworld in Florida, for example, may need salt supplements. The severity of cystic fibrosis is variable and, as discussed later, varies with the particular mutation in CFTR which is its cause.

Management of CF currently includes chest percussion and postural drainage to relieve bronchial and bronchiolar obstructions, administration of antibiotics to treat infection, and replacement of pancreatic enzymes. The survival of patients with CF has greatly improved over the last 40 years, with the median age of survival now being around 29 years. Individuals born with CF today may expect to survive about 40 years, even with no further improvements in therapy. Thus CF is no longer a disease confined to children and young adults.

Mutation Analysis

More than 400 mutations in the CF gene have been described, although only a tiny number have been characterised functionally. They may be grouped into four different classes (Fig. 12.6, Welsh and Smith, 1993). Class I mutations produce premature truncation of the protein and result from splice-site abnormalities, frameshifts caused by deletions and insertions, or nonsense mutations. Class II mutations affect the correct trafficking of the protein and result in a dramatic loss of CFTR expression at the cell surface. Class III mutations include those in which the regulation of the channel is defective, while class IV mutations result in channels with altered conduction properties. Although this classification provides a useful framework, it is not completely distinct because some mutations fall into more than one class. The 12 most common mutations that occur worldwide are listed in Table 12.1. Missense mutations occur throughout the entire coding region of *CFTR*, but are most heavily concentrated in TM1, TM6, the NBDs, and the loop linking TM10 and TM11, suggesting that these are structurally or functionally important regions of the protein. Their study has helped elucidate the relationship between channel structure and function.

The most common mutation, found in ~70% of northern European Caucasian patients, is the deletion of a single amino acid, a phenylalanine at position 508, in the first NBD (ΔF508). This Class II mutation is associated with a minimal cAMP-activated epithelial chloride conductance and produces a severe form of CF. At 37°C, the mutant protein is processed incorrectly and, as a result, does not reach the plasma membrane. Chloride channels in the apical membrane of epithelial cells are thus completely absent or drastically reduced. Interestingly, at lower temperatures (25–30°C) the mutant protein is processed correctly and delivered to the membrane, although the channels show a reduced open probability (Denning *et al.*, 1992).

Many mutations which cause the rarer CF genotypes occur within the NBDs. Like ΔF508, some of these lead to defective processing of the protein and reduced surface expression of CFTR. Since mutations in NBD1, but not

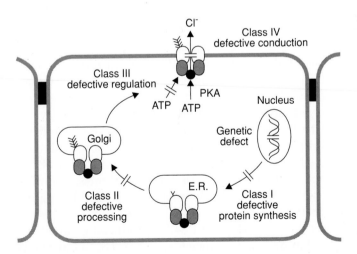

FIGURE 12.6 DIFFERENT CLASSES OF MUTATIONS IN CFTR
Glycosylation is indicated by the branched structure. After Welsh and Smith (1993).

TABLE 12.1   FREQUENT CYSTIC FIBROSIS MUTATIONS

Name	Frequency (%)	Mutation	Consequence
ΔF508	67	Deletion of 3 bp between nt 1652 and 1655 in exon 10	Deletion of Phe at codon 508
G542stop	3.4	G→T at nt 1756 in exon 11	Gly→Stop at codon 542
G551D	2.4	G→A at nt 1784 in exon 11	Gly→Asp at codon 551
W1282stop	2.1	G→A at nt 3978 in exon 20	Trp→Stop at codon 1282
3905insT	2.1	Insertion of T after nt 3905 in exon 20	Frameshift
N1303K	1.8	C→G at nt 4041 in exon 21	Asn→Lys at codon 1303
3849+10 kbC→T	1.4	C→T in a 6.2-kb *Eco*RI fragment 10 kb from 5′ junction of intron 19	Aberrant splicing
R553stop	1.3	C→T at nt 1789 in exon 11	Arg→Stop at codon 553
621+1G→T	1.3	G→T at nt 1 from 5′ junction of intron 4	Splice mutation
1717-1G→A	1.1	G→A at nt 1 from 3′ junction of intron 10	Splice mutation
1078delT	1.1	Deletion of T at nt 1078 in exon 7	Frameshift
2789+5G→A	1.1	G→A at 5 nt from 5′ end of intron 14b	Splice mutation

From Tsui (1992). bp = base pair, nt = nucleotide.

the equivalent mutations in NBD2, result in a reduced channel density, NBD1 appears to be more critical for protein maturation and membrane targeting than NBD2 (Gregory *et al.*, 1991).

Class III mutations are correctly processed but the channel fails to open on cAMP stimulation. A relatively common CF mutation, for example, is a glycine-for-aspartic acid substitution at position 551 (G551D) which lies within the linker region of NBD1. The equivalent mutation in NBD2 (G1349D) is also associated with CF. Both these mutations result in significantly decreased nucleotide binding (Logan *et al.*, 1994). ATP binding to wild-type CFTR is half-maximal at 3–5 mM ATP, a concentration which is similar to that found within the cell. This suggests that CFTR activity may be coupled to physiological nucleotide levels and thus that even a small change in nucleotide binding might have a marked effect on channel activity.

In contrast to the lack of functional channels associated with the NBD mutations described earlier, other (class IV) mutations reduce the epithelial Cl⁻ conductance by a different mechanism. A mild form of CF is found in about 2% patients and results from one of several missense mutations which are associated with altered Cl⁻ channel properties (Sheppard *et al.*, 1993). When these mutations were engineered into CFTR and expressed in mammalian cells, two of them (R334Y and R347P) produced channels with a greatly reduced single-channel conductance (Fig. 12.7). These mutations are located in the sixth membrane-spanning domain, suggesting that the TM6 may form part of the pore. The third mutation (R117H), located at the external surface

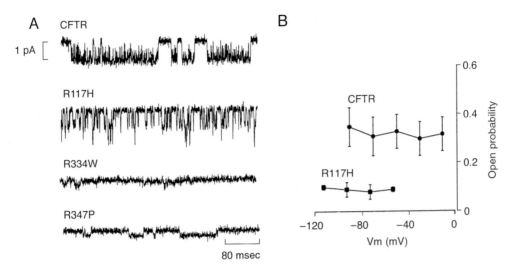

FIGURE 12.7 MILD MUTATIONS ALTER CFTR PERMEABILITY
AND GATING
(A) Single channel currents recorded from inside-out patches excised from
HeLa cells transfected with wild-type or mutant CFTR, as indicated. The
intracellular solution contained 1 mM ATP and 75 nM protein kinase A.
(B) Relationship between membrane potential and open probability for
wild-type or R117H CFTR. From Sheppard *et al.* (1993).

of TM2, had little effect on the single-channel conductance but nevertheless
substantially reduced the epithelial chloride conductance. This may be explained by the fact that this mutation produced a profound decrease in the
channel open probability (Fig. 12.7B). The consequence of all three mutations
is a marked decrease, but not a total loss, of the whole-cell chloride conductance. The residual current may explain the mild form of the disease. It is
not known how much the Cl⁻ conductance must be reduced for the CF
phenotype to occur, although it is clearly less than 50% because heterozygotes
appear asymptomatic.

The intracellular loop that connects TMs 10 and 11 of CFTR (ICL4) also
exhibits a high frequency of CF mutations; at least 19 different mutations
have been identified in this loop. This region of the protein is also highly
conserved between ABC transporters, suggesting that it may have an important functional role. Mutagenesis studies have shown that most CF mutations
in ICL4 disrupt processing of the protein and result in a dramatic reduction
in channel density (Cotten *et al.*, 1996). In addition, channel gating is affected,
which has led to the suggestion that ICL4 may couple the activity of the
NBDs to gating of the channel pore.

It is difficult to define a clear relationship between genotype and phenotype for CFTR mutations because the clinical phenotype shows considerable
variability even between individuals carrying the same mutation. This variability may be a consequence of differences in genetic background or environ-

mental factors, or reflect the fact that some CF patients are compound hetero-zygotes. Nevertheless, the different classes of CFTR mutation do show some correlation with the extent of pancreatic function. Thus, class I and II muta-tions are generally associated with severely compromised pancreatic function, class III mutations often (but not always) cause pancreatic insufficiency and individuals carrying Class VI mutations are usually pancreatic sufficient.

Summary of CFTR mutations

To summarise, mutations which give rise to cystic fibrosis may disrupt CFTR function in different ways: by preventing expression of a full-length transcript, by reducing cell-surface expression of the protein, by impairing channel regulation, or by altering the single-channel properties. Those which result in the total loss of functional channel activity, either because the protein is processed incorrectly and is missing from the surface membrane or because it is present but completely inactive, give rise to a severe form of the disease. Mutations which result in a reduced Cl^- current are associated with a milder form of the disease. These may produce a reduction in the single-channel conductance or channel open probability or a reduced level of protein expres-sion. Mutations may also have self-compensating effects, as is the case for P574H, which causes a markedly decreased channel density but an enhanced open probability (60% higher than normal) of those channels that make it to the membrane (Champigny et al., 1995). As a consequence, there is a small but significant Cl^- current and the mutation only causes a mild form of CF. Mild mutations are dominant in conferring disease phenotype. This is because a compound heterozygote carrying one allele with a severe mutation (e.g., ΔF508) and another with a mild mutation (e.g., R117H) will have significant residual channel activity (Welsh and Smith, 1993).

How Do Mutations in CFTR Cause Cystic Fibrosis?

In this section we consider how the defective Cl^- secretion which is the hallmark of CF gives rise to the symptoms of the disease. This is most easily done by first looking at a tissue in which CFTR is involved principally in fluid absorption, as in the sweat gland, and then a tissue such as the exocrine pancreas, where it is primarily involved in fluid secretion. We will then consider the lung.

Sweat

The lack of CFTR leads to a high concentration of NaCl in the sweat, which, as we have seen, forms the basis of the diagnostic 'sweat' test. Sweat is used to cool the body by the evaporative loss of water. In normal sweat glands, NaCl and water enter the lumen of the duct in the distal coil by filtration. NaCl is then reabsorbed in the proximal ducts to produce a dilute

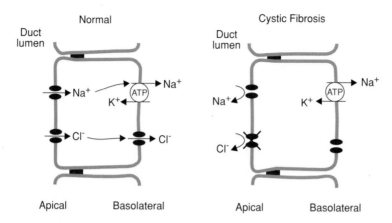

FIGURE 12.8 MODEL OF ABSORPTION IN THE SWEAT DUCT
NaCl transport across the sweat duct epithelia cell is mediated by the combined activity of a basolateral Na/K-ATPase and apical Na^+ and Cl^- channels. When CFTR is absent, Cl^- can no longer enter the cell from the duct lumen. This leads to a secondary loss of Na^+ uptake.

solution (Fig. 12.8). This is achieved by the activity of the Na/K-ATPase in the basolateral membrane, which creates an electrochemical gradient for Na^+ uptake across the apical membrane, with Cl^- following passively through CFTR channels. The ductal epithelium is impermeable to water, which exits via the duct to the surface of the skin. When sweat production is increased, circulating adrenaline acts on β-receptors to elevate cyclic AMP, thereby stimulating CFTR and increasing salt uptake.

In CF, epithelial Cl^- absorption is defective. As a consequence, Na^+ uptake causes a membrane depolarisation (because there is no accompanying Cl^-), which reduces the driving force for Na^+ entry and limits its further reabsorption. Consequently, NaCl levels are elevated in the sweat. Because of the Cl^- impermeability of the apical membrane, the potential gradient across the ductal cells is also more negative in CF epithelia (as is also the case in other epithelia).

Pancreas

It is well documented that the level of pancreatic function is associated with different CF genotypes. Mutations are classified as severe or mild according to whether they are associated with very low (<1% of normal), or close to normal, levels of pancreatic enzyme secretion into the small intestine. Around 90% of CF mutations produce pancreatic insufficiency. This is caused by a failure of fluid secretion from the pancreatic duct cells, which results in decreased washing out of pancreatic enzymes and mucins from the ducts.

The study of fluid secretion from the exocrine pancreas has an illustrious history. It is stimulated by secretin, which was the first endocrine agent to be discovered in a series of classical experiments by Bayliss and Starling between 1902 and 1904. They coined the name hormone (from the Greek, "I arouse to activity") to describe the class of molecules they discovered. Studies of the mechanism of action of secretin-like molecules also led Sutherland to the discovery of cyclic AMP and the concept of an intracellular second messenger. We now know that secretin, via elevation of cAMP, stimulates the activity of CFTR, and thereby fluid secretion. Figure 12.9 shows the current model for how this is achieved.

The pancreatic duct cells secrete a watery fluid, rich in sodium bicarbonate. HCO_3^- is generated within the cell from CO_2 and water by the action of carbonic anhydrase. The luminal membrane contains a HCO_3/Cl exchanger that transports the HCO_3^- out of the cell down its concentration gradient in exchange for Cl^-. The Cl^- recycles into the lumen via CFTR Cl^- channels and thus maintains the electrochemical gradient. Some Cl^- ions also remain in the lumen and are secreted. The protons which are formed together with HCO_3^- when CO_2 is split, leave the cell via the Na/H antiporter in the basolateral membrane and the pH of the blood in the capillaries leaving the ducts is therefore decreased. The Na^+ concentration gradient is maintained by the activity of the Na/K-ATPase and a NaK2Cl cotransporter, and the entering potassium is recycled through K^+ channels: all three proteins are located in the basolateral membrane. In the secreting state, the result of these ion fluxes is the transport of both Cl^- and HCO_3^- into the lumen. This creates a state of lumen negativity and allows Na^+ ions to move through the spaces

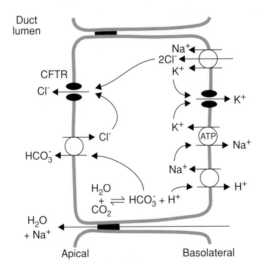

FIGURE 12.9 MODEL OF SECRETION IN THE PANCREATIC DUCT
The activity of CFTR allows Cl^- efflux from the epithelial cells of the pancreatic ducts into the duct lumen and, by enhancing the activity of the HCO_3/Cl exchanger, also promotes the efflux of HCO_3^-. This transport of negative ions leads to the movement of Na^+ across the epithelium by a paracellular route. Consequently water follows by osmosis. For a full explanation see text.

between the cells from the basolateral side into the lumen of pancreatic duct. The increase in the ion concentration in the lumen causes the osmotic movement of water from the blood to the lumen and results in a watery secretion. Secretin activates CFTR further, leading to increased apical Cl^- flux into the lumen and thus driving both the HCO_3/Cl exchanger and HCO_3^- production.

In CF, the apical membrane Cl^- channels are either absent or defective so the Cl^- permeability is reduced, and secretion of Cl^- and HCO_3^- does not occur. Although the pancreatic enzymes are released from the acinar cells in response to hormonal stimulation, they do not enter the duodenum, both because of the lack of a fluid vehicle to carry them and because mucous plugs the pancreatic duct. This results in two problems. First, a lack of pancreatic digestive enzymes, which is treated by giving the patient oral supplements of pancreatic enzymes. Second, the lack of fluid secretion in the pancreas, together with the presence of digestive enzymes, may result in local auto-digestion, leading to destruction of the exocrine tissue and pancreatitis. Meconium ileus may similarly result from diminished secretion in the small intestine.

The Lung

The epithelium of the small airways consists chiefly of ciliated epithelial cells, interspersed with mucous-secreting goblet cells, which are interrupted at intervals by submucosal glands (Fig. 12.10). A thin layer of fluid covers the surface of the epithelium. This is moved up the airways by directional

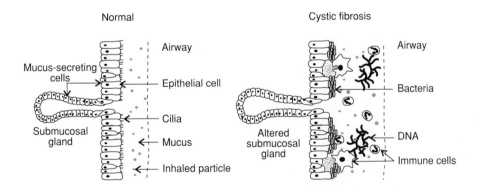

FIGURE 12.10 MODEL OF NORMAL AND CF EPITHELIUM
Normal epithelium. A thin wet mucus solution traps inhaled particles and bacteria and is moved up the airways by the cilia of the surface epithelia cells. Most fluid is secreted by the submucosal glands. CF epithelium. In the absence of CFTR, little fluid is secreted by the submucosal glands, so the mucus is thick and difficult to remove. Defensins are inactivated by the higher salt concentration. Bacteria proliferate and attract immune cells leading to inflammation and damage of healthy tissue. DNA released from dying cells increases the stickiness of the mucus.

beating of the cilia, carrying with it any small particles or bacteria that have been deposited in the small airways. This process has been referred to as the mucociliary escalator. As the surface area of the airways diminishes markedly from the lungs to the tracheae, the airways surface fluid must move faster in the upper airways. In CF patients, the fluid layer is more viscous and contains a higher concentration of mucous than that of normal individuals and it is consequently harder to move. It therefore accumulates in the lower airways. Mucous plugs in the small airways of CF patients not only impair breathing, they also trap bacteria so that infection is established more readily. These infections lead to inflammation and a gradual destruction of the small airways that ultimately leads to respiratory failure. Lung disease accounts for most deaths from CF, and gentle pounding on the chest (chest percussion) to clear mucous from the clogged airways is central to the management of the disease. The hyperviscosity of the airways fluid in CF is not simply the result of dehydration. It is also due to the presence of DNA released from dead cells and bacteria. Hence some CF patients are treated with aerolised DNAse.

The thickness of the fluid layer that covers the epithelia of the lung and airways is determined by a fine balance between fluid uptake and secretion. This is osmotically driven by ion transport, with fluid secretion accompanying Cl^- efflux into the airways and fluid reabsorption accompanying Na^+ uptake. Both process are modified in CF. First, the lack of functional CFTR means that Cl^- efflux is reduced, which leads to a decrease in fluid secretion. Second, Na^+ uptake increases because the inhibitory influence of CFTR on the activity of the epithelial Na^+ channel (ENaC) is lost, as described earlier. This leads to enhanced fluid uptake. It is thought that both these changes combine to alter the quantity and/or composition of the airways surface fluid.

Although there is clear experimental evidence for changes in cAMP-activated Cl^- transport (attributable to CFTR) and in amiloride-sensitive Na^+ uptake (attributable to ENaC) in CF epithelia, the precise way in which they contribute to the pathogenesis of the disease is far from resolved. One difficulty is that the main site of complications in CF is the small airways, which are hard to access. While the epithelial cells of the upper airways, such as the nasal passages and the tracheae, have been studied it is by no means certain that their properties will resemble those of the lower airways epithelia. Indeed, it is thought that the electrolyte composition of the airway surface fluid is modified in the upper airways. Given that the pulmonary complications in CF are the most severe, it may also be surprising to discover that CFTR is hard to detect in the lung. In adult humans, CFTR is expressed most strongly in the submucosal cells, and little is found in the surface epithelium of the airways or the alveolae. The mechanism of Cl^- secretion by submucosal gland cells has yet to be resolved, but it is thought to resemble that found in the pancreatic ducts (Fig. 12.9) with two exceptions. First, the HCO_3/Cl exchanger may not be present in the apical membrane of the gland cells, nor the Na/H exchanger in the basal membrane. Second, the airway epithelium is believed to be relatively impermeable to water, so that the water movement that accompanies Cl^- efflux is likely to be mediated by water channels in the gland cell membranes. It is not yet known whether ENaC and CFTR are coexpressed in the same cells thoughout the airways or have different cellular distributions; but clearly CFTR can only modify ENaC activity in those cells in both channels are expressed.

Importance of the composition of the airway surface fluid

The thin layer of fluid that coats the surface of the airways has a much higher NaCl concentration in CF patients that in normal subjects. Direct measurements have shown Na^+ and Cl^- concentrations in normal tracheal fluid of 82 and 84 mM as compared with values of 121 and 129 mM in CF individuals (Joris *et al.*, 1993). The hypotonicity of normal airway fluid probably results from a significant NaCl reabsorption in excess of water. The NaCl concentration and tonicity of the airway fluid of CF patients are similar to those of the blood, suggesting that reabsorption and/or secretion of fluid is impaired. The higher salt concentration reduces the capacity of the lung to deal with bacterial infection.

The major cause of death in CF is lung disease. This results from chronic bacterial infections that lead to progressive destruction of the lung parenchyma and, eventually, respiratory failure. *Pseudomonas aeruginosa* and *Staphylococcus aureus* are the major culprits, but *Burkholderia cepacia* is also important. The lung is an ideal environment for bacteria, being warm, wet and not exposed to ultraviolet light, and the lung is exposed to a multitude of potential pathogens with every breath. Normal individuals do not suffer from infection because the lung epithelial cells secrete antibacterial agents that are effective against a broad spectrum of bacteria. These agents are also secreted by CF epithelia but are inactivated by the high NaCl concentration of the airway surface fluid in CF patients (Smith *et al.*, 1996a). This enables bacteria to multiply, producing a chronic pulmonary infection that is established within the first few months of life. As a consequence, the body mounts an inflammatory response, and breakdown products of inflammatory cells, such as DNA and cytoskeletal proteins, are major components of the mucous which characterises CF. The combination of infection and inflammation progressively leads to chronic lung damage.

Goldman and colleagues (1997) have isolated one antimicrobial factor from airway surface fluid, a peptide called human β-defensin 1. It has broad spectrum activity against gram-negative bacteria, including *P. aeruginosa*, and is inactivated by NaCl concentrations within the range found in the airways surface fluid of CF patients. Defensins kill bacteria by virtue of their ability to form lytic pores in bacterial cell membranes, and are discussed in more detail in Chapter 23.

Infertility

Cystic fibrosis is associated with infertility in men. Up to 80% of men with congenital bilateral absence of the vas deferens have mutations in *CFTR*, although many of them have no obvious other symptoms. The most common of these mutations are ΔF508, G551D, G542stop, W1283stop, N1303K and

R117H. Around 50% of men who suffer from obstructive azoospermia—the failure to produce sperm because of obstruction of the vas deferens and/or epididymis—also carry point mutations in *CFTR* or possess a 5-thymidine variant of intron 8, which results in less efficient splicing and is predicted to lead to reduced CFTR expression (Jarvi *et al.*, 1995).

Gene Therapy

CF is a recessive disease and heterozygotes who produce only 50% of the wild-type protein are asymptomatic. Thus it may be predicted that the transfer and expression of the normal gene into CF epithelia would correct the disorder. This possibility was first addressed using a mouse model of CF, in which the CFTR gene had been disrupted (knocked out). Like their human counterparts, heterozygous mice are asymptomatic, but homozygous animals show markedly reduced epithelial Cl$^-$ secretion and exhibit some of the characteristics of human CF. The secretory defect in the nasal and tracheal epithelium was corrected by delivery of CFTR cDNA–liposome complexes to the airways of CF mice (Hyde *et al.*, 1993).

Gene therapy trials with human patients are now in progress which attempt to correct the defect in the lungs, where the most serious complications in CF occur. The lungs have the advantage that they are an extracellular surface and can be readily and selectively accessed using aerosols. Attempts at adenoviral-mediated CFTR gene transfer have been complicated by a significant inflammatory reaction induced by the vector. By contrast, no adverse immune reaction has been reported for liposome-mediated CFTR gene transfer, and partial restoration of the Cl$^-$ permeability in nasal biopsies of CF patients treated with the gene has been observed (Caplen *et al.*, 1995). Considerable work is still required, however, to determine whether gene transfer will be viable as a routine therapy for CF. In particular, it is unclear whether the submucosal glands in the small airways (which are the major site of CFTR expression and of CF pathology) will be accessible to gene therapy. Nor is it clear whether it is actually necessary to correct CFTR expression in these cells to obtain a clinical improvement. Indeed, the current hope is that enhanced expression of CF in the surface airway epithelia may be sufficient to correct the salt composition of the airway surface fluid and thereby prevent infection and the subsequent events that lead to lung damage. Clinical trials are required to determine if this is the case.

Geographic and Ethnic Variations

The frequency of CF varies among different ethnic groups and is highest in individuals of northern European extraction. Significant numbers are also found in southern Europeans, American blacks and Ashkenazi Jews, but it is extremely rare in Orientals and African blacks. The high incidence of the disease in Caucasian populations (1 in ~22 carry the gene) has led to the suggestion that there may be a selective advantage in being a carrier of CF. One possibility is that carriers may be more resistant to the effects of diarrheal diseases, such as cholera. *Vibrio cholerae*, the bacterium responsible for cholera, produces a toxin that causes massive fluid secretion in the gut and leads to severe diarrhoea and death from dehydration. Cholera toxin acts on the G-protein G_s to stimulate adenylate kinase to produce cyclic AMP. This leads to activation of CFTR, and the resulting chloride secretion causes the osmotic flow of water from the cells into the gut lumen. Individuals with a lower complement of CFTR may secrete less chloride and thus potentially be less susceptible to dehydration. Although cholera is no longer a disease of northern Europe, being mainly confined to third world countries, this was not always the case. One of the most notable successes of public hygiene was the removal of the handle of a water pump in Broad Street, London, by Dr. John Snow, which contained an outbreak of cholera, and confirmed his contention that the disease was spread in the water. Furthermore, *Escherichia coli* continues to be a major cause of secretory diarrhoea in northern Europe, and resistance to its effects may still confer a selective advantage.

The relative frequency of cystic fibrosis in different populations may help to define their genetic relationship to one another. A map of the distribution of the different CF mutations in Europe shows that the ΔF508 mutation is concentrated in the north, being highest in Denmark and England, whereas other mutations are more common in the south. Indeed it appears that some mutations radiate out from the southeast of Europe towards the northwest, a trend which geneticists suggest reflects the movement of early people throughout Europe. Clinically, an important implication of these studies is that simply screening for the most common mutations may miss as many as 20% of them because of differences in the ethnic and geographic background of individuals within a population. Consequently, diagnosis of CF is still dependent on the 'sweat test.'

CHAPTER 13

EPITHELIAL Na+ CHANNEL

The amiloride-sensitive epithelial sodium channel (ENaC) mediates the transport of sodium across the apical membranes of epithelial cells. It serves as the limiting step for Na^+ reabsorption from the kidney distal tubule, the distal colon and the ducts of several exocrine glands. As a consequence, it plays an important role in the control of blood volume and blood pressure. This is emphasised by the fact that, in man, mutations in ENaC can result in hereditary hypertension or hypotension. ENaC is also involved in taste perception, regulates fluid secretion and absorption in the lung and controls the low Na^+ concentration of the endolymph in the cochlea of the ear. Related channels are found in neurones, but their functional role is unknown.

Epithelial cells are polarised and their proteins are distributed asymmetrically between the apical and basolateral membranes. The transport of Na^+ across epithelia requires two proteins: a Na^+ uptake protein located on the apical membrane and the Na/K-ATPase, which resides in the basolateral membrane. These two proteins work in concert (Fig. 13.1). In some epithelia, such as those of the distal tubule, Na^+ enters the cell passively through ENaC channels in the luminal membrane. It is then extruded across the basolateral membrane by the activity of the Na/K-ATPase, which provides the driving force for the movement of Na^+ across the cell. K^+ entering the cell via the Na/K-ATPase leaves via one of many types of K^+ channels and transporters.

Basic Properties

The most distinctive characteristic of ENaC is that it is blocked by low concentrations of amiloride or triamterene, half-maximal inhibition of the channel requiring less than 50 nM amiloride. Electrophysiological studies suggest that there may be a family of amiloride-sensitive Na^+ channels in epithelial tissues, which can be distinguished by their sensitivity to amiloride, selectivity to Na^+ and single-channel conductance (Eaton *et al.*, 1995). Only the classical type of channel, which we will call ENaC, is considered here. It is important to remember, however, that other kinds of amiloride-sensitive channels do exist.

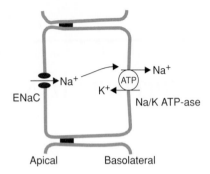

FIGURE 13.1 MODEL FOR EPITHELIAL Na⁺ TRANSPORT
Na⁺ transport across epithelia is mediated by the combined activity of a
basolateral Na/K-ATPase and ENaC channels in the apical membrane.

Studies on native membranes have shown that ENaC has a very low K^+
permeability ($P_{Na}/P_K > 10$), whereas Li^+ is even more permeable than Na^+.
At room temperature, the single channel conductance is ~5 pS in 100 mM
Na^+ and the kinetics are slow, with mean open times of 1–2 sec. There is no
marked voltage dependence of gating, although the open probability does
increase slightly with hyperpolarisation. ENaC is modulated by a variety of
hormones, which mediate their effects via phosphorylation, GTP-binding
proteins and lipid metabolites (Eaton *et al.*, 1995).

Structure

The epithelial Na^+ channel is a hetero-oligomer formed of three homologous
subunits, named α, β and γ, which share about 30% identity at the amino
acid level. The α-subunit (αENaC) codes for a 698 amino acid protein with
a molecular mass (M_r) of 78 kDa (Canessa *et al.*, 1993). The β- and γ-subunits
encode proteins of 638 and 650 amino acids with predicted M_r of 72 kDa and
75 kDa, respectively (Canessa *et al.*, 1994). All three subunits are glycosylated
and have M_r of around 90 kDa when glycosylated.

Hydropathy plots predict that each of the different ENaC subunits consist
of two transmembrane domains linked by a large extracellular loop of ~500
amino acid residues which contains a cysteine-rich domain, several potential
glycosylation sites and the amiloride-binding site (Fig. 13.2). The N and C
termini are both short and there is good evidence that they lie on the intracellular side of the membrane.

ENaC channels are composed of several subunits

Neither the β- nor γ-subunit gives rise to Na^+ currents when expressed
individually, while expression of the α-subunit alone only induces a very

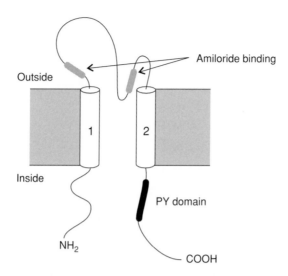

FIGURE 13.2 *PUTATIVE TOPOLOGY OF ENaC SUBUNITS*
Putative membrane topology of ENaC, with the positions of the main
functional domains indicated. The parts of the channel involved in pore
formation are still not fully resolved, but there is evidence that a region
of the TM1-TM2 linker just prior to TM2 is involved. The PY domain is
a regulatory element that acts to reduce the channel open probability and
enhance the rate of removal of the channel from the membrane surface.

small amiloride-sensitive Na^+ current. By contrast, coexpression of all three
ENaC subunits results in large currents with properties resembling those of
the native epithelial sodium channel (Fig. 13.3A; Canessa *et al.*, 1994). Firsov
and colleagues (1996) have shown that this is because assembly of the different
ENaC subunits within the endoplasmic reticulum is required for correct
targeting of the channel to the plasma membrane. When each subunit is
expressed individually the protein is translated but it does not reach the
surface membrane (Fig. 13.3B). There is evidence that $\alpha\beta$- and $\alpha\gamma$-subunits
can also form functional Na^+ channels, with distinctive properties (McNicholas and Canessa, 1997). Although the expression of channels composed of
only two types of ENaC subunit is much less than that of $\alpha\beta\gamma$ENaC channels,
they may contribute to the diversity of amiloride-sensitive Na^+ channels
observed in native membranes (Eaton *et al.*, 1995).

The stoichiometry of the channel remains unknown, although at least
one of each of the α-, β- and γ-subunits obviously is required. A δ-subunit
has also been isolated which is not epithelial but is expressed mainly in testis
and ovary (Waldmann *et al.*, 1995a). It has properties similar to those αENaC
and associates with β- and γ-subunits to form functional channels. Since
βENaC and γENaC do not share the same tissue distribution as δENaC, it
is possible that additional, as yet unidentified, subunits serve this function
in vivo.

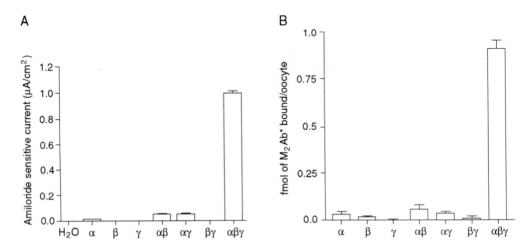

FIGURE 13.3 ENaC IS COMPOSED OF α-, β- and γ-SUBUNITS
(A) Amplitude of the amiloride-sensitive Na+ current in Xenopus oocytes
injected with mRNA encoding α, β or γ ENaC subunits either alone or
in combination. The maximum current is observed when all three subunits
are expressed. From Canessa et al. (1994). (B) Surface density of different
ENaC subunits observed when expressed alone, or in combination, in
Xenopus oocytes, detected by antibody binding to an extracellular epitope
(M_2Ab^*). The maximum amount of surface expression is observed when
all three subunits are present. From Firsov et al. (1996).

ENaC Belongs to a Family of Related Ion Channels with Diverse Functions

Three other groups of channels have been cloned which probably belong to
the same superfamily as ENaC (Fig. 13.4). The first of these, cloned from
neurones of the snail Helix aspersa, is a ligand-gated channel referred to FaNaC
because it is opened by the peptide FMRFamide (phenylalanine-methionine-
arginine-phenylalanine; Linguelia et al., 1995). Although the overall sequence homology is low (16% identity), this channel shares the same overall
structural organization as the ENaC subunits. Like ENaC, it is also Na+
selective and is blocked by amiloride. The three α,β,γ ENaC subunits also
share significant sequence homology with a family of Caenorhabditis elegans
genes involved in sensory touch transduction (Hamill and McBride, 1996).
These genes are also often known as degenerins because dominant mutations
are associated with neuronal degeneration. They include mec-4, mec-10 and
deg-1.
    The third type of ENaC-related channels are the acid-sensing ion channels
(ASICs), which are expressed in brain and sensory neurones (Waldmann and
Lazdunski, 1998). As these cation channels are activated by H+, it has been
postulated that they may be involved in the non-adapting pain that accompanies tissue acidosis. However, although ASIC1 opens when the pH falls below

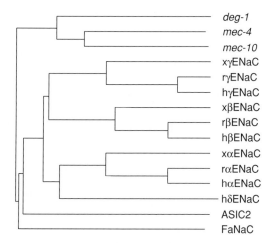

FIGURE 13.4 ENaC GENE FAMILY LINEAGES
deg1, mec4, mec10 are *C. elegans* degenerins; ENaC, epithelial sodium channel (h = human, r = rat, x = *Xenopus*); ASIC2, mammalian brain sodium
channel; FNaC, *H. aspersa* FMRF-amide channel. The horizontal branch
lengths are proportional to the mean number of differences per residue
along each branch. From North (1996).

6.9, other members of this family require a far more acidic pH for activation—
for example, ASIC2 is half-maximally activated by pH 4.9. Thus it is not clear
whether protons serve as physiological regulators of ASIC activity or if there
are other, as yet unidentified, ligands for these channels.

Chromosomal Location

The gene for αENaC is located on chromosome 12p13 (Table 13.1; Voilley *et
al.*, 1994). The β- and γ-subunits map to adjacent regions of chromosome
16p12, being located within 400 kb of each other (Voilley *et al.*, 1995). This
proximity suggests that they may be derived rather recently by gene duplica-

TABLE 13.1   **HUMAN EPITHELIAL Na⁺ CHANNEL GENES**

Gene	Protein	Chromosome location	Major sites of expression	Reference
SCNN1A	αENaC	12p13	Kidney, lung, colon	Voilley *et al.* (1994)
SCNN1B	βENaC	16p13–p12	Kidney, lung, colon	Voilley *et al.* (1995)
SCNN1G	γENaC	16p13–p12	Kidney, lung, colon	Voilley *et al.* (1995)
SCNN1D	δENaC	1p36.3–p36.2	Brain, testis, ovary, pancreas	Waldmann *et al.* (1995a)
ACCN1	ASCI2	17q11.2–q12	Brain	Waldmann *et al.* (1996); García-Anoveros *et al.* (1997)

tion from a common ancestor, a view which is supported by the fact that in the colon the β- and γ-subunits are subject to the same transcriptional control, whereas transcription of αENaC is regulated differently (Renard *et al.*, 1995). The chromosomal clustering of the genes for the βENaC and γENaC subunits is reminiscent of that found for the subunits of the GABA$_A$ receptor (Chapter 18).

Regulation of ENaC Expression

As expected if the functional channel is composed of α, β and γ ENaC, all three subunits are coexpressed in native tissues. They are found at high levels in the apical membranes of epithelial cells in the distal tubules of the kidney, the distal colon, the stomach, the trachea, bronchi and bronchioles of the lung, and the secretory ducts of the salivary and sweat glands (Canessa *et al.*, 1993, 1994; Voilley *et al.*, 1994; Renard *et al.*, 1995).

The expression of the α-, β- and γ-subunits is differentially controlled (Renard *et al.*, 1995). In the colon, transcription of both β- and γ-subunits is enhanced by steroids, whereas expression of αENaC is constitutive and is unaffected by steroids. By contrast, expression of all three subunits is upregulated by glucocorticoids in the lung. In the adult, circulating levels of glucocorticoids are sufficiently high to induce maximal expression of ENaC subunits. It is possible, however, that elevation of glucocorticoids mediates the marked upregulation of αENaC expression that occurs just before birth (see also page 245; Voilley *et al.*, 1994; Renard, 1995). In the rat kidney, βENaC and γENaC mRNA levels are constitutively high and aldosterone has no effect: there is a marginal effect on αENaC mRNA abundance (Renard *et al.*, 1995).

Regulation of ENaC Activity

In the epithelial cells of the airways, ENaC is inhibited by intracellular cyclic AMP. This inhibition does not result from a direct action of cAMP on the channel, or from protein kinase A (PKA)-dependent channel phosphorylation, but is instead mediated indirectly, via the effect of cAMP on the cystic fibrosis transmembrane conductance regulator (CFTR), a protein which regulates ENaC activity. This was first shown by Stutts and colleagues (1995), who examined the effect of agents that elevate intracellular cAMP in MDCK cells transfected either with $\alpha\beta\gamma$ENaC, or with $\alpha\beta\gamma$ENaC together with CFTR. Whereas cAMP stimulated Na⁺ currents in the former case, coexpression with CFTR resulted in smaller basal Na⁺ currents and a small decrease in response to cAMP. This argues that CFTR mediates the inhibitory response of ENaC to cAMP, probably via a PKA-dependent phosphorylation of CFTR itself (see Chapter 12). Subsequently, the regulatory effects of CFTR were confirmed both when $\alpha\beta\gamma$ENaC and CFTR were coexpressed in *Xenopus* oocytes (Mall *et al.*, 1996) and when the purified proteins were incorporated into artificial lipid bilayers (Ismailov *et al.*, 1996). In these later studies, cAMP was without effect on the activity of $\alpha\beta\gamma$ENaC, suggesting that the stimulation observed when ENaC was expressed in MDCK cells may be due to an additional protein, endogenously present in MDCK cells, which upregulates the channel in response to cAMP. The mechanism by which CFTR regulates ENaC activity has not been fully established, but single-channel recordings from artificial bilayers suggest that the channel open probability is reduced.

Correlating Structure and Function

The amiloride-binding site

Several pieces of evidence support the idea that amiloride acts as an open channel blocker, plugging the pore and physically preventing ion flux. Among these are the observations that amiloride block is voltage dependent and that it is influenced by permeating cations. All three ENaC subunits are involved in forming the amiloride binding site, because mutations in any one of them can markedly reduce the affinity of amiloride block (Schild *et al.*, 1997). These mutations are located in a short stretch of amino acids that precedes the second transmembrane domain (TM), which presumably lines the outer mouth of the channel pore (Fig. 13.2). The most dramatic change in amiloride sensitivity was observed when βG525 or γG537 was mutated, suggesting that these residues are particularly important for binding.

The pore

The structure of the ENaC pore is far from resolved. One suggestion is that the pore is formed from a hairpin segment, analogous to the H5 segment of K^+ channels, which loops down into the membrane prior to the second TM. There is evidence, however, that two residues in the second transmembrane domain (αS589 and αS593) are also involved in lining the channel pore (Wald-mann *et al.*, 1995b). Replacement of serine 593 with threonine, for example, increased the single-channel Na^+ current amplitude from 4.6 to 7 pS. This suggests a topology reminiscent of Kir channels in which both the H5 segment and the second TM contribute to the pore (see Chapter 8). All three types of ENaC subunit contribute to the pore, because mutations in α, β or γ ENaC alter the single-channel conductance and generate binding sites for channel block by divalent cations (Schild *et al.*, 1997).

Regulatory elements

The C terminus of all three ENaC subunits contains a highly conserved proline-rich sequence (PPPxY) known as the PY motif. This is involved in interactions with proteins that regulate the activity of ENaC (Schild *et al.*, 1996; Staub *et al.*, 1996). One of these is α-spectrin, a cytoskeletal structural protein which binds to αENaC. Another is a cytosolic protein known as Nedd4 that binds to the PY motif of each ENaC subunit. Deletion of the PY motif, which occurs in patients with Liddle's syndrome, enhances ENaC currents. This suggests that the interaction with Nedd4 may play an important physiological role in channel regulation. The mechanism by which Nedd4 reduces ENaC currents is unknown. One possibility is that it reduces the channel open probability. The idea currently favoured, however, is that Nedd4 is involved in the internalization of the protein, thereby reducing the number of channels in the membrane.

ENaC is unusual in that its activity is also increased by the action of an extracellular protease (Vallet *et al.*, 1997). This protease is known as the channel-activating protease (CAP) and is expressed in kidney, gut, lung, skin and ovary. Sequence analysis indicates that it is a secreted and/or membrane-anchored protein, which suggests that ENaC activity may be regulated by a protease expressed in the same cell. The physiological role of CAP remains to be elucidated.

DISEASES ASSOCIATED WITH ENaC CHANNELS

Disease of Blood Pressure Regulation

Mutations in ENaC may produce either hereditary hypotension or hypertension. To understand why this is the case, we must first consider the role of ENaC in the regulation of blood pressure.

All vertebrates must maintain a constant extracellular fluid volume in the face of marked variations in water and salt uptake. A chronic increase in blood volume produces hypertension and is accompanied by an increase in the interstitial fluid volume, which results in oedema. Conversely, decreases in blood volume cause hypotension. The plasma Na⁺ concentration is the most important determinant of blood volume and is tightly controlled by regulating the amount of Na⁺ uptake across the epithelia of the gut and kidney. This is principally achieved by regulating the activity of ENaC, which thereby plays an essential role in the regulation of blood volume and blood pressure. This is dramatically illustrated by the effects of hereditary mutations in ENaC subunits. As discussed later, such mutations give rise to familial forms of both hypertension and hypotension, the former being associated with enhanced ENaC activity and the latter with reduced ENaC activity.

The kidney glomeruli filter ~120 ml of plasma per min (approximately twice the blood volume per hour!), retaining the red cells and plasma proteins in the capillaries and allowing salts and water to pass through into the kidney tubule. Almost all of the sodium and much of the water which are filtered are subsequently reabsorbed as the filtrate passes down the tubule (Fig. 13.5A). What remains is excreted as urine. About 90% of the filtered sodium is constitutively recovered in the proximal tubule and loop of Henle: the remainder is reabsorbed in a regulated fashion in the lower parts of the nephron, principally by the cells of the distal tubule and the cortical collecting ducts (Fig. 13.5B). The epithelial sodium channel plays a key role in this regulated Na⁺ uptake. Sodium absorption is accompanied by a corresponding

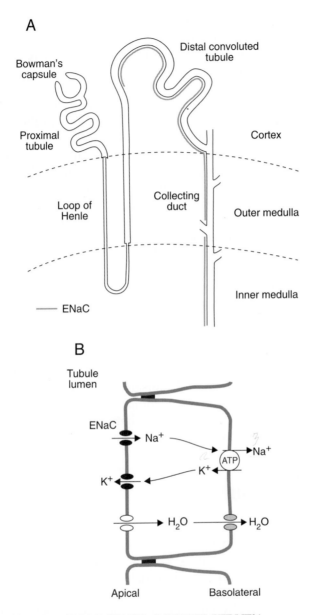

A

Bowman's capsule

Distal convoluted tubule

Proximal tubule

Cortex

Loop of Henle

Collecting duct

Outer medulla

Inner medulla

—— ENaC

B

Tubule lumen

ENaC

Na⁺

Na⁺

ATP

K⁺

K⁺

H₂O

H₂O

Apical

Basolateral

FIGURE 13.5 THE ROLE OF ENaC IN THE KIDNEY

(A) Anatomy of a single kidney tubule. Regions which express ENaC in the apical membranes are indicated by the gray line. (B) Salt transport in the late distal convoluted tubule and cortical collecting duct. The activity of the basolateral Na/K-ATPase creates a Na^+ concentration gradient that causes the passive uptake of Na^+ from the tubule lumen via ENaC. This leads to the osmotic uptake of water. Potassium entering the cell via the Na/K ATPase leaves via Kir1.1 (ROMK) in the apical membrane, leading to K^+ secretion. This K^+ loss is facilitated by depolarization of the apical membrane as a consequence of Na^+ uptake.

osmotic uptake of water, to maintain a constant extracellular Na$^+$ concentration. Thus Na$^+$ absorption leads to expansion of the plasma volume and, when sustained, raises the blood pressure. This explains why enhanced ENaC activity leads to hypertension, whereas reduced activity results in hypotension.

Although the Na/K-ATPase provides the energy for Na$^+$ reabsorption (Fig. 13.5B), all of the regulation of Na$^+$ uptake occurs at the level of ENaC. Sodium uptake in the principal cells of the cortical collecting duct is under the control of several hormones. The most important of these is aldosterone, which promotes Na$^+$ uptake both by enhancing the channel open probability and by increasing the number of channels in the apical membrane (by stimulating their synthesis). An increase in Na$^+$ uptake has secondary consequences: in particular, K$^+$ secretion into the tubule lumen is stimulated because the apical membrane depolarizes and so increases the driving force for K$^+$ efflux (the membrane potential is now positive to E_K). In addition, more K$^+$ enters the cell due to the enhanced activity of the Na/K-ATPase. This explains why excess ENaC activity in Liddle's syndrome is associated with hypokalaemia (reduced plasma K$^+$) and, conversely, why reduced ENaC activity is accompanied by hyperkalaemia.

In addition to its effect on water channels (see Chapter 19), the antidiuretic hormone arginine-vasopressin (AVP) also enhances Na$^+$ uptake across epithelial cells of the renal cortical collecting ducts and the colon. The current view is that this is principally achieved by insertion of fresh ENaC channels into the plasma membrane from an intracellular pool of vesicles which sits just below the membrane, an effect that is mediated by elevation of intracellular cyclic AMP. In addition, effects on the channel open probability have been postulated.

Liddle's Syndrome

In 1963, Liddle and co-workers described a rare form of inherited salt-sensitive hypertension in a family in which there was onset of severe hypertension prior to age 20 in multiple siblings. Liddle's syndrome is characterised by a very high rate of renal Na$^+$ uptake, despite low levels of aldosterone (pseudo-hyperaldosteronism). As a consequence, individuals suffer from severe hypertension and secondary hypokalaemia and metabolic acidosis. The disorder can be corrected by renal transplantation (Botero-Velez et al., 1994), indicating that it is associated with a kidney defect. Amiloride or triamterene, coupled with restriction of dietary Na$^+$ intake, is an effective treatment for both the hypertension and hypokalaemia.

Liddle's syndrome is caused by mutations in the β- or γ-subunits of ENaC which result in constitutive channel hyperactivity. These mutations are either missense mutations in the PY motif (Hansson et al., 1995a), or are substitutions or deletions which result in the introduction of novel stop codons and lead to deletion of the C terminus (Fig. 13.6; Shimkets et al., 1994; Hansson et al., 1995b). Expression studies have been used to determine the effects of these

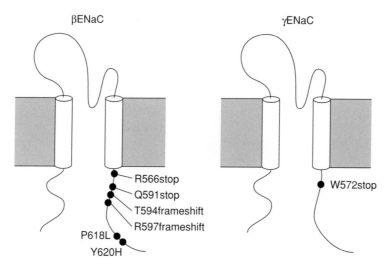

FIGURE 13.6 MUTATIONS ASSOCIATED WITH LIDDLE'S SYNDROME

Putative membrane topology of β and γ ENaC with the mutations associated with Liddle's syndrome marked. All mutations occur in the C terminus of the molecule.

mutations on ENaC currents. In the original Liddle kindred, the mutation is a C to T base-pair substitution in the β-subunit which results in the introduction of a stop codon at arginine 564. This deletes 75 of the 85 amino acids in the C terminus. Coexpression of the mutant β-subunit with wild type α- and γ-subunits produced a three- to five-fold increase in the macroscopic amiloride-sensitive Na^+ current when compared with the wild-type channel (Schild *et al.*, 1995; Fig. 13.7). Similar results were found with other Liddle mutations. The increase in current caused by these mutations results both from an increase in the number of functional channels (Fig. 13.7) and from an increase in the open probability of existing channels (Firsov *et al.*, 1996). The unitary conductance, ion selectivity and sensitivity to amiloride block are unaffected.

It was the study of the natural mutations in families with Liddle's syndrome which led to the recognition that the PY motif (PPPxY) is a key regulator of ENaC. In particular, the tyrosine (Y) plays a critical role, because mutation of this residue to alanine enhanced ENaC currents to the same extent as did truncation of the entire C terminus (Fig. 13.8; Snyder *et al.*, 1995). Mutation of all three prolines to alanine also increased the Na^+ current. In part, the enhanced current results because truncation of the C terminus causes an increased amount of the protein in the plasma membrane. This appears to be due to an inability to remove the truncated ENaC from the membrane, probably because deletion of the PY motif prevents interaction with proteins such as Nedd4. A different mechanism is needed to explain

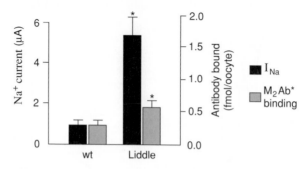

FIGURE 13.7 LIDDLE'S MUTATIONS RESULT IN AN ENHANCED Na⁺ CURRENT AND AN INCREASED DENSITY OF ENaC CHANNELS

Macroscopic Na⁺ currents (black bars) and surface expression (hatched bars) of wild-type and mutant $\alpha\beta\gamma$ENaC. Oocytes were injected with mRNA encoding wild-type α- and γ-ENaC together with either wild-type βENaC or βENaC containing the R564stop Liddle's mutation. Surface expression was detected by antibody binding to an extracellular epitope on the β-subunit. From Firsov *et al.* (1996).

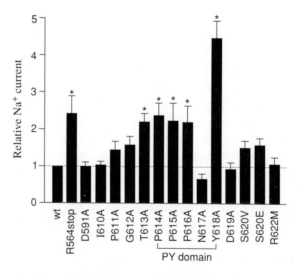

FIGURE 13.8 MUTATIONS WITHIN THE PY DOMAIN CAUSE INCREASED ENaC CURRENTS

Relative current amplitude recorded from oocytes injected with mRNA encoding wild-type α- and β-ENaC together with either wild-type or mutant β-subunits. The greatest enhancement of current is seen when residues T613, P614, P615, P616 or Y618 are mutated. From Schild *et al.* (1996).

why mutations and deletions in the PY motif also increase the channel open probability.

Liddle's syndrome is inherited in an autosomal dominant fashion and patients heterozygous for the disease may be expected to produce both wild-type and mutant subunits. Coexpression of mutant and wild-type β-subunits of ENaC, together with wild-type αENaC and βENaC, results in channel activity that is greater than that of wild type, consistent with the autosomal dominant nature of the disease (Snyder *et al.*, 1995). However, the current was less than that produced when only mutant β-subunits were expressed, suggesting that the disease will be milder in heterozygotes. Although mutations in both β- and γ-subunits have been identified in Liddle's patients, to date none have been reported in αENaC. This may be because mutations in αENaC are phenotypically silent, since mutagenesis studies have found that truncation of the C terminus of αENaC produces only a very small enhancement of the Na$^+$ current (Schild *et al.*, 1995).

Liddle and colleagues (1963) wrote that the syndrome they described was due to "a disorder in which renal tubules reabsorb ions with an abnormal facility despite suppression of aldosterone secretion". The combination of genetics, molecular biology and electrophysiology has shown that their idea was correct. Furthermore, the Liddle's mutations have been important in defining the functional role of the C terminus of the channel. Thus they have shown that the terminal part of the C terminus is not needed for correct targeting of the channel to the membrane or for functional channel activity. Instead, it appears to reduce the channel open probability and to be involved in the reinternalization of the protein.

Pseudohypoaldosteronism Type-1

Pseudohypoaldosteronism type 1 (PHA-1) is a rare inherited disorder in which the kidney fails to respond to aldosterone (Cheek and Perry, 1957). It is characterised by marked hypotension and dehydration of newborn babies and infants. These patients also show salt-wasting, high serum K$^+$ levels and metabolic acidosis, despite having high levels of aldosterone in the circulation. A number of organs can be affected, including the kidney, colon and sweat glands, all of which demonstrate salt loss. In most families, the disease shows an autosomal recessive pattern of inheritance, with varying clinical severity. Patients are usually treated with salt supplementation, which can sometimes be discontinued after infancy.

Genetic analysis indicated that PHA-1 was linked to the regions on chromosomes 12 or 16 where the different ENaC subunits are located, and it was subsequently found that mutations in either the α, β or γ ENaC subunits produce the disease (Fig. 13.9) (Chang *et al.*, 1996; Strauknieks *et al.*, 1996a,b). Some of these are frameshift mutations that cause premature termination of translation, while others are missense mutations. All cause a marked *reduction* in ENaC activity (Fig. 13.10), which results in a marked decrease in Na$^+$ absorption by the kidney. This stimulates renin and aldosterone secretion,

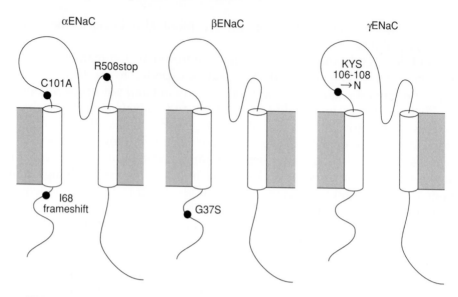

FIGURE 13.9 MUTATIONS ASSOCIATED WITH PSUEDOHYPOALDOSTERONISM TYPE-1
Putative membrane topology of α, β and γ ENaC with the mutations associated with PHA-1 marked.

but salt reabsorption cannot be augmented as ENaC is not functional. The high Na+ concentration in the tubular fluid causes water to be osmotically retained in the tubule lumen, leading to diuresis and dehydration: in normal individuals, a similar diuresis can be produced by amiloride.

A reduction in Na+ uptake in the distal tubule cells leads to a secondary failure to secrete K+ because the activity of the Na/K-ATPase is impaired by low intracellular Na+ (Fig. 13.6A). This accounts for the hyperkalaemia which is a feature of PHA-1.

Respiratory Distress in the Newborn

During fetal life, the lungs are filled with fluid. This causes expansion of the fetal lungs and contributes to the formation of amniotic fluid (25–50% depending on fetal age). Most of the fluid is secreted by the alveolar epithelium, which extrudes Cl− ions into the pulmonary lumen, providing an osmotic force for water efflux (Olver and Strang, 1974). At birth, the water must be removed as the lung switches to an air-conducting system. This is achieved by the activation of a Na+ uptake system, and the absorption of Na+ leads to an associated osmotic uptake of water (Strang, 1991). The Cl− secretion also declines following birth. Na+ uptake is localised to the type II pnemocytes of the lung alveolae and is mediated by amiloride-sensitive channels with properties most closely resembling those of ENaC (Kemp and Olver, 1996).

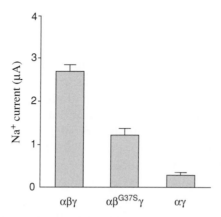

FIGURE 13.10 PHA-1 MUTATIONS RESULT IN A REDUCED
ENaC CURRENT
Macroscopic Na$^+$ currents were recorded from oocytes injected with
mRNA encoding wild-type α and γENaC together with either wild-type
βENaC or βENaC containing the G37S mutation associated with PHA-1.
From Chang et al. (1996).

Clearly, Na$^+$ and water uptake must be timed to coincide with birth. This
is achieved by the enhanced surface expression of ENaC in the weeks prior
to birth and a marked increase of ENaC activity at parturition. In rats, expression of αENaC mRNA increases dramatically just before birth. Although
similar data is not available for man, it is known that αENaC is not expressed
in human fetal lung (20–25 weeks) but is expressed at high levels in adults
(Voilley et al., 1994). It is therefore likely that the rise in glucocorticoid hormones that occurs prior to birth triggers αENaC expression. A similar increase
in aquaporin expression and water permeability in the lung also occurs
around birth (Carter et al., 1997; Chapter 19). Activation of ENaC at parturition
results from the elevated adrenaline levels which occur in the baby during
labour. In sheep, adrenaline levels begin to rise and water uptake is initiated
between 150 and 50 min before delivery (Brown et al., 1983). As might perhaps
be expected, infants delivered by elective caesarean section, who do not
experience the endocrine changes associated with labour, take longer to clear
their lungs of water and experience a higher incidence of respiratory complications in the postnatal period.

The importance of ENaC-mediated Na$^+$ uptake at birth is demonstrated
by the fact that mice in which αENaC expression in the airway epithelia has
been abolished die within 40 hr of birth of respiratory distress caused by a
failure to clear their lungs of liquid (Hummler et al., 1996). In addition,
amiloride treatment delays the clearance of liquid from the lungs of newborn
guinea pigs (O'Brodovich et al., 1990). Whether diminished ENaC activity
contributes to respiratory distress syndrome (RDS) in premature human infants remains controversial. It is widely believed that the principal cause of
RDS is lack of surfactant, a lipid secreted by the lung epithelium which lowers

the surface tension of the alveoli and prevents their collapse. However, other factors may also be important since RDS continues to be a major cause of mortality in premature infants despite surfactant replacement therapy. Insufficient ENaC expression due to immature development may be one such factor. It is well known that glucocorticoids are important in lung development and that the combined administration of glucocorticoid and the thyroid hormone triiodothyronine (T3) causes maturation of the immature fetal lung. This is mediated not only by stimulation of surfactant secretion, but also by induction of Na⁺ uptake (Barker *et al*, 1990). Because T3 potentiates the stimulatory action of steroids on αENaC expression (Champigny *et al.*, 1994), and both hormones normally rise *in vivo* prior to birth, the failure of premature babies to clear their lungs of fluid may reflect a low level of ENaC expression. In support of this idea, a reduced Na⁺ absorption is observed in the nasal epithelium of premature infants suffering from respiratory distress syndrome (Barker *et al.*, 1997). This can be measured easily, because defective Na⁺ uptake causes a decrease in the amiloride-sensitive potential difference (PD) across the nasal epithelium. An abnormal nasal PD is also observed at birth in transient tachypnea of the newborn, a condition in which babies delivered at term have wet lungs despite normal surfactant levels, and in neonates delivered by caesarian section without prior labour (Gowen *et al.*, 1988). Both these conditions, therefore, also appear to be associated with low ENaC activity.

Cystic Fibrosis

Cystic fibrosis (CF) results from mutations in the cystic fibrosis transmembrane conductance regulator (CFTR), which is both an epithelial chloride channel and acts as a regulator of ENaC activity. Patients with cystic fibrosis fail to secrete Cl⁻ from epithelial cells of the airways, sweat ducts and exocrine glands. In addition, they show abnormally high levels of Na⁺ uptake in airway epithelia. Cystic fibrosis in general, and the function of CFTR as a Cl⁻ channel, are discussed in detail in Chapter 12. Here, we briefly consider how the loss of ENaC regulation contributes to the pathogenesis of the disease.

In 1986, Boucher and colleagues demonstrated that the airway epithelia of CF patients had an abnormally high Na⁺ permeability. It was subsequently shown that this was a consequence of the loss of CFTR regulation of ENaC activity (Stutts *et al.*, 1995). As discussed earlier, CFTR mediates the inhibitory effect of cAMP on ENaC activity. The absence of CFTR in the surface membrane, as occurs with the most common CF mutation—a deletion of phenylalanine 508—will therefore result in enhanced ENaC activity. The increased Na⁺ absorption will lead to an accompanying uptake of water, thus drying the lungs. This may enhance mucus accumulation and predispose the lung to the infections that ultimately cause the death of CF patients. Amiloride applied as an aerosol reduces the excessive Na⁺ absorption and causes a partial amelioration of the decline in pulmonary function.

Neuronal Degeneration

A family of *C. elegans* genes involved in touch sensitivity, which includes *mec-4*, *mec-10* and *deg-1*, share significant sequence homology with the different ENaC subunits (Hamill and McBride, 1996). Dominant mutations in these genes cause neuronal degeneration. Consequently, their gene products are referred to as degenerins. All of the mutations in *mec-4* that cause neurodegeneration occur in the same residue, an alanine at position 442 (A442) which is located at the extracellular face of the second transmembrane domain. Mutation of A442 causes selective degeneration of certain sensory neurones, possibly by hindering pore closure (Hong and Driscoll, 1994). The affected neurones are observed to swell to several times their normal diameter and develop large vacuoles before they die due to cell lysis. This suggests that enhanced channel activity may be integral to the degeneration and that the water influx accompanying increased salt uptake is responsible for the observed cell swelling.

While dominant mutations in *mec-4* and *mec-10* cause neurodegeneration, recessive mutations result in the loss of mechanosensitivity. Because *mec-4* and *mec-10* show homology with ENaC, it has been suggested that their gene products may form part of a mechanosensitive channel involved in touch sensitivity. Genetic interactions have identified another gene, *mec-6*, which is required for *mec-4* and *mec-10* function. Thus, the mechanosensitive channel of *C. elegans* may be composed of three types of subunit, in a manner analogous to the α,β,γ subunits of ENaC (Huang and Chalfie, 1994). This channel is only part of a much larger protein complex because 13 genes are actually needed for touch cell function. Since recessive mutations in *mec-4* and *mec-10* result in the loss of touch sensitivity, it seems likely that they produce a reduction or complete loss of channel activity.

A degenerin-like channel, initially known as MDEG or BNaC1, has also been cloned from human brain (Waldmann *et al.*, 1996; García-Anoveros *et al.*, 1997). This channel has been renamed ASIC2, as it is now known to belong to the H^+-gated cation channel family. It did not express in *Xenopus* oocytes or mammalian cell lines unless a point mutation was introduced at glycine 430, the position corresponding to the site of mutations which cause neurodegeneration in *C. elegans*. Cells expressing the mutant form of ASIC2 exhibited an amiloride-sensitive Na^+ current and died rapidly. The role of ASIC2 in human brain is unknown but it is clear that it would cause neuronal degeneration if it were to become permanently activated. Possibly, therefore, it may be implicated in human neurodegenerative disease.

CHAPTER 14

LIGAND-GATED Ca^{2+} CHANNELS

Two major classes of intracellular ligand-gated Ca^{2+} channels have been identified: the ryanodine receptors (RyR) and the inositol 1,4,5-triphosphate receptors (IP_3R). These mediate the release of Ca^{2+} from intracellular stores. Mutations in the ryanodine receptor of skeletal muscle cause malignant hyperthermia in man and pig. Disease-causing mutations in the IP_3 receptor have not been described in man, but defective regulation of channel activity may cause Lowe's oculocerebrorenal syndrome.

Because the RyR and IP_3R channels are located in intracellular membrane systems, it is difficult to study their single-channel properties *in situ*. Instead, single-channel currents have usually been recorded from artificial lipid bilayers following incorporation of the purified protein, or fusion of vesicles prepared from intracellular membranes. An alternative approach has been to monitor the activity of the channel indirectly by measuring changes in intracellular Ca^{2+}, or Ca^{2+}-activated currents, or by measuring Ca^{2+} release from membrane vesicles. Two techniques that allow the measurement of intracellular ion channels *in situ* have been reported. The first of these capitalizes on the fact that the endoplasmic reticulum (ER) membrane is continuous with that of the nucleus, which makes it possible to record the activity of ion channels in native ER membranes by patch-clamping isolated nuclei (Stehno-Bittel *et al.*, 1995). The second technique allows channel activity to be recorded from intracellular membranes of intact cells and is somewhat of a technical *tour-de-force*. It uses two concentric patch electrodes: an outer electrode, used to penetrate the plasma membrane of the cell, and an inner electrode, whose tip is exposed only after penetration, which is used to form a seal on the intracellular membrane (Jonas *et al.*, 1997). It is to be expected that both these methods will be used more widely in the future.

RYANODINE RECEPTORS

Ca^{2+} release from the sarcoplasmic reticulum (SR) of skeletal muscle is mediated by a Ca^{2+} channel which is usually referred to as the ryanodine receptor, because it binds the plant alkaloid ryanodine with high affinity. This property was used to purify and clone the protein. Reconstitution of the purified skeletal muscle ryanodine receptor into lipid bilayers confirmed that it acts as a Ca^{2+} channel. Ryanodine receptors are not only found in the sarcoplasmic reticulum membranes of skeletal muscle, they also occur in the intracellular membranes of other tissues such as brain and epithelial cells where they mediate Ca^{2+} release from intracellular stores. All ryanodine receptors are activated by an increase in the cytoplasmic Ca^{2+} concentration and this property forms the basis of the phenomenon known as Ca^{2+}-dependent Ca^{2+} release.

Structure

The skeletal muscle ryanodine receptor (RyR1) is one of the largest ion channels cloned to date. It is so large, in fact, that its gross structure can be resolved by cyroelectron microscopy (Radermacher *et al.*, 1994; Serysheva *et al.*, 1995). The purified receptor is a tetramer composed of four identical subunits arranged in a quatrefoil or cloverleaf shape. It consists of a large cytoplasmic assembly which measures ~270 × 270 × 125 Å and a smaller transmembrane assembly that protrudes 65 Å from the lower surface of the tetramer (Fig. 14.1). A central channel, 20–30 Å in diameter, appears to run down the centre of the transmembrane assembly and may correspond to the pore of the RyR. Intriguingly, this channel is plugged at the cytoplasmic end of the transmembrane domain. However, in some images, the central channel appears to be radially connected to four peripheral channels, one in each monomer, which extend from partway across the membrane to its cytoplasmic edge. The cytoplasmic assembly has a striking lacelike structure. When viewed from the cytoplasmic side, its centre contains a depression with a diameter of ~50 Å, which may represent the outer mouth of the Ca^{2+} channel pore and into which the four radial channels possibly open.

Each RyR1 monomer has a molecular mass of 565 kDa and is 5037 amino acids long (Takeshima *et al.*, 1989). The putative topology is given in Fig. 14.2. Most of the protein lies outside the membrane in the cytoplasm and forms the foot process observed to span the gap between the SR and T-tubule membranes in electron micrographs of skeletal muscle (Fig. 14.3). The membrane-spanning regions, which contain the pore, are located in the C terminus of the molecule. The number of transmembrane domains (TMs) is still debated and there are several contending models, which postulate between 4 and 12 TMs (Takeshima *et al.*, 1989; Otsu *et al.*, 1990; Franzini-Armstrong and Protasi, 1997).

Three types of ryanodine receptors have been cloned to date: RyR1, RyR2, and RyR3 (Table 14.1). RyR1 is mainly expressed in skeletal muscle, RyR2 in brain and cardiac muscle and RyR3 is found in epithelia, parts of the brain and, to a lesser extent, skeletal muscle. Each RyR2 subunit is 4968 residues in length and shows 66% sequence identity with RyR1 (Nakai *et al.*, 1990). The RyR3 receptor is 4872 amino acids long and 67% identical with RyR1

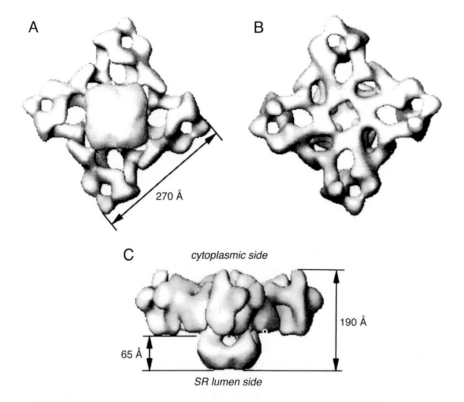

FIGURE 14.1 ELECTRON MICROSCOPY OF THE PURIFIED
RYANODINE RECEPTOR
Schematic representation of the three-dimensional structure of the ryano-
dine receptor in its closed state, viewed from the lumen (A), the cytoplasm
(B), and the side (C). From Serysheva *et al.* (1995).

(Hakamata *et al.*, 1992). The location of the different human RyR genes is
given in Table 14.1. Alternatively spliced variants of both RyR1 and RyR2
have been identified.

Functional Roles

The ryanodine receptors RyR1 and RyR2 play important roles in the contraction of skeletal and cardiac muscle, respectively, as they mediate the release
of Ca^{2+} from the intracellular store known as the sarcoplasmic reticulum,
which is required to activate the contractile proteins. In both types of muscle,
depolarisation of the T-tubule membrane initiates SR Ca^{2+} release, but the
mechanism of excitation–contraction coupling differs in heart and skeletal
muscle. The cardiac ryanodine receptor (RyR2) is activated by Ca^{2+} ions
entering the cell from the extracellular solution through voltage-gated L-type
Ca^{2+} channels. The amount of Ca^{2+} that enters is not sufficient to activate the

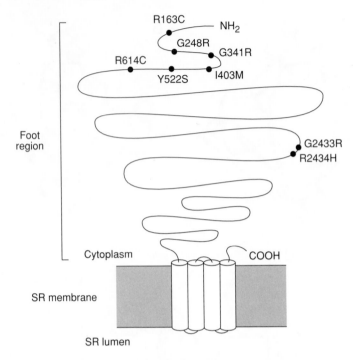

FIGURE 14.2 PUTATIVE TOPOLOGY OF THE RYANODINE RECEPTOR, RYR1

Putative membrane topology of RYR1. Both N and C termini are thought to be cytosolic, but the number of transmembrane domains is debated. Residues associated with malignant hyperthermia are indicated.

contractile proteins directly but it triggers the release of much larger amounts of Ca^{2+} from the SR by opening RyR2 channels. A different mechanism of excitation–contraction coupling has been adopted in skeletal muscle, possibly to ensure more rapid Ca^{2+} release and thus faster contraction (see also Chapter 9). Although the voltage-gated L-type Ca^{2+} channel in the plasma membrane also plays a key role in this process, Ca^{2+} entry from the extracellular solution is not required. Instead, the α_1-subunit of the voltage-gated Ca^{2+} channel acts as the voltage sensor for the ryanodine receptor, linking depolarization of the T-tubule membrane to opening of the ryanodine receptor in the adjacent SR membrane (Fig. 14.4). Franzini-Armstrong and colleagues (1997) have explored the structure of the junction between the T-tubule and SR membranes. These studies have revealed that at specialised regions, known as triads, the two membranes come together in close apposition. They are separated by a gap of ~120 Å, which is bridged by electron-dense structures known as foot processes (Fig. 14.3). The foot processes have the same size and quatrefoil shape as purified RyR1 and are thus believed to correspond to ryanodine receptors. Freeze-fracture electron microscopy has shown that every other foot process is positioned exactly in register with a group of four particles (a tetrad) in the opposing T-tubule membrane. Each tetrad is

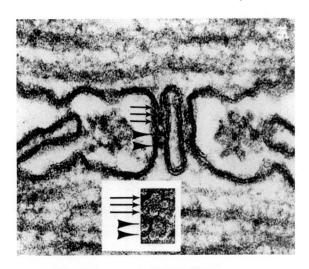

FIGURE 14.3 STRUCTURE OF THE TRIAD
Electron microscope picture of the triad junction between the T-tubule membrane (centre) and the SR membrane (left and right) of skeletal muscle. Arrows indicate the electron-dense structures (foot processes) which bridge the gap between the two sets of membranes. The inset shows that when viewed from above, the foot appears to be composed of four subunits. From Ferguson *et al.* (1984).

considered to be a cluster of four voltage-gated Ca^{2+} channels, because they are absent in dysgenic muscle, which lacks functional voltage-gated Ca^{2+} channels.

It is believed that the voltage-gated Ca^{2+} channel and RyR1 do not interact directly and that additional components are involved, but their identity has not yet been established. RyR1 retains the ability to be activated by Ca^{2+} and the fact that alternate RyRs are not associated with tetrads suggests that this effect may be of physiological importance. Thus it has been postulated that

TABLE 14.1   HUMAN LIGAND-GATED Ca^{2+} CHANNEL GENES

Gene	Chromosome location	Tissue distribution	Reference
RYR1	19q13.1	Mostly skeletal muscle,	MacKenzie *et al.* (1990)
RYR2	1	Cardiac muscle, brain	Otsu *et al.* (1990)
RYR3	15q14–q15	Epithelia, some brain regions, some skeletal muscles	Sorrentino *et al.* (1993)
ITPR1	3p25–p26	Brain (Purkinje cells), uterus, vascular smooth muscle	Yamada *et al.* (1994)
ITPR2	12p11	Brain	Yamamoto-Hino *et al.* (1994)
ITPR3	6p21	Pancreatic islets, kidney, gastrointestinal tract	Yamamoto-Hino *et al.* (1994)

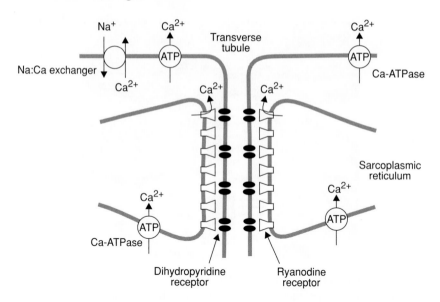

FIGURE 14.4 MODEL FOR RYANODINE RECEPTOR/L-TYPE Ca²⁺ CHANNEL INTERACTION
Structure of the junction between the T-system and sarcoplasmic reticulum membranes of skeletal muscle with proteins involved in EC coupling indicated. Four dihydropyridine receptors (L-type Ca²⁺ channels) form a tetrad that associates with alternate RYRs: only two Ca²⁺ channels in each tetrad are shown for clarity. Also shown are proteins responsible for returning the cytosolic Ca²⁺ to its resting level following contraction (the plasma membrane and SR Ca-ATPases, and the NaCa exchanger).

Ca^{2+}-dependent Ca^{2+} release, subsequent to voltage-activated release, might serve to amplify the amount of Ca^{2+} released in skeletal muscle.

Properties

Although the ryanodine receptor serves as the conduit for Ca^{2+} release from intracellular Ca^{2+} stores, it does not discriminate strongly between different cations, and the relative Ca^{2+}/K^+ permeability of RyR1 is only ~6 (Smith *et al.*, 1988). Both native and cloned RyRs exhibit multiple conductance states. The most common is 100–150 pS with 50 mM luminal Ca^{2+}. In both these respects, RyRs differ from the voltage-gated Ca^{2+} channels (Chapter 9). Studies of the permeability of RyRs to ions of different size suggest that the selectivity filter lies towards the lumenal side of the membrane and has a minimum diameter of ~7 Å (Tinker and Williams 1993). The location of the pore within the protein is not clear.

The ryanodine receptor is modulated by a variety of second messengers and intracellular ligands. The most important of these is cytosolic Ca^{2+}. The relationship between channel activation and cytoplasmic Ca^{2+} is bell-shaped, with maximum activation around 5 μM $[Ca^{2+}]_i$ (Fig. 14.5). For this reason it has been proposed that the RyR has two Ca^{2+}-binding sites: a high-affinity site that induces channel opening and a low-affinity site which inhibits the channel. The precise location of these Ca^{2+}-binding sites in the molecule remains controversial (Franzini-Armstrong and Protasi, 1997). The ability of high Ca^{2+} concentrations to inhibit Ca^{2+} release may be of functional signifi-cance since it provides a mechanism for limiting the amount of Ca^{2+} released and so prevents a small amount of activator Ca^{2+} from emptying the Ca^{2+} stores in a regenerative positive feedback fashion. The mechanism by which Ca^{2+} release is terminated is, however, controversial and other hypotheses have also been suggested, including the local depletion of Ca^{2+} within the SR in the vicinity of the RyR.

Other ligands of physiological importance are ATP and cyclic ADP-ribose. Millimolar concentrations of ATP enhance the sensitivity to Ca^{2+} by shifting the relationship between channel activation and cytoplasmic $[Ca^{2+}]_i$ to lower concentrations. In sea urchin eggs, cyclic ADP ribose activates the ryanodine receptor and thereby plays an important role in the response to fertilization. The role of cyclic ADP ribose in mammalian cells is still unclear. The cardiac RyR is also the substrate for several protein kinases, including protein kinase

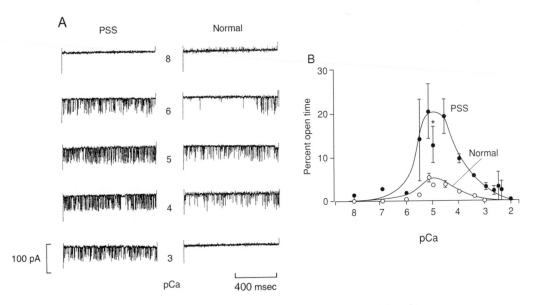

FIGURE 14.5 Ca^{2+} SENSITIVITY OF NORMAL AND MH Ca^{2+} RELEASE CHANNELS

Ca^{2+} dependence of single Ca^{2+} release channels isolated from normal (right) or porcine stress syndrome susceptible (PSS, left) pig skeletal mus-cle and reconstituted into a lipid bilayer. pH 6.8. (A) Single-channel cur-rents recorded with 200 mM K^+ as the permeant ion. Ca^{2+} was added to the *cis* side of the bilayer. (B) Open probability as a function of the *cis* Ca^{2+} concentration. From Shomer *et al.* (1994).

A (PKA). Phosphorylation by PKA enhances Ca^{2+} release and may contribute to the increased contractile force observed in cardiac muscle in response to catecholamines.

Pharmacology

The ryanodine receptor is the target for a large number of pharmacological agents, many of which have complex effects (Zucchi and Ronca-Testoni, 1997). In general, drugs that enhance RyR function act by shifting the relationship between channel activation and cytosolic Ca^{2+} to lower Ca^{2+} concentrations while leaving the single-channel conductance unaffected. The classical example is caffeine, which enhances the Ca^{2+} sensitivity of RyR1 so that significant channel opening, and thus Ca^{2+} release, occurs at resting (nM) Ca^{2+} levels. Volatile anaesthetics, such as halothane and enflurane, also enhance the activity of both cardiac and skeletal muscle RyRs. As these effects are observed at concentrations which are lower than their minimally effective alveolar concentrations, they may be of clinical importance. The cardiac glycoside digoxin activates RyR2 (but not RyR1) at therapeutic concentrations, an action which may contribute to the inotropic action of the drug.

A number of drugs block ryanodine receptors. These include ruthenium red, which inhibits both RyR1 and RyR2 and is used as a research tool, and local anaesthetics, such as procaine and tetracaine, which block RyR1. Dantrolene is a postsynaptic muscle relaxant and is the chief drug used to treat malignant hyperthermia (see later). At therapeutic concentrations (10 μM) it blocks Ca^{2+} release from the SR of skeletal muscle and it is likely that this constitutes its main mechanism of action. There is some evidence that dantrolene may not interact directly with RyR1.

Proteins Associated with RyR Channels

The ryanodine receptor interacts with a number of proteins, including triadin, calsequestrin, calmodulin, and FKBP12. Triadin is a 95-kDa SR membrane protein that interacts with the ryanodine receptors of skeletal and cardiac muscle. It has been proposed that triadin may link the channel to calsequestrin, a low-affinity, high-capacity Ca^{2+}-binding protein found in the SR lumen. The Ca^{2+}-binding protein calmodulin binds directly to RyR and decreases channel opening at concentrations (2 μM) likely to occur *in vivo*.

FKBP12 (FK506-binding protein of 12 kDa) associates with the ryanodine receptor of skeletal muscle in a one-to-one stoichiometry so that four FKBP12 molecules are bound to each RyR1 tetramer. Coexpression of FKBP12 with RyR1 modifies the gating of the cloned channel by stabilising the fully open conductance level and reducing the number of subconductance states, effects that are reversed on addition of the immunosuppressant drug FK506, which binds to FKBP12 (Brillantes *et al.*, 1994). Immunosuppressant drugs have similar effects on RyR channels in native membranes (Ahern *et al.*, 1994).

This may be attributed to the removal of the regulatory effect of FKBP12, as immunosuppressants have no effect on cloned RyR1 channel function (Brillantes *et al.*, 1994). A related protein, FKBP12.6, is found in cardiac muscle; it associates with RyR2 and modifies its activity in a similar way. Cardiac hypertrophy has been reported in some patients during treatment with high doses of FK506 (tacrolimus); it is possible that this results from alteration of RyR2 function (Atkison *et al.*, 1995).

DISEASES ASSOCIATED WITH RYANODINE RECEPTORS

Malignant Hyperthermia

Malignant hyperthermia (MH) is an autosomal dominant neuromuscular disorder in which common inhalation anaesthetics or muscle relaxants trigger a potentially fatal state, characterized by accelerated skeletal muscle metabolism, muscle contractures, hyperkalaemia, arrhythmias, respiratory and metabolic acidosis, and a rapidly rising body temperature. The latter may increase by as much as 1°C every 5 min and can exceed 43°C. The condition affects about 1 in 20,000 adults. It is usually only detected when the patient is given a volatile anaesthetic such as halothane or enflurane or a depolarizing muscle relaxant such as suxamethonium. The MH-induced increase in body temperature can be fatal unless the patient is immediately treated, and dantrolene sodium, which blocks SR Ca^{2+} release, is kept in all operating theatres for MH emergencies.

MH results from a sustained increase in intracellular Ca^{2+} ($[Ca^{2+}]_i$) in skeletal muscle that activates both metabolic and contractile activity (Mickelson and Louis, 1996). The former results in respiratory and metabolic acidosis and the latter produces the elevation in body temperature.

Pigs provide the key

Malignant hyperthermia is also found in pigs and, until recently, was widespread in the pig population of the United Kingdom. In pigs, MH is known as porcine stress syndrome (PSS), as it is usually precipitated by stress. The PSS pig has proved to be an invaluable animal model for studying the molecular basis of malignant hyperthermia. In contrast to normal muscle, in PSS muscle fibres halothane induces a rise in resting $[Ca^{2+}]_i$. The rate and extent of Ca^{2+}-induced Ca^{2+} release from the SR of skeletal muscle are also higher in PSS pigs than normal animals, although the Ca^{2+} dependence of release and the resting $[Ca^{2+}]_i$ appear to be normal (Ohta *et al.*, 1989). Analysis of single-channel currents recorded following the incorporation of purified porcine RyRs into lipid bilayers has shown that PSS-RyR1 channels have longer open times and shorter closed times than normal channels (Shomer *et al.*, 1994). This produces an increase in the channel open probability (Fig. 14.5), which accounts for the enhanced Ca^{2+}-induced Ca^{2+} release observed in flux studies. The single-channel conductance is unaffected. Together, these studies convincingly demonstrate that MH in pigs results from a defect in the ryanodine receptor of skeletal muscle. Conclusive proof that is the case came with the demonstration that PSS results from a single mutation in the RyR1 gene, an arginine-to-cysteine substitution at position 615 (Fujii *et al.*, 1991). When muscle cells were transfected with RyR1 carrying the R615C mutation, clinical doses of halothane caused a rapid increase in $[Ca^{2+}]_i$, which was not observed when the normal gene was expressed (Otsu *et al.*, 1994). Arginine 615 lies within the large cytoplasmic foot structure (Fig. 14.2), but it is not understood how its mutation affects the gating of the channel or induces PSS.

PSS is of considerable economic importance because it can be triggered by exercise, sex (in boars), parturition, the stress of transporting the pigs to market, or even by the conditions in which the animals are kept (e.g., overcrowding). Not only do the afflicted pigs die, but the meat also becomes very tough and is then unsaleable. All affected pigs carry the same mutation and derive from a single founder animal. The high incidence of PSS in the United Kingdom arose because pigs were selectively bred for lean meat and decreased back fat, attributes that unfortunately turned out to be associated with the gene for malignant hyperthermia. The PSS gene has now been almost completely bred out of the U.K. pig population by giving each pig 3% halothane to breathe and removing those pigs that developed muscle rigidity, and a rise in temperature of 2°C within 5 min, from the breeding pool.

Malignant hyperthermia in man

In man, malignant hyperthermia is a genetically heterogeneous disease. In a number of families it maps to chromosome 19q13.1, the same locus as the gene encoding RyR1. In some of these families, mutations in *RYR1* have been identified and the search for others will be aided by the description of the structure of the *RYR1* gene (Phillips *et al.*, 1996). In other families, however, *RYR1* is not linked to MH and the cause of the disease remains unknown. Muscle biopsy and *in vitro* testing for a reduced contractile threshold to halothane and caffeine remain the standard tests for MH susceptibility since the disease is genetically heterogeneous and not all mutations have been identified.

The genetic basis of MH was first identified in the pig (see earlier) and the corresponding mutation (R614C) was subsequently found in about 2% of human MH families. A number of other mutations have also been identified (Fig. 14.2; Table 14.2). In Europe, the most common of these is G341R, which is found in around 10% of MH families. All mutations which cause MH, or central core disease (see later) are clustered in two regions of the cytoplasmic foot domain of RyR1. The functional consequences of most of these mutations have not yet been determined, and analysis of skeletal muscle biopsies has

TABLE 14.2   MUTATIONS IN *RYR1*

Phenotype	Codon change	Reference
MH, CCD	R163C	Quane *et al.* (1993)
MH	G248R	Gillard *et al.* (1992)
MH	G341R	Quane *et al.* (1994)
MH, CCD	I403M	Quane *et al.* (1993)
MH, CCD	Y522S	Quane *et al.* (1994)
MH	R614C	Gillard *et al.* (1991)
PSS	R615C (pig)	Fujii *et al.* (1990)
MH	G2433R	Keating *et al.* (1994)
MH, CCD	R2434H	Zhang *et al.* (1993)

not been as informative in man as in the pig because the data are fewer and the results sometimes contradictory. This may be due in part to the difficulty in obtaining tissue of the required quality and in part to genetic heterogeneity in MH. However, all data agree that both RyR1 channel activity and Ca^{2+} release are consistently stimulated by lower concentrations of halothane and caffeine in MH muscle than in normal muscle.

Central Core Disease

Central core disease (CCD) is a rare autosomal dominant nonprogressive myopathy that presents in infancy as proximal muscle weakness and hypertonia. Diagnosis is by muscle biopsy, which reveals that regions of type 1 skeletal muscle fibres (known as 'central cores') are depleted of mitochondria and oxidative enzymes. The disease is often associated with a predisposition to malignant hyperthermia and this observation led to the demonstration that CCD is linked to the *RYR1* locus on chromosome 19q13.1 and results from mutations in *RYR1* (Table 14.1; Quane *et al.*, 1993; Zhang *et al.*, 1993). Thus both CCD and MH appear to be allelic disorders of the same gene. It is not clear how the different phenotypes arise, especially since the same mutation can give rise to MH in some individuals and CCD in others. However, since all CCD patients are MH-susceptible, it is possible that additional factors are necessary for the development of central core disease. One possible explanation for the mitochondrial atrophy observed in CCD is that RyR mutations lead to enhanced intracellular Ca^{2+} release. At the periphery, the excess Ca^{2+} will be extruded by surface membrane Ca^{2+} pumps, but in the centre of the fibre Ca^{2+} levels may be buffered by mitochondria. Accumulation of Ca^{2+} within the organelle may lead to mitochondrial degeneration and thus to the development of 'central cores.'

Dyspedic Mice

Dyspedic mice, in which the RyR1 gene has been disrupted (knocked out), lack functional ryanodine receptors in their skeletal muscles and exhibit complete loss of excitation–contraction coupling (Takeshima *et al.*, 1994). The lack of 'foot processes' at the T-tubule–SR junction gave the mice their name and provided conclusive proof that the RyR indeed constitutes the foot process. Like the dysgenic mouse (Chapter 9), dyspedic mice die at birth from respiratory failure. Studies of myotubes isolated from dyspedic mice confirmed that they do not twitch in response to electrical stimulation, but also showed that caffeine induces a small contracture. The latter effect is attributable to the presence of RyR3 in skeletal muscle, which can support Ca^{2+}-dependent Ca^{2+} release but is unable to couple to the voltage-gated Ca^{2+} channel. Dyspedic myotubes have also provided information about the relationship between the ryanodine receptor and the L-type Ca^{2+} channel (Nakai *et al.*, 1996). Surprisingly, dyspedic mice have Ca^{2+} currents which are ~30 times smaller than wild-type animals. The amount of Ca^{2+} current gating charge, however, is close to normal, which suggests that dyspedic mice have the normal complement of Ca^{2+} channels and that their smaller Ca^{2+} currents result from a reduction in either the channel open probability or the single-channel conductance. Injection of dyspedic myotubes with cDNA encoding RyR1 restored the magnitude of the L-type Ca^{2+} current to normal levels. This suggests that the ryanodine receptor is able to modulate the activity of the L-type Ca^{2+} channel, although how it might do so is unknown. By contrast, RyR2 was unable to restore either excitation–contraction coupling or the amplitude of the Ca^{2+} current, suggesting that it is unable to receive or send signals to the L-type Ca^{2+} channel.

IP₃ RECEPTORS

Inositol 1,4,5-trisphosphate (IP₃) is an important second messenger that mediates the effects of a number of hormones, neurotransmitters, and growth factors (Berridge, 1993). It is produced by the action of phospholipase C, which splits the membrane lipid phosphotidylinositol 4,5-bisphosphate into diacylglycerol and IP₃ (Fig. 14.6). The former remains associated with the plasma membrane, but IP₃ is released into the cytosol and diffuses to intracellular Ca^{2+} stores where it triggers the release of Ca^{2+} by binding to a ligand-gated Ca^{2+} channel. The term IP₃ receptor is used to refer to these IP₃-gated Ca^{2+} channels, which structurally resemble the ryanodine receptors discussed earlier. This structural homology accounts for the many functional similarities which exist between IP₃ and RyRs.

IP₃-gated Ca^{2+} release plays an important role in excitation–contraction coupling in smooth muscle by mobilising Ca^{2+} release from intracellular stores in response to hormonal stimulation (e.g., acetylcholine). In the nervous system, IP₃ receptors are involved in olfaction and they mediate the synaptic transmission and hormonal secretion stimulated by certain neurotransmitters. They are also involved in the regulation of cell growth, cell division and, probably, apoptosis.

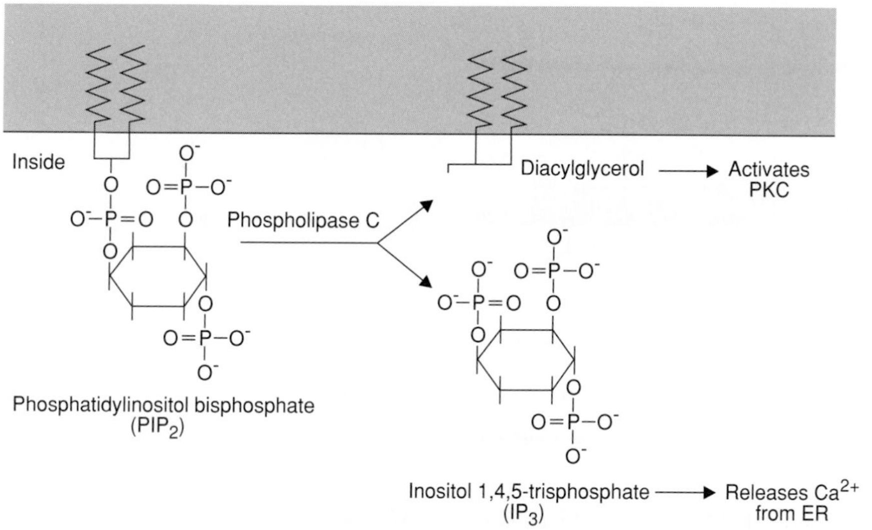

FIGURE 14.6 MECHANISM OF IP₃ GENERATION

Cleavage of PIP_2 by the enzyme phospholipase C leaves diacylglycerol attached to the membrane and releases IP_3. Both substances are able to serve as second messengers and activate effector molecules.

Basic Properties

The properties of the native IP_3-gated Ca^{2+} channel have been inferred from IP_3-stimulated Ca^{2+} fluxes from membrane vesicles or, more directly, analysed by recording single-channel currents following fusion of membrane vesicles with artificial lipid bilayers (Fig. 14.7A; Bezprozvanny and Ehrlich, 1994). With 55 mM *trans* (intraluminal) Ca^{2+}, the single-channel conductance (γ) is 53 pS, while under physiological conditions the estimated γ is ~0.5 pS, about fourfold lower than that of the RyR. Like the ryanodine receptor, the IP_3 receptor has a low cation selectivity ($P_{Ba}/P_K \sim 6$) and an open probability which is independent of membrane potential.

Inositol 1,4,5-trisphosphate binds to the purified receptor with a K_d of ~100 nM, when measured in the presence of 1 mM EDTA using 2.5 nM radiolabeled IP_3 (Supattapone *et al.*, 1988). It is considered to be the primary ligand for the IP_3-gated Ca^{2+} channel. Channel activity is also strongly modulated by the intracellular Ca^{2+} concentration and because IP_3 has little effect in the absence of $[Ca^{2+}]_i$, Ca^{2+} is sometimes referred to as a coagonist. Figure 14.7B shows that the Ca^{2+} dose–response curve is bell shaped, channel activity being first enhanced and then suppressed as Ca^{2+} is increased (Bezprozvanny *et al.*, 1991). Maximal activation is achieved with ~300 nM $[Ca^{2+}]_i$. The Ca^{2+} dependence of the IP_3 receptor suggests it might function as a Ca^{2+}-gated channel if $[Ca^{2+}]_i$ were varied at a constant IP_3 concentration. It also suggests

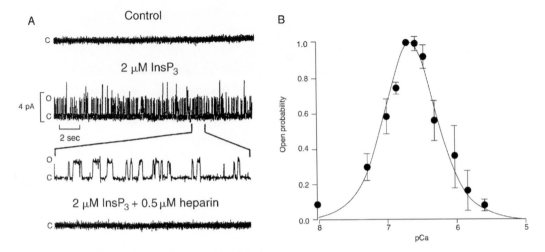

FIGURE 14.7 IP₃ RECEPTORS ARE ACTIVATED BY IP₃ AND Ca²⁺
(A) Single-channel currents through cerebellar IP₃-gated channels recorded following incorporation into an artificial lipid bilayer. Barium
(55 mM) was used as the permeant ion and all traces were recorded in a
solution containing 0.2 μM Ca²⁺ and 0.5 mM ATP at a membrane potential
of 0 mV. IP₃ channels were activated by IP₃ and blocked by heparin.
From Bezprozvanny and Erlich (1994). (B) Ca²⁺ dependence of the open
probability of IP₃-gated channels. From Bezprozvanny *et al.* (1991).

that the Ca²⁺ flux through the IP₃ receptor will produce first a positive, and
then a negative, feedback regulation of its own release, giving rise to a rapid
and transient increase of [Ca²⁺]$_i$. Put another way, at Ca²⁺ concentrations
below 300 nM, Ca²⁺ release will stimulate further IP₃ receptor activation but
as soon as Ca²⁺ rises above 300 nM, channel activity will be inhibited and
calcium release suppressed. This effect of Ca²⁺ on the gating of the IP₃ receptor
is similar to that observed for the ryanodine receptor.

The IP₃-gated Ca²⁺ channel is modulated by a range of physiological and
pharmacological agents (Bezprozvanny and Ehrlich, 1994). One of the most
important of these is ATP, which acts as an allosteric activator of channel
activity. The mechanism of ATP action does not involve phosphorylation:
rather, ATP binds directly to the protein to gate the channel. However, phosphorylation of type 1 IP₃ receptors by protein kinase A reduces channel
activity by shifting the relationship between Ca²⁺ and channel activation to
higher Ca²⁺ levels (it has no effect on ligand binding). Cameron and colleagues
(1995) have shown that FKBP12 also binds to IP₃ receptors and modifies Ca²⁺
fluxes in a manner similar to that found for FKBP12 modulation of RyR.
Inhibitors of IP₃ receptors include arachidonic acid, caffeine and heparin; the
last binds to the channel with high affinity and was used as a ligand to purify
the protein from brain.

Cloning of the IP$_3$ Receptor

The IP$_3$ receptor (IP$_3$R1) was cloned by first purifying an IP$_3$-binding protein from rat cerebellum and then screening a cDNA library with probes based on the partial sequence of this protein (Supattapone *et al.*, 1988; Mignery *et al.*, 1989). At about the same time, IP$_3$R1 was also cloned serendipitously, by screening an expression library with an antibody raised against a protein that was enriched in the cerebellum of normal mice and substantially reduced in the cerebellum of the Purkinje cell-deficient mice mutants known as *staggerer, nervous,* and *Purkinje-cell degeneration* (Furuichi *et al.*, 1989). This protein was subsequently found to be IP$_3$R1. The receptor was found to be present in all tissues tested, but is particularly abundant in cerebellar Purkinje cells and cerebral cortex.

Alternative splicing of the IP$_3$R1 gene produces several receptor subtypes, which are differentially expressed. Other genes encode IP$_3$R2 and IP$_3$R3 receptors that are similar in sequence and structure to IP$_3$R1 (60–70% identity). The gene encoding IP$_3$R1 is located on human chromosome 3p25–26, that for IP$_3$R2 on chromosome 12p11, and that for IP$_3$R3 on chromosome 6p21 (Table 14.1).

Structure

Electron microscope images indicate that, like the ryanodine receptors, the IP$_3$ receptor is a tetramer. Since each monomer has a single IP$_3$-binding site, there are four sites in each IP$_3$R complex. Both homomeric and heteromeric IP$_3$ receptors occur *in vivo* (Monkawa *et al.*, 1995). IP$_3$R1 is strongly expressed in the cerebellum, and because the level of other IP$_3$ receptors is very low, it is likely to exist predominantly as a homotetramer. On the other hand, both IP$_3$R1 and IP$_3$R2 are expressed in liver, and antibodies directed at either of these receptors are able to coimmunoprecipitate the other protein from liver membranes. Thus, IP$_3$R1 and IP$_3$R2 are able to form a heteromeric complex in liver.

IP$_3$R1 is 2749 amino acids long and has a predicted M_r of 313 kDa. It shares significant sequence homology with the RyRs in the transmembrane domains but is otherwise very different, indicating that IP$_3$R and RyR belong to different gene families. Each IP$_3$R monomer consists of three domains, which have distinct functional roles (Fig. 14.8). The N terminus contains the binding site for IP$_3$, the C-terminal domain forms the Ca^{2+} channel and is important for tetramerization, and there is a very long coupling domain (residues 651–2275) containing binding sites for modulatory agents, which links the N- and C-terminal domains. Binding of IP$_3$ to the N-terminal domain induces a large conformational change in the coupling region which opens the Ca^{2+} channel. Modulators of IP$_3$R activity mediate their effects on channel activity by binding to the coupling domain and interfering with this process.

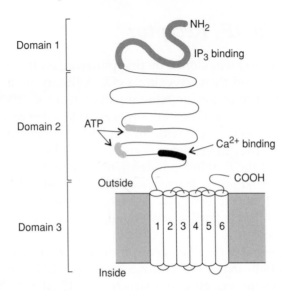

FIGURE 14.8 PUTATIVE TOPOLOGY OF THE IP₃ RECEPTOR
Putative membrane topology of IP₃R1. Domain 1 denotes the IP₃-binding domain, domain 3 indicates the transmembrane domains which form the channel pore, and domain 2 is involved in coupling IP₃ binding to channel activation. It also serves as the binding site for regulatory molecules such as ATP and Ca²⁺.

Mutagenesis studies have demonstrated that the IP₃-binding site is located within the first 650 amino acids of IP₃R1 and that within this region, three positively charged residues—arginine 265, lysine 508, and arginine 511—are critical (Yoshikawa *et al.*, 1996). These residues presumably interact with the negatively charged phosphate groups of IP₃. The central regulatory domain contains consensus sequences for ATP binding and for phosphorylation. A Ca²⁺-binding site with a K_d (0.8 μM) similar to that expected for the stimulatory-binding site also lies within this domain (residues 2124–2146; Sienaert *et al.*, 1996), and residues 1564–1585 are involved in calmodulin binding (Yamada *et al.*, 1995). There are six putative transmembrane domains which lie clustered in the C-terminal part of the molecule and which contribute to the Ca²⁺ channel.

DISEASE ASSOCIATED WITH IP₃ RECEPTORS

No human diseases associated with mutations in any of the genes encoding IP₃ receptors have been identified to date, but a number of naturally occurring mutations in mice produce disease. The *opisthotonos* (*opt*) mouse, which has a deletion in the IP$_3$R1 gene, displays epileptic-like activity (Street *et al.*, 1997), and a less severe, but similar, ataxic phenotype is observed in the mutant mice *staggerer, nervous*, and *Purkinje-cell degeneration*, which express very low levels of IP$_3$R1. In addition, mice in which the IP$_3$R1 gene has been disrupted (knocked out) either die *in utero* or exhibit ataxia and epileptic seizures at about 9 days after birth and die before weaning (Matsumoto *et al.*, 1996). Thus IP$_3$R1 is essential for normal brain function.

IP$_3$ is an important second messenger that mediates the effects of many neurotransmitters and hormones. Consequently, changes in the level of the IP$_3$—either as a result of altered synthesis or metabolism—will influence IP$_3$ receptor gating and may be expected to have profound physiological effects. Lowe's oculocerebrorenal syndrome is an example of such a disease. Manipulation of IP$_3$ levels may also provide a therapeutic target for diseases in which neurotransmitters which elevate IP$_3$, or the receptors for these transmitters, are affected.

The Opisthotonos Mouse

The autosomal recessive *opisthotonos* (*opt*) mutant mouse develops seizures approximately 14 days postnatally, which progress to severe and frequent convulsions with increasing age. Death occurs within 3–4 weeks of birth. The phenotype results from the deletion of two exons in the gene encoding IP$_3$R1 (Street *et al.*, 1997). Although the mouse IP$_3$R1 gene is alternatively spliced, all of the *opt* mRNA transcripts produced lack the nucleotides encoding amino acids 1732–1839, which lie within the coupling domain of the protein. Several important modulatory sites are contained within this region, including a potential PKA phosphorylation site and an ATP-binding domain. Although the mutant protein is expressed at lower levels than normal, Ca^{2+}-induced Ca^{2+} release could still be elicited in Purkinje neurones, and the mechanism by which the mutation produces the epileptic phenotype is not yet clear.

Manic Depression and Li$^+$ Treatment

Manic depression is characterised by dramatic mood swings between mania and depression. It can be controlled effectively by lithium, the therapeutic blood concentration being between 0.5 and 1 mM. It has been argued that the therapeutic effects of Li$^+$ result from its potent inhibitory effect on inositol phosphate metabolism (Berridge *et al.*, 1989). Inhibition of inositol monophosphate phosphatase by Li$^+$ will cause depletion of inositol and reduce the synthesis of inositol lipids required for formation of IP$_3$ (Fig. 14.9). Inositol depletion will be particularly marked in those cells which are unable to replenish inositol by uptake from external sources; this is the case for brain neurones where the inositol supply is limited by the blood–brain barrier. By blocking the supply of inositol, Li$^+$ will impair IP$_3$ generation in response to

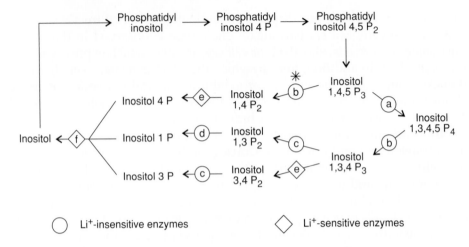

○ Li⁺-insensitive enzymes   ◇ Li⁺-sensitive enzymes

*FIGURE 14.9 LITHIUM INHIBITION OF PHOSPHOINOSITIDE
METABOLISM*
The diagram shows the pathway for metabolism of IP$_3$. The Li⁺ insensitive
and Li⁺ sensitive enzymes are indicated. The asterisk shows the enzyme
which is defective in Lowe's oculocerebrorenal syndrome. a, Inositol
1,4,5 trisphosphate 3-kinase; b, Inositol polyphosphate 5-phosphatase;
c, Inositol polyphosphate 4-phosphatase; d, Inositol polyphosphate 3-
phosphatase; e, Inositol polyphosphate 1-phosphatase; f, Inositol monophosphate phosphatase. After Berridge *et al.* (1989).

neurotransmitters and dampen the effects of receptor hyperactivity, which
is thought to be responsible for the mood swings found in manic depression.
    A side effect of Li⁺ therapy is fine tremor, and at toxic concentrations
(>2.5 mM) Li⁺ causes seizures and death. These effects resemble those reported for mice in which the IP$_3$ receptor has been knocked out (Matsumoto
et al., 1996). Thus it is possible that a complete loss of IP$_3$ receptor activation,
as a consequence of the inability to manufacture sufficient IP$_3$, may contribute
to the toxic effects of Li⁺.

Lowe's Oculocerebrorenal Syndrome

Lowe's oculocerebrorenal syndrome is an X-linked disease which affects the
kidneys, the brain and the lens of the eye. Patients exhibit profound mental
retardation, blindness as a result of cataracts, glaucoma, and microphthalmos
(literally, small eyes). The renal defect manifests principally as proteinuria
and rickets. Lowe's oculocerebrorenal syndrome is not caused by mutations
in the IP$_3$ receptor itself, but instead results from defective regulation of Ca²⁺
release due to impairment of IP$_3$ metabolism.
    Linkage studies initially showed that the gene responsible for Lowe's
oculocerebrorenal syndrome is located on the long arm of chromosome X
(Xq25–26), and positional cloning subsequently revealed that it encodes a

112-kDa protein of 968 amino acids that shows 71% similarity (53% identity) to the human inositol polyphosphate-5-phosphatase (Attree *et al.*, 1992). This suggests that the protein may be a related phosphatase, and thus also involved in the metabolism of IP$_3$. It is expressed in brain, skeletal muscle, heart, kidney, lung and placenta. The loss of the functional protein which occurs in Lowe's oculocerebrorenal syndrome may be expected to lead to the elevation of cytosolic IP$_3$, and thus possibly to an accompanying increase in $[Ca^{2+}]_i$, which could account for the profoundly deleterious effect of the mutations.

CHAPTER 15

ACETYLCHOLINE RECEPTORS

Acetylcholine receptors (AChR) come in two major subtypes, nicotinic and muscarinic, which are named for the agonists which activate them: nicotine and muscarine. Muscarinic AChR are seven transmembrane receptors which do not possess an intrinsic ion channel and mediate their effects by activation of second messenger systems. We will not consider them further here. Nicotinic acetylcholine receptors (nAChR) are ligand-gated ion channels. They are expressed in both muscle and nerve and play a key role in fast synaptic transmission both at neuronal–neuronal synapses within the nervous system and at the neuromuscular junction. Mutations in muscle nAChR subunits produce the congenital fast and slow channel syndromes, while mutations in neuronal nAChRs have been found in nocturnal frontal lobe epilepsy. Myasthenia gravis results from autoantibodies to muscle nAChR and is discussed in Chapter 21.

Basic Properties

Synaptic transmission

To understand the physiological role of muscle nAChRs and how their mutation gives rise to the slow channel syndromes, it is helpful to have some knowledge of neuromuscular transmission. At the neuromuscular junction the pre- and postsynaptic membranes come together in close apposition, being separated by a gap of ~20–30 nm known as the synaptic cleft (Fig. 15.1). The arrival of a nerve impulse in the presynaptic nerve terminal leads to an influx of Ca^{2+} through voltage-gated Ca^{2+} channels, which triggers the exocytosis of ACh-containing synaptic vesicles. The ACh that is released diffuses across the synaptic cleft and interacts with nAChRs in the postsynaptic muscle membrane, which lie immediately opposite the release sites in the nerve terminal. Binding of ACh to its receptor opens an intrinsic ion channel and results in an influx of cations (mainly Na^+, but also some Ca^{2+}). The current that flows through the open channels, the excitatory postsynaptic current (epsc) causes a local depolarization of the muscle membrane that is

A

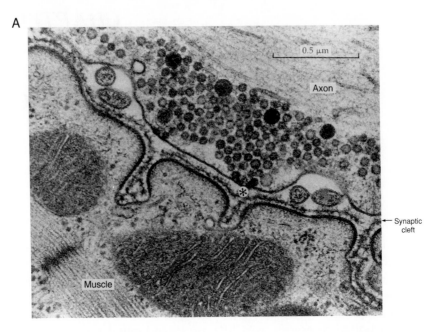

B

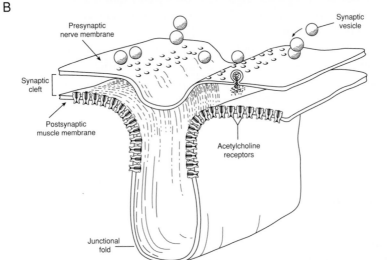

FIGURE 15.1 STRUCTURE OF THE NEUROMUSCULAR JUNCTION
(A) Electron micrograph of a neuromuscular junction. Many synaptic
vesicles are clustered in the terminal of the presynaptic axon. The asterisk
marks an active zone where vesicles are being released. These lie opposite
the junctional folds in the membrane of the postsynaptic muscle fibre.
From Heuser (1977). (B) Diagram of the neuromuscular junction. After
Hille (1992).

known as the endplate potential (epp). The amplitude of the epp is determined by the number of synaptic vesicles which are released. A single nerve impulse releases between 100 and 300 vesicles and depolarises the membrane from its resting value of ~-90 mV to ~0 mV. Miniature epps result from the spontaneous release of just one, or at most a few, vesicles and are <1 mV in amplitude. If it is sufficiently large, the epp activates voltage-gated Na^+ channels in the extrasynaptic membrane and triggers an action potential in the muscle fibre that leads to a muscle twitch. The action of ACh is terminated by its unbinding from the nAChR. Once ACh has unbound it is rapidly cleared from the synaptic cleft as a result of hydrolysis by the enzyme acetylcholinesterase or by diffusion.

The difference between the amplitude of the epp and the depolarization which is required to elicit an action potential is known as the 'safety factor.' In human muscle, the threshold of the muscle action potential is fairly high, so that even a moderate reduction in epp amplitude may cause a failure of neuromuscular transmission and consequent muscle weakness. Neuromuscular transmission in human patients may be investigated using electromyography, a technique in which the compound muscle action potential evoked by nerve stimulation is recorded.

Many of the principles of synaptic transmission at neuromuscular junction also apply to those neuronal–neuronal synapses involving nAChRs. The main differences are that whereas a muscle fibre is innervated by a single axon, a nerve cell may receive many thousands of inputs. These can be inhibitory as well as excitatory and, in addition to ACh, may involve many different type of transmitters and postsynaptic receptors. Each of the innervating axons may release just a few vesicles in response to a nerve impulse (in contrast to the 100–300 released at the neuromuscular junction) so that the postsynaptic potential may be only a few hundreds of microvolts, far smaller than the epp. The output of the postsynaptic neuron is therefore an integrated response to all of the different inputs, and summation of many excitatory postsynaptic potentials (epsps) may be necessary to evoke an action potential.

Desensitization

Desensitization refers to the closure of the nAChR for long periods, despite the continued presence of bound agonist. Desensitization is of little consequence for normal neuromuscular transmission because it is more likely that a receptor with bound ACh will open, or that ACh will unbind, than that the bound receptor will desensitize. If ACh unbinding is slowed, however, as is found in some forms of congenital slow channel syndromes (see later), then receptor desensitization may occur during repetitive stimulation. This is manifest as the depression of the epsp with successive stimuli and recovery can take seconds to minutes. It is also possible that low levels of circulating nicotine in habitual smokers induce desensitization of neuronal nAChR and thereby contribute to nicotine tolerance (see later).

Pharmacology

Many drugs interact with nAChRs. Nicotine, the best known agonist, is extracted from the leaves of the tobacco plant *Nicotiana tabacum*. Other agonists include carbachol and succinylcholine (suxamethonium). The latter dissociates only slowly from the nAChR and consequently induces a long-lasting epp that triggers a brief train of muscle action potentials and muscle twitches. This is followed by a depolarizing neuromuscular block that results from the inactivation of voltage-gated Na^+ channels. Probably the best known antagonist of nAChRs is curare, which was used by South American Indians as an arrow poison. Its active ingredient is *d*-tubocurarine. α-Bungarotoxin, isolated from the venom of banded krait, is another classical antagonist. Its affinity for the nAChR is so high that it was used for biochemical isolation of the receptor (Changeux *et al.*, 1970) and for determining the number, distribution and lifetime of nAChRs at the neuromuscular junction. The tranquiliser chlorpromazine acts as an open channel blocker.

A digression on kinetics

To understand how disease-causing mutations produce changes in nAChR channel function, it is helpful to consider the mechanism of channel activation. Much insight into this process has been provided by analysis of the single-channel kinetics. Consider Fig. 15.2, which shows a typical recording of

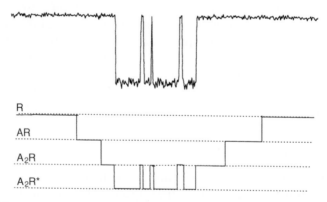

FIGURE 15.2 SINGLE AChR CHANNEL CURRENTS
(A) Single-channel currents though an nAChR channel, evoked by a low concentration of ACh in the pipette solution. (B) Diagrammatic model illustrating how the channel kinetics are influenced by the number of ACh molecules bound. Once two ACh molecules (A) have bound to the nAChR (A_2R), the channel undergoes a conformational change that results in opening of the pore (A_2R^*). The channel may then flicker briefly between open and closed states before finally closing for an extended period. The rapid closings and reopenings are associated with transitions between the open and closed states of the activated receptor. The long closings occur when the ACh molecule dissociates from its receptor.

single nAChR channel currents activated by the application of ACh. It is evident that channel openings occur in bursts, which are separated by much longer closed intervals. The long closed intervals between the bursts correspond to the time when ACh is not bound to its receptor, while the short closings within each burst of openings correspond to the time when ACh is bound but the channel remains closed. Each nAChR must bind two ACh molecules before the channel opens. Thus, a very simplified scheme for nAChR activation is

$$R \underset{\substack{k_{-2} \\ A}}{\overset{\substack{A \\ k_2}}{\rightleftharpoons}} AR \underset{\substack{k_{-1} \\ A}}{\overset{\substack{A \\ k_1}}{\rightleftharpoons}} A_2R \underset{\alpha}{\overset{\beta}{\rightleftharpoons}} A_2R^*$$

closed (3)	closed (2)	closed (1)	open

where A is acetylcholine, R is the receptor and α, β, k_1, k_{-1}, and so on are the rate constants for the transitions between the open and closed states. For simplicity, desensitized states are not included here. The model may be further simplified if it is assumed that the rate of ACh binding and unbinding at each site is identical: in this case, $k_2 = k_1$ and $k_{-2} = k_{-1}$. It is more likely, however, that the two sites are not equivalent. As explained in Chapter 3, the mean time spent in a given state is determined by the reciprocal of the sum of the rate constants for leaving that state. Thus, in the above scheme, the mean open time is determined by a single rate constant and is $1/\alpha$. The mean time spent in closed state 1 (A_2R), during which ACh is bound but the channel is not open, is given by $1/(\beta + k_{-1})$. The mean number of openings in a burst is given by $1 + \beta/k_{-1}$. If we ignore the very small contribution from the brief closures (A_2R) within a burst of openings, the mean burst duration is $(1 + \beta/k_{-1})/\alpha$. It therefore depends on the opening rate (β), the closing rate (α), and the rate of ACh dissociation (k_{-1}). Because β and k_{-1} are of comparable value, the ACh receptor oscillates several times between the open state and the closed state before ACh dissociates. Mutations may produce changes in channel function by affecting one or more of the rate constants in the above scheme, as discussed later. Kinetic analysis of the single-channel currents can elucidate which rate constant(s) is altered and so provide insight into the relationship between the structure of the nAChR and its gating.

Structure

Subunit composition

The nAChR are pentamers of homologous subunits arranged in a ring around a central ion pore (Fig. 15.3). There are five types of muscle subunits, α_1, β_1, γ, δ and ε, but only four of these are found in the channel complex as there are always two α_1-subunits. The adult muscle nAChR channel comprises $2\alpha_1$, β_1, ε, δ, while that of embryonic and denervated muscle is composed of $2\alpha_1$, β_1, γ, δ (Mishina et al., 1986). The different subunits have molecular

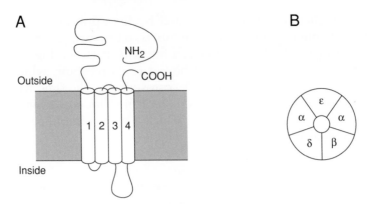

FIGURE 15.3 PUTATIVE TOPOLOGY OF THE nAChR
(A) Membrane topology of a single nAChR subunit deduced from hydropathy analysis in conjunction with mutagenesis and antibody studies.
(B) The receptor is a pentamer: in adult skeletal muscle the δ subunit is
replaced by an ε subunit.

masses ranging between 45 and 60 kDa, giving a total molecular mass of
around 290 kDa for the nAChR complex. The subunits are the products of
different genes and share between 37% and 57% sequence identity, suggesting
that they derive from duplication of a single ancestral gene early in evolution.

Neuronal nAChR are composed of just two types of subunit, α and β,
probably in a 2:3 stoichiometry (Cooper *et al.*, 1991). There are at least seven
different neuronal α-subunits (named α_2 to α_8) and three different neuronal
β-subunits (β_2-β_4) (McGehee and Role, 1995), which show differential tissue
expression within the brain (Wada *et al.*, 1988). An additional α-subunit (α_9),
is found in the cochlea (Elgoyhen *et al.*, 1994). Some subunits, such as α_7,
α_8, and α_9, can form functional homopentameric channels in heterologous
expression systems, but *in vivo* it is likely that they exist in association with
each other, or with β-subunits. Neither α_5 nor α_6, in combination with any
β-subunit, has been shown to produce ACh-activated currents; this does not
imply, however, that these subunits do not participate in functional channel
formation as it is conceivable that they may form heteromeric channels with
other α-subunits. The chromosomal location of the genes encoding the different human nAChR subunits is given in Table 15.1.

Subunit composition influences AChR channel properties

Nicotinic AChR channels formed from different subunit combinations
vary in their gating kinetics, single-channel conductance, Ca^{2+} permeability,
sensitivity to ACh and affinity for pharmacological drugs. For example, α-
bungarotoxin binds with high affinity to α_1-, α_7-, and α_8-subunits but does
not interact with α_4-subunits, and homomeric α_7 nAChR channels are far
more permeable to Ca^{2+} than other nAChR channels (P_{Ca}/P_{Na} ~20).

In mammalian muscle, the functional properties of the nAChR change
after birth. The single-channel conductance and Ca^{2+} permeability increase,

TABLE 15.1   CHROMOSOMAL LOCATION OF HUMAN AChR GENES

Gene	Protein	Chromosome location (human)	Tissue	Reference
CHRNA1	α_1	2q24–q32	Neuromuscular junction	Beeson et al. (1990)
CHRNB1	β	17p12–p11	Neuromuscular junction	Beeson et al. (1990)
CHRNG	γ	2q33–q34	Fetal, dennervated muscle	
CHRND	δ	2q33–q34	Neuromuscular junction	Lobos et al. (1989); Beeson et al. (1990)
CHRNE	ε	17p13.1	Neuromuscular junction (adult)	Lobos (1993)
CHRNA2	α_2	8	Neuronal	Anand and Lindstrom (1992)
CHRNA3	α_3	15q24	Neuronal	Raimondi et al. (1992); Anand and Lindstrom (1992)
CHRNA4	α_4	20q13.2–q13.3	Neuronal	Steinlein et al. (1994)
CHRNA5	α_5	15q24	Brain	Raimondi et al. (1992)
CHRNA7	α_7	15q13–q14	Neuronal	Orr-Urtreger et al. (1995)
CHRNB2	β_2	1p21	Neuronal	Anand and Lindstrom (1992)
CHRNB3	β_3	8p11.2	Neuronal	Koyama et al. (1994)
CHRNB4	β_4	15q24	Neuronal	Raimondi et al. (1992); Anand and Lindstrom (1992)

while the duration of a burst of openings decreases, producing a reduction in the duration of the mepp and epp. This results from a developmental switch in the expression of the nAChR subunit genes, so that the γ-subunit of the fetal nAChR is replaced by an ε-subunit, thus giving the adult form of the receptor (Mishina et al., 1986). The opposite change in subunit composition and functional properties is observed after denervation of adult skeletal muscle (γ replaces ε). Expression of the ε-subunit during development, and repression of γ-subunit expression, are regulated by a nerve-dependent trophic factor. Genetic deletion of the ε-subunit gene has deleterious consequences: homozygous 'knockout' mice develop impaired neuromuscular transmission, progressive muscle weakness and muscle atrophy, which results in premature death (Witzemann et al., 1996).

Postsynaptic Localization

nAChR are distributed diffusely over the surface of embryonic muscle but are found at high density at the postsynaptic region of adult muscle fibres. This clustering is achieved by a number of different proteins that act in concert to anchor the receptor to the cytoskeleton. One of these is rapsyn, a 43 kDa protein which copurifies with the nAChR under certain conditions. When this protein is eliminated by gene targeting, the homozygous mice have impaired movements and die within hours of birth, as a result of the failure of AChR to cluster at the postsynaptic density (Gautam et al., 1995). Other proteins involved in the aggregation and anchoring of ACh receptors

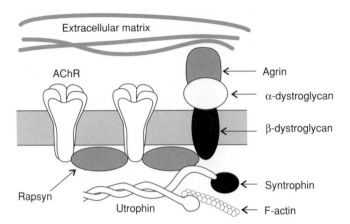

FIGURE 15.4 ANCHORING OF nAChR AT THE NEUROMUSCULAR JUNCTION

Rapsyn anchors the nAChR to a large complex of proteins which interact with the extracellular matrix and the cytoskeleton. After Sunada and Campbell (1995).

at the endplate include agrin, dystroglycans, utrophin and the syntrophins (Sunada and Campbell, 1995). One model suggests that agrin, an extracellular matrix protein produced by the nerve, mediates AChR clustering and that the α- and β-dystroglycans serve to anchor agrin to the postsynaptic membrane and form a scaffold to which the AChR is tethered by rapsyn (Fig. 15.4).

Correlating Structure and Function

The muscle nicotinic acetylcholine receptor channel has the distinction of being the first ion channel to have its amino acid sequence determined, by Shosaku Numa and his colleagues in 1982 (Noda *et al.*, 1982, 1983). Its structure has been extensively studied by biochemical methods, site-directed mutagenesis and cryoelectron microscopy. In this section, therefore, we focus on the structure of the muscle nAChR.

Hydropathy analysis first suggested that each nAChR subunit consists of a large extracellular N terminus of ~210 amino acids, four transmembrane domains (M1 to M4) and a short extracellular C terminus (Fig. 15.3). This topology has now been confirmed by determining whether antibodies directed against particular regions of the protein bind to the outside or inside surface of the cell membrane. Extensive studies, described later, have shown that the ACh-binding site lies within the N-terminal extracellular domain, at the interface between the αβ- and αγ (or αε)-subunits, and that the pore is lined by the five M2 domains, one contributed by each subunit.

Three-dimensional structure

The density of nAChR in the postsynaptic membranes of the electric organ of the *Torpedo* ray is so high that they appear as an almost crystalline array (Fig. 15.5A). Consequently, this organ has been used to provide a source of nAChR for ultrastructural analysis. When membranes containing nAChR are isolated and resuspended in low salt solution, some of them form tubular two-dimensional crystalline arrays. These have been examined at 9-Å resolution by diffraction analysis of high resolution electron microscope pictures (Unwin, 1993, 1995). In order to capture the channel in the open state, ACh was sprayed onto the membranes in aerosol form and the membranes frozen within milliseconds of the arrival of the transmitter. Multiple images were then averaged to visualise the structure.

These studies have shown that the five subunits are assembled in a pseudosymmetric ring around a central pore, rather like the staves of a barrel. Each nAChR projects about 60 Å above the membrane on its synaptic side and ~15 Å on the cytoplasmic side (Fig. 15.5B). The central channel has a wide (20–25 Å) extracellular mouth that narrows to a waist with a diameter of ~10 Å across the bilayer and then dilates again on the cytoplasmic side of the membrane. Within the extracellular domains, two cavities were observed, 30 Å above the plane of the membrane. These were interpreted as the binding pockets for ACh.

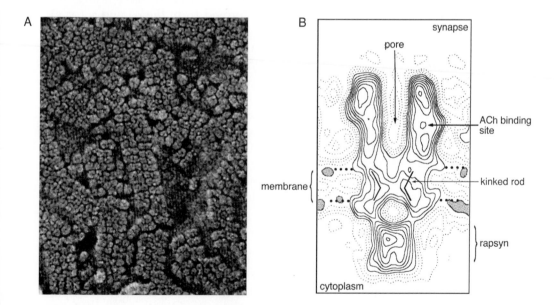

FIGURE 15.5 AChR AT THE ELECTRIC ORGAN OF TORPEDO
(A) Postsynaptic membrane of *Torpedo* electroplaque showing the semi-crystalline array of nAChRs. Freeze-fracture electron micrograph, viewed face-on (×216,000). From Heuser and Salpeter (1979). (B) Cross-section of the nAChR obtained from cryoelecton microscopy. The dense kinked rods correspond to the M2 domains and the dotted lines indicate the estimated limits of the lipid bilayer. The square structure at the base of the channel is not part of the nAChR but is an associated 43 kDa protein (rapsyn) that tethers the nAChR to the cytoskeleton. From Unwin (1993).

The ligand-binding site

One approach to identifying those nAChR residues that contribute to the ligand-binding site has been to use ligands which compete with ACh for its binding site on the receptor. A radiolabeled form of such a ligand can be covalently (irreversibly) attached to the nAChR, and the residues which it labels can then be determined by proteolytic digestion and sequencing of the protein. Using this approach, six residues in the α-subunit were labeled: three tyrosines, at positions 93, 190 and 198 (*Torpedo* numbering); a tryptophan (W149); and two adjacent cysteines, C192 and C193 (Dennis *et al.*, 1988; Galzi *et al.*, 1990; Changeux *et al.*, 1992). This suggests that these residues, located in the extracellular N terminus of the α-subunit, contribute to the ACh-binding site. Further support for this idea comes from the finding that mutation of any of these six residues caused a decrease in the apparent affinity of the receptor for ACh (Karlin and Akabas, 1995; Galzi *et al.*, 1991). Thus it appears that three widely separated regions of the α-subunit (Y93, W149 and Y190-W193) come together to form the binding site. Negatively charged residues on both δ-(D180, E189) and γ-subunits also contribute to agonist binding, as indicated by both affinity-labeling and mutagenesis studies (Czajkowski and Karlin, 1991, 1995; Czajkowski *et al.*, 1993).

Figure 15.6 illustrates the current picture of the two ligand-binding sites of the muscle nAChR. One binding site is formed from the α- and γ (or ε)-subunits and the other lies at the interface between the α- and the δ-subunit. The adjacent cysteines (at positions 192 and 193) within each binding site form a disulphide bond. How does acetylcholine interact with its binding site? One idea is that the positively charged quaternary ammonium group of ACh interacts with the negatively charged aspartate residue in the δ (or γ)-subunit and that the other end of the molecule simultaneously binds to the side chains of the tyrosine, tryptophan and cysteine residues in the adjacent α-subunits via hydrophobic interactions and hydrogen bonding (Czajowski *et al.*, 1993; Karlin and Akabas, 1995). It is suggested that this results in shortening of the distance between these two regions of the binding site, so producing a conformational change which ultimately leads to channel opening (see later).

The pore

The pore of the ACh channel was first explored by examining its permeability to organic molecules of different sizes and charge (Huang *et al.*, 1978). The results of these studies indicated that the diameter of the narrowest part of the channel is ~6.5 Å. This is considerably larger than the pore of Na^+ or K^+ channels (see Chapters 5 and 6), and partially explains the lower cation selectivity of the AChR channel. Since the diameter of both a water molecule, and of a Na^+ ion, is ~3 Å, it also means that Na^+ should be able to permeate the AChR channel in a partially hydrated state.

There is considerable evidence that the second transmembrane domain (M2) lines the pore of the nAChR channel. Because the M2 domains of the five subunits have very similar sequences, this implies that the pore will be

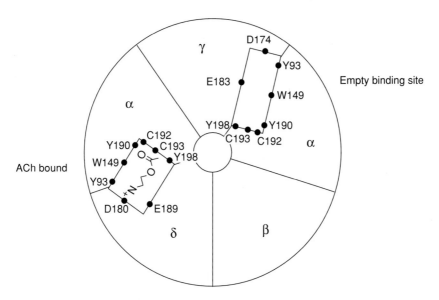

FIGURE 15.6 MODEL OF THE ACh-BINDING SITE OF THE nAChR
The receptor is viewed from above, with the subunits surrounding a
central channel. The ACh-binding sites are represented by the rectangles
between α-and γ- and α- and δ-subunits. Residues implicated in ACh
binding are indicated, numbered according to their position in the mouse
α_1 subunit. ACh is shown bound to the α-δ site, which is contracted
compared to the unoccupied α-γ site. Modified from Karlin (1993).

lined by rings of similar or identical residues along its length. Single-channel
measurements on mutated nAChRs have shown that three rings of negatively
charged residues are of major importance in determining the selectivity and
conductance properties of the channel. These rings surround both the internal
and the external entrances of the pore and lie just outside the M2 domain
(Fig. 15.7). Counting from the N terminal (intracellular) end of M2, they lie
at positions -4, -1 and 20 (Fig. 15.7). Mutations at position -1 have the
largest effects on permeability and conductance, suggesting that this region
forms part of the selectivity filter of the pore. The anionic rings at positions
-4 and 20 probably serve to attract cations into the vicinity of the pore.

Site-directed mutagenesis of the innermost ring of negatively charged
residues (position -4), reduced the outward current through the channel,
but had little effect on the inward current (Imoto *et al.*, 1988). Conversely,
mutation of negatively charged residues in the extracellular ring (position
20) lowered the inward current without altering the outward current. Not
only do these results imply that the two rings of charge are located on opposite
sides of the membrane, they also demonstrate that each ring is unable to
influence ions on the other side of the membrane. They therefore define the
extracellular and intracellular mouths of the pore. These flanking rings of
negative charge may serve to attract cations into the channel entrances and

A

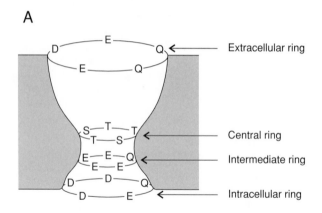

B

```
            -4       -1     2                 9            13                  20
ACh α₁  Torpedo    D  S  G  -  E  K  M  T  L  S  I  S  V  L  L  S  L  T  V  F  L  L  V  I  V  E
ACh β   Torpedo    D  A  G  -  E  K  M  S  L  S  I  S  A  L  L  A  V  T  V  F  L  L  L  L  A  D
ACh γ   Torpedo    Q  A  G  G  Q  K  C  T  L  S  T  S  V  L  L  A  Q  T  I  F  L  F  L  I  A  Q
ACh δ   Torpedo    E  S  G  -  E  K  M  S  V  A  I  S  V  L  L  A  Q  A  V  F  L  L  L  T  S  Q
                              -------------------------------------------------------
                          in                      M2 segment                          ex
```

FIGURE 15.7 M2 DOMAINS LINE THE PORE OF THE nAChR
(A) Model of the M2 domains showing the three rings of negative charge
and the ring of polar residues that contribute to the pore. (B) Aligned
amino acid sequences of the M2 domain and its flanking regions in the
Torpedo α-, β-, γ-, and δ-subunits. The M2 domain is indicated by the
dashed line. Amino acids indicated in bold type are exposed to the channel
lumen. Residues are numbered according to their position relative to the
N-terminal (intracellular) end of the M2 domain (K = 0).

to repel anions: as they are made more negative the local concentration of
cations will rise and the single-channel conductance will increase accordingly.

An additional set of negative charges is found on the inside of the membrane just before the beginning of the M2 domain at position −1. Removal
of negative charges at this position substantially reduces both the inward
and outward currents (Imoto *et al.*, 1988; Konno *et al.*, 1991). These charges
therefore seem to be more important in determining the rate of ion transport
through the pore than the other two anionic rings. Mutations within this
intermediate ring also influence the relative permeability to different cations,
with selectivity varying according to the total amount of local charge. This
suggests that the intermediate ring of negative charges may contribute to the
selectivity filter of the nACh channel.

An additional ring of polar residues at position 2 also contributes to the
permeability properties of the channel and may form part of the selectivity
filter (Imoto *et al.*, 1991; Villaroel *et al.*, 1992). When this residue is mutated,
the single-channel conductance depends on the side-chain volume of the

amino acid with which it is replaced. The results suggest that the residues at position 2 participate unequally in the formation of a narrow constriction in the channel pore, with the δ-subunit being the main determinant of the single-channel conductance.

A dramatic demonstration that the nature of the amino acids which line the nAChR pore determines the channel selectivity is provided by experiments carried out by Galzi and colleagues (1992). They were able to convert the neuronal α_7 nAChR, which forms homomeric channels, from a cation-selective channel to an anion-selective one by making three separate mutations in each of the five subunits: replacement of the negatively charged glutamate at position −1 by the neutral alanine, the substitution of valine at position 13 by threonine and the introduction of a proline residue between positions −2 and −1. Interestingly, a proline is found at the latter position in all anion-selective channels (see also Chapter 17).

A different approach to identifying those residues that contribute to the nAChR pore has been to mutate residues in the M2 segment one at a time to cysteine and subsequently determine the susceptibility of the mutant channel to block by thiol reagents, which interact specifically with cysteine residues (Akabas et al., 1994). This method is known as cysteine-scanning mutagenesis. Since some thiol reagents are small enough to enter the pore and interact with exposed side chains, this approach enables those amino acids which are exposed to the channel lumen to be identified. Ten residues within M2 were found to face the channel lumen (Fig. 15.7B). The pattern of exposure suggests the M2 segment crosses the membrane partly as an α-helix and partly in an extended conformation (positions 8–10).

Summary

The studies described above have led to the following model of the ACh channel in its open state. The M2 domains line the pore of the nAChR channel. The outer part of the pore is funnel shaped with a diameter of 25–30 Å and protrudes some distance from the membrane surface. The narrowest region of the pore is very short, being only 3–6 Å long. This region also acts as the selectivity filter. It has a diameter of ~7 Å and is lined by two rings, one of negatively charged residues and the other of polar residues. Additional rings of negatively charged residues surround both the internal and the external entrance of the pore. These serve to attract cations into the channel entrance and so influence the conductance of the channel. This picture is in good agreement with that derived from analysis of cyroelectron micrographs of the purified receptor.

Gating

The way in which ligand binding results in opening of the ACh channel pore remains contentious. Because the binding site involves more than one subunit, a conformational change could be produced by the movement of one subunit relative to another. This conformational change must propagate through the protein to open the channel because the binding site for ACh and the channel gate are separated by a distance of as much as 50 Å. Cryoelectron microscope images of the nAChR imaged in the presence and absence of ACh have shed some light on this problem (Unwin, 1995). In the closed state, the α-helices lining the channel pore appear as dense rods with a kink at their middle that defines the narrowest region of the pore (Fig. 15.5B). Activation causes these rods to twist relative to one another, which moves the kink out of the axis of the pore and so increases its diameter. Interpretation of these images has provided a model for channel opening (Fig. 15.8). The α-helices are presumed to be the M2 segments and the kink is suggested to be produced by the leucine residues (at position 9). In the closed state, these leucines, which are large and hydrophobic, protrude into the pore from each of the five subunits and occlude ion flow. Binding of ACh causes a reorientation of the protein so that the leucine residues are rotated out of the pore, thus opening the ion pathway. Evidence both for and against this model has been obtained experimentally.

Summary of structure–function studies

The relationship between the structure and function of the nAChR is probably one of the best understood of any ion channel. It constitutes the paradigm for most ligand-gated ion channels (the exceptions being glutamate and P_2X receptors). The nAChR channel is a pentamer of similar but non-identical subunits arranged pseudosymmetrically around a central ion channel. The ACh-binding pocket is found in the extracellular domains at the interface between α- and non-α-subunits ($\gamma,\delta,\varepsilon$). The M2 domains of each subunit line the walls of the pore and contribute to the selectivity filter. In general, the effects of nAChR mutations that give rise to human disease can

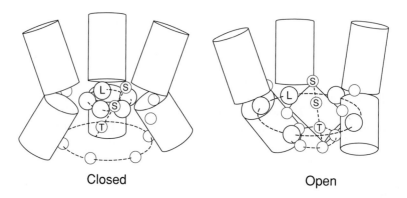

Closed   Open

FIGURE 15.8 MODEL FOR GATING OF nAChR
Closed and open conformations of the M2 segment with the leucine and polar residues indicated. In the closed state the point of the kink faces into the channel lumen, blocking ion flux, while in the open state the kink is swung out of the way and aligned with the walls of the pore. From Unwin (1995).

be explained on the basis of our current knowledge of the channel structure. However, there have been a few surprises, which have shed some light upon the function of regions of the channel whose role was not well understood.

DISEASES OF SKELETAL MUSCLE AChRs

The congenital myasthenic syndromes (CMSs) are inherited disorders of neuromuscular transmission. They may result from a deficiency of acetylcholinesterase (the enzyme which breaks down ACh), presynaptic abnormalities which influence the amount of transmitter released, postsynaptic abnormalities associated with a reduction in the number of functional nAChRs, or kinetic abnormalities of the nACh receptor itself (Vincent *et al.*, 1997). Here we focus on those congenital myasthenic syndromes which result from mutations in the genes which encode the various nAChR subunits.

Slow Channel Syndrome

Slow channel syndrome (SCS) is a rare congenital myasthenic syndrome that often, but not invariably, presents at birth or in childhood. It is inherited in an autosomal dominant fashion, but many cases appear to arise sporadically. The hallmarks of the disease are muscle weakness, rapid fatigue, progressive muscle atrophy and the generation of repetitive muscle action potentials in response to a single nerve stimulus. The last may be demonstrated by electromyography. Electrophysiological studies of muscle biopsies taken from SCS patients have shown that the decay phase of the endplate potential (epp) and miniature endplate potential (mepp), and their underlying synaptic currents, are slowed, suggesting prolonged activation of nAChR channels. In some cases, this idea has been confirmed by single-channel recordings. The mepp amplitude is often reduced and degenerative changes in the endplate region of the muscle have also been described, including loss of junctional nAChRs and destruction of the junctional folds. In contrast to myasthenic gravis (see Chapter 21), the symptoms of SCS are exacerbated by acetylcholinesterase inhibitors and the patients do not respond to immunotherapies.

Slow channel syndrome results from one of a spectrum of mutations in the muscle nAChR channel. These mutations have been found in all four adult nAChR (α, β, δ, and ε) subunits but they occur within different domains (Fig. 15.9). Several of them (εT264P, εL269F, βV266M, and βL262M) are located within the second transmembrane domain (M2), while αN217K lies in M1. By contrast, αG135S and αV156S lie in an extracellular region of the N terminus that is believed to contribute to ACh binding, and S269I lies in the extracellular loop linking transmembrane domains 2 and 3. All SCS

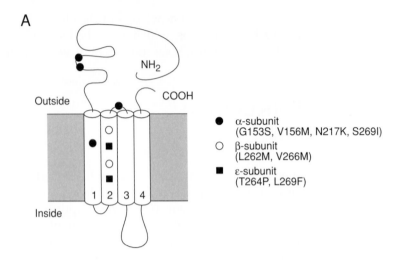

A

Outside

NH₂

COOH

Inside

1 2 3 4

● α-subunit
   (G153S, V156M, N217K, S269I)
○ β-subunit
   (L262M, V266M)
■ ε-subunit
   (T264P, L269F)

B

ACh α₁ human	M T L S I S V L L S L T V F L L V I V
ACh β human	M G L S I F A L **L T** L T **V** F L L L I V
ACh γ human	T S V A I S V L L A Q S V F L L L I S
ACh δ human	C T V A I N V L L A **Q** T V F L F L V A
ACh ε human	C T V S I N V L L A Q **T** V F L F **L** I A

in   *ex*

FIGURE 15.9 *MUTATIONS ASSOCIATED WITH SLOW CHANNEL SYNDROME*
(A) Putative membrane topology of a stereotypical nAChR subunit with SCS mutations indicated. (B) Aligned amino acid sequences of the M2 domain of the human α-, β-, γ-, and δ-subunits. Residues that are mutated in slow channel syndrome are indicated in bold type.

mutations result in prolonged channel activation by ACh, but they may do so by different molecular mechanisms.

Mutations affecting channel opening

Site-directed mutagenesis studies have shown that several of the SCS mutations in M2 lead to prolonged channel openings in response to ACh (Ohno *et al.*, 1995; Engel *et al.*, 1996). Figure 15.10 illustrates an example. These residues therefore appear to be involved in stabilizing the open state of the channel by slowing the rate of channel closure (i.e., α in the reaction scheme depicted on page 273). Most of the M2 mutations are also associated with an unusually high rate of spontaneous channel openings (i.e., openings in the absence of agonist), suggesting that they may also enhance the rate of channel opening.

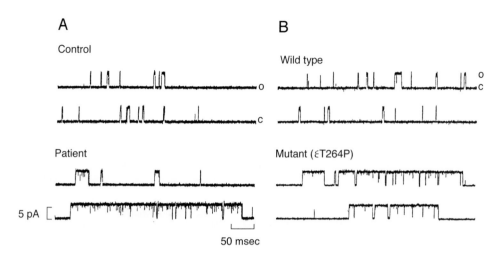

FIGURE 15.10 SOME SCS MUTATIONS INCREASE THE MEAN OPEN TIME
(A) Single nAChR channel currents recorded from muscle biopsies taken from a normal individual (top) and a patient with slow channel syndrome caused by the εT264P mutation (bottom). (B) Single-channel currents recorded from cloned wild-type (top) or εT264P mutant (bottom) nAChR channels expressed in HEK cells. From Ohno *et al.* (1995).

Mutations affecting receptor occupancy

A number of SCS mutations do not alter the mean open time of the nAChR channel but instead produce prolonged bursts of openings (Fig. 15.11). The increase in the number of openings within a burst suggests that these mutations markedly decrease the rate at which ACh unbinds from its receptor and so allow the channel to open repeatedly before ACh dissociates. One of these mutations (G153S) lies in an extracellular part of the α-subunit close to the ACh-binding site (Sine *et al.*, 1995). A detailed kinetic analysis has indicated that the main effect of this mutation is to decrease the rate of ACh dissociation (k_{-1} in the simple reaction scheme depicted earlier). Possibly, this is the result of local structural changes in the vicinity of the ACh-binding site produced by the mutation. An increase (~20-fold) in the apparent affinity of the nAChR for acetylcholine is also found for the αN217K mutation (Wang *et al.*, 1997). This mutation lies within the first transmembrane domain and the discovery that it causes a reduction in the rate of ACh unbinding was somewhat unexpected because the ACh-binding site lies ~30 Å above the plane of the membrane. It therefore appears that N217 is involved in the allosteric linkage between agonist binding and channel gating and that it may contribute to the external entrance of the pore.

Expression of the γ-subunit

In some SCS patients, aberrant expression of the fetal γ-subunit in the postsynaptic membrane is also found, which is probably related to the de-

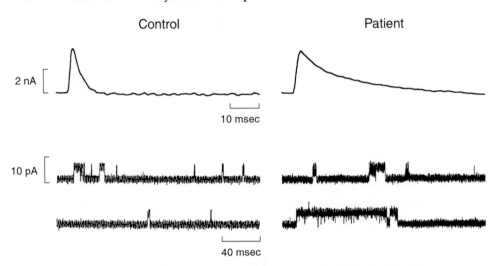

FIGURE 15.11 SOME SCS MUTATIONS PRODUCE PROLONGED BURSTS OF OPENINGS AND SLOW POSTSYNAPTIC POTENTIALS
Miniature excitatory postsynaptic currents (top) and single nAChR channel currents (bottom) recorded from muscle biopsies taken from a normal individual (left) and from a patient with slow channel syndrome caused by the αG153S mutation (right). nAChR currents were activated by inclusion of ACh in the extracellular (patch pipette) solution. From Sine *et al.* (1995).

struction and regeneration of the junctional folds. nAChRs containing a γ-(rather than ε)-subunit have longer bursts of openings, which will tend to prolong the epp and contribute to the SCS phenotype.

How do the SCS mutations cause myasthenia?

The increased duration of channel openings, or bursts of openings, characteristic of SCS AChR channels produces a prolonged excitatory postsynaptic current and thus epps and mepps that decay more slowly. Temporal summation of these epps may be expected at physiological rates of stimulation, and will cause a prolonged depolarization of the muscle membrane. The consequent inactivation of voltage-gated Na^+ channels will lead to failure of muscle excitability. This explains the muscle weakness and rapid fatigue experienced by patients suffering from SCS. A similar 'depolarization block' is observed with acetylcholinesterase inhibitors or with AChR agonists such as suxamethonium which dissociate slowly.

If ACh dissociation is slowed, as with the αN217K and αG153S mutations, then the probability of receptor desensitization is enhanced. This suggests that an increased fraction of mutant AChR channels may become desensitized at physiological rates of stimulation, so contributing to muscle fatigue. The prolonged epp also causes enhanced Ca^{2+} entry, which may account for the progressive destruction of the postsynaptic neuromuscular junction. Abnormal channel openings in the absence of ACh may also contribute to the 'endplate myopathy,' and could help explain why the SCS mutations which cause spontaneous openings are often associated with a more severe phenotype. Acetylcholinesterase inhibitors will clearly exacerbate SCS since they further prolong the epp: however, long-lived nAChR channel blockers may be of some benefit.

Fast Channel Syndrome

Congenital myasthenia may also be caused by nAChR mutations which produce a reduction in ACh affinity and decrease the rate of channel opening (Ohno *et al.*, 1996). This disorder has been referred to as fast channel syndrome (FCS). Its clinical features resemble those of slow channel syndrome, but FCS can be distinguished in muscle biopsies by the reduced agonist-binding affinity, the very low frequency of channel reopenings during ACh occupancy of the receptor, the very small miniature endplate potentials, and the lack of degenerative changes in the postsynaptic region of the muscle.

   To date, FCS has only been reported in one family, where it is inherited in a recessive fashion. It results from mutation of the proline at position 121 in the nAChR ε-subunit to leucine. εP121 lies close to residues which participate in agonist binding and, because proline produces kinks in proteins, its mutation may be expected to produce a local conformational change that influences the ACh-binding site. This is probably why the affinity for ACh at one of its two binding sites was reduced in the mutant channel (Fig. 15.12). That only one site was affected is consistent with the fact that one binding site is formed from $\alpha_1\varepsilon$-subunits and the other from $\alpha_1\delta$-subunits. Single-

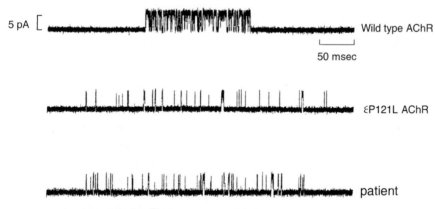

FIGURE 15.12 *εP121L DECREASES THE FREQUENCY OF nAChR OPENINGS*
Single-channel currents recorded from cloned wild-type and εP121L mutant nAChR channels expressed in HEK cells, compared with those recorded from the muscle endplate of a patient with fast channel syndrome caused by the εP121L mutation.

channel recordings also revealed that the εP121L mutant channels open much less frequently and do not exhibit the long bursts of openings characteristic of wild-type channels (Fig. 15.12). Consequently, the channel open probability is markedly reduced. The low open probability arises both because of the reduced affinity for ACh and because of a marked reduction in the rate of channel opening (so fewer channel openings occur during receptor occupancy). The ability of εP121L to influence both ACh binding and channel opening argues that the agonist-binding site is closely coupled to that which influences channel gating.

Mutation of εP121L decreases the safety margin for neuromuscular transmission by reducing the response of the nAChR to acetylcholine. Because fewer receptors are activated, and those which are activated exhibit shorter burst durations, the excitatory postsynaptic current is reduced and the epp is smaller. It may therefore fail to elicit an action potential.

DISEASES OF NEURONAL AChRs

Nocturnal Frontal Lobe Epilepsy

Epilepsy is characterised by bursts of synchronised electrical discharges that interfere with normal neuronal function and commonly cause seizures. It is genetically heterogeneous and may result from defects in K^+ or Ca^{2+} channels, as described elsewhere (Chapters 6 and 9), as well as nAChR channels. Autosomal dominant nocturnal frontal lobe epilepsy (ADNFLE) is a rare form of partial epilepsy which usually presents in childhood. It is associated with frequent violent, but fortunately brief, seizures that are usually confined to the frontal lobe and occur almost exclusively during light sleep. Mild cases of the disease are sometimes misdiagnosed as nightmares.

In one Australian kindred, the gene for ADNFLE was shown to map to chromosome 20q13.2–13.3, at the same locus as the gene encoding the neuronal nAChR α_4-subunit (CHRNA4). A single mutation in the CHRNA4 gene was detected in affected individuals—substitution of a conserved serine at position 248 by phenylalanine (S248F; Steinlein et al., 1995). This residue is located within the second transmembrane domain of the nAChR channel. It is thought to lie within the pore as it forms part of the binding site for the noncompetitive inhibitor chlorpromazine, which acts as an open channel blocker. α_4-Subunits are expressed in many brain regions, where they coassemble with β_2 subunits to form AChRs. When the α_4-subunits containing the S248F ADNFLE mutation were coexpressed with β_2-subunits in Xenopus oocytes, the response to ACh desensitized faster and showed slower recovery from desensitization than that of the wild-type channel (Wielland et al., 1996). Single-channel current recordings showed that mutant α_4,β_2 channels also exhibited a greater frequency of a low conductance state (Kuryatov et al., 1997). The net effect of the ADNFLE mutation is thus to reduce nAChR function by decreasing the channel open time, reducing the single-channel conductance and increasing the rate of desensitization. Why this should result in epilepsy is not altogether clear, but one possibility is that α_4-subunit-containing AchRs might mediate the release of an inhibitory neurotransmitter; a reduction in nAChR function would then result in the enhanced excitability of post-synaptic neurones and lower the seizure threshold.

Several intriguing issues remain to be explained. Why, for example, does epilepsy affect only the frontal lobe when *CHRNA4* is widely expressed throughout the cerebral cortex? And why do seizures occur almost exclusively during sleep?

Neuronal Degeneration in C. elegans

A gain-of-function mutation (*u662*) in the gene encoding a nicotinic ACh α-subunit (*deg-3*) leads to degeneration of a subset of neurones in the worm *C. elegans* (Treinin and Chalfie, 1995). This mutation substitutes an asparagine for an isoleucine at position 293, which lies within the pore-forming M2 domain. Channel hyperactivity may underlie the neuronal degeneration observed with this mutation since nicotinic AChR antagonists suppress *u662* mutant phenotypes. Within the pore region, the *deg3* α-subunit is most similar to the neuronal α_7-subunit of the rat, which is highly permeable to Ca^{2+}. This suggests that in *C. elegans*, neuronal death may result from the enhanced cation influx and osmotic swelling which accompanies channel hyperactivity.

Nicotine Abuse

Tobacco consumption has been estimated to cause nearly 20% of deaths in developed countries, and smokers are more than 20 times more likely to develop lung cancer than individuals who do not smoke (Peto *et al.*, 1992). Despite being aware of its harmful effects, most smokers find it extremely difficult to give up tobacco. Current evidence suggests that the addictive effects of tobacco are principally due to nicotine. It is likely that neuronal nAChRs are the principal target for the drug and that their activation stimulates dopaminergic neurones of the mesolimbic system which mediate 'reward' pathways in the brain. Dani and Heinemann (1966) have suggested a simple model to explain nicotine addiction. They argue that smoking a cigarette delivers sufficient nicotine to active nAChRs that, directly or indirectly, stimulate dopamine release within the mesolimbic system and thereby induce a pleasurable effect. With continued use, nicotine builds up to a low steady-state level that causes receptor desensitization. The pleasurable effect of subsequent cigarettes is therefore reduced and, if nicotine levels remain elevated, the number of nAChRs is increased to compensate for receptor desensitization. During periods of abstinence, such as at night or when trying to stop smoking, steady-state nicotine levels drop and some or all of the desensitized nAChRs recover.

nAChRs containing β_2-subunits appear to mediate most of the effect of addictive effects of nicotine. Functional disruption ('knock-out') of the β_2-subunit gene resulted in the loss of high-affinity binding sites for nicotine in the brain of homozygous knockout ($\beta_2^{-/-}$) mice (Picciotto et al., 1995). In addition, nicotine did not stimulate dopaminergic neurones or cause striatal dopamine release in mice lacking β_2-subunits, as it does in normal mice (Picciotto et al., 1998). Self-administration of nicotine was also attenuated in $\beta_2^{-/-}$ mice.

It is well documented that, in addition to its addictive effects on the brain, nicotine enhances learning and memory. As the performance of $\beta_2^{-/-}$ mice in associative memory tests was *not* enhanced by nicotine, β_2-containing nAChRs may also mediate the memory-enhancing effects of nicotine (Picciotto et al., 1995). It is worth noting that the concentrations of nicotine used in these studies are similar to those attained in plasma by cigarette smoking: 150–500 nM from a single puff. Interestingly, there is some evidence that nicotine may improve memory retention in Alzheimer's disease, a condition which is associated with the loss of ACh-containing neurones. One finding that is as yet unexplained is why the $\beta_2^{-/-}$ mice performed better than their wild-type littermates in associative memory tests conducted in the absence of nicotine.

CHAPTER 16

GLUTAMATE RECEPTORS

Glutamate receptors come in two types: ionotropic and metabotropic. The ionotropic receptors are ligand-gated ion channels, whereas the metabotropic receptors do not have an integral ion channel and mediate their effects by activation of a second messenger cascade. We shall only be concerned with the former type here. Traditionally, ionotropic glutamate receptors have been classified according to their preferred agonists into AMPA (α-amino-3-hydroxy-5-methyl-4-isoxazoleproprionate), kainate and NMDA (*N*-methyl-D-aspartate) types. All three types of channel are also, of course, activated by glutamate, which is the physiological agonist and the major excitatory neurotransmitter in the brain. Rapid synaptic transmission is mediated by the activation of AMPA and kainate receptors. In contrast, NMDA receptor (NMDAR) activation results in slow synaptic potentials which are involved in various forms of activity-dependent synaptic plasticity. NMDARs also play important roles in brain development, learning and memory. In this chapter, we will consider the NMDA and non-NMDA (kainate and AMPA) glutamate receptors separately.

Mutations in one non-NMDA type of glutamate receptor (GluR) subunit (Rδ2) cause epileptic seizures and degeneration of Purkinje neurones in the *Lurcher* mouse. Autoantibodies to another non-NMDA type of subunit (GluR3) have been implicated in Rasmussen's encephalitis; this is discussed in Chapter 21. In addition, overstimulation of glutamate receptors is thought to be involved in the neuronal cell death produced by ingested toxins, hypoxia, and ischaemia and has also been implicated in a number of neurodegenerative diseases. The precise contribution of glutamate excitotoxicity to neuronal death in most of these disorders has, however, not yet been fully established.

NON-NMDA RECEPTORS

Basic Properties

AMPA receptors (AMPARs) are activated by AMPA, quisqualate and glutamate and are blocked by CNQX (6-cyano-7-nitroquinoxaline-2,3-dione). They are permeable to Na^+ and K^+ but, generally, show low Ca^{2+} permeability. Because they exhibit fast activation kinetics and often undergo rapid desensitization in the maintained presence of agonist, AMPARs mediate the fast component of the excitatory postsynaptic potential (epsp). The AMPA response peaks within ~1 ms and decays within 30 ms. AMPARs are half-maximally activated by ~0.5 mM glutamate. Since the estimated glutamate concentration in the synaptic cleft rises to around 1 mM during synaptic release, some AMPARs may not open in response to a presynaptic depolarization. Kainate receptors show many similar properties to AMPARs but are selectively gated by kainate and domoic acid (domoate). Like AMPARs, they are blocked by CNQX and their activation gives rise to a fast epsp.

Structure

The first glutamate receptor subunit to be identified (GluR1) was isolated by expression cloning in 1989 (Hollman *et al.*, 1989), and more than 16 different non-NMDA glutamate receptor subunits have now been cloned (Fig. 16.1; Table 16.1). Unfortunately, different laboratories have given these subunits different names so there is considerable confusion in the literature. For example, GluR1 to GluR4 are also known as GluRA to D and as α-1 to α-4. This chapter uses the first of these classifications (i.e., GluR1). The chromosomal

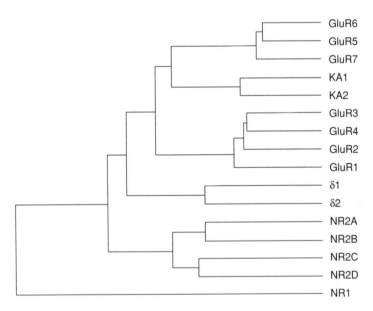

FIGURE 16.1 RELATIONSHIP BETWEEN THE GENES ENCODING THE DIFFERENT GLUTAMATE RECEPTOR SUBUNITS
The line lengths are proportional to the mean number of differences per residue along each branch. After Seeburg (1995).

TABLE 16.1 MAMMALIAN GLUTAMATE RECEPTOR GENES

Gene	Protein	Chromosome location (human)	Reference
GRIA1	GluR1	5q31.3–q33.3	Puckett *et al.* (1991)
GRIA2	GluR2	4q32–33	McNamara *et al.* (1992)
GRIA3	GluR3	Xq25–26	McNamara *et al.* (1992)
GRIA4	GluR4	11q22–23	McNamara *et al.* (1992)
GRIK1	GluR5	21q21.1–22.1	Eubanks *et al.* (1993)
GRIK2	GluR6	6	Paschen *et al.* (1994)
GRIK3	GluR7	1p34–p33	Puranam *et al.* (1993)
GRIK4	KA1	11q22.3	Szpirer *et al.* (1994)
GRIK5	KA2	19q13.2	Szpirer *et al.* (1994)
	δ1		Lomeli *et al.* (1993)
	δ2		Lomeli *et al.* (1993)
GRIN1	NR1	9q34.3	Karp *et al.* (1993); Takano *et al.* (1993)
GRIN2A	NR2A	16p13	Takano *et al.* (1993)
GRIN2B	NR2B	12p12	Mandich *et al.* (1994)
GRIN2C	NR2C	17q25	Takano *et al.* (1993)
	NR2D		Ishii *et al.* (1993)

location of the genes encoding the glutamate receptor subunits is given in Table 16.1.

There are four closely related AMPA receptor subunits: GluR1 to GluR4. They have molecular masses of between 96 and 101 kDa, are ~900 amino acids long, and share ~70% identity at the amino acid level. Injection of mRNA encoding a single type of AMPAR subunit into *Xenopus* oocytes gives rise to currents gated (in order of potency) by AMPA, glutamate and kainate. The same affinity sequence was found for all four cloned receptors in ligand-binding studies, and as it resembles that found for AMPA-binding sites in the brain GluR1–4 are known as AMPA receptors. Five different kainate receptor subunits have been cloned. Three of these show ~40% homology with GluR1–4 and have therefore been called GluR5, GluR6 and GluR7. Expression of GluR5 mRNA in mammalian cells gives rise to homomeric channels which can be activated by domoate, kainate, glutamate and AMPA, in that order of potency. GluR6 also expresses functional homomeric channels that can be activated by kainate and glutamate, but not by AMPA. GluR7 has not been reported to form functional channels. Two other kainate receptor subunits, KA1 and KA2, have been described that also fail to form functional homomeric channels, but which bind kainate with very high affinity. These show ~70% sequence identity with each other, but only around 40% sequence identity with the GluR5–7 kainate receptor family, and somewhat less with the AMPAR subunits. No currents are detected when KA1 or KA2 are expressed alone, but coexpression of KA2 with either GluR5 or GluR6 leads to the formation of functional channels with novel properties. Finally, several other non-NMDAR subunits have been identified, including the δ1 and δ2 subunits, which have not yet been shown to form functional channels in heterologous expression systems (but see page 310).

Tissue distribution

In situ hybridization studies have demonstrated that there are marked differences in the level of expression of the different non-NMDAR subunits within the brain (Hollmann and Heinemann, 1994). Almost all cells express GluR2, with the notable exception of the Bergman glial cells in the cerebellum, the type II cells of the hippocampus, the globus pallidus and the brainstem. GluR1,3 and 4 also show a widespread distribution. High levels of expression of GluR5 are found in dorsal root ganglion neurones; of GluR6 in cerebellar granule cells and CA3 hippocampal neurones; and of GluR7 in neocortex. KA1 is found in high levels in the CA3 and dentate gyrus regions of the hippocampus, whilst KA2 has a much more widespread distribution.

Topology

Each GluR channel is thought to be made up of four subunits (Rosenmund *et al.*, 1998). By analogy with the AChR channels, it was originally proposed that each GluR subunit would have four transmembrane domains (TMs). This idea was later revised when it was found to be incompatible with data obtained from glycosylation and antibody studies (Hollmann *et al.*, 1994; Wo and Oswald, 1995). It is now clear that GluR subunits have only three TMs, with a large extracellular N-terminal domain and a cytosolic C terminus (Fig. 16.2). The cytosolic region linking the first and second TMs loops back into the membrane to line the pore in a manner reminiscent of the pore loops of K^+ channels; in this case, however, the loop is inserted from the inner side of the membrane rather than the extracellular side. The loop linking the first and second TMs is called TM2 and the second and third transmembrane domains are called TM3 and TM4, respectively. This rather strange nomenclature derives from the original hypothesis that each GluR subunit contained four TMs.

Functional Diversity

Native AMPA and kainate receptors differ in their pharmacological sensitivity, rate of desensitization and Ca^{2+} permeability (Fig. 16.3). This functional diversity is not simply the result of the several different genes which encode AMPA and kainate receptors. It is also created by mRNA editing, alternative splicing and the formation of heteromeric receptors.

Heteromerization

Expression of a single type of AMPA receptor mRNA in *Xenopus* oocytes produces functional channels, demonstrating that GluR1–4 subunits are capable of forming homomeric receptors (Fig. 16.3). Native AMPARs, however, appear to be predominantly heteromeric. This is evident from the fact that most native AMPA receptors are only poorly permeable to Ca^{2+}, whereas

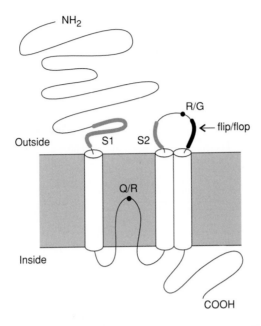

FIGURE 16.2 PUTATIVE TOPOLOGY OF GLuR SUBUNITS
Membrane topology of a single GluR subunit deduced from a combination
of hydropathy analysis, antibody-binding studies and site-directed mutagenesis. Residues and domains indicated in the text are marked. S1 and
S2 are domains involved in agonist binding; the flip/flop domain is an
alternatively spliced module; and the Q/R and R/G sites are subject to
mRNA editing (the first letter gives the amino acid encoded by the genomic
DNA and the second letter indicates the edited residue).

homomeric GluR1, 3, and 4 receptors show significant Ca^{2+} permeability.
Coexpression of GluR2 with GluR1, 3, or 4, however, generates AMPARs
with properties resembling those of native channels (Hollman *et al.*, 1991).
The low-affinity kainate receptors, GluR5, 6 and 7, do not form heteromeric
complexes with any of the four different AMPAR subunits. However, GluR5
and GluR6 can coassemble with KA2 (Fig. 16.3). Somewhat surprisingly,
GluR7 does not interact with either KA1 or KA2. Clearly, it is important to
determine which of the many possible combinations of AMPA and kainate
receptor subunits comprise native glutamate receptors. One particularly elegant approach to this problem has been to combine single-cell PCR with
patch-clamp recordings, thus enabling correlation of mRNA transcripts with
the electrophysiological properties of the native receptor in a single neurone
(Geiger *et al.*, 1995).

mRNA editing

The cloning of the AMPA receptor subunits was followed by the striking
finding by Sommer and his colleagues that the genomic sequence of GluR2

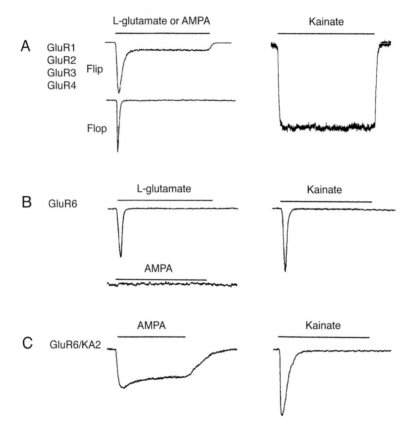

FIGURE 16.3 AMPA AND KAINATE RECEPTORS EXHIBIT DIFFERENT KINETICS AND AGONIST SENSITIVITY
The time course of the currents elicited by glutamate (or AMPA) and by kainate varies between the different types of subunits and between the flip and flop variants of GluR1–4. From Seeburg (1993).

differs from that of the cDNA at a single residue within TM2 (Sommer *et al.*, 1991). This was subsequently shown to result from editing of the mRNA, which resulted in the substitution of an arginine (R) residue for glutamine (Q). Hence this position is designated the Q/R site. Editing takes place within the nucleus, prior to removal of the introns. It is carried out by an enzyme that operates upon a double-stranded RNA structure, formed by base pair interactions between the exon that contains the critical edited residue and a sequence within a downstream intron known as the ECS site (Higuchi *et al.*, 1993). mRNA editing is highly efficient: almost 100% of GluR2 mRNA is edited. GluR5 and GluR6 are also edited, but with lower efficiency.

mRNA editing has dramatic consequences (Sommer *et al.*, 1991). When glutamine is present at the Q/R site, the channel is permeable to Ca^{2+} ions. In contrast, the presence of a positively charged arginine residue at this position renders the channel impermeable to Ca^{2+} (Fig. 16.4). Homomeric

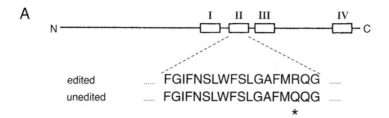

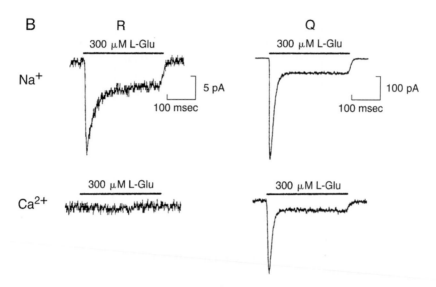

FIGURE 16.4 mRNA EDITING AT THE Q/R SITE OF AMPA
RECEPTORS INFLUENCES Ca²⁺ PERMEABILITY
(A) Linear diagam of GluR2 showing the position of the Q/R site (in
bold type) in TM2. The edited and unedited sequences are given below.
(B) Whole-cell currents evoked by 300 μM glutamate in high Na⁺ or high
Ca²⁺ extracellular solution in a cell expressing homomeric GluR2 with
either arginine (R, left) or glutamine (Q, right) at the Q/R editing site.
From Burnashev *et al.* (1992a).

GluR2 AMPA receptors are therefore not Ca^{2+} permeable. Heteromeric GluRs
containing a single GluR2 subunit are also Ca^{2+} impermeable (Hollmann *et
al.*, 1991). This explains why most native AMPARs, which are heteromers of
GluR2 and either GluR1, 3, or 4, are impermeable to Ca^{2+}. The Q/R site also
influences the shape of the current–voltage relation and the sensitivity to
channel blockers. Homomeric GluR4 channels exhibit pronounced inward
rectification at negative membrane potentials and outward rectification at
positive potentials (Fig. 16.5; Verdoorn *et al.*, 1991). They are also blocked by
the spider toxin argiotoxin. In contrast, AMPARs containing GluR2 subunits
have a more linear current–voltage relationship and are insensitive to argiotoxin.

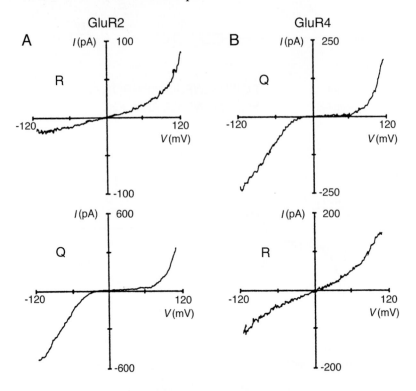

FIGURE 16.5 *mRNA EDITING AT THE Q/R SITE OF AMPA RECEPTORS ALTERS THE SHAPE OF THE I-V RELATION*
(A) Current–voltage relations recorded for homomeric wild-type GluR2 (top) or GluR2 in which arginine 586 has been mutated to glutamine (below). (B) The reverse experiment, showing current–voltage relations recorded for homomeric wild-type GluR4 (top) or GluR4 in which glutamine 587 has been mutated to arginine (below). From Verdoorn *et al.* (1991).

Neurones appear to control the Ca^{2+} permeability of their native AMPA receptors by regulating the level of GluR2 expression, rather than by varying the efficiency of editing at the Q/R site. For example, CA3 hippocampal neurones express high levels of GluR2 and generally have GluRs with low Ca^{2+} permeability, whereas the Bergmann glia cells of the cerebellum, which do not express detectable GluR2, exhibit the highest Ca^{2+} permeability (Geiger *et al.*, 1995).

mRNA editing is not confined to the Q/R site. In some AMPAR subunits, a second site within the extracellular loop linking TM3 and TM4 is also edited (Fig. 16.2; Lomeli *et al.*, 1994). All AMPAR subunits contain an arginine codon at this site, but the mRNA encoding GluR2, 3, and 4 may be edited to glycine (hence it is called the R/G site). Around 80–90% of GluR2 and GluR3 mRNA is edited, but only ~50% of GluR4. Edited subunits recover faster from desensitization than unedited subunits (Fig. 16.6). Unedited subunits are therefore expected to respond only to transmitter released by the first few

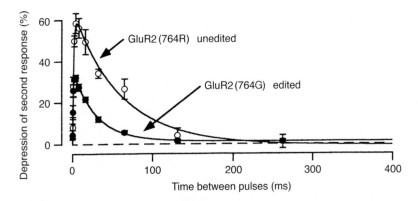

FIGURE 16.6 FUNCTIONAL EFFECTS OF mRNA EDITING AT THE R/G SITE OF AMPA RECEPTORS
When two glutamate pulses are applied close together, the response to the second pulse is depressed, by an amount which is dependent on the pulse interval. This depression is greater if the R/G site is unedited (i.e., R) because the receptor takes longer to recover from desensitization. Heteromeric GluR1/GluR2 receptors were used. From Lomeli *et al.* (1994).

impulses of a fast train of presynaptic action potentials, in contrast to edited subunits which recover rapidly from desensitisation and therefore respond to all impulses in the train.

Alternative splicing

The diversity of non-NMDA receptors is further enhanced by alternative splicing. It turns out that each of the GluR1 to GluR4 channels may exist in either a 'flip' or a 'flop' version (Sommer *et al.*, 1990a). This results from the use of alternative exons to code for a region of 38 amino acids that precedes TM4 and is located extracellularly (Fig. 16.2). The flip/flop module influences the rate of receptor desensitization (Mosbacher *et al.*, 1994). It is not the sole determinant of the desensitization rate, however, as the subunit type is also important. Studies of cloned AMPARs have shown that the influence of the flip/flop splice variant is specific to each subunit and is strongest for GluR4.

Both the flip and flop versions of GluR1–GluR4 are widely expressed in the brain, but show differential tissue and developmental distribution (Sommer *et al.*, 1990). This is particularly marked in the hippocampus, where adult pyramidal CA3 neurones primarily express the flip module but CA1 neurones predominantly express the flop module. There is also differential expression of flip and flop modules during development (Monyer *et al.*, 1991). The flip module dominates prior to birth and persists throughout life, whereas the flop form is only expressed postnatally. Thus the properties of the AMPARs in the early brain may differ from those of the adult brain, because of their different molecular structure.

Correlating Structure and Function

The pore

The diameter of the narrowest part of the pore of both AMPA and kainate receptors lies between 0.7 and 0.8 nm (Burnashev *et al.*, 1996), and is thus rather larger than that found for NMDA receptors (see later). As described earlier, mutation of the Q/R site within TM2 dramatically alters the Ca^{2+} permeability of AMPA and kainate receptors, thus providing clear evidence that the TM2 loop contributes to the pore. Additional mutagenesis studies, reviewed by Hollmann and Heinemann (1994), provide further support in favour of this idea. There is some evidence that TM1 may also contribute to the pore, as mutation of two residues in this domain results in altered GluR permeability properties (Kohler *et al.*, 1993). This is perhaps not surprising, since the adjacent transmembrane domains contribute to the pore of both the voltage-gated and inwardly rectifying K^+ channels, as described in Chapters 6 and 8.

The ligand-binding site

The structure of the ligand-binding site of the glutamate receptor is now known with reasonable accuracy as a result of a combination of site-directed mutagenesis and X-ray crystallographic studies. To identify regions involved in agonist binding, Stern-Bach and colleagues (1994) made chimeras between the AMPAR subunit GluR3 and the kainate receptor subunit GluR6. They observed that exchange of the first 150 amino acids upstream of TM1 (which they called the S1 domain) and the M3-M4 loop (the S2 domain) was required to alter agonist selectivity. The 3-dimensional structure of the glutamate-binding site of the GluR2 receptor was subsequently solved by construct-ing a cDNA consisting of the S1 domain linked to S2 domain, expressing the cDNA in bacteria and crystallizing the recombinant protein in the presence of the GluR agonist kainate (Armstrong *et al.*, 1998). This revealed that the binding site is formed from two kidney-shaped domains, one of which is mainly composed of S1 and the other of S2, which are arranged with their concave sides facing each other. This creates a central pocket into which kainate nestles. The crystal structure is strikingly similar to that of a glutamine-binding protein found in *Escherichia coli* (Hsaio *et al.*, 1996).

Summary of structure–function studies

To recapitulate, AMPA and kainate subunits consist of three transmem-brane domains (TMs 1, 3 and 4) with a large extracellular N terminus and an intracellular C terminus. The pore is lined by a hairpin loop (TM2) which enters the membrane from the cytosolic side after TM1. The ligand-binding site comprises two extracellular domains (S1 and S2), located in the N termi-nus, immediately prior to TM1, and in the M3–M4 loop, respectively. An edited residue (Q/R) forms part of the selectivity filter and determines whether or not the channel is permeable to Ca^{2+}. Most AMPA receptors are heteromers and contain one or more edited GluR2 subunits, which render them Ca^{2+} impermeable. Those receptors lacking GluR2 possess high Ca^{2+} permeability.

NMDA RECEPTORS

Basic Properties

NMDA receptors are activated by NMDA and by glutamate. They are insensitive to the AMPAR antagonist CNQX, but D-2-amino-5-phosphonovalerate (APV) acts as a competitive inhibitor, and some agents such as phencyclidine and MK-801 act as voltage-dependent channel blockers, entering and plugging the channel pore. NMDARs are permeable to Na^+ and K^+ and, unlike most non-NMDARs, have a high Ca^{2+} permeability. They also have much slower kinetics and show little desensitization. This means that the time course of the NMDAR current is determined by the unbinding of glutamate and contrasts with AMPARs, where rapid desensitisation makes a marked contribution to the receptor kinetics. NMDA receptors are also about a thousand times more sensitive to glutamate than AMPARs, being half-maximally activated by ~5 μM glutamate. Because the ambient glutamate concentration is around 1–3 μM, and NMDARs show little desensitization, depolarisation of the postsynaptic membrane may be sufficient to open the NMDAR even in the absence of presynaptic glutamate release.

A unique property of the NMDAR is that it requires the simultaneous presence of two agonists for activation: glutamate and glycine. The first clue to this fact came when Johnson and Ascher (1987) discovered that the amplitude of the glutamate-activated NMDAR current was dramatically increased by micromolar concentrations of glycine. It is now established that the reason many investigators observe a small glutamate current in the absence of added glycine is because of background contamination from the water or salts used to make the saline solution. When special precautions are taken to make a glycine- or glutamate-free solution, neither agonist is able to open the channel on its own. *In vivo*, the ambient glycine concentration is sufficient for NMDAR activation, so that channel opening is determined by the amount of glutamate released from the presynaptic terminal.

Another distinguishing property of NMDARs, and one of considerable physiological significance, is that they are blocked by external Mg^{2+} (Nowak *et al.*, 1984). The block is voltage dependent and is largest at hyperpolarised potentials, so that the current first increases and then falls again as the membrane is hyperpolarised (Fig. 16.7). This is because the positively charged Mg^{2+} ion is attracted into the pore at negative membrane potentials, but is unable to pass all the way through. Consequently, it plugs the pore and impedes Ca^{2+} flux. The levels of Mg^{2+} found in plasma are sufficient to block

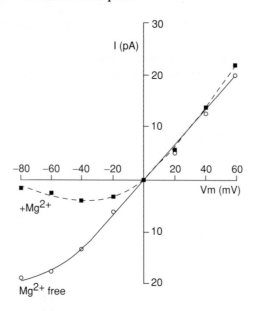

FIGURE 16.7 Mg^{2+} CAUSES A VOLTAGE-DEPENDENT BLOCK OF
NMDA RECEPTORS
Voltage dependence of the whole-cell glutamate-activated current measured in the absence (●) or presence (■) of 0.5 mM Mg^{2+}. The current–
voltage relation deviates from linearity in the presence of the cation because Mg^{2+} produces a voltage-dependent block. From Nowak *et al.* (1984).

NMDARs. This means that NMDARs will be blocked at the resting membrane
potential, so that no current will flow even when glutamate levels are high.
If, however, the membrane is depolarised for any length of time the block
will be relieved, allowing Ca^{2+} influx through the NMDAR.

The influx of Ca^{2+} through NMDARs can trigger lasting changes in synaptic efficacy. One such change, known as long-term potentiation (LTP), is
involved in learning (McKernan and Shinnick-Gallagher, 1997). LTP describes
the enhanced response of the postsynaptic neurone to a single stimulus
which follows a fast train of impulses, or two closely spaced stimuli. This
enhancement persists for long periods (hence it is known as long-term potentiation or LTP). At glutamergic synapses, both AMPARs and NMDARs are
generally present in the postsynaptic membrane (Fig. 16.8). Glutamate released by a single nerve impulse arriving at the presynaptic terminal triggers
a fast epsp mediated by AMPARs, but the membrane depolarization is not
sufficient to relieve the Mg^{2+} block of NMDARs and thus does not elicit LTP.
If two or more epsps occur in close succession, however, they may summate
to produce a membrane depolarization large enough to relieve the Mg^{2+} block
and allow Ca^{2+} influx through NMDARs. Exactly how the resulting rise in
intracellular Ca^{2+} produces enhanced synaptic efficacy is a topic of intensive
investigation.

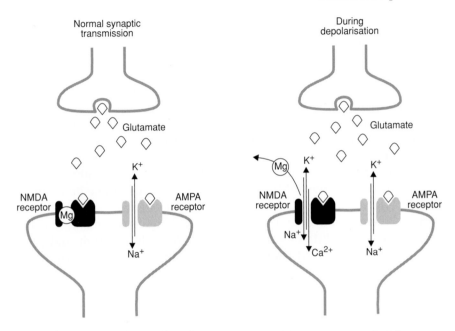

FIGURE 16.8 MODEL FOR LONG-TERM POTENTIATION (LTP)
Excitatory glutaminergic synapse with both AMPA and NMDA receptor channels in the postsynaptic membrane. The presynaptic neurone releases glutamate on stimulation. This causes rapid activation of AMPA receptors, which have fast kinetics. At low stimulation rates, NMDA receptors do not conduct because Mg^{2+} blocks the pore, but at high stimulus frequencies the membrane depolarization caused summation of the AMPAR responses is sufficient to relieve the resting Mg^{2+} block of the NMDAR. This leads to Ca^{2+} influx and LTP.

Structure

Two major gene families which encode NMDA receptor subunits have been identified to date: NR1 and NR2 (Fig. 16.1; Table 16.1). The former was isolated by expression cloning from rat brain (Moriyoshi *et al.*, 1991). It has a molecular mass of ~103 kDa, is ~900 amino acids long, and shows ~25% amino acid identity with the AMPA and kainate receptor subunits. NR1 is subject to alternative splicing, and at least eight splice variants have been identified (Zukin and Bennett, 1995). These differ in regions of their N and/or C termini that are involved in channel regulation by polyamines, Zn^{2+} and protein kinase C. Additional NMDA receptor subunits were identified by screening brain-derived cDNA libraries with nucleotide sequences conserved throughout known GluRs (Monyer *et al.*, 1992). These subunits, termed NR2A to NR2D, are closely related to each other, but only distantly related to other types of GluR subunit, including NR1. In particular, NR2 subunits have a very much longer C-terminal sequence which follows TM4. They are also larger, having molecular masses which range from ~130 to ~170 kDa. The chromosomal location of the genes encoding NR1 and NR2 subunits is given in Table 16.1.

Expression of the different NR2 subunits by themselves does not produce functional GluRs, and although NR1 produces homomeric receptors in *Xenopus* oocytes, their properties are not identical with those of native NMDARs. Coexpression of NR1 with any of the NR2 subunits produces functional channels which more closely resemble the native receptors (Kutsuwada *et al.*, 1992). Moreover, subunit-specific antibodies demonstrate that NR1 and NR2 subunits can be coimmunoprecipitated from rat brain membranes (Sheng *et al.*, 1994a). Thus, native NMDA receptors are believed to be heteromers of NR1 and NR2 subunits. The binding of two glutamate and two glycine molecules is required to open native NMDAR channels (Clements and Westbrook, 1991). Since glutamate interacts with the NR2 subunit and glycine binds to NR1, as described later, this suggests that the native NMDAR may be a tetramer composed of two NR1 and two NR2 subunits.

Tissue distribution

In situ hybridization studies have revealed that NR1 is abundantly expressed throughout almost all brain regions, consistent with the idea that it is a primary component of the NMDA receptor (Moriyoshi *et al.*, 1991). By contrast, the different NR2 subunits show differential tissue and developmental expression (Monyer *et al.*, 1994; Hollmann and Heinemann, 1994). Coassembly of NR1 with different types of NR2 subunits produces NMDARs with different functional properties, and suggests that part of the variation in the functional properties of NMDARs in different brain regions may arise because they contain different NR2 subunits. The different splice variants of NR1 are also differentially expressed and may contribute to the regional variation in NMDAR properties (Laurie and Seeburg, 1994).

Topology

Like the AMPAR subunits, NMDAR subunits are thought to consist of three transmembrane-spanning domains, a hydrophobic segment (TM2) that forms a hairpin loop within the membrane and lines the channel pore, a large extracellular N terminus that comprises about half of the protein mass, and an intracellular C-terminal domain that contains a number of phosphorylation sites (Fig. 16.2). There are also two extracellular segments, located in the N terminus and M3–M4 loop, that are predicted to form the agonist-binding site.

Correlating Structure and Function

The pore

The narrowest part of the NMDA receptor channel has a diameter of ~0.55 nm (Villarroel *et al.*, 1995), which is smaller than that of other cation-selective channels; that of the nAChR, for example, is ~0.75 nm (Chapter 15). NMDARs also differ from most other ligand-gated channels in being more permeable to Ca^{2+} than to monovalent cations. This selectivity is largely determined by a single uncharged residue, an asparagine, which occupies a position in the M2 domain of NR1 and NR2 corresponding to the Q/R site of AMPA receptor subunits. In NMDA receptors, this site is sometimes termed the N site (unlike the Q/R site it is not subject to mRNA editing). Replacing the critical asparagine in NR1 with a positively charged arginine residue abolished the Ca^{2+} permeability of the channel—presumably because of electrostatic effects (Burnashev et al., 1992b). Thus, like the AMPA receptors, the pore of NMDARs is formed from the TM2 loop. Both NR1 and NR2 subunits contribute to the pore, but the critical asparagine residue in NR2 is not located at the N site (residue 595) defined by sequence homology with NR1 (Wollmuth et al., 1996). Instead, it is shifted towards the C terminus by one position (N596).

The structure of the M2 segment has also been investigated by cysteine-scanning mutagenesis (Kuner et al., 1996) (see Chapter 4 for an explanation of this method). The results of these experiments have been interpreted to indicate that the M2 segment loops up into the membrane from the cytoplasmic side, with the N-terminal (ascending) limb forming an α-helical structure and the C-terminal (descending) limb adopting an extended form. The critical asparagine is positioned at the tip of the loop and forms a constriction in the pore which constitutes the selectivity filter.

Native NMDA receptors are blocked by extracellular Mg^{2+} in a voltage-dependent manner (Mayer et al., 1984; Nowak et al., 1984). Studies of cloned NMDARs have shown that the strength of this block depends on the subunit combination. For example, channels formed from NR1 and NR2A are blocked more strongly than channels formed from NR1 and NR2C subunits (Monyer et al., 1992; Kutsuwada et al., 1992). The N site also appears to be involved in this block because mutation of the critical asparagine in either NR1 or NR2 influences Mg^{2+} block.

Ligand binding

Site-directed mutagenesis of NR1 and NR2 has revealed that, as is the case for the AMPAR subunits, agonist binding involves two main regions of the protein, one located in the extracellular region preceding TM1 (S1) and the other in the M3-M4 loop (S2) (Fig. 16.2). Mutation of residues F390, Y329 and F466 in NR1, which lie within the S1 domain, markedly reduces the affinity for glycine (Kuryatov et al., 1994). Two residues in the M3-M4 loop (F735 and F736) are also important determinants of the glycine binding site (Hirai et al., 1996). None of these mutations affected the glutamate sensitivity of NR1, however. In contrast, mutation of equivalent residues within the S1 and S2 domains of NR2B strongly reduced the ability of glutamate, but not glycine, to gate the NMDAR (Laube et al., 1997). Taken together, these results suggest that in the native NMDA receptor the NR1 subunit is mainly responsible for glycine binding whereas the NR2 subunit is the main determinant of glutamate binding.

Modulation

NMDARs are modulated by a number of agents, including polyamines, Zn^{2+}, and protein phosphorylation by serine/threonine and tyrosine kinases. The different NR1 splice variants differ in their sensitivity to these forms of modulation. For example, tyrosine kinases of the *src* family potentiate NR1/NR2A currents but are without effect on NMDARs containing other types of NR2 subunits (Kohr and Seeburg, 1996). This effect is mediated by the C-terminal domain of NR2A.

Like other ligand-gated channels, the localisation and anchoring of NMDARs at the postsynaptic region are facilitated by interaction with other proteins. In particular, the C-terminal domain of both NR1 and NR2 subunits interacts with a postsynaptic density protein known as PSD-95 (Kornau *et al.*, 1995). An ever-increasing number of other cytosolic proteins are also found to interact with the NMDAR.

GLUTAMATE CHANNELS AND DISEASE

Numerous studies have indicated that overstimulation of Ca^{2+}-permeable glutamate receptors has neurotoxic effects. John Olney has coined the term *'excitotoxicity'* to describe this process. Excessive activation of NMDA, AMPA, or kainate receptors is associated with a large increase in intracellular calcium, which triggers a cascade of events that can lead to neuronal death. The rise in intracellular calcium comes from Ca^{2+} influx, both through glutamate receptors themselves and through voltage-gated Ca^{2+} channels activated by the depolarisation which results from GluR activation. The reason why elevation of $[Ca^{2+}]_i$ causes cell death is less clear, but has been postulated to include the activation of Ca^{2+}-dependent proteases, disturbance of cellular metabolism, accumulation of toxic free radicals, and macromolecular degradation. Overactivation of glutamate receptors is also associated with enhanced neuronal excitation and may lead to recurrent excitation of neuronal circuits and thus to seizures. In turn, this may enhance glutamate release and lead to further autoexcitation and, ultimately, destruction of the neural network. Excitotoxicity can result from ingestion of glutamate agonists and has also been implicated in the neuronal death caused by ischaemia, and in neurodegenerative disorders. Glutamate is therefore something of a Jekyll and Hyde molecule—essential for normal synaptic function, but with the potential to cause neuronal destruction.

Ingestion of Natural Excitotoxins

A number of glutamate analogues are produced by plants, including domoate, β-N-oxalylaminoalanine, and β-N-methylamino-L-alanine. Their consumption can cause severe neurological disorders in man (Olney, 1994).

Domoic acid is a kainate receptor agonist which is found at high concentrations in shellfish, particularly mussels, that have fed on domoate-rich phytoplankton. It is responsible for sporadic outbreaks of food poisoning—such as that which occurred in western Canada in 1987 when more than 100 people were affected. In addition to gastrointestinal symptoms, some of them suffered from neurological sequelae, which included disorientation, memory loss, seizures, and coma. A high proportion of the most severely affected individuals were elderly. Autopsies on the brains of two people who died revealed they had disseminated lesions in the central nervous system with extensive loss of hippocampal neurones. Some survivors also sustained brain damage as they continued to show memory loss and mental confusion 18 months after being poisoned; one 84-year-old man developed temporal lobe epilepsy a year after intoxication (Cendes et al., 1995). The symptoms of domoate poisoning are indicative of hippocampal damage and are consistent with the presence of high-affinity domoate-binding sites in this brain region.

The seeds of the chickling pea Lathyrus sativus contain the potent AMPA receptor agonist β-N-oxalylaminoalanine (BOAA or β-ODAP), and their regular consumption for several months can cause lathyrism (Spencer et al., 1986). This crippling neurological disorder is endemic in certain parts of India and Bangladesh, and may become more widespread during famine, when other foods are less available. It is one of the oldest neurotoxic diseases known. The concentration of BOAA in Lathyrus seeds is increased by zinc deficiency, and a higher incidence of lathyrism may be found when the soil is depleted of micronutrients (as can occur during the monsoon rains). Lathyrism is characterised by severe muscle rigidity and spastic paralysis of the lower limbs. There is little or no recovery, even after chickling pea consumption ceases.

It has been postulated that a neurological disorder endemic to the western Pacific island of Guam may also be caused by ingestion of glutamate receptor agonists (Spencer et al., 1987). During periods of famine, the Guam islanders used to consume large quantities of flour made from the seeds of the cycad Cycas circulnalis, which contain high concentrations of β-N-methylamino-L-alanine (BMAA). In recent years this practice has ceased and the incidence of Guam disease has concomitantly dropped. BMAA has neurotoxic effects that can be blocked by NMDA antagonists: it is therefore believed to cause excitotoxic neuronal death by stimulation of NMDARs. Some of the features of Guam disease resemble those of amyotrophic lateral sclerosis (ALS) and of Parkinson's disease, which has raised the question of whether glutamate toxicity might also be involved in these neurodegenerative disorders.

A number of fungi, including the fly agaric *Amanita muscaria*, contain ibotenic acid, a nonselective GluR agonist that is active on all three types of glutamate receptor. Consumption of the fungus causes dizziness, drowsiness, delirium and euphoria. Erratic behavior is sometimes observed, which is followed by deep sleep.

Aspartame (Nutrasweet), used widely as a synthetic sweetener, is hydrolysed in the gut to produce aspartate, which binds to GluRs with an affinity similar to that of glutamate. It can cause neuronal cell death when administered orally to infant mice, but it is argued that human consumption is likely to be too low to cause neurotoxic effects.

Finally, we should not forget that glutamate itself (as monosodium glutamate, or MSG) is one of the most widely used of all food additives. Excessive consumption of MSG can lead to the well-known Chinese restaurant phenomenon, characterised by a splitting headache, dizziness and nausea. Studies of immature animals have shown that ingestion of MSG is associated with neuronal death, and that neuroendocrine cells which lie outside the blood–brain barrier are particularly vulnerable. In some cases, a single feeding was sufficient to cause rapid and irreversible neuronal destruction. It is therefore not surprising that there remains continuing concern that MSG consumption in man, particularly by human infants, may lead to later neuroendocrinopathies (Olney, 1994). It is perhaps worth remembering that the neurotoxic effects of glutamate were discovered as a result of investigations into the use of MSG as a food additive.

Ischaemic Neuronal Death

The brain depends on a continual supply of oxygen and glucose, and thus on a continuous blood supply, for its survival. Interruption of blood flow for only a few minutes results in neuronal degeneration and death. Such ischaemia may occur globally, as in cardiac arrest, or be confined to the local vasculature as in stroke, the third most common cause of death in North America. Certain neuronal populations, such as the CA1 pyramidal cells of the hippocampus, are more vulnerable to ischaemic injury than others. Interestingly, neuronal death does not occur instantaneously, and histological evidence of cell damage is not observed until 48–72 hr after the circulation has been restored. The question of why this is the case, and how ischaemia results in neuronal death, has been the subject of intensive investigation for many years.

It is currently believed that neuronal death is triggered by a rise in glutamate during the ischaemic period, caused by a combination of enhanced release and reduced uptake. This results in an excessive activation of glutamate receptors that continues after the circulation has been restored. Inhibitors of AMPARs appear much more effective at preventing neuronal death than NMDAR blockers, even when given as late as 24 hr after the ischaemic period. There is thus a growing consensus that AMPA receptor activation is of key importance for ischaemic-induced cell death (Pellegrini-Giampietro *et al.*,

1997). Changes in GluR expression may contribute to the mechanism of this effect, because GluR2 mRNA is downregulated in vulnerable neurones in response to transient forebrain or global ischaemia (Pellegrini-Giampietro *et al.*, 1992; Gorter *et al.*, 1997). This leads to the formation of AMPARs with increased Ca^{2+} permeability. An enhanced sensitivity to glutamate is therefore expected to follow an ischaemic episode and may contribute to neuronal death. The mechanism of this downregulation has not yet been established.

It is worth noting that those brain regions that are most vulnerable to ischaemia do not normally have Ca^{2+}-permeable AMPA receptors, whereas the more insensitive neurones express lower levels of GluR2 and show higher Ca^{2+} permeability. The latter may therefore have mechanisms for dealing with the increased Ca^{2+} influx, which are not present in sensitive neurones.

Excitotoxicity and Neurodegeneration

The discovery of the excitotoxic effects of glutamate fuelled speculation that a similar mechanism might underlie neuronal cell death in chronic neurodegenerative diseases such as Alzheimer's disease, Huntingdon's disease and amyotrophic lateral sclerosis (ALS). Several studies have examined whether the subunit composition of AMPARs is altered in these states to produce GluR with higher Ca^{2+} permeability. A reduction in GluR2 expression has been observed in spinal motor neurones in ALS patients, which would be consistent with an enhanced Ca^{2+} permeability and excitotoxicity (Virgo *et al.*, 1996). The level of GluR2 expression is also decreased in pyramidal neurones of the parahippocampal gyrus in patients with schizophrenia (Eastwood *et al.*, 1995). Whether these changes in GluR2 levels are the principal cause, or simply a consequence, of these disease, however, remains unknown.

Defective Q/R editing causes neurodegeneration

A decrease in GluR2 editing at the Q/R site of GluR2 might be expected to predispose to neuronal degeneration, by increasing the Ca^{2+} permeability of AMPARs. No such editing deficiency has been reported in man, but its effect is clearly seen in a mouse mutant in which GluR2 editing has been abolished by genetic deletion of the intronic ECS sequence required for operation of the editing enzyme (see earlier, page 296) (Brusa *et al.*, 1995). This results in a GluR2 protein that is not edited at the Q/R site. As expected, the AMPA receptors of heterozygous GluR2$^{+/\Delta ECS}$ knockout mice have a higher Ca^{2+} permeability than normal (Fig. 16.9). Although the heterozygous animals appear normal for the first 2 weeks after birth, thereafter they develop spontaneous and recurrent epileptic seizures and die before they are 3 weeks old. Postmortem examination reveals selective degeneration of CA3 hippocampal neurones, reminiscent of the hippocampal lesions induced by kainate.

Neurodegeneration in the *Lurcher* mouse

When recombinant δ1 or δ2 GluR subunits are expressed in heterologous systems, glutamate-activated currents are not observed. The functional role

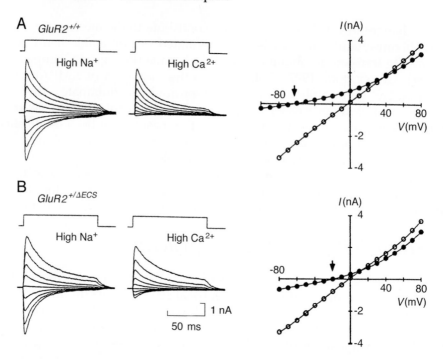

FIGURE 16.9 A DEFICIENCY IN GLuR2 EDITING CAUSES
INCREASED Ca^{2+} PERMEABILITY
Glutamate-activated currents (left) and corresponding current–voltage
relations (right) recorded in high Na^+ (O) or high Ca^{2+} (●) extracellular
solution from neurones of wild-type (A) and GluR2$^{+/\Delta ECS}$ mutant (B) mice.
Voltage steps (−80 to +80 mV) were applied as indicated. The presence
of inward Ca^{2+} currents and a more depolarised reversal potential in high
Ca^{2+} solution demonstrates the higher Ca^{2+} permeability of the mutant
neurones, in which GluR2 mRNA editing has been abolished. From Brusa
et al. (1995).

of the δ-subunits therefore remained a mystery until studies demonstrated
that the δ2-subunit is essential for normal brain function. The mutant mouse
Lurcher (Lc) results from a mutation in the δ2-subunit of the glutamate receptor, which is inherited in a semidominant fashion (Zuo et al., 1997). Homozygous Lc/Lc mice die shortly after birth because of a dramatic loss of midbrain
and hindbrain neurones during late embryogenesis. Heterozygous mice exhibit selective death of cerebellar Purkinje cells, which results in the ataxic
gait that gives the mouse mutant its name. The mutation that causes the Lc
phenotype is an alanine-for-threonine substitution within M3 of the GluR
δ2-subunit. It appears to cause a gain of function since Purkinje neurones of
Lc mice have a much higher resting conductance and a more depolarised
resting potential than those of normal animals (Fig. 16.10). This is a consequence of a constitutive inward current which is insensitive to NMDA and
AMPA antagonists, consistent with the idea that ligand binding is not re-

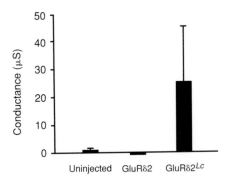

FIGURE 16.10 *THE* **LURCHER** *MUTATION CAUSES A CONSTITUTIVE INWARD CURRENT*
The average Na$^+$ conductance at -60 mV recorded from oocytes injected with GluRδ2 expressing the *Lurcher* mutation (GluRδ2^{Lc}) is much greater than that observed for wild-type GluRδ2. Currents were recorded in the absence of glutamate. From Zuo *et al.* (1997).

quired to open the mutant channel. Expression of δ2 mRNA is concurrent with the development of the *Lc* phenotype and immediately precedes large-scale neuronal death. Thus, neuronal death appears to result from the increased inward current through the mutant receptor.

Olivopontocerebellar degeneration

Olivopontocerebellar atrophy (OPCA) is an adult-onset progressive neuro-degenerative disorder of man characterized by ataxic movements, dysar-thria, and, often, nystagmus. It is associated with widespread progressive neuronal degeneration affecting the cerebellar cortex, inferior olive, and pon-tine nuclei (hence its name). Patients may also show rigidity and akinesia due to degeneration of the striatum and substantia nigra. In one subgroup of patients with OPCA the disorder is caused by a deficiency of the enzyme glutamate dehydrogenase, which is involved in glutamate metabolism (Plait-akis *et al.*, 1982, 1984). These patients exhibit elevated levels of resting blood glutamate and abnormally high levels of glutamate following ingestion of the amino acid. It has therefore been suggested that the deficit in glutamate dehydrogenase may lead to a buildup of glutamate in the brain close to glutamate receptors and thereby precipitate excitotoxic neuronal death. This idea is supported by the finding that mice in which the glutamate transporter GLT-1 has been deleted genetically not only clear glutamate from the synaptic cleft more slowly, but also suffer from epilepsy and show increased suscepti-bility to brain injury (Tanaka *et al.*, 1997). The homozygous mice also die prematurely.

CHAPTER 17

GLYCINE RECEPTORS

Inhibition at central synapses is mediated by glycine and γ-aminobutyric acid (GABA). Glycine is predominantly used in the spinal cord and brain stem, whereas GABA is more commonly used in the brain. Glycine acts by binding to a specific receptor (GlyR) and opening an intrinsic chloride channel. The increase in Cl^- permeability that results prevents the membrane depolarization and neuronal firing induced by excitatory neurotransmitters. Mutations in the glycine receptor give rise to startle disorders in man and animals.

Basic Properties

The primary physiological agonist for the glycine receptor (GlyR) is glycine, but taurine and β-alanine also act as agonists. All three amino acids open the same anion-selective ion channel, which has a halide selectivity sequence of $I^- > Br^- > Cl^-$ and is also significantly permeable to bicarbonate. It exhibits multiple conductance states, which range from 10 to 90 pS in amplitude. Strychnine acts as a competitive antagonist of glycine binding, thereby reducing the activity of inhibitory neurones and producing overexcitation, muscle spasms, and convulsions. It has been used as a rat poison for centuries.

  Glycine receptors are also modulated. Low concentrations of zinc ions (1–10 μM) potentiate the currents about threefold, while higher concentrations may cause inhibition. Since zinc is present within many presynaptic secretory vesicles and is coreleased with the neurotransmitter on stimulation, this modulation is of physiological significance. Phosphorylation also modulates GlyRs, with protein kinase A (PKA) phosphorylation enhancing, and protein kinase C (PKC) decreasing, the response to glycine (Kuhse *et al.*, 1995).

Structure

The glycine receptor was the first neurotransmitter receptor protein to be isolated from the mammalian central nervous system, by Pfeiffer and his colleagues in 1982 (Pfeiffer *et al.*, 1982). The adult GlyR is a pentameric complex composed of three α- and two β-subunits, which have M_r of 48 kDa (α_1) and 58 kDa (β). Each type of subunit consists of a large and highly conserved N-terminal extracellular domain, four putative transmembrane domains, and a short extracellular C terminus (Fig. 17.1). An additional 93 kDa protein, gephyrin, copurifies with the glycine receptor and resides at the cytoplasmic side of the membrane (Prior *et al.*, 1992). The name gephyrin is derived from the Greek word for bridge and alludes to the fact that the protein is thought to anchor GlyRs to the subsynaptic cytoskeleton. An 18 amino acid motif within the cytoplasmic loop linking the third and fourth transmembrane domains (TMs) of the GlyR β-subunit forms the binding site for gephyrin (Meyer *et al.*, 1995). This motif is flanked by consensus sequences for phosphorylation by PKC and tyrosine kinases, which may therefore be

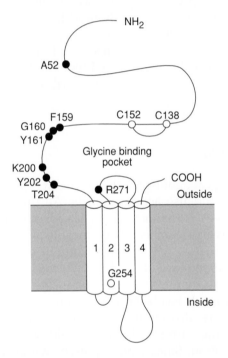

FIGURE 17.1 PUTATIVE TOPOLOGY OF THE GlyR α_1 SUBUNIT
Domains and residues indicated in the text are marked. The numbering follows that of the human α_1 subunit. A disulphide bridge links cysteines 138 and 152. Residues determining ligand-binding affinities are indicated by filled circles: residues G160, K200, and Y202 are involved in strychnine binding; F159, Y161, and T204 are important determinants of agonist specificity and affinity. Mutation of A52, which causes the *spasmodic* mouse phenotype, results in a small decrease in glycine binding. R271 is mutated in human startle disease: this produces a change in glycine affinity and single-channel conductance. Residue G254 is involved in determining the conductance of the main open state. After Kuhse *et al.* (1993).

implicated in the regulation of GlyR–gephyrin interactions. The α-subunit does not bind gephyrin.

The α- and β-subunits of the glycine receptor share significant sequence homology with each other ($\sim$35%) and with the subunits of the nACh and GABA$_A$ receptors. This similarity is especially strong within the four trans-membrane domains and a cysteine-rich motif lying within the extracellular domain. Three different genes encoding GlyR α-subunits (α_1–α_3) have been identified in man (4 in mouse). They display $\sim$80% overall sequence identity. Alternative splicing contributes to further heterogeneity; for example, both the α_1- and α_2-subunits have two isoforms. The human α_1 gene is located on chromosome 5p32 (Ryan et al., 1992) and that encoding the α_2-subunit is on chromosome Xp21.2–22.1 (Grenningloh et al., 1990). Both contain nine exons. Only a single type of β-subunit gene has been identified to date, which is located on human chromosome 4q32 (Handford et al., 1996).

When expressed in Xenopus oocytes, recombinant α_1-subunits form func-tional homomeric channels with properties similar to those of native channels, being opened by glycine, taurine and β-alanine and blocked by strychnine and picrotoxin (Schmieden et al., 1987). Thus the α-subunit possesses both agonist- and antagonist-binding sites. Expression of the β-subunit alone does not result in either significant glycine-activated currents or binding sites for [^{3}H]strychnine, indicating that the β-subunit binds neither agonists nor antagonists with high affinity. Coexpression of α_1- and β-subunits markedly enhanced the magnitude of whole-cell currents. This suggests an important role of the β-subunit may be to facilitate membrane targeting of the glycine receptor or to prevent receptor degradation. Possibly this is achieved by interaction with gephyrin, which selectively binds to the β-subunit.

Expression

The α_1- and α_2-subunits are subject to developmental regulation; α_1-subunits increasing and α_2-subunits decreasing after birth (Malosio et al., 1991). Expres-sion of α_3-subunits also increases postnatally. The neonatal form of the glycine receptor is probably a homopentamer of α_2-subunits while the adult form is primarily composed of $\alpha_1\beta$-subunits and, to a lesser extent, $\alpha_3\beta$-subunits. This change may underlie the shorter open time of adult GlyRs and the faster decay time of inhibitory postsynaptic currents.

In situ hybridization studies have shown that α_1-subunits are expressed in the spinal cord, brain stem and colliculi, whereas α_2-subunits are found in the hippocampus, cerebral cortex and thalamus, and low levels of the α_3-subunit are detected in cerebellum, olfactory bulb and hippocampus (Malosio et al., 1991). The β-subunit is expressed throughout the entire brain and spinal cord.

GlyRs are localised at postsynaptic membrane specializations opposite glycinergic nerve terminals. Gephyrin links the GlyRs to the underlying cytoskeleton and is required for GlyR clustering (Kirsch et al., 1993).

Correlating Structure and Function

Agonist- and antagonist-binding sites

Recombinant GlyRs consisting solely of α_1-subunits have a Hill coefficient for glycine activation of 4.2, which suggests that binding of at least five glycine molecules is required to activate the channel and is consistent with its pentameric structure. The Hill coefficient for glycine activation of $\alpha_1\beta$ GlyRs was 2.5, suggesting that only three glycine molecules are needed for activation and supporting the idea that three α_1-subunits and two β-subunits (which do not bind glycine) comprise the channel complex.

Taurine and β-alanine have a dual action on the glycine receptor: they act as channel agonists, but at lower concentrations they may also inhibit the response to glycine. This suggests there may be two functional amino acid binding sites. In agreement with this idea, mutation of R271 strongly reduces the ability of taurine and β-alanine to act as agonists, without altering their antagonistic actions.

Figure 17.1 shows residues known to be important for the binding of agonist and antagonists to the α_1-subunit of the glycine receptor. Two regions appear to be of particular importance for ligand recognition. These lie in different regions of the protein and are postulated to form loops which come together to form the ligand-binding site. One loop is believed to be formed by a disulphide bridge between two cysteine residues located at positions 198 and 209 in the α_1-subunit. Proteolytic digestion of [3H]strychnine-labelled GlyR first showed that residues 170–220 contribute to the high-affinity strychnine-binding site (Ruiz-Gomez et al., 1990) and mutagenesis studies subsequently narrowed down the key amino acids within this region to K200, Tyr202 and T204 (Vandenberg et al., 1992; Rajendra et al., 1995b). Mutation of cysteines 198 and 209 also abolishes agonist and antagonist binding, probably by disrupting the tertiary structure of the binding loop. A second region important for high-affinity binding of both agonists and antagonists comprises residues 159–161, which are proposed to form a β-turn (Kuhse et al., 1990; Vandenburg et al., 1992; Schmieden et al., 1993). In particular, residue 160 is crucial for high-affinity glycine binding, while adjacent residues determine agonist specificity. Thus replacement of the phenylalanine at position 159 by tyrosine both enhanced the affinity of the channel for glycine and, remarkably, also rendered it sensitive to GABA (wild-type GlyRs are unaffected by GABA). It is worth noting that the regions of the GlyR involved in ligand binding are in homologous positions to those residues which determine ligand binding to ACh and GABA$_A$ receptors, suggesting that the ligand-binding pockets of all three receptors share a common molecular architecture.

Modulatory sites

Volatile anaesthetics such as enflurane and isoflurane act by binding to GlyR and most GABA$_A$Rs and markedly potentiating their activity. Alcohol acts in a similar way. By contrast, the activity of the GABA$_A$R ρ_1-subunit is reduced by alcohol and anaesthetics (Chapter 18). This difference was exploited to identify the site at which both agents act, by making chimeric molecules between the GlyR α_1-subunit and the GABA ρ_1 receptor (Mihic et al., 1997). A stretch of 45 amino acids required for the potentiation of receptor activity by anaesthetics and alcohol was identified in this way. Site-directed mutagenesis was then employed to locate two residues within TM2 of the GlyR α_1-subunit, T264 and S267I, which are involved in the potentiating effects of ethanol. When the latter residue was mutated to tyrosine, potentiation by the anaesthetic enflurane was also abolished. It is not known whether these residues are involved in the binding site for alcohol and volatile anaesthetics, form part of the channel gate, or contribute to the allosteric linkage between binding and gating.

The pore

The second putative transmembrane domain (M2) of the glycine receptors shows significant sequence homology to that of cation-selective ligand-gated ion channels. It is therefore believed that the M2 domains of the five GlyR subunits come together to form the walls of the ion pore. Evidence in favour of this idea is that mutation of specific residues in the M2 domain affects the inhibitory action of picrotoxinin, which is believed to act as an open-channel blocker and plug the pore (Pribilla et al., 1992). Mutation of glycine 221, which lies at the cytoplasmic end of the second transmembrane domain, to alanine also alters the single-channel conductance (Bormann et al., 1993).

The M2 domains of the anion-selective GlyRs and GABA$_A$Rs resemble those of their cation-selective counterpart, the AChR, but differ at three important positions (Fig. 17.2). First, the negatively charged ring of amino acids that lies within the pore of the AChR is replaced by a hydrophobic ring of alanine residues. Second, an additional ring, of prolines, is located adjacent to this alanine ring in GlyRs and GABA$_A$Rs, but is absent in AChRs. Third, GlyRs possess a ring of neutral residues at the outer mouth of the pore, in place of the cationic ring found in AChRs. There is good evidence that these changes are responsible for the ability to discriminate between cations and anions. As discussed in Chapter 15, it is possible to convert the nACh receptor from a cation-selective to an anion-selective channel simply by adding a proline at position 259, replacing glutamine 260 with alanine and substituting threonine for valine at position 274 (Galzi et al., 1992).

A characteristic of glycine receptors is that they do not exhibit a single conductance state but instead display a range of conductance states (Fig. 17.3). The single-channel conductance of the main conductance state of the homomeric α_1-subunit GlyRs is 86 pS. When coexpressed with the β-subunit, however, the main conductance is only 44 pS. This is similar to that found for native glycine receptors and confirms that native GlyR are $\alpha_1\beta$ heteroli-

A

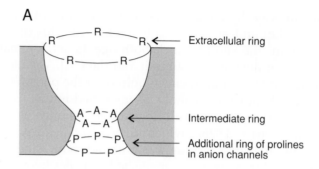

B

	$\ast\ast$	$\ast$
Glycine α_1 rat	DAAPARVGLGITTVLTMTTQSSGSRA	anion-selective
GABA α_1 rat	ESVPARTVFGVTTVLTMTTLSISARN	anion-selective
ACh α_1 mouse	DSG-EKMTLSISVLLSLTVFLLVIVE	cation-selective

in   M2 segment   *ex*

FIGURE 17.2 THE M2 DOMAINS LINE THE PORE OF ANION-SELECTIVE CHANNELS
(A) Rings of similar amino acid residues in the M2 domain of GlyRs
thought to be important for anion selectivity. (B) Comparison of the amino
acid sequence of the M2 domains of the α_1 subunits of the rat GlyR, rat
GABA$_A$R, and mouse AChR. There are three main differences between
the M2 domains of the anion-selective GlyR and the cation-selective AChR
(starred): a ring of positively charged arginine residues lines the outer
mouth of the GlyR pore, an intermediate ring of hydrophobic alanine
residues is found within the pore, and there is an additional ring of prolines.

gomers. A residue within the second transmembrane domain (G254) is important for determining the main state conductance of the channel.

Summary of structure–function studies

The GlyR consist of five similar subunits ($3\alpha,2\beta$) arranged to form a
ring around a central pore. Each subunit consists of a large extracellular N
terminus, which contains the ligand-binding site, and four transmembrane
domains. This pore is formed from the M2 domains of each subunit. Since the
glycine-binding site is distant from the pore, long-range allosteric interactions
must couple agonist binding to channel gating. The GlyR is anchored at the
postsynaptic region by gephyrin, which binds to the cytosolic domain of the
β-subunit.

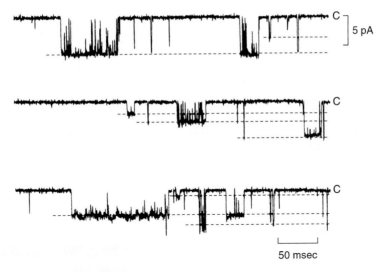

5 pA

C

C

C

50 msec

FIGURE 17.3 GlyR CHANNELS HAVE MORE THAN ONE CONDUCTANCE STATE
Single-channel currents recorded from outside-out patches excised from cells expressing homomeric α_1-GlyR. The dotted lines indicate the different conductance levels. From Borman *et al.* (1993).

GLYCINE RECEPTORS AND DISEASE

Startle Disease

Mutations in the gene encoding the α_1-subunit of the glycine receptor, *GLRA1*, give rise to startle disease (Shiang *et al.*, 1993). Startle disease (hyperekplexia) is an autosomal dominant neurological disorder characterized by muscle spasm in response to an unexpected stimulus, which manifests as facial grimacing, hunching of the shoulders, clenching of the fists and exaggerated jerks of the limbs. In some patients, sudden unexpected noises or lights can cause them to become rigid and even fall over. Because their arms are held stiffly by their sides and they make no attempt to protect themselves, such falls may cause multiple injuries. Umbilical and inguinal hernias are common. Infants also exhibit a sustained increase in muscle tone (hypertonia) between attacks, but this declines with age. The symptoms resemble those observed with sublethal doses of the glycine antagonist strychnine. Startle disease is very rare and its prevalence is not known with certainty.

Genetic linkage studies initially mapped human startle disease to chromosome 5q32 (Ryan *et al.*, 1992), a region which was known to contain the gene encoding the α_1-subunit of the glycine receptor *GLRA1*. A subsequent study demonstrated the presence of two missense mutations in *GLRA1* (Shiang *et al.*, 1993). These resulted in substitution of a leucine or glycine residue for the arginine at position 271, which lies in the extracellular loop linking the first and second transmembrane domains (Fig. 17.1). Both these mutations produce a dramatic (up to 400-fold) decrease in the glycine-activated current (Fig. 17.4). The reduction in current is not due to a decrease in the number of glycine receptors because the number of binding sites per cell is not decreased. Instead, the mutant receptors exhibit a marked (>200-fold) reduction in glycine sensitivity (Fig. 17.5): strychnine affinity is unaffected (Langosch et al., 1994; Rajendra et al., 1994). Single-channel recordings reveal that the single-channel currents also exhibit a greater percentage of the lower (~16 pS) conductance levels. These two effects account for the smaller whole-cell currents. An additional, and surprising, finding was that taurine and

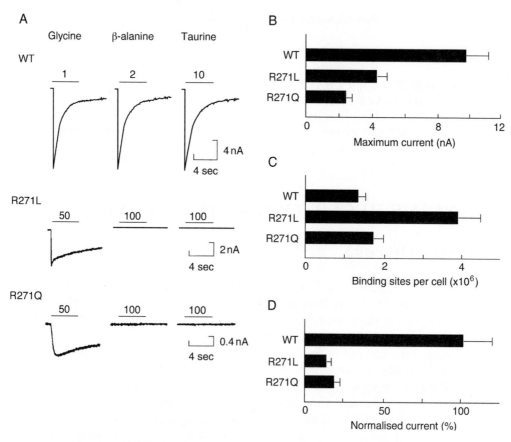

FIGURE 17.4 EFFECTS OF STARTLE MUTATIONS ON AGONIST-ACTIVATED CURRENTS
(A) Currents elicited by glycine, β-alanine, or taurine in cells expressing wild-type GlyR or the startle mutations, R271L and R271Q. (B) Maximal current observed for cells expressing wild-type or mutant GlyRs. (C) Maximum number of glycine-binding sites in cells expressing wild-type or mutant GlyRs. (D) Normalised current, obtained by dividing the maximal current by the number of binding sites, for cells expressing wild-type or mutant GlyRs. From Rajendra et al. (1995a).

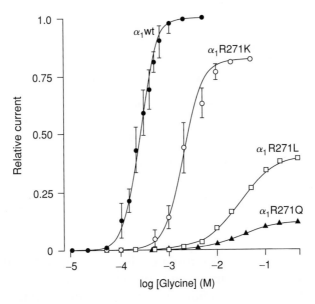

FIGURE 17.5 GLYCINE CURRENTS ARE REDUCED BY THE STARTLE DISEASE MUTATIONS, R271L AND R271Q

Relationship between glycine concentration and current amplitude in oocytes injected with wild-type or mutant GlyR α_1-subunits. The currents are expressed as a fraction of their value at a saturating glycine concentration. From Langosch *et al.* (1994).

β-alanine, which normally act as channel agonists, acted as competitive antagonists of the mutant channel and inhibited the effects of glycine (Rajendra *et al.*, 1995a). Binding of these amino acids appears to be normal and the reduction in current must result from an uncoupling of agonist binding and channel opening. Thus, R271 appears to play a critical role in linking agonist binding to opening of the channel pore.

Additional startle mutations are found in the M2–M3 loop (Y279C and K276E). They produce a similar effect on the properties of α_1GlyR channels to that of the R271 mutation (Lynch *et al.*, 1997). A fifth mutation (I224N) is found in the intracellular loop linking TMs 1 and 2. This mutation greatly impairs the efficiency of α_1GlyR expression, a fact which probably accounts for its ability to cause startle disease. It also impairs the ability of taurine to act as an agonist (Lynch *et al.*, 1997). The functional effects of the startle mutations suggest that both the M1–M2 and M2–M3 linkers may be involved in transducing the allosteric change that occurs on agonist binding into channel activation, and led Lynch and his colleagues to test this hypothesis by making a range of mutations within each of these regions. Their results support the idea that the M1–M2 and M2–M3 loops act in parallel to transduce ligand binding into channel activation (Lynch *et al.*, 1997).

Glycine is the major inhibitory transmitter in the spinal cord and glycinergic interneurones are important for normal spinal cord reflexes, muscle tone

TABLE 17.1   MAMMALIAN GlyR CHANNEL GENES

Gene	Protein	Chromosome location (human)	Major sites of expression	Reference
GLRA1	α_1GlyR	5q32	Spinal cord, brain stem, colliculi	Grenningloh et al. (1987); Ryan et al. (1992)
GLRA2	α_2GlyR	Xp22.1–p21.2	Forebrain (e.g., hippocampus, cortex thalamus)	Grenningloh et al. (1990)
GLRA3	α_3GlyR	4q33–q34	Cerbellum, olfactory bulb, hippocampus	
GLRB	βGlyR	4q32	Brain, spinal cord	Handford et al. (1996)

and the pattern of motoneurone firing during movement. It is therefore not surprising that a reduction in glycine current amplitude leads to excessive and uncontrolled movements. Since the disease is dominant, it may be expected that both normal and mutant α_1-subunits will be coexpressed *in vivo*. The dominant nature of the disease is explained by the fact that mutant subunits can coassemble with wild-type subunits to form heteroligomeric GlyRs with intermediate affinity for glycine. Thus glycinergic tone in the spinal cord will be reduced, but not abolished. Other inhibitory pathways may compensate for the glycinergic deficit in startle diseases: for example, GABAergic tone appears to be increased in startle syndromes and increased numbers of $GABA_A$-binding sites have been found in myoclonic cattle. Startle disease is usually treated with drugs such as pentobarbitone or clonazepam, which enhance $GABA_A$ transmission.

Mice Mutants

Three mutations in the glycine receptor of the mouse cause recessively inherited disorders which resemble human startle disease. These are *spa* (*spastic*), *spd* (*spasmodic*), and *oscillator*. The *spasmodic* phenotype is caused by point mutation in the α_1-subunit of the glycine receptor (Saul et al., 1994). This mutation results in substitution of an alanine for the serine at position 52, which lies within the extracellular N-terminal domain (Fig. 17.1). As a result, the concentrations of glycine, β-alanine and taurine required to produce half-maximal current activation are higher (Fig. 17.6). Strychnine affinity, however, is unaffected. The phenotype develops around 2–3 weeks postnatally, at about the time expression of the α_1-subunit reaches its normal level.

The *oscillator* mouse has a microdeletion at the end of the third transmembrane domain of the α_1-subunit of the glycine receptor, which produces a frameshift and premature truncation of the protein (Buckwalter et al., 1994). This results in a dramatic decrease in strychnine binding, indicative of a loss of GlyR function. The mice exhibit fine motor tremor and muscle spasms,

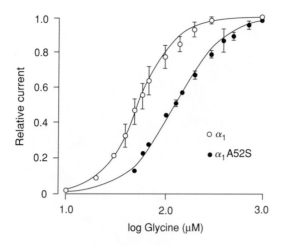

FIGURE 17.6 GLYCINE CURRENTS ARE REDUCED BY THE
SPASMODIC MUTATION, A52S
Relationship between glycine concentration and current amplitude in
oocytes injected with wild-type or A52S GlyR α_1-subunit. Currents were
normalised to that obtained at a saturating concentration of glycine. From
Saul *et al.* (1994).

which develop around 2 weeks after birth, coincident with the expression of
the α_1-subunit. Death usually occurs by ~3 weeks of age. The startle reflex
of the *oscillator* mouse is more severe than that of *spasmodic* (Fig. 17.7), which
can be explained by the fact that the *oscillator* mutation causes an almost total
loss of GlyR function, while the *spasmodic* mutation disrupts, but does not
completely prevent, GlyR function.

Homozygous *spastic* mice appear normal at birth, but at about 2 weeks
of age they develop an exaggerated startle response, myoclonus, tremor,
rigidity, and an impaired righting reflex. This is associated with a marked
reduction in the density of glycine receptors in the adult brain and spinal
cord (Becker *et al.*, 1992). The residual GlyRs, however, function normally.
The disorder results from the insertion of a LINE-1 element in intron 5 of
the gene encoding the β-subunit of the glycine receptor (Mülhardt *et al.*, 1994;
Kingsmore *et al.*, 1994). This transposon disrupts the gene and causes aberrant
splicing of the GlyR β-subunit mRNA, which results in an excess of truncated
transcripts (which are predicted to be nonfunctional) and a severe reduction
of full-length transcripts (to <10% of normal). The reduction in the glycine
receptors found in *spastic* indicates that the β-subunit is essential for glycine
receptor assembly and/or synaptic targeting *in vivo*. Furthermore, as the
onset of the *spastic* phenotype coincides with the developmental switch from
α_2- to α_1-subunits, it also implies that *in vivo* GlyRs comprising only α_2-
subunits are functionally normal.

When a transgene encoding the GlyR β-subunit is expressed in the *spastic*
mouse background, the normal phenotype is rescued (Hartenstein *et al.*, 1996).

FIGURE 17.7 THE STARTLE RESPONSE OF THE OSCILLATOR MOUSE IS MORE SEVERE THAN THAT OF THE SPASMODIC MOUSE
Startle reflexes triggered in response to a camera flash in a six-month-old spasmodic mouse (background) and a 24-day-old oscillator mouse (foreground). The spasmodic mouse reacts with a stiff tail and hindlimbs. The muscle spasms of the oscillator mouse are far more severe and cause it to fall over. Prior to the flash both mice exhibited normal movement.

These animals show ~25% of the normal level of β-subunit mRNA expression, so this must be sufficient for the normal functioning of glycinergic synapses.

Cattle and Horses

A disease resembling the mice and human startle diseases is found in poll Hereford cattle and Peruvian Paso horses, where it is known as myoclonus (Grundlach *et al.*, 1988). The symptoms of cattle myoclonus are more pronounced than those of human startle disease and the consequences are more dramatic: the calves often spasm *in utero* and are born with dislocated hips, so that they are unable to stand. The condition is associated with reduced glycine receptor expression in the spinal cord, but the molecular basis of the disease has not yet been identified (both α and β GlyR subunits are good candidates).

GABA$_A$ RECEPTORS

The neurotransmitter γ-aminobutyric acid (GABA) interacts with two major types of receptor: the GABA$_A$ (GABA$_A$R) and the GABA$_B$ receptors. GABA$_A$ receptors are anion-selective ligand-gated channels which mediate fast synaptic inhibition in the CNS. They are widely distributed throughout the brain and spinal cord. Like the glycine receptors, the ligand-binding site and the anion channel form a single functional unit. Binding of GABA to its receptor opens an integral Cl$^-$ channel, producing an increase in membrane conductance that results in inhibition of neuronal activity. The GABA$_A$ receptor is also the target for a wide range of clinically important drugs, including antiepileptic agents, anxiolytics (anti-anxiety drugs), sedatives, hypnotics, muscle relaxants and anaesthetics. There are no human diseases that have been shown unequivocally to arise from mutations in the genes encoding GABA$_A$ receptor subunits. Nevertheless, because GABA is the major inhibitory neurotransmitter in the brain and GABA$_A$ receptors are present on most neurones, GABA$_A$Rs are good candidate genes for a number of neurological disorders. In particular, there is accumulating evidence that deletion of the GABA$_A$R β_3-subunit may be the cause of Angelman syndrome.

GABA may also act on one of a number of GABA$_B$ receptors. These do not possess an intrinsic ion channel, although they may modulate the activity of separate ion channels via second messengers or GTP-binding proteins. They will not be considered further here.

Basic Properties

Like the glycine receptors, GABA$_A$ receptors are anion channels, with the permeability sequence SCN$^-$ > I$^-$ > Br$^-$ > Cl$^-$ $\gg$ F$^-$. They are also permeable to bicarbonate ions but are essentially impermeable to cations. GABA$_A$ receptors exhibit multiple conductance levels, which range from 10 to 45 pS in amplitude.

Depending on the Cl$^-$ equilibrium potential, the opening of GABA$_A$R Cl$^-$ channels may produce either a hyperpolarization or a depolarization. Early in development, activation of GABA$_A$ receptors in CNS neurones produces membrane depolarization because of the elevated intracellular Cl⁻ concentration (Ben Ari *et al.*, 1997). In many cases, this depolarization is sufficient to reach the threshold for action potential initiation and thus in immature neurones GABA may act as an excitatory neurotransmitter. By contrast, in adult neurones, GABA$_A$R activation almost invariably causes either a membrane hyperpolarization or no potential change. In either case, the increase in membrane Cl⁻ conductance serves to clamp the membrane at E_{Cl} and so produces inhibition of neuronal activity.

Pharmacology

GABA$_A$ receptors have a very complex pharmacology. In addition to GABA, they possess binding sites for barbiturates, benzodiazepines, anaesthetics, alcohol and some steroids. All of these agents potentiate the response to GABA, but do not open the Cl⁻ channel themselves. They bind to sites distinct from the GABA-binding site and act by increasing the channel open time and/or enhancing the frequency of channel openings: the single-channel conductance is unaffected. Gonadal steroids (progesterone, oestrogen) and their derivatives serve as physiological ligands, so that the GABA response varies according to the hormonal state of the animal. The other agents are of pharmaceutical importance. Their ability to depress the central nervous system, impair motor function and induce sedation or loss of consciousness reflects the widespread distribution of GABA$_A$Rs throughout the brain.

Benzodiazepines used to be widely prescribed as anxiolytics, sedatives, muscle relaxants, anticonvulsants and hypnotics. Their long-term use is now discouraged, however, following concern over their ability to cause physical dependence. Currently they are prescribed principally for short-term use— for example, to help patients in a crisis by reducing anxiety and insomnia. There are many different types of benzodiazepine, which have varying duration of action. They include the well-known tranquillizers diazepam (Valium) and temezepam.

A number of drugs cause convulsions because they inhibit GABA$_A$Rs. Among these are picrotoxin and bicuculline. The latter acts as a competitive antagonist: it binds to the GABA-binding site but does not open the Cl⁻ channel. In contrast, picrotoxin is a noncompetitive inhibitor which reduces channel activity by binding to a site different from that at which GABA acts. Its mode of action is still not completely clear. One idea is that it blocks the channel pore, while another suggests that it interferes with the coupling between agonist binding and channel opening (Xu *et al.*, 1995).

In summary, drugs which interact with the GABA$_A$R fall into two main categories: antagonists and potentiators. The former act as convulsants while the latter depress the central nervous system and are used as sedatives, anaesthetics and anticonvulsants.

Structure

GABA$_A$ receptors constitute a very large gene family. Six major types of GABA$_A$ receptor subunits have been cloned: α, β, γ, δ, ε, and π. In most cases, multiple subtypes within each of these classes have been described (Table 18.1). In the rodent, for example, there are six types of α, three β, three γ, and one δ subunit, and further diversity is created by alternative splicing. The subunits have molecular masses of between 50 and 60 kDa. There is 30–40% sequence homology between the different types of subunit and 70–80% homology within each type of subunit.

The GABA$_A$ receptor is a multimeric protein with a total molecular mass of 230–270 kDa. Although there is as yet no direct evidence, by analogy with GlyR and AChR channels, GABA$_A$Rs are believed to be composed of five subunits. The potential number of subunit combinations is therefore very large but, as with most other multimeric channels, current evidence suggests that only a limited number of combinations are found *in vivo*. Immunoprecipitation studies using antibodies directed against specific subunits have shown that most GABA$_A$ receptors contain a single type of α-subunit, in combination with one kind of β-subunit and a γ_2-subunit (McKernan *et al.*, 1991; Benke *et al.*, 1991). Thus, most native GABA$_A$ receptors are likely to be ternary structures composed of α-, β-, and γ-subunits. These subunits are believed to assemble as a pentamer, comprising a total of two α- and two β-subunits with a single γ-subunit (Tretter *et al.*, 1997). The α-subunit determines the different affinities for allosteric ligands shown by the GABA$_A$ receptor sub-

TABLE 18.1 **MAMMALIAN GABA CHANNEL GENES**

Gene	Protein	Chromosome location (human)	Reference
GABRA1	$\alpha1$	5q31.1–33.2	Buckle *et al.* (1989); Wilcox *et al.* (1992); Kostrzewa *et al.* (1996)
GABRA2	$\alpha2$	4p12–p13	Buckle *et al.* (1989); McLean *et al.* (1995)
GABRA3	$\alpha3$	Xq28	Buckle *et al.* (1989)
GABRA4	$\alpha4$	4p14–q12	McLean *et al.* (1995)
GABRA5	$\alpha5$	15q11–q13	Knoll *et al.* (1993)
GABRA6	$\alpha6$	5q31.1–33.2	Kostrzewa *et al.* (1996)
GABRB1	$\beta1$	4p12–p13	Buckle *et al.* (1989); McLean *et al.* (1995)
GABRB2	$\beta2$	5q31.1–33.2	Kostrzewa *et al.* (1996)
GABRB3	$\beta3$	15q11–q13	Wagstaff *et al.* (1991)
GABRG1	$\gamma1$	4p14–q21.1	Wilcox *et al.* (1992); McLean *et al.* (1995)
GABRG2	$\gamma2$	5q31.1–33.2	Wilcox *et al.* (1992)
GABRG3	$\gamma3$	15q11–q13	Greger *et al.* (1995)
GABRD1	$\delta1$	1p	Sommer *et al.* (1990b)
GABRE1	ε	Xq28	Wilke *et al.* (1997)
	$\rho1$	6q14–q21	Cutting *et al.* (1992)
	$\rho2$	6q14–q21	Cutting *et al.* (1992)

types. The β-subunit markedly enhances functional channel expression and appears to be an essential structural component of the GABA$_A$R complex. The γ-subunit is required for benzodiazepine sensitivity, a property of the native receptor, and also enhances the single-channel conductance and open probability.

The δ-, ε-, and π-subunits do not form homomeric receptors but can coassemble with $\alpha\beta$-subunits. The role of the δ-subunit is uncertain. The presence of the ε-subunit renders the GABA$_A$R complex insensitive to the potentiating effect of anaesthetics. It is expressed at high levels in heart and placenta (Davies et al., 1997). The π-subunit is expressed in the reproductive system, being particularly abundant in uterus (Hedblom and Kirkness, 1997). It does not express by itself, but when combined with the β_3-subunit alters the sensitivity to the steroid pregnalone.

An additional type of GABA receptor subunit, the ρ-subunit (ρ_1) has also been cloned. This is principally expressed in the retina, although it is detected at lower levels in the brain and lung (Cutting et al., 1991). Unlike the different GABA$_A$ subunits it readily forms homomeric channels in *Xenopus* oocytes. These channels are blocked by picrotoxin, but are unaffected by barbiturates. As these properties resemble those of native GABA$_A$Rs in the retina, it is believed that the ρ_1 receptor may function as a homomer *in vivo*. It corresponds pharmacologically to the GABA$_C$ receptor subtype.

Topology

Each type of GABA$_A$R subunit consists of a large N-terminal extracellular domain of $\sim$200 amino acids, four putative transmembrane domains and a short extracellular C terminus (Fig. 18.1). A cytosolic loop of variable length links TMs 3 and 4. The GABA$_A$R β-subunits possess a gephyrin-binding motif similar to that found in GlyRs, and coexpression with gephyrin modulates the subcellular targeting of GABA$_A$Rs carrying the β_3-subunit (Kirsch et al., 1995). This suggests that clustering of GABA$_A$ receptors in the postsynaptic membrane may be regulated by gephyrin, as in the case of GlyRs (Chapter 17).

Tissue expression

The genes encoding the GABA$_A$ receptor subunits are differentially expressed, depending on both developmental age and tissue type (Lüddens et al., 1995). Some cell types express most types of GABA$_A$ mRNAs, while others possess only a limited number. Each of the six different α-subunit genes has a unique expression pattern in the brain, so there must be at least six different types of GABA$_A$ receptor. The α_1-subunit is the most prevalent and is expressed ubiquitously. It is usually found in combination with β_2- and γ_2-subunits, suggesting that $\alpha_1\beta_2\gamma_2$ complexes are the most common type of GABA$_A$ receptor in brain. Another common combination, found in spinal cord motor neurones, appears to be $\alpha_2\beta_3\gamma_2$. Other subunits are expressed less widely. For example, α_6-subunits are only found in cerebellar granule cells. In some, but not all, regions of the rat brain there is a developmental switch

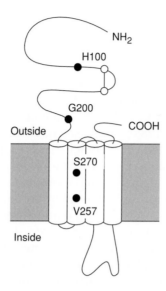

FIGURE 18.1 PUTATIVE TOPOLOGY OF GABA$_A$ RECEPTORS
Domains and residues indicated in the text are marked. The numbering
follows that of the rat α_1-subunit sequence. A disulphide bridge is formed
between cysteines 139 and 153. The filled circles indicate residues involved
in benzodiazepine binding. A histidine at position 100 is required for
benzodiazepine binding (α_4- and α_6-subunits, which do not bind benzodiazepines, have an arginine at this position). Residue 200 (glycine in α_1,
glutamate in other α-subunits) determines the differential sensitivity to
different benzodiazepines. Valine 257 lies close to the narrowest region
of the channel pore, and in insects, is involved in picrotoxin and cyclodiene
insecticide sensitivity. Serine 270 has been implicated in sensitivity to
alcohol and volatile anaesthetics.

from embryonic to adult types of GABA$_A$ receptor. GABA$_A$ receptors are also
found outside the CNS, for example, they occur in pancreatic α-cells, where
they may play a role in glucagon release.

Chromosomal Location

The chromosomal location of the different GABA$_A$R subunits is given in Table
18.1. It is noteworthy that members of this gene family are generally organised
in clusters. Thus, the genes encoding α_1-, α_6-, β_2-, and γ_2-subunits are clustered
on chromosome 5q31.1–q33.2, those encoding α_2-, α_4-, β_1-, and γ_1-subunits
are localised to chromosome 4p14–q12, the α_5, β_3, and γ_3 genes are found on
chromosome 15q11–q13, and the ρ_1 and ρ_2 genes map to chromosome 6q14–
q21. The exceptions are the genes encoding the α_3- and δ-subunits, which are
not clustered and lie on chromosomes Xq28 and 1p, respectively.

Correlating Structure and Function

Ligand-binding sites

When expressed in *Xenopus* oocytes, recombinant α_1-, α_2-, and α_3-subunits form functional homomeric channels with properties similar to those of native channels: being opened by GABA, potentiated by pentobarbital and blocked by picrotoxin (Blair *et al.*, 1988). Likewise, the β-subunits can form functional homomeric GABA-activated Cl$^-$ channels. Thus both α- and β-subunits must possess GABA-binding sites. The structure of the GABA-binding site is not fully solved, but mutational analysis has identified two domains in the amino-terminal region of the β-subunit which are important for activation by GABA. Each of these includes a crucial tyrosine and threonine residue (Amin and Weis, 1993). Interestingly, tyrosine residues are also key components of both the ACh- and the glycine-binding sites (Chapters 15 and 17).

The γ_2-subunit is essential for benzodiazepine sensitivity, because marked potentiation of the GABA-induced current is only produced by benzodiaze-pines when the recombinant GABA$_A$ receptor contains a γ_2-subunit (Pritchett *et al.*, 1989). Mice in which the γ_2 gene has been inactivated (knocked out) are also insensitive to benzodiazepines and show a dramatic decrease ($\sim$95%) in the number of benzodiazepine-binding sites in the brain (Günther *et al.*, 1995). A phenylalanine at position 77 and a methionine at position 130 of the γ_2-subunit are important determinants of benzodiazepine binding (Wingrove *et al.*, 1997).

There is evidence that the α-subunit also contributes to the benzodiazepine-binding site. First, benzodiazepines are able to photoaffinity label the α (but not the γ_2) subunit (Stephenson *et al.*, 1990). Second, in GABA$_A$Rs containing α-, β-, and γ_2-subunits the affinity for different benzodiazepines was found to vary with the α-subunit (Pritchett *et al.*, 1989). Mutagenesis studies have shown that two residues within the α-subunit are of key importance for benzodiazepine sensitivity (residues 100 and 200 in the α_1-subunit). The first of these determines the affinity for this class of drugs in general and the other regulates the selectivity of the GABA$_A$R to the different types of benzodi-azepine (Pritchett and Seeburg, 1991; Wieland *et al.*, 1992). Both residues are located within the extracellular N-terminal domain (Fig. 18.1) and may form part of the high-affinity-binding site for benzodiazepines. It is perhaps sig-nificant that residue 200 is located in an analogous position to residues impli-cated in the binding of acetylcholine to the nAChR (Chapter 15).

The potentiation of GABA$_A$ currents by alcohol and volatile anaesthetics involves amino acids within the second and third transmembrane domains (TMs). Mutation of serine 270 in TM2 of the α_1 or α_2 GABA$_A$R subunits to leucine largely abolished the potentiatory effects of alcohol and volatile anaesthetics such as enflurane and isoflurane (Fig. 18.2; Mihic *et al.*, 1997). Similar results were obtained with mutations in TM3 of the α_2 (A291Y)- or β_1 (M286Y)-subunits. It is not known whether these residues actually form part of the binding site for enflurane and isoflurane, or whether they influence binding and/or gating allosterically.

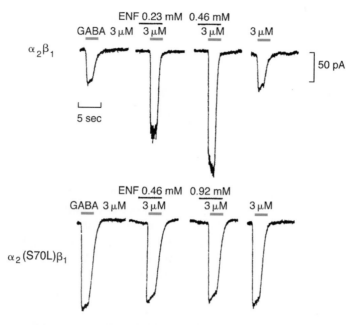

FIGURE 18.2 VOLATILE ANAESTHETICS INTERACT WITH GABA$_A$ RECEPTORS

Currents recorded from HEK cells coexpressing wild-type GABA$_A$ $\alpha_2\beta_1$ receptors or mutant α_2(S270L)β_1 receptors. Wild-type receptors show significant potentiation of the GABA response by the anaesthetic enflurane (ENF), unlike the mutant receptor. From Mihic *et al.* (1997).

The pore

Like the glycine and nicotinic ACh receptors, the pore of the GABA$_A$ receptor is believed to be formed from the second TM. Studies similar to those described for these channels (Chapters 15 and 17) have provided evidence in support of this idea. For example, Xu and Akabas (1996) replaced each amino acid in TM2 of α_1GABA$_A$R, in turn, with a cysteine residue and then examined the effects of thiol reagents, which interact specifically with the side chains of cysteine residues, on the mutant channel. The rationale behind these experiments is that thiol reagents will react with residues lining the channel pore and thereby block the conduction pathway. In this way, Xu and Akabas (1996) confirmed that TM2 does indeed line the channel pore and that it has a predominantly α-helical structure. When applied to the extracellular side of the channel, the positively charged thiol reagent MTSEA was able to access residues as deep as α_1T261, suggesting that the selectivity filter lies towards the cytoplasmic end of the channel pore. Picrotoxin, which may block GABA$_A$ receptors by physically plugging the pore, prevented thiol reagents from reacting with residue α_1V257 but not with residues located more extracellularly (Xu *et al.*, 1995). This would be consistent with the narrowest region of the GABA$_A$R pore lying towards the cytoplasmic end of the channel, as is the case for nAChR (Chapter 15).

DISEASES ASSOCIATED WITH GABA$_A$ CHANNELS

Angelman Syndrome

Angelman syndrome (AS) is characterised by severe mental retardation, absence of speech, paroxysmal laughter, 'puppet-like' ataxic movements and seizures. Affected individuals also have craniofacial abnormalities and hypopigmentation. The incidence is estimated to be around 1 in 20,000 and the vast majority of patients appear to be sporadic cases. AS is a classical example of genomic imprinting and in 70–80% of patients is associated with deletion of the q11–q13 region of maternal chromosome 15 (Nicholls, 1993). As explained in Chapter 2, genomic imprinting refers to the fact that the expression of certain genes is dependent upon whether they are inherited from the mother or from the father. The gene responsible for AS is switched off on the paternal chromosome so the disease phenotype is only expressed when the gene is deleted on the maternal chromosome. AS can also arise by uniparental disomy—the inheritance of two copies of a genetic locus from only one parent (in this case, the father).

The region of chromosome 15 deleted in AS is usually very large and contains many genes, making it difficult to identify the gene (or genes) responsible for the disorder. The best candidate to date is the gene encoding the GABA$_A$R β_3-subunit (GABRB3), which maps to the same region of chromosome 15 and is deleted (along with other genes) in AS patients (Wagstaff et al., 1991). Although the GABA$_A$R α_5 gene is located adjacent to GABRB3, it is not deleted in all patients and thus cannot be the cause of the disorder (Knoll et al., 1993). The available data suggest that deletion of GABRB3 may contribute to the pathogenesis of Angelman syndrome, although the possible role of other genes within this region cannot yet be excluded. However, the neurological symptoms of the disorder are readily explained by a loss of GABA$_A$R receptors. Furthermore, in one patient, a marked reduction in benzodiazepine receptor density in the cerebellum, a region in which GABRB3 is known to be expressed, has been reported (Odano et al., 1996). Since the AS gene is imprinted, GABRB3 must express preferentially on the maternal chromosome if it is indeed the cause of AS. Unfortunately, there are currently conflicting reports regarding the preferential expression of GABRB3. Thus, it remains unclear whether or not GABRB3 is indeed involved in AS.

Approximately 20% of AS patients show no evidence of chromosomal deletions and the reason for their phenotype is unknown. The possibility that they possess microdeletions or mutations in GABRB3 deserves investigation. Deletion of the same region of chromosome 15 is also involved in a quite separate disorder called Prader–Willi syndrome (Nicholls, 1993). The phenotype is quite different from that of AS and is generated by deletion of the paternal (rather than maternal) chromosome. The critical region of the deletion is also distinct and does not include *GABRB3*.

Animal models

A cluster of three GABA$_A$ receptor subunit genes, encoding the α_5-, β_3-, and γ_3-subunits, which lie adjacent to each other on the same chromosome, are absent in the pink-eyed cleft palate (p^{cp}) mouse (Nakatsu *et al.*, 1993). This mutant mouse strain is characterised by greatly decreased pigmentation, a cleft palate and neurological disorders. Most affected animals die within a few days of birth because they fail to feed. A very few mice (<5%) survive longer; these animals do not have a cleft palate, but they exhibit ataxia and tremor. Mice that fail to express either the γ_3-subunit alone, or both α_5 and γ_3 GABA$_A$R subunits, do not exhibit obvious neurological symptoms (Culiat *et al.*, 1994). However, mice in which the β_3-subunit alone has been eliminated are severely affected, with a phenotype resembling that of the p^{cp} mouse (Homanics *et al.*, 1997). Most animals (>90%) die as neonates with craniofacial abnormalities, while those that survive have a reduced life span, suffer from epileptic seizures (Fig. 18.3), and show behavioural abnormalities such as hyperactivity and hypersensitivity to stimuli. The similarities between the symptoms of the GABA$_A$Rβ_3 knockout mouse and those of Angelman syndrome offer support for the idea that the human disorder is caused by deletion of *GABRB3*.

The γ2 Knockout Mouse

Mice in which the γ_2-subunit was 'knocked out' have provided insight into the importance of this subunit in the GABA$_A$ receptor complex (Günther *et al.*, 1995). These mice do not express the γ_2-subunit, but other types of GABA$_A$ subunits are expressed at similar levels to wild-type mice. They lack high-affinity benzodiazepine-binding sites in the brain and their behaviour is unaffected by diazepam, confirming that the γ_2-subunit forms an essential part of the benzodiazepine-binding site. Electrophysiological analysis revealed that the GABA$_A$ receptors of homozygous knockout ($\gamma_2^{-/-}$) animals also exhibit reduced GABA sensitivity and enhanced inhibition by Zn^{2+}. Since $\gamma_2^{-/-}$ mice were normal at birth, the γ_2-subunit is not essential for embryonic development. However, most animals died within a few days of birth. Those which survived were hyperactive, showed severe growth retardation, and later developed an abnormal gait and righting reflexes; they also died prematurely. This argues that the γ_2-subunit is essential for normal GABA$_A$R function (remember that most GABA$_A$ receptors contain γ_2-subunits). It is not known whether this is because the γ_2-subunit enhances the response to GABA or because it confers sensitivity to some putative endogenous ligand of the

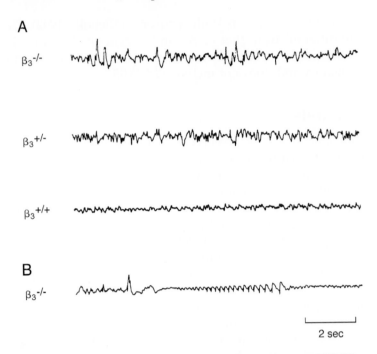

FIGURE 18.3 ELECTROENCEPHALOGRAPHIC RECORDINGS
(A) Recordings taken at 2 months of age from a homozygous GABA$_A$Rβ_3
knockout mouse ($\beta_3^{-/-}$), a heterozygous mouse ($\beta_3^{+/-}$), and a wild-type
littermate ($\beta_3^{+/+}$). The $\beta_3^{-/-}$ recordings are abnormal. (B) Seizure in a $\beta_3^{-/-}$
mouse at 5 months of age. From Hominacs *et al.* (1997).

benzodiazepine-binding site. Heterozygous animals showed no obvious differences from wild-type animals.

Alcohol Intolerance in the Rat

The alcohol nontolerant (ANT) rat has been selectively bred for enhanced
motor impairment in response to alcohol. It is also abnormally sensitive to
benzodiazepine agonists, such as diazepam, which cause a marked impairment of postural reflexes. These effects result from the conversion of GABA$_A$
receptors in cerebellar granule neurones from a diazepam-insensitive to a
diazepam-sensitive form (Fig. 18.4). The ANT phenotype is caused by a point
mutation in the α_6-subunit of the rat GABA$_A$ receptor, which substitutes a
glutamine for the arginine at position 100 (R100Q) (Korpi *et al.*, 1993). This
residue lies within the extracellular N terminus (Fig. 18.1), and other experiments described earlier, have demonstrated that it is critical for benzodiazepine binding. Expression studies confirmed that GABA currents though
α_6(R100Q)$\beta_2\gamma_2$ receptors are markedly enhanced by diazepam, in contrast to
wild-type receptors (Fig. 18.4).

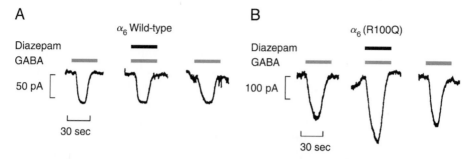

FIGURE 18.4 *A POINT MUTATION IN GABA$_A$R α_6 CAUSES BENZODIAZEPINE SENSITIVITY IN THE ANT RAT*
Currents recorded from HEK cells coexpressing wild-type GABA$_A$ $\alpha_6\beta_2\gamma_2$ (A) or mutant α_6(R100Q)$\beta_2\gamma_2$ (B) receptors. The response of the mutant receptor is potentiated by the benzodiazepine diazepam, in contrast to that of wild-type channels. From Korpi *et al.* (1993).

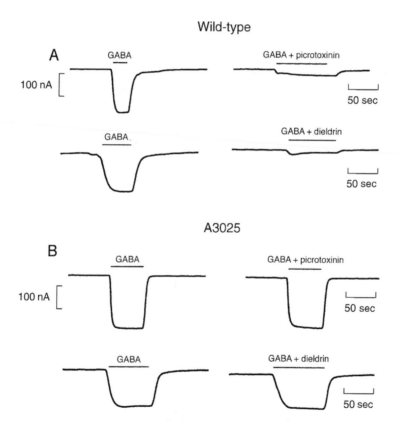

FIGURE 18.5 *A POINT MUTATION IN TM2 OF THE INSECT GABA$_A$R CONFERS INSECTICIDE RESISTANCE*
GABA-activated currents recorded from oocytes expressing either wild-type (A) or mutant (B) *Drosophila* GABA receptors. The mutant receptor (A302S) is insensitive to both the insecticide dieldrin and to picrotoxin, which block the wild-type channel. From ffrench-Constant *et al.* (1993).

Resistance to Insecticides

GABA is a major inhibitory transmitter in invertebrates, as well as in vertebrates. Insect ionotropic GABA receptors are blocked by picrotoxin, but they differ from vertebrate GABA$_A$Rs, in their sensitivity to GABA analogues and allosteric modulators (Hosie *et al.*, 1997). Cylodiene insecticides, such as dieldrin, act by inhibiting insect GABA receptors, and compete for the same binding site as picrotoxin. Resistance to cyclodiene pesticides is widespread, being found in ~300 arthropod species and accounting for >60% of reported cases of insecticide resistance. In most cases that have been examined, resistance is caused by a point mutation in the insect GABA receptor that results in substitution of an alanine (or glycine) residue for the serine at position 302, which lies within the second transmembrane domain (ffrench-Constant *et al.*, 1993). As Fig. 18.5 shows, this mutation markedly reduces the channel sensitivity to picrotoxin and the cylodiene insecticide dieldrin. This results from disruption of the binding site and is associated with a marked reduction in antagonist affinity. The residue equivalent to S302 in the vertebrate GABA$_A$ α_1-subunit is valine 257, which is thought to lie within the pore, just above the channel gate on the extracellular side (Xu and Akabas, 1996). This suggests that cylodiene insecticides inhibit wild-type channels either by physically plugging the pore or by influencing channel gating allosterically.

CHAPTER 19

WATER CHANNELS

Almost all cell membranes exhibit some water permeability. In many cells, water movement takes place by simple diffusion across the lipid bilayer or through ion channels that are open under resting conditions. However, tissues with unusually high water permeabilities possess specialised water-selective channels. At least ten water channels, known as aquaporins 0 to 9, have been identified to date.

Aquaporin 1

The first water channel to be characterised was isolated serendipitously, by virtue of the fact that it copurified with a Rhesus blood group antigen. This 28 kDa protein was initially called CHIP28 (an acronym for channel-forming integral protein) but is now more commonly known as aquaporin 1 (AQP1). It was found to be expressed at high levels in red blood cells and in renal tubules, but its function remained a mystery until the gene was cloned (Preston *et al.*, 1992). It was then found that *Xenopus* oocytes injected with mRNA encoding AQP1 swelled rapidly and finally burst when placed in a solution of low osmotic strength (Fig. 19.1). This effect was prevented by mercurial compounds, which are known to block water channels, and was not observed in control oocytes. Confirmation that AQP1 is indeed a water channel, rather than simply a regulator of oocyte water movements, was provided by purification of the AQP1 protein and its reconstitution into liposomes, where it was shown to be selective for water and to exclude urea, protons, hydroxyl ions, ammonium ions and salts (Zeidel *et al.*, 1992).

AQP1 is only one of a number of water-selective channels known as aquaporins (van Os *et al.*, 1994; Nielsen and Agre, 1995). These channels have a common molecular architecture which consists of six hydrophobic domains that are considered to cross the membrane as α-helices (Fig. 19.2A). The N and C termini are located intracellularly. The primary sequence suggests that the protein includes two internal repeats which are orientated at 180° to each other in the membrane. The first repeat corresponds to the segment including transmembrane domains (TMs) 1 to 3, while the second repeat comprises

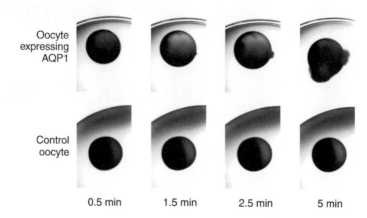

Oocyte expressing AQP1

Control oocyte

0.5 min   1.5 min   2.5 min   5 min

FIGURE 19.1 WATER PERMEABILITY IS ENHANCED BY AQP1 EXPRESSION
Effect of exposure to a solution of low osmotic strength on *Xenopus* oocytes injected with water (control, below) or mRNA encoding AQP1 (CHIP28, above). Water moves into the oocyte by osmosis to equalise the ionic concentrations on either side of the cell membrane. In oocytes expressing AQP1 the water permeability is higher so that the oocytes rapidly swell. Photographs supplied by Dr. Peter Agre.

TMs 4–6. Both repeats contain a highly conserved short sequence [asparagine (N), proline (P), and alanine (A)] known as the NPA motif, which is important for water permeability: one of these is located in the linker between TMs 2 and 3 and the other in the region linking TMs 5 and 6. Mutation of these amino acids alters the permeability of water channels, suggesting that they form critical parts of the channel pore. The mercurial sensitivity characteristic of water channels is also associated with residues lying in these regions (specifically, alanine 73 and cysteine 189). This has led to the idea that the conserved domain located intracellularly between TMs 2 and 3, and its extracellular counterpart between TMs 5 and 6, may dip down into the membrane to form part of the central pore of the water channel (Fig. 19.2B; Jung *et al.*, 1994b). In this 'hourglass' model of the channel the NPA motifs are predicted to form the selectivity filter. The proposed structure, in which the pore is formed from a short hydrophobic loop, is reminiscent of that discussed earlier for other ion channels.

There is evidence that the water channel is composed of four subunits (Cheng *et al.*, 1997). Both radiation inactivation analysis and site-directed mutagenesis studies suggest that each individual subunit contains its own water pore and may function independently as a water channel. Thus it is currently envisaged that the water channel consists of four monomers, with four water pores located in the centre of the structure. This idea of a multibarrelled channel is reminiscent of that proposed for some types of voltage-gated chloride channel (see Chapter 10).

Cryoelectron microscopy (electron crystallography) of AQP1 in native membranes or lipid bilayers has revealed the three-dimensional structure of

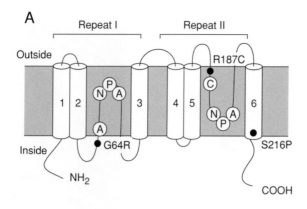

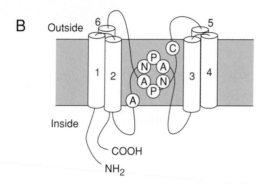

FIGURE 19.2 PUTATIVE MEMBRANE TOPOLOGY OF AQUAPORINS

(A) Deduced membrane topology of AQPs. The NPA motif and the residues involved in interaction with mercurial compounds (A37, C189) are indicated by white circles. Some of the mutations associated with NDI discussed in the text are also shown (●). (B) Hourglass model of AQP1 structure, showing how the NPA motifs come together to form the pore. After Jung *et al.* (1994b).

the protein at 6–7 Å resolution (Cheng *et al.*, 1997; Walz *et al.*, 1997). The individual AQP1 monomers are ~60 Å high by 30 Å in diameter and they assemble as tetramers. Each monomer has six membrane-spanning α-helices which lie at a tilt to the plane of the membrane (Fig. 19.3). The six α-helices are assembled like the staves of a barrel around a central vestibule which has a diameter of ~8 Å at its narrowest point, and probably leads to the pore of the water channel.

Functional role of AQP1

In situ hybridization and immunocytochemistry indicate that AQP1 is abundantly expressed in renal tubules, where it is confined to the cells of the proximal tubule and thin descending loop of Henle, and in the vasa recta

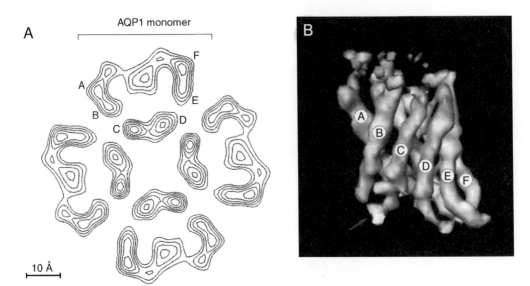

FIGURE 19.3 CRYOELECTRONMICROSCOPE PICTURES OF AQP1
(A) AQP1 tetramer viewed perpendicular to the bilayer and depicted as
a contoured section through a three-dimensional density map. The six
helices of a single monomer are indicated (A–F). From Cheng *et al.* (1997).
(B) Representation of the six-helix barrel of the AQP1 monomer viewed
parallel to the plane of the membrane. The six α-helices are labeled. From
Walz *et al.* (1997).

(Nielsen *et al.*, 1995). These are regions associated with high constitutive water
permeability, and AQP1 is responsible for the retrieval of up to 90% of the
water filtered by the kidney glomerulus. Aquaporin 1 is also found in many
other absorptive and secretory epithelia, including the choroid plexus epithelium, the bile duct, the gall bladder epithelium, and the ciliary and lens
epithelium of the eye. This suggests that AQP1 mediates water movements
associated with the formation of cerebrospinal fluid, the aqueous humour of
the eye, bile secretion and concentration, and the removal of water from the
lens and cornea. AQP1 may also contribute to vascular water permeability
since it is found in many capillary endothelia, including those of the lung
and the descending vasa recta in the kidney. It has been suggested that
movement of water across the alveolar epithelium to offset evaporative losses
during respiration may be mediated by AQP1 (in the vascular endothelium)
and by AQP5 (in alveolar type I pneumocytes) (King *et al.*, 1996). AQP1 may
also facilitate water absorption from the lung at birth.

Intuitively, it seems likely that mutations in AQP1 which severely com-
promise channel function would be fatal. This is not the case, however, as
individuals homozygous for AQP1 mutations which result in the lack of
functional water channels are clinically normal (Preston *et al.*, 1994). AQP1
is located on chromosome 7p14, coincident with the gene that codes for

the Colton blood group antigen, and the different antigens result from a polymorphism at residue 45 (alanine or valine) which is located in the first extracellular loop of AQP1. A few very rare individuals have been found who do not express either the Colton A or the Colton B antigen; AQP1 was also not detected in their red blood cells. These people turned out to homozygous for mutations in AQP1, which result in either major disruption or complete loss of the protein. Since no clinical symptoms are associated with AQP1 mutations, clearly some other water channel or transport mechanism must compensate for the deficiency in AQP1 in tissues such as the kidney. In contrast to man, genetic deletion of the AQP1 gene in mice severely impairs the ability to produce a concentrated urine (Ma *et al.*, 1998).

Aquaporin 2

The second water channel to be cloned, aquaporin 2 or AQP2, is 42% identical to AQP1 at the amino acid level (Fushimi *et al.*, 1993). Human AQP2 is a 271 amino acid protein with a molecular mass of 29 kDa and a predicted transmembrane structure similar to that of AQP1 (Sasaki *et al.*, 1994). The gene is located on chromosome 12q13 and contains four exons (Table 19.1).

AQP2 is expressed exclusively in the collecting duct of the kidney and plays a fundamental role in the production of a concentrated urine (Fig. 19.4). The concentration of the urine is regulated by varying the extent to which water is absorbed from the collecting ducts of the kidney tubule. The membranes of the collecting duct cells are relatively impermeable to water. Water uptake is achieved by the regulated insertion of AQP2 channels into the apical membranes of the principal cells of the collecting duct, thereby increasing their water permeability (Fig. 19.5). This process is under the control of arginine-vasopressin (AVP). Binding of AVP to its receptor on the basal membrane of the principal cells results in activation of adenylate cyclase (via the heterotrimeric G-protein G$_s$) and thus in elevation of cyclic AMP. This stimulates protein kinase A (PKA) and initiates a chain of (as yet, poorly understood) events that ultimately result in the translocation of vesicles containing water channels to the apical membrane. Here, they fuse, inserting AQP2 channels into the surface membrane and increasing its water permeability. The vesicles containing water channels are subsequently retrieved by endocytosis and are then continuously recycled for as long as AVP continues to be present. Removal of AVP allows the continued uptake of AQP2 channels by endocytosis, but there is no longer any concomitant insertion, so that water uptake ceases. AQP2 is expressed exclusively in the apical membranes of the collecting duct cells, and the efflux of water across their basal membranes is mediated by another type of aquaporin, AQP3. In addition to stimulating the insertion of AQP2 in the apical membrane, AVP also exerts a longer-term effect on water permeability by increasing the level of AQP2 expression.

The link between AQP2 and the production of a concentrated urine was firmly established when it was discovered that the AQP2 gene is mutated in

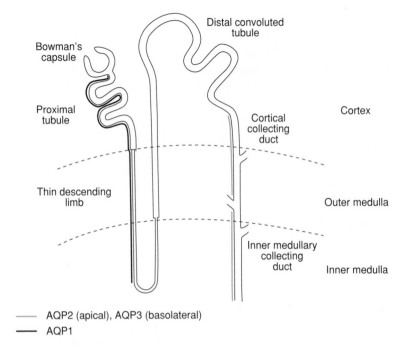

FIGURE 19.4 EXPRESSION PATTERN OF DIFFERENT
AQUAPORINS IN THE KIDNEY
The structure of a single kidney tubule. The distribution of the different
aquaporins along the tubule is indicated.

some patients with familial nephrogenic diabetes insipidus (NDI; Deen *et al.*,
1994a). This is an inherited disease in which water uptake by the kidney
tubules is impaired. The disorder manifests within the first few weeks of life.
The most typical symptoms are the excretion of large amounts of hypotonic
urine and excessive thirst. In early infancy these may not be noticed, and the
disease is often recognised by signs of dehydration, such as poor feeding, poor
weight gain, irritability and fever. Without treatment, the severe dehydration
associated with NDI can lead to mental retardation or death.

In most cases, familial NDI is caused by a mutation in the vasopressin
receptor. In some families, however, the disease results from a mutation in the
AQP2 gene and is inherited in an autosomal recessive fashion. Six missense
mutations (G64R, N68S, T126M, A147T, R187C, and S216P) and two nucleotide deletions (at codons 310 and 369) have been identified to date (Fig.
19.2A). Both deletions shift the reading frame, the latter producing premature
termination of translation and a truncated protein while the former produces
a longer protein. When the missense mutations were engineered in AQP2
and expressed in *Xenopus* oocytes, either no increase or only a very small
increase, in water permeability was observed (Deen *et al.*, 1995a; Mulders *et
al.*, 1997). This is because incorporation of the mutant protein into the plasma
membrane is impaired. Coinjection of mRNAs encoding mutant and wild-
type AQP2 did not affect water permeability, which is consistent with the

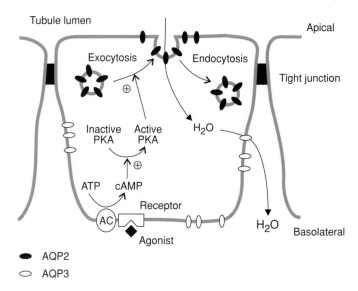

FIGURE 19.5 ROLE OF AQP2 IN URINE CONCENTRATION
Cellular events involved in water uptake in the principal cells of the
kidney collecting duct. Agonists such as AVP stimulate the production
of cyclic AMP, which promotes the fusion of membrane vesicles containing
AQP2 channels with the apical membrane. Water entering the cell through
apical AQP2 channels exits via AQP3 channels in the basolateral membrane. This gives rise to a transcellular uptake of water.

autosomal recessive inheritance of the disease. As it is believed that water
channels exist as tetrameric complexes, this argues that the mutant AQP2
subunits either do not complex with wild-type subunits or that the wild-type
subunit is able to offset the deleterious effect of the mutant subunit.

Both prolonged hypokalaemia or lithium treatment result in nephrogenic
diabetes insipidus. Reduced AQP2 levels are also found in animal models
of both these syndromes (Marples *et al.*, 1995, 1996), further emphasizing the
clinical importance of this water channel. In contrast, AQP2 was found to be
upregulated in the apical membranes of renal-collecting duct cells in an
animal model of congestive heart failure with fluid retention (Nielsen *et
al.*, 1997b).

Interestingly, AQP2 can be detected in the urine, and in normal people
its excretion is increased by dehydration or AVP (Kanno *et al.*, 1995). By
contrast, in patients with mutations in AQP2, the protein was not detected
in the urine even in response to dehydration. This suggests that urinary
excretion of AQP2 may provide a simple way to detect its presence in the
collecting duct membranes. This could be useful diagnostically, for example,
to distinguish between central and nephrogenic NDI or to determine whether
secondary NDI (such as that caused by drugs or hypokalaemia) results from
a decrease in the expression of AQP2 or its delivery to the plasma membrane.

Other Aquaporins

In addition to AQP1 and AQP2, a number of other aquaporins have been identified (AQP0–AQP9; King and Agre, 1996; Deen and van Os, 1998). They exhibit very different water permeabilities, with AQP4 having the highest water permeability (Yang *et al.*, 1997). The major site of AQP0 expression is the lens of the eye, where it constitutes more than 50% of the membrane protein of the lens fibre cells. AQP3 is strongly expressed in the basolateral membranes of the principal cells of the kidney collecting ducts. It thus provides an exit pathway for the water that enters the cell across the apical membranes through AQP2. The brain constitutes the primary site of AQP4 expression, but it is also found in the basolateral membranes of the renal-collecting ducts. In the brain, high levels are found in the glial cells of the paraventricular and supraoptic nuclei (in the hypothalamus), which are involved in thirst and in the regulation of vasopressin release. Thus AQP4 may contribute to osmosensitivity. AQP4 is also found in glial cells bordering the subarachnoidal space, ventricles and blood vessels (Nielsen *et al.*, 1997a), and in the lung. AQP5 is expressed in the apical membranes of the acinar cells within salivary and lacrimal glands, in cornea, and in lung; it may therefore be involved in the formation of tears and saliva (Raina *et al.*, 1995).

The three aquaporins expressed in the lung—AQP1, AQP4, and AQP5—are found in the capillary endothelium, the basolateral membrane of the airway epithelium and the apical membrane of the type 1 alveolar cells, respectively. Expression of all three genes is switched on a few days prior to birth, rises sharply just after birth and remains high throughout life (Umenishi *et al.*, 1996). In conjunction with the epithelial Na^+ channel (Chapter 13), they mediate the osmotic reabsorption of lung fluid that takes place at parturition.

The human genes for AQP0, AQP2 and AQP5 are clustered together on the long arm of chromosome 12, at position q13 (Table 19.1). Mutations of clinical importance in these genes have not yet been identified. A deficiency of AQP0 (originally called major intrinsic protein or MIP) in mice, however, results in congenital cataracts (Shiels and Bassnett, 1996).

TABLE 19.1   **MAMMALIAN AQUAPORIN GENES**

Gene	Chromosome location	Tissue distribution	Reference
AQP0	12q13	Lens	Saito *et al.* (1995)
AQP1	7p14	Kidney, erythrocytes, eye, lung	Deen *et al.* (1994b)
AQP2	12q13	Kidney collecting duct	Sasaki *et al.* (1994); Saito *et al.* (1995)
AQP3	9p13	Kidney	Ishibashi *et al.* (1995)
AQP4	18q11.2–q12.1	Brain	Jung *et al.* (1994a); Lu *et al.* (1996)
AQP5	12q13	Cornea, lung, salivary, and lacrimal glands	Lee *et al.* (1996); Raina *et al.* (1995)

Aquaporins Belong to the MIP Superfamily

The aquaporins belong to an ancient and diverse family of channel-forming proteins known as the major intrinsic protein (MIP) family (Chepelinsky *et al.*, 1994). Some members of this family also serve as water channels. Their physiological roles include mediating water uptake in plant roots (TobRB7) and conferring the water permeability of the tonoplast membrane which surrounds plant vacuoles (γ-TIP). Mutation of the *mod* gene, which encodes a plant aquaporin, eliminates self-incompatibility in *Brassica* and so allows viable self-pollination (Ikeda *et al.*, 1997). Other members of the MIP family do not appear to act as water channels, but instead are permeable to ions or other substances. Thus, various bacterial proteins which transport glycerol (GIF) or dicarboxylic acids (Nod26) also belong to the MIP family. Aquaporin 3 is permeable to both water and glycerol, although, interestingly, they appear to move through separate permeation pathways since inhibitors which abolish water movements do not affect glycerol translocation (Echevarría *et al.*, 1996).

CHAPTER 20

GAP JUNCTION CHANNELS

Gap junctions act as pathways for communication between the cytoplasm of adjacent cells, allowing the passive exchange of ions and small molecules (<1–2 kDa). In mammals, gap junctions are found in virtually all cell types, and are particularly abundant in liver and heart.

Gap junctions provide a means of coordinating the activity of adjacent cells or groups of cells. In excitable cells, gap junctions act as electrical synapses, enabling the fast transmission of electrical impulses between cells. They are important in synchronizing the electrical activity (and hence contraction) of cardiac and smooth muscle and the output of certain neuronal circuits, and their ability to relay information rapidly between pre- and postsynaptic neurones is utilised in escape behaviours such as the tail flip of the crayfish. They also synchronise exocytosis from endocrine cells, such as the islets of Langerhans. In non-excitable cells, the function of gap junctions is to permit the exchange of nutrients and regulatory signals between cells, such as intermediary metabolites and second messengers (Ca^{2+}, cyclic AMP and IP_3 are all permeant). They are involved in buffering the extracellular K^+ concentration in the brain: the local increase in K^+ which results from neuronal activity is removed by uptake into glial cells and is then siphoned away to the capillaries by passing from glial cell to glial cell through gap junctions. Gap junctions also play a critical role in the control of cell growth, differentiation and embryonic development (Warner et al., 1984). It is perhaps not surprising, given their many functions, that mutations in the genes encoding gap junction proteins cause diseases as diverse as peripheral neuropathy, abnormal cardiac development and congenital deafness.

Structure

At gap junctions, the plasma membranes of adjacent cells come into close apposition and are separated by a gap of only 1–2 nm (Fig. 20.1A). In freeze-fracture electron microscope pictures, gap junctions appear as hexagonal, semi-crystalline arrays of 8.5-nm particles which span the gap between neigh-

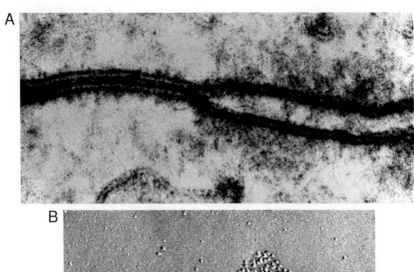

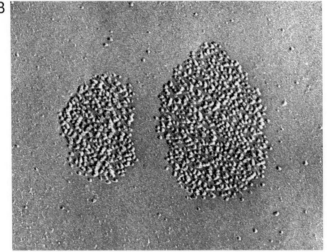

FIGURE 20.1 GAP JUNCTION MORPHOLOGY
(A) A thin section electron micrograph of a gap junction (From Brightman and Reese, 1969). (B) Freeze-fracture replica of two gap-junctional plaques from the moth *Manduca sexta*. Each plaque is made up of a cluster of connexons. Magnification ×75,400. Micrograph supplied Nancy Lane.

bouring cells (Fig. 20.1B). These particles correspond to individual gap junction channels, which associate in large numbers to form the gap junction. Structurally, gap junction channels are unique in that they span two cell membranes. Each channel is formed by the association of two hemichannels, known as connexons, one of which is contributed by each of the associating cells (Fig. 20.2). Each connexon is in turn composed of six subunits (called connexins) which are arranged in a ring around a central pore. Thus the gap junction channel is a dodecamer, formed from 12 connexins. Hydropathy plots suggest that the connexin has four transmembrane domains (TMs) linked by one cytoplasmic and two extracellular loops, with a short intracellu-

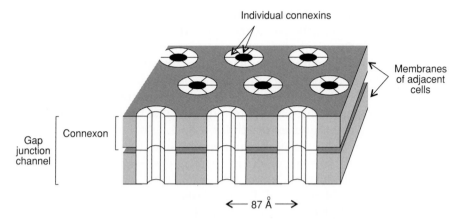

FIGURE 20.2 MODEL OF CONNEXON INTERACTIONS
Schematic representation showing how the connexons in each membrane
interact with their partners in the adjacent cell to form a gap junction
channel.

lar N terminus and a cytoplasmic C terminus of variable length (Fig. 20.3).
The extracellular loops are involved in linking the two hemichannels together:
they bridge the gap between the membranes of participating cells and contact
the aqueous fluid on two sides—both internally within the pore and exter-
nally in the gap. Each loop contains three conserved cysteine residues. There

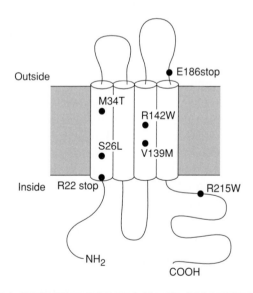

FIGURE 20.3 PUTATIVE TOPOLOGY OF CONNEXIN 32
Topology of Cx32 deduced from hydropathy analysis in conjunction with
mutagenesis and antibody studies. Some of the residues associated with
the Charcot-Marie-Tooth disease mentioned in the text are indicated.

is evidence that these cysteines are involved in the formation of disulphide bridges both within and between the two extracellular loops (they do not appear to participate in intermolecular interactions, i.e., those between individual connexins). The extracellular loops also contain a number of conserved glycine and proline residues. Because these amino acids form bends in molecules, they are likely to be important for determining the folding pattern of the extracellular loops. It has also been suggested that the hydrophobic groups found in the longer second extracellular loop might interact with those of the partner connexon in the opposite membrane and so link the two halves of the gap junction channel together. There is good evidence that the first TM contributes to the channel pore (Oh *et al.*, 1997; Zhou *et al.*, 1997). The amphipathic nature of the third TM suggests it may also form part of the lining of the pore, but there is as yet no clear evidence in support of this idea. Mutagenesis studies suggest that the cytoplasmic loop and the C-terminal domain are involved in the regulation of channel gating by protons (Morley *et al.*, 1996). Voltage-dependent gating, by contrast, involves residues within the N terminus, the first TM and the first extracellular loop (Verselis *et al.*, 1994).

It is possible to isolate gap junctions relatively intact, which has allowed the three-dimensional structure of the channel to be examined using high-resolution electron-microscopy and X-ray scattering, most recently at 7 Å resolution (Fig. 20.4A) (Unwin and Zampighi 1980; Unwin and Ennis, 1984; Unger *et al.*, 1997). These studies suggest that each connexin consists of four transmembrane domains, each of which adopts an α-helical conformation. The six connexins fit together to form an annulus that is tilted at an angle of 7° to the central axis of the channel. The connexon they form is about 70 Å tall and 65 Å wide, with a central pore that has a minimum diameter of ~16 Å within the membrane and that widens to ~20 Å in the extracellular region of the junction. The two hemichannels are not connected in register. Instead, one connexon is rotated by ~30° with respect to the other so that a connexin in one hemichannel makes contact with two of those in its partner connexon (Fig. 20.4B). The closing of the gap junction channel by Ca^{2+} has been suggested to result from the partial untwisting of the connexins so that the space between them is occluded, rather like the iris diaphragm of a camera (Fig. 20.4C). This rotation reduces the degree of tilt, moving small polar groups out of the lumen and replacing them with bulky residues which block the pore. According to this model, therefore, the gating of the channel involves changes in the packing of connexins but not in the structure of the individual subunits.

Gap Junctions Are Heterogeneous

Connexins come in many different types: at least 13 different kinds of rodent connexin have been cloned to date (Kumar and Gilula, 1996; White and Bruzzone, 1996). They are named according to their predicted molecular mass and the animal species in which they were first identified. All connexins

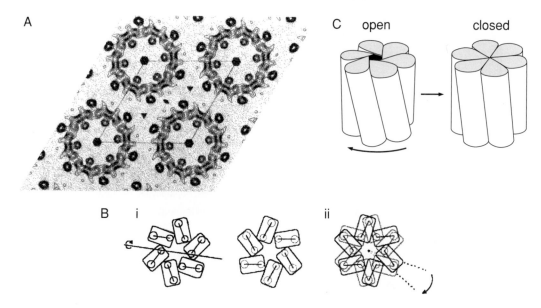

*FIGURE 20.4 HIGH RESOLUTION STRUCTURE OF THE GAP
JUNCTION CHANNEL*
(A) Projected density map of rat Cx43, showing six-fold symmetry. From
Unger *et al.* (1997). (B) Schematic model of the packing of connexons
within a gap junction. (i) Each connexin is represented by a rectangle and
the transmembrane helices are depicted as circles. (ii) The two partner
connexons are rotationally staggered so that each subunit makes contact
with two others in the opposing membrane. From Unger et al (1997).
(C) Model of a gap junction depicting the transition from the open to the
closed state. From Unwin and Zampighi (1980).

possess a similar gene structure and are about 50% identical at the amino
acid level (the most divergent region is the C terminus, which varies in
length). They belong to a superfamily of connexin genes which has two
distinct lineages, α (class I) and β (class II), typified by connexins 32 (Cx32,
liver) and 43 (Cx43, heart), respectively (Fig. 20.5). The different connexons
differ in their tissue distribution, single-channel conductance, gating properties and selectivity to ions and large molecules. The latter is clearly of particular physiological importance. Interestingly, there appears to be no correlation
between the single-channel conductance of a particular gap junction and its
permeability to large molecules (Veenstra *et al.*, 1995). This argues that the
gap junction does not act as a simple aqueous pore. Instead, the electrostatic
effects of charged amino acids situated close to the mouth of the pore influence
the channel selectivity and conductance.

Expression of the various connexins varies between tissues and is subject
to developmental regulation. Most isoforms are expressed in more than one
tissue and a single cell can express more than one type of connexin. It is
possible for heterotypic as well as homotypic gap junctions to form, by the

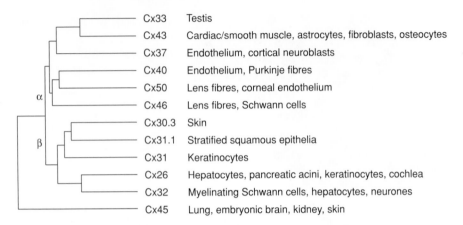

Cx33	Testis
Cx43	Cardiac/smooth muscle, astrocytes, fibroblasts, osteocytes
Cx37	Endothelium, cortical neuroblasts
Cx40	Endothelium, Purkinje fibres
Cx50	Lens fibres, corneal endothelium
Cx46	Lens fibres, Schwann cells
Cx30.3	Skin
Cx31.1	Stratified squamous epithelia
Cx31	Keratinocytes
Cx26	Hepatocytes, pancreatic acini, keratinocytes, cochlea
Cx32	Myelinating Schwann cells, hepatocytes, neurones
Cx45	Lung, embryonic brain, kidney, skin

FIGURE 20.5 CONNEXIN LINEAGES
Relationship between the different members of the connexin family. The horizontal branch lengths indicate the percentage of amino acid identity. From Paul (1995).

association of cells expressing two different types of connexon (White and Bruzzone, 1996). Heterotypic gap junctions may exhibit novel properties, as is the case when Cx32 hemichannels form gap junctions with Cx26 hemichannels (Barrio *et al.*, 1991). Not all connexon interactions are viable, however; for example, Cx40 and Cx46 do not form functional gap junctions. The second extracellular domain appears to be important for compatibility between connexons (White *et al.*, 1994). The hemichannels can also be heteromeric, that is composed of different connexins. An important consequence of the heterogeneity of connexin interactions is that it leads to diversity in the strength of coupling between cells, and in its regulation, depending on the type of connexins involved.

Studying Gap Junction Channels

The study of gap junction channels is rather more difficult than that of other ion channels as it is necessary to examine the extent of electrical coupling between pairs of cells by voltage clamping each cell. The 'paired-oocyte' system has proved useful for studying cloned connexons (Swenson *et al.*, 1989). In this technique, the vitelline membranes are removed from two *Xenopus* oocytes expressing heterologous connexons (the vitelline membrane is an additional membrane surrounding the oocyte which lies outside the plasma membrane). The oocytes are gently pushed together so that the plasma membranes touch, and gap junctions then form spontaneously. Junctional coupling is studied by voltage clamping both oocytes at the same holding potential and then applying voltage steps to one oocyte. The current required to hold the other oocyte at the same voltage is equal in magnitude but opposite in sign to that flowing through the gap junction channels. The presence of endogenous connexons in the oocyte can complicate interpretation of such experiments but this problem can be resolved by injecting antisense DNA to downregulate expression of endogenous channels to undetectable levels.

Regulation of Gap Junction Channels

The degree of coupling between participating cells is highly plastic. Long-term regulation, such as that which occurs during development, is achieved by changes in the level of connexin expression. For example, Cx26, but not Cx32, is expressed in embryonic brain, whereas after birth there is reciprocal upregulation of Cx32 and downregulation of Cx26 (Dermietzel et al., 1989). Short-term regulators of gap junction coupling influence channel gating. They include second messengers such as cyclic AMP, which opens gap junction channels and so enhances coupling, and ions such as Ca^{2+} or H^+, which rapidly close gap junctions and decrease coupling. Variations in the membrane potential of coupled cells, or in the phosphorylation state of the cell, also influence gap junction gating. Different connexins vary in their sensitivity to these short-term regulators.

The mechanism by which short-term regulators modulate gap junction activity is still not fully worked out: indeed, with the exception of H^+, voltage and certain kinases there is little evidence that they interact directly with the connexin. Protons induce a rapid closing of Cx43 (cardiac) gap junctions. Two regions of the protein appear to be involved in this process: the C-terminal part of the molecule (particularly residues 261–300 and 374–382) and a region lying within the cytoplasmic loop, close to the end of the second TM, which includes a histidine residue at position 95 (Ek-Vitorin et al., 1996). One model suggests that the C terminus acts as a blocking particle which interacts with a receptor close to the second TM to plug the pore in a manner analogous to the ball-and-chain blocking particle model proposed for voltage-dependent K^+ channel inactivation (see Chapter 6). The C terminus is also involved in the regulation of gap junction gating by protein kinases. Phosphorylation of serine or tyrosine residues in this domain results in channel closure or alteration of gap junction properties (Moreno et al., 1994). The closing of gap junction channels by intracellular Ca^{2+} requires concentrations greater than those observed physiologically, but may be an important safety measure under pathological conditions to ensure that cells are isolated from damaged neighbours, since cytosolic Ca^{2+} rises in dying or damaged cells.

Chromosomal Location

The human gene encoding Cx26 (*GJB2*) lies on chromosome 13 (13p11-q12), that for Cx32 (*GJB1*) on chromosome X (Xq13.1), and that for Cx43 (*GJA1*) on chromosome 6 (6q21-q23.2) (Mignon et al., 1996; Corcos et al., 1992; Hsieh et al., 1991; Willecke et al., 1990).

DISEASES OF GAP JUNCTION CHANNELS

Charcot-Marie-Tooth Disease (Cx32)

Charcot-Marie-Tooth disease (CMT) is the most common inherited peripheral neuropathy, affecting about 1 in 2500 individuals. The disease causes progressive degeneration of the peripheral nerves. It presents in childhood or adolescence, generally beginning with a weakness in the legs, which causes difficulty in walking, and progressing later to the arms. CMT is also characterized by foot deformities, muscle wasting, ataxia, decreased tendon reflexes and distal sensory loss. The disease is both pathologically and genetically heterogeneous. Traditionally, two main forms of CMT have been distinguished, based on electrophysiological differences (Vance, 1991). CMT1 is associated with a decreased nerve conduction velocity which results from demyelination of the peripheral nerves, whereas CMT2 is a non-demyelinating disease in which nerve conduction velocity is nearly normal. The CMT1 form of the disease is genetically heterogeneous and has been linked to chromosomes 17 (CMT1A), 1 (CMT1B) and X (CMTX). The first two result from mutations in the genes encoding peripheral myelin protein and myelin protein zero, respectively (Patel and Lupski, 1994). The X-linked form of Charcot-Marie-Tooth disease, however, is a disease of gap junction channels. It shows incomplete dominant inheritance, with heterozygous females being affected less severely than hemizygous males. X chromosome inactivation in heterozygous females probably accounts for the fact that the dominance is not complete. The phenotype may vary from mild, in which the patient has a normal gait, to a severe form which may necessitate the use of a walking stick or wheelchair.

CMTX results from mutations in the gene encoding connexin 32 (Bergoffen *et al.*, 1993). Nearly 90 of these have now been identified and they occur in all domains of the channel (Fig. 20.3; Table 20.1). Although not all mutations have been examined electrophysiologically, it appears that they fall into two main groups: those in which the protein never reaches the plasma membrane and those where the protein reaches the membrane but forms connexins with altered functional properties. The former primarily result from deletions, insertions and nonsense mutations which introduce premature stop codons and give rise to a severe phenotype. Missense mutations may be associated with either mild or severe phenotypes, according to whether they partially or completely disrupt channel function (Ionasescu *et al.*, 1996).

CMTX mutations provide insight into the structure of gap junction channels

One mutation, a leucine-for-serine substitution at position 26 (S26L) in Cx32, has proved useful in defining the region of the protein that forms the

TABLE 20.1   SOME Cx32 MUTATIONS ASSOCIATED WITH
CHARCOT–MARIE–TOOTH DISEASE

Mutation	Phenotype	Reference
W3S	Moderate/Severe	Ionasescu *et al.* (1996)
W3R	Mild	Ionasescu *et al.* (1996)
V13L		Bone *et al.* (1995)
R22G	Moderate/Severe	Ionasescu *et al.* (1996)
R22stop	Severe	Ionasescu *et al.* (1996)
S26L	Moderate	Oh *et al.* (1997)
M34T		Oh *et al.* (1997)
Y65C		Bone *et al.* (1995)
W77S	Mild	Ionasescu *et al.* (1996)
G80R	Mild	Ionasescu *et al.* (1996)
V95M		Bone *et al.* (1995)
E102G	Mild	Ionasescu *et al.* (1996)
111–116 del	Mild/Moderate	Ionasescu *et al.* (1996)
W133R		Bone *et al.* (1995)
V139M		Bone *et al.* (1995); Bergoffen *et al.* (1993)
R142W	Mild	Ionasescu *et al.* (1996)
L156R		Bone *et al.* (1995); Bergoffen *et al.* (1993)
R164W	Moderate/Severe	Ionasescu *et al.* (1996)
E186stop	Severe	Ionasescu *et al.* (1996)
C217stop	Severe	Ionasescu *et al.* (1996)
R220stop		Bone *et al.* (1995); Ionasescu *et al.* (1996)

pore of the gap junction channel (Oh *et al.*, 1997). This residue lies within
the first transmembrane domain (Fig. 20.3). When the mutated connexin is
expressed, the gap junction channels conduct Cs^+ ions normally but their
permeability to large molecules is dramatically reduced. Glycerol, for example, is no longer permeant. The conductance to Li^+ ions is also decreased.
As both the hydrodynamic radius of glycerol and the radius of the hydrated
Li^+ ion are ~3 Å, the radius of the mutant channel must be reduced to less
than 3 Å (the normal radius is ~6–7 Å). Leucine has a larger side chain than
serine and its substitution for serine at position 26 may introduce a constriction at the mouth of the channel and so lower the pore diameter and permeability to large molecules. Importantly, the mutant channels are likely to be
impermeable to Ca^{2+} and to second messengers such as cAMP and IP_3, which
have diameters greater than 3 Å. The demonstration that mutation of S26
affects the single-channel conductance suggests that the first TM contributes
to the pore of the gap junction channel. A similar conclusion has been reached

by cysteine scanning mutagenesis (Zhou *et al.*, 1997) (see Chapter 4 for an explanation of this method).

Another CMTX mutation that lies within TM1 (M34T) forms functional channels which have a near normal 70 pS conductance in the fully open state: however, most of the time the channel resides in a low conductance (15 pS) substate so that the macroscopic conductance is greatly reduced (Oh *et al.*, 1997). The latter also shows a marked shift in voltage dependence. Although M34 is predicted to face the channel lumen, the similarity in the single-channel conductance of the fully open state of the mutant channel with that of the wild-type channel is not surprising as the side chain of threonine is actually smaller than that of methionine.

Some CMTX missense mutations produce non-functional gap junctions because the protein is not inserted correctly into the plasma membrane. It has been reported that several of these mutations (eg., V139M, R215W) have a dominant negative effect, as in coexpression studies they are able to combine with wild-type Cx32 and suppress functional gap junctional formation (Omori *et al.*, 1996). This effect is unlikely to be of physiological or clinical significance, however, because the Cx32 gene is located on the X chromosome: thus, both males and females will have only a single functional copy of the gene in any one cell. More significantly, connexin 32 expressing the mutation R142W, which lies within the third TM, prevented gap junction coupling when it was coexpressed with Cx26 (Bruzzone *et al.*, 1994). Since Cx32 forms heteromeric connexons with Cx26, the effect presumably results from a dominant negative action of the mutant Cx32. This suggests that mutant Cx32 may interfere with the formation of Cx26 junctions in those tissues where both Cx32 and Cx26 are expressed, such as hepatocytes, pancreatic acinar cells and mammary gland cells.

Why does the lack of functional Cx32 give rise to Charcot-Marie-Tooth disease?

The Cx32 protein is expressed at high levels in myelinated peripheral nerve, where it appears to be located in the Schwann cells at the nodes of Ranvier and at Schmidt-Lanterman incisures. In these regions, the myelin is not completely compacted but instead there is a thin layer of cytoplasm between each of the enveloping turns of the Schwann cell. This suggests that Cx32 may form channels within a single Schwann cell (rather than between different cells) and connect the folds of cytoplasm between adjacent turns of myelin (Fig. 20.6). Nutrients and other substances would therefore not need to diffuse through the long thin cytoplasmic spirals of the Schwann cell to reach the innermost layers of the cell. Failure of the gap junctions may therefore lead to impaired Schwann cell function and thus to demyelination. The CMTX mutation S26L, in which the pore diameter is narrowed, is particularly illuminating in this respect. Its ability to cause CMT indicates that electrical coupling is not sufficient to prevent neuropathy and that the

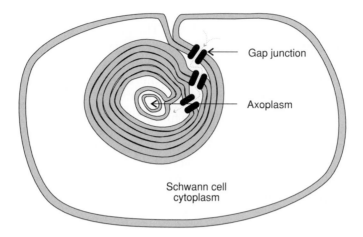

FIGURE 20.6 CONNEXIN 32 MAY FORM CONNECTIONS BETWEEN SCHWANN CELL LAYERS
Diagram illustrating a single Schwann cell wrapped many times around an axon. Gap junctions are depicted connecting adjacent layers of cytoplasm at Schmidt-Lantermann incisures. They therefore provide a short-cut for the diffusion of nutrients, regulatory molecules and trophic substances to the inner part of the Schwann cell. After Paul (1995).

supply of nutrients and second messengers (between 3 and 7 Å in radius) is critical for the integrity of the Schwann cell. Likewise, a decrease in the flux of such molecules may be expected with those mutations that reduce the channel open probability. Axonal degeneration also occurs in CMTX, suggesting that the Schwann cell gap junctions may also be involved in providing nutrients and trophic signals to the axon itself.

One somewhat surprising finding is that Cx32 is also present in liver, epithelial cells and brain, yet the effects of CMTX appear to be confined to the peripheral nervous system. The reason for this is unknown. One possibility is that different connexins may be able to substitute for Cx32 in other tissues; alternatively, the loss of coupling may not be deleterious in most cells.

Cardiac Malformations (Cx43)

Connexin 43 is strongly expressed in the heart, where it is involved in junctional coupling between adjacent cardiac cells. This coupling is not only important for the spread of electrical signals throughout the heart but also for cardiac development. Mice in which the Cx43 gene has been 'knocked out' show a defective cardiac anatomy, caused by a failure to establish the normal left-right asymmetry of the heart during embryonic development. As a consequence, the passage of blood from the right ventricle to the pulmonary arteries is impeded and the mice die shortly after birth from a failure to oxygenate the blood (Reaume *et al.*, 1995). The phenotype of the Cx43 knock-out mouse resembles a human disorder known as familial visceroatrial hetero-taxia (VAH), which is characterised by multiple malformations, including gross cardiac malformation. This raises the question of whether VAH is caused by mutations in Cx43. There is currently controversy on this issue. Mutations in Cx43 have been identified in some patients with familial VAH (Britz-Cunningham *et al.*, 1995). However, a number of subsequent studies of other patients with familial defects of cardiac laterality have failed to find associated Cx43 mutations (Splitt *et al.*, 1995). One possible explanation for these disparate findings is that visceroatrial heterotaxia shows significant genetic heterogeneity and that Cx43 mutations are not a major cause of the disorder.

Non-Syndromic Deafness (Cx26)

The most common inherited sensory disorder is deafness. It is commonly classified into syndromic deafness, which is associated with other symptoms, and non-syndromic deafness, in which severe hearing defects are found in the absence of other symptoms. About 70% of all cases of prelingual deafness are non-syndromic. The disorder shows marked genetic heterogeneity, but in some families it maps to chromosome 13q11–12, the location of the gene encoding connexin 26 (Kelsell *et al.*, 1997). Analysis of these families has revealed over 60 mutations in Cx26 that result in either dominant or recessive forms of non-syndromic deafness. The most common mutation is a deletion of the glycine residue at position 30 (G30), which results in the introduction of a premature stop codon and is expected to produce a non-functional protein (Denoyelle *et al.*, 1997). This mutation is inherited in a recessive fashion and since heterozygotes have normal hearing, or only suffer mild high-frequency hearing loss in adult life, it appears that a 50% reduction in Cx26 expression has little functional effect. Dominant mutations in Cx26 that cause non-syndromic deafness have also been identified, but are much rarer: one example is a methionine-to-threonine substitution at position 34 (M34T), which is located in the first TM (Kelsell *et al.*, 1997). These mutations may act in a dominant negative fashion to inactivate heteromeric (wild-type and mutant) connexons in heterozygous individuals.

Connexin 26 is expressed in the cochlea, including the stria vascularis, the cochlear duct and the cochlear nerve. The mechanism by which the lack of functional Cx26 leads to hearing loss remains obscure. Likewise, it is not clear why there are no other observable phenotypes in affected individuals because Cx26 is expressed in many other tissues, including the liver and pancreas.

Infertility in Mice (Cx37)

At birth, female mammals already possess their entire complement of oocytes. These are arrested in meiosis and lie surrounded by granulosa cells within primordial follicles. During each menstrual cycle one or more follicles enlarge, the granulosa cells proliferating and forming a fluid-filled cavity surrounding the oocyte. After ovulation, the granulosa cells form the corpus luteum. Connexin 37 is present in the gap junctions which couple the oocyte to the surrounding granulosa cells. Female mice which lack Cx37 do not possess mature follicles, fail to ovulate and inappropriately develop many corpus lutea (Simon *et al.*, 1997a). Male mice appear unaffected. Thus, Cx37 coupling between oocyte and granulosa cells is required for normal oogenesis, ovulation and corpus luteum formation. The abnormalities observed in Cx37 knockout mice are similar to those observed in the human disease known as karyotypically normal spontaneous ovarian failure. The aetiology of this disorder is unknown.

Summary

We have seen that gap junction channels form a pathway between the cytoplasm of adjacent cells. They are composed of two hemichannels (connexons), each of which is made up of six subunits (connexins). Multiple types of connexins exist and gap junctions may be formed from identical or different types of connexin. Mutations in one type of connexin, Cx32, produce CMTX, a disease in which the peripheral nerves degenerate. This appears to result because the gap junctions in the Schwann cells which surround the axons are no longer functional. Gross cardiac malformations result from mutations in Cx43 and congenital deafness from mutations in Cx26.

SUMMARY

CHAPTER 21

AUTOANTIBODIES TO ION CHANNELS

A number of autoimmune disorders are caused when the body produces antibodies to its own ion channels. Of these, myasthenia gravis is the most common. It is one of the best characterised of autoimmune diseases and is produced by autoantibodies to the nicotinic acetylcholine receptor. Autoantibodies to the P/Q-type of voltage-gated Ca^{2+} channels give rise to Lambert–Eaton myasthenic syndrome, antibodies to voltage-gated K^+ channels are responsible for acquired neuromyotonia, and Rasmussen's syndrome is associated with antibodies to the GluR3 glutamate receptor subunit. In this chapter, we consider how these autoantibodies produce disease. A brief introduction to immunology is also provided for those unfamiliar with the subject.

BASIC IMMUNOLOGY

Two types of molecule are involved in recognition of a foreign protein (antigen): the immunoglobulins (or antibodies) and the T-cell antigen receptors. We will only be concerned with the former here. Immunoglobulins are glycoproteins produced by the B lymphocytes of the immune system and five classes are recognised in higher mammals: IgG, IgA, IgM, IgD and IgE. The major immunoglobulin in normal human serum is IgG, which constitutes 70–75% of the total immunoglobulin pool. Approximately 15–20% is IgA and 10% is IgM, while IgD and IgE make up less than 1% of the total pool.

All immunoglobulins share the same characteristic Y-shaped structure (see Fig. 21.1), which consists of two antigen-binding arms (Fab) linked via a common stalk (Fc). The latter binds to the Fc receptor of lymphocytes and may also activate complement (see Chapter 23). Antibodies do not bind to all of a foreign protein but instead to specific domains, which are known as antigenic determinants or epitopes. Each antibody binds to one epitope only. However, if two different proteins—such as two types of Ca^{2+} channel—share the same (or similar) epitope, a single type of antibody can react with both proteins.

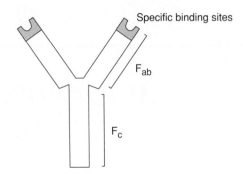

FIGURE 21.1 MODEL OF IMMUNOGLOBULIN STRUCTURE
The two Fab arms, which bind antibody, are joined by an Fc domain, which binds complement.

When circulating immunoglobulins encounter their specific epitope on an antigen, they bind to it via one of their Fab arms. The strength of this interaction is determined by the snugness with which the epitope fits into the antigen-binding site. The binding of an immunoglobulin to its antigen on the surface of the target cell may have several effects. First, the C1q component of the complement system may attach to the Fc stalk of the antibody and thereby initiate a chain of reactions which results in the lysis and death of the target cell. The mechanism of complement-mediated cell lysis is discussed in Chapter 23. Specific types of lymphocytes (killer cells) may also attach to the Fc stalk of the antibody. This triggers the lymphocyte to secrete lytic pores known as perforins (see Chapter 23), or degradative enzymes and cytokines, which destroy the target cell. Since each antibody has two antigen-binding sites, it is also possible for a single molecule to interact with two antigens simultaneously. This occurs in myasthenia gravis and in Lambert-Eaton myasthenic syndrome, in which autoantibodies can cross-link adjacent acetylcholine receptors or Ca^{2+} channels, respectively (see later). Cross-linking of adjacent ion channels by the two Fab arms of an immunoglobulin increases the rate of normal internalization of the target protein/antibody complex and thereby decreases the channel density on the cell surface. Finally, antibody binding may have acute effects on channel function as exemplified by the direct inhibition of AChR channel activity produced by the IgG of some patients with myasthenia gravis.

The role of the immune system is to react against foreign antigens. It must therefore be able to distinguish between foreign and self-proteins. In normal circumstances antibodies are not produced against autologous (host) proteins: autoimmune disease results when this self-tolerance fails and anti-bodies are produced against epitopes found on autologous cells. These auto-antibodies may be the primary cause of disease or they may be produced as a secondary consequence of tissue damage, resulting from a disease with a different primary aetiology. It is worth noting that the brain is protected against circulating antibodies, which normally do not cross the blood–brain barrier. This explains why antibodies to voltage-gated K^+ channels affect the peripheral, but generally not the central, nervous system. In rare cases, however, disruption of the blood–brain barrier may permit pathogenic antibodies to reach brain antigens, as in the case of Rasmussen's syndrome.

Autoimmune diseases of ion channels share several distinctive features. Among these are the presence of circulating antibodies to the ion channel; a reduction in the number or function of the ion channel in the target tissue; the amelioration of symptoms by plasma exchange, which removes the offending antibodies; and the ability to transfer the phenotype of the disease to mice by injection of immunoglobulins isolated from the serum of affected patients. In addition, immunization of experimental animals against the purified ion channel protein produces an animal model of the disease. Certain autoimmune diseases are the indirect result of a neoplastic tumor, and are referred to as paraneoplastic disorders: the best characterized example is Lambert-Eaton myasthenic syndrome, which is associated with small cell carcinoma of the lung.

AUTOANTIBODIES AFFECTING NEUROMUSCULAR TRANSMISSION

It is perhaps initially surprising to discover that autoimmune diseases are associated with antibodies directed against all of the major ion channels involved in neuromuscular transmission: the presynaptic voltage-gated Ca^{2+} and K^+ channels and the postsynaptic ligand-gated nicotinic acetylcholine receptors (nAChRs). Full details of the properties of these ion channels—and how mutations in the genes that encode them give rise to disease—may be found in the relevant chapters. A detailed description of neuromuscular transmission is also provided in Chapter 15. This chapter focuses on how autoantibodies to ion channels at the neuromuscular junction cause autoimmune disorders.

When a nerve impulse arrives in the presynaptic terminal it activates voltage-gated Ca^{2+} channels (probably of the P/Q type), producing a rise in intracellular Ca^{2+} that triggers the exocytosis of synaptic vesicles containing acetylcholine (ACh). The amount of ACh released is determined by the Ca^{2+} concentration in the presynaptic terminal and therefore depends on the magnitude of the voltage-gated Ca^{2+} current. In turn, this is influenced by the duration of the membrane depolarization and thus by the amplitude of the voltage-gated K^+ current which underlies membrane repolarization. Once released, ACh diffuses across the synaptic cleft and binds to nAChRs in the postsynaptic membrane. This induces a conformational change which results in the opening of an intrinsic ion channel. The current that flows through the open nAChR causes a depolarization of the postsynaptic membrane known as the endplate potential (epp), which, if it is sufficiently large, triggers an action potential in the muscle fibre and thereby initiates muscle contraction. The concentration of ACh in the synaptic cleft is rapidly lowered both by the action of acetylcholinesterase, which hydrolyses the transmitter, and by diffusion. Clearly, interference with either the voltage-gated Ca^{2+} or K^+ currents, which govern ACh release, or with the nAChR that mediates its postsynaptic action, will influence neuromuscular transmission. A reduction in the number of Ca^{2+} channels, or nAChRs, decreases transmission and leads to muscle weakness, whereas a reduction in K^+ channel density increases transmitter release and facilitates repetitive muscle activity.

The 'safety threshold' for neuromuscular transmission is the difference between the amplitude of the epp and the depolarization required to elicit an action potential. The critical threshold for action potential firing in human muscle is fairly high, so that even a moderate reduction in epp amplitude may cause a failure of neuromuscular transmission and consequent muscle weakness. Although a reduction in epp amplitude (fatigue) occurs during repetitive activity at the normal neuromuscular junction, the epp amplitude does not fall below the safety threshold. If the efficiency of synaptic transmission is already compromised, however, (as in myasthenia gravis), fatigue may be sufficient to cause a failure of transmission. In other species, the safety threshold is lower, and neuromuscular transmission only fails when the epp amplitude is reduced to very low levels.

Myasthenia Gravis

Myasthenia gravis is one of the best understood autoimmune diseases, with a prevalence in northern Europe of 7–9 cases per 100,000 people. Two-thirds of patients are women, in whom the disease often develops in early adult life, with a peak age of onset during the third decade; both men and women can also develop the disease in later life. The hallmark of myasthenia gravis is a profound weakness of the skeletal muscle, which increases with exercise. It is caused by autoantibodies directed against the nAChR in the postsynaptic membrane of skeletal muscle (Vincent, 1980). These antibodies lead to a loss of functional nAChR, primarily by a combination of complement-mediated lysis and an increased rate of receptor degradation. Consequently, the end-plate potentials are substantially reduced in amplitude. In turn, this leads to muscle weakness and fatigue.

Clinical features

The muscle weakness characteristic of myasthenia gravis usually manifests first in the facial and eye muscles and many patients present with double vision and ptosis (drooping upper eyelids). They also may have difficulty in holding their head up, and in chewing and swallowing (which increases during a meal). Weakness in the limbs develops progressively and particularly affects the arms and the proximal muscles in the legs, so that the patient has difficulty in standing up or in climbing stairs. A distinctive feature of myasthenia gravis is that muscle weakness is exacerbated by exercise—

indeed in milder cases it may only be obvious after physical exertion—and it increases in severity throughout the day. The symptoms may be aggravated by stress or infection but are relieved by rest. Respiratory insufficiency can be life-threatening, but with modern treatments mortality is infrequent.

Many patients with myasthenia gravis also have thymus abnormalities. This is most common in early onset disease (<40 years of age), in which antibodies to the nAChR may be synthesized by B lymphocytes in the thymus. Thymectomy can be beneficial in these individuals. Around 10% of patients have a thymus tumour; curiously, removal of the tumor does not usually alleviate the myasthenia.

Mild forms of myasthenia gravis are treated with anticholinesterase drugs, such as pyridostygmine, which inhibit the breakdown of acetylcholine. This increases the lifetime of the transmitter in the synaptic cleft and so facilitates neuromuscular transmission. Thymectomy and immunosuppressive therapy are used where anticholinesterases are unable to control the myasthenia. In severe acute cases, particularly when respiratory insufficiency is present, plasma exchange may be required; but it is only a temporary measure.

Myasthenia gravis is associated with loss of AChR

Analysis of muscle biopsies from myasthenia gravis patients revealed that the epps were abnormally small, or rapidly decreased in amplitude with repetitive stimulation (Elmqvist *et al.*, 1964), and that mepp amplitudes were also substantially smaller (Fig. 21.2). These changes do not reflect a reduction

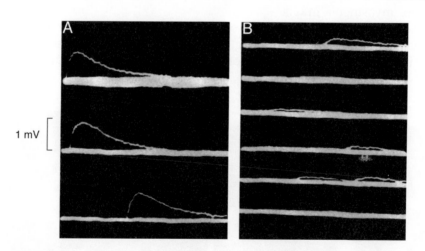

FIGURE 21.2 MINIATURE ENDPLATE POTENTIALS RECORDED FROM NORMAL AND MYASTHENIC HUMAN MUSCLE
Miniature endplate potentials (mepps) recorded from the endplate region of intercostal muscle biopsies from a control (A) and myasthenia gravis (B) patient. All mepps were recorded in the presence of an AChE inhibitor to increase their amplitude (which explains the prolonged time course). From Vincent (1997).

in the number of synaptic vesicles released, or in the amount of ACh each vesicle contains. Rather, they result from a reduction in the number of functional nAChR in the postsynaptic membrane. This was demonstrated by a marked decrease in the binding of α-bungarotoxin (a snake toxin which binds to nAChR with very high affinity) to human myasthenia gravis muscle (Fambrough et al., 1973). It is believed that transmission block occurs in human muscle when the number of nAChR is reduced to ~60% of normal (Vincent, 1980).

Anti-nAChR antibodies

The involvement of anti-nAChR antibodies in myasthenia gravis was first clearly demonstrated when rabbits that had been immunised with purified nAChRs developed extreme flaccid paralysis and respiratory failure (Patrick and Lindstrom, 1973). As in the human disease, the paralysis was transiently reversed by anti-acetylcholinesterase drugs (Fig. 21.3). Additional support for the pathogenic role of anti-nAChR antibodies is provided by the fact that it is possible to transfer the disease to a mouse by injection of IgG isolated from an affected human patient (Toyka et al., 1977) and that IgG and complement components are found at the neuromuscular junction in myasthenia gravis (Engel et al., 1977). Anti-nAChR antibodies have also been observed in the serum of more than 85% of patients with myasthenia gravis (Vincent and Newsom-Davis, 1985). These are detected by the ability of serum, or an IgG fraction, to immunoprecipitate nAChRs tagged with ^{125}I-labelled α-bungarotoxin. A positive anti-nAChR assay is now the main diagnostic criterion for myasthenia gravis.

Adult nAChRs are composed of α-, β-, ε-, δ-subunits while fetal AChRs consist of α-, β-, γ-, δ-subunits (Fig. 21.4). Fetal nAChRs are present throughout muscle development and do not disappear until ~33 weeks after birth, by

FIGURE 21.3 nAChR AUTOANTIBODIES PRODUCE MYASTHENIA
(A) Rabbit showing flaccid paralysis 5 days after injection with purified nAChRs. (B) The same rabbit 1 min after intravenous injection of 0.3 mg of the acetylcholinesterase inhibitor edrophonium. From Patrick and Lindstom (1973).

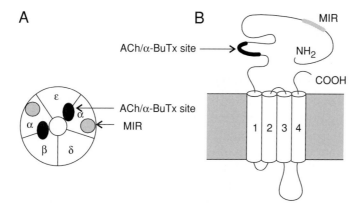

FIGURE 21.4 IMMUNOGENIC REGIONS OF THE nAChR
(A) Adult nAChR seen from above. The main immunogenic region (MIR) is hatched. The ACh-binding sites (black) lie at the interface between the α- and δ (or ε)-subunits. In the fetal receptor the ε-subunit is replaced by a γ-subunit. The fetal-specific antibodies associated with arthrogrypopsis multiplex congenita interfere with ACh action and are proposed to overlap with the αγ-binding site. (B) AChR seen from the side.

which time the γ-subunit is replaced by the ε-subunit. The main immunogenic region of the nAChR is a complex epitope on the extracellular surface of the α-subunits of the acetylcholine receptor, which includes residues 65–76 (Fig. 21.4; Tzartos *et al.*, 1991). It is distinct from the ACh-binding site and was initially defined by its reactivity with a high proportion of monoclonal antibodies raised against the *Torpedo* AChR. Parts of the β-, δ-, and γ (fetal)-subunits are also significantly immunogenic. nAChR antibodies found in myasthenia gravis patients are heterogeneous: although most seem to bind to the main immunogenic region on the α-subunit, some antibodies interact with epitopes on other subunits. There appears to be no correlation between disease severity and the type of anti-nAChR antibody.

The marked loss of nAChRs observed in patients with myasthenia gravis is produced by several mechanisms. The most important of these is complement-mediated lysis of the postsynaptic membrane following antibody binding, which leads to destruction of the endplate region of the muscle (Engel *et al.*, 1977). Cross-linking of nAChRs followed by their accelerated degradation (Drachman *et al.*, 1978) and, in rare cases, direct inhibition of function have also been demonstrated.

Antibodies to fetal nAChR

Fetal or neonatal muscle weakness was found in ~10% of pregnancies in mothers with myasthenia gravis, as a result of the transfer of anti-nAChR antibodies across the placenta. These antibodies are directed against both the adult and fetal acetylcholine receptors. The condition ameliorates after birth as the levels of antibody in the baby decline. It is less common now that more mothers with myasthenia gravis are well controlled.

Arthrogryposis multiplex congenita

In rare cases, maternal antibodies block the function of fetal nAChRs. This leads to paralysis of the fetus and, because fetal movements are essential for normal muscle development, may produce severe developmental abnormalities. Infants are born with muscle weakness, muscle wasting and multiple contractures of the extremities. The condition is known as arthrogryposis multiplex congenita (AMC) and often leads to fetal or neonatal death. It is sometimes also associated with limb deformities, small palate, poor lung development and polyhydramnios (excess amniotic fluid due to reduced swallowing by the fetus). Plasma exchange may be helpful if carried out during the critical period when fetal nAChRs are present.

Not all cases of AMC are associated with maternal myasthenia gravis. In some cases, the mother herself may be asymptomatic, as exemplified by the case of a woman whose first child was healthy, but who subsequently lost six babies from AMC (Vincent et al., 1995). In this patient, high titres of antibodies were able to inhibit fetal, but not adult nAChR (Fig. 21.5), thus explaining why the mother was unaffected. The rapidity of the inhibition, and the fact that the antibodies also severely reduced α-bungarotoxin binding, suggest that they may act by inhibiting ACh binding to the fetal receptor: as explained in Chapter 15, one of the two ACh-binding sites lies at the interface between the α- and γ (or ε)-subunit.

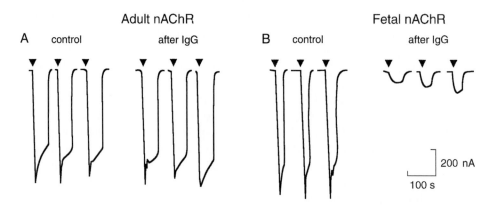

FIGURE 21.5 MATERNAL IgG INHIBITS FETAL BUT NOT ADULT nAChR

Whole-cell ACh-induced currents recorded from oocytes expressing fetal-type (α, β, γ, δ) or adult-type (α, β, ε, δ) nAChR before and 40 min after application of IgG purified from the serum of a woman who had lost six babies as a result of placental transfer of these antibodies. The arrows indicate the addition of ACh. From Vincent et al. (1995).

Patients without AChR antibodies

In about 15% of patients with myasthenia gravis, no anti-AChR antibodies can be detected. Nevertheless, they appear to suffer from an autoimmune disorder, because their serum and plasma immunoglobulins (Igs) transiently inhibit nAChR function (Mossman *et al.*, 1986). There is no evidence that their Igs bind to the nAChR itself and the cause of the transient inhibition is unknown. One possibility is that these patients' immunoglobulins bind to some other protein that indirectly modulates AChR function, for example, by enhancing receptor desensitization.

Lambert–Eaton Myasthenic Syndrome

Clinical features

Lambert-Eaton myasthenic syndrome (LEMS) is a presynaptic disorder of neuromuscular transmission which is caused by the production of antibodies to voltage-gated Ca^{2+} channels at the motor nerve terminals. This results in a marked reduction in acetylcholine release, the failure of neuromuscular transmission and muscle weakness. Muscle weakness is most common in the limbs, so that patients complain that their legs feel stiff or weak and they find difficulty in walking: indeed, in some individuals the symptoms of the disease may be sufficiently severe to render them bedbound. Tendon reflexes are also weak or absent. Unlike myasthenia gravis, in LEMS muscle weakness does not increase with exercise; in fact, muscle strength and tendon reflexes are briefly *enhanced* during the first few seconds of maximal effort. LEMS is also associated with symptoms indicative of disturbance of the autonomic nervous system, including decreased salivation and sweating, constipation and impotence. In most, but not all, patients the onset of the disease is gradual.

In around 60% of patients, LEMS is associated with small cell carcinoma of the lung. It is believed that voltage-gated Ca^{2+} channels present in the tumour trigger the production of autoantibodies that cross-react with Ca^{2+} channels at the nerve terminal. Consistent with this idea, LEMS antibodies downregulate Ca^{2+} channels in cells isolated from small cell carcinomas (Johnston *et al.*, 1994). In addition, removal of the carcinoma, or radiotherapy, leads to neurological recovery, as is expected if the tumour is driving autoantibody production. LEMS often presents before radiological evidence of small cell carcinoma, and provides a useful indicator of the underlying tumour, which enables its earlier treatment. It is therefore important to investigate all patients presenting with LEMS for small cell carcinoma of the lung.

In many LEMS patients, 4-aminopyridine therapy may be beneficial. This drug blocks K^+ channels at the presynaptic terminal. It therefore lengthens the action potential duration, prolonging Ca^{2+} entry and thus augmenting transmitter release. The effects of 4-aminopyridine can be quite spectacular—a patient unable to stand initially may be able to walk nearly normally within an hour of treatment with the drug (Murray and Newsom-Davis, 1981).

Autoantibodies

Evidence for the autoimmune nature of LEMS was first provided by studies showing that injection of mice with IgG derived from LEMS patients reproduced many of the features of the disease (Lang *et al.*, 1983). *In vitro* studies of neuromuscular transmission in these mice showed that at low stimulation rates the amplitude of the epp was much lower than that of control animals because of a reduction in acetylcholine release (Fig. 21.6A). At high frequencies of nerve stimulation, there was some facilitation of the epps in LEMS animals, indicating that the final stimulus in a train caused more transmitter release than the first (Fig. 21.6B). This facilitation is probably the explanation for the short-term increase in strength observed in LEMS patients after vigorous muscle activation. It results from the fact that the reduced density of Ca^{2+} channels at LEMS nerve terminals means that at low stimulation rates Ca^{2+} entry is insufficient to stimulate the normal level of transmitter release, whereas at higher stimulation rates release is facilitated by the accumulation of intracellular Ca^{2+}.

At the neuromuscular junction, transmitter release occurs at discrete sites called active zones. In freeze-fracture electron micrographs, the active zones are characterized by intramembranous particles, which are organised into linear arrays resembling tramlines (Fig. 21.7A). These particles are thought to constitute voltage-gated Ca^{2+} channels. LEMS antibodies disrupt the regular tramlines of active zone particles and reduce their number, both in patients and in mice injected with LEMS immunoglobulin (Fukunaga *et al.*, 1982, 1983) (Fig. 21.7B). These morphological changes correlate with a reduction in the amplitude of the endplate potential, as is expected if LEMS antibodies cause the internalization and downregulation of Ca^{2+} channels at the motor nerve terminals (Engel, 1991). There is evidence that the antibodies cross-link adjacent Ca^{2+} channels, thus producing an aggregation of membrane

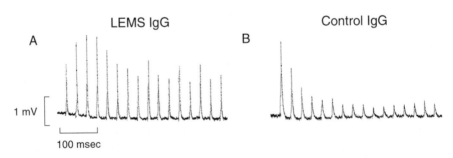

FIGURE 21.6 LEMS ANTIBODIES AFFECT NEUROMUSCULAR TRANSMISSION

Trains of endplate potentials recorded in response to nerve stimulation from the muscle of a mouse pretreated with IgG isolated from LEMS patients (A) or pretreated with control IgG (B). From Lang *et al.* (1983).

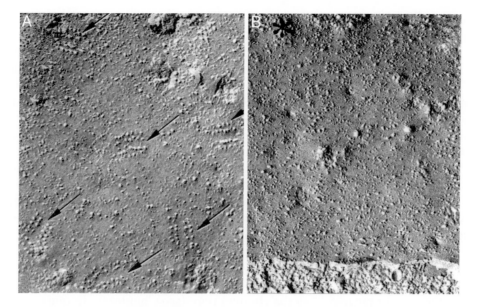

FIGURE 21.7 LEMS ANTIBODIES DISRUPT PRESYNAPTIC Ca^{2+} CHANNEL ORGANIZATION
Freeze-fracture electron micrographs of the presynaptic membrane of mouse diaphragm muscle, viewed face on. Mice were pretreated with control IgG (A) or IgG isolated from LEMS patients (B). Numerous active zone particles arranged in tramlines (arrows) are present in (A), while in (B) active zone particles are disorganized and fewer in number. The active zone particles are believed to correspond to P/Q-type Ca^{2+} channels. Photograph supplied by Andrew Engel.

particles and disrupting the linear arrays, and that this is followed by internalization and degradation of the antibody–channel complex (Peers *et al.*, 1993).

A number of studies have addressed the question of which type of Ca^{2+} channel constitutes the primary target for LEMS antibodies. As explained in Chapter 9, there are several different types of voltage-gated Ca^{2+} channels. Of these, N-, L- and P/Q-type Ca^{2+} channels are present in presynaptic nerve terminals. The P/Q-type of Ca^{2+} channel appears to be the most important for acetylcholine release at the mammalian neuromuscular junction, since specific inhibitors of P/Q-type, but not N- and L-type, Ca^{2+} channels block neuromuscular transmission (Protti *et al.*, 1996). Approximately 90% of patients with LEMS have antibodies to P/Q-type Ca^{2+} channels, as demonstrated by their ability to immunoprecipitate radiolabeled P/Q-type Ca^{2+} channels extracted from human cerebellum (Motomura *et al.*, 1995). By contrast, only about 30% of patients have antibodies to N-type Ca^{2+} channels and even fewer have antibodies to L-type Ca^{2+} channels. Some studies have also shown a linear relationship between the titre of P/Q-type Ca^{2+} channels and disease severity in individual patients (Motomura *et al.*, 1997). Furthermore, LEMS antibodies block Ca^{2+} uptake into mammalian cells transfected with P/Q-type, but not with N-, L- or R-type Ca^{2+} channels (Lang *et al.*, 1998). Thus, P/Q-type Ca^{2+} channels appear to be the primary target for LEMS antibodies. The ability of LEMS sera to immunoprecipitate N- and L-type, as well as P/Q-type, Ca^{2+} channels (Motomura *et al.*, 1997) is now believed to reflect immunologic cross-reactivity (i.e., different types of voltage-gated Ca^{2+} channels have a similar epitope) or the possibility that in some patients the underlying small cell carcinoma may express multiple types of Ca^{2+} channel.

The P/Q-type Ca^{2+} channel is composed of α_{1A}-, α_2-, β-, and δ-subunits (see Chapter 9 for full details). The α_1-subunit is functionally the most important as it acts as the channel pore, the voltage sensor and the receptor for many drugs. The other subunits have auxiliary roles. The β-subunit is entirely cytosolic and is therefore unlikely to be a target for circulating autoantibodies, but theoretically the extracellular domains of α_{1A}-, α_2-, or δ-subunits could serve as potential epitopes for LEMS antibodies. The α_{1A}-subunit has four homologous repeats (I–IV), each of which consists of six putative transmembrane segments and a loop between segments 5 and 6 which dips back down into the membrane and lines the channel pore (see Chapter 9). In one study, antibodies to the S5–S6 linker in repeats II and IV were identified in 30% and 20% of LEMS patients respectively, suggesting that these regions of the protein may serve as targets for autoantibodies (Takamori *et al.*, 1997). Other studies have suggested that some LEMS antibodies may interact with the β_2-subunit of voltage-gated Ca^{2+} channels (Rosenfeld *et al.*, 1993), but these antibodies would not be pathogenic since the β-subunit is cytosolic.

In addition to suppressing neuromuscular transmission, LEMS antibodies also interfere with transmitter release from parasympathetic and sympathetic neurones, by downregulation of voltage-gated Ca^{2+} channels (Waterman *et al.*, 1997). This effect accounts for the autonomic dysfunction observed in LEMS patients. P/Q-type channels also are found at high density in the cerebellum. Fortunately, however, antibodies do not generally cross the blood–brain barrier and therefore cannot access neurones of the central nervous system. Consequently, the effect of LEMS antibodies is usually confined to Ca^{2+} channels of the peripheral nervous system. However, a small proportion of patients with paraneoplastic cerebellar ataxia have antibodies to voltage-gated Ca^{2+} channels (Mason *et al.*, 1997).

In a small number of patients with LEMS, antibodies to N- or P/Q-type Ca^{2+} channels cannot be detected (Motomura *et al.*, 1997). However, as their clinical symptoms respond to plasma exchange and immunosuppressive drugs, their disease is likely to be antibody-mediated. In these patients, however, the offending antibodies appear to be directed at other proteins of the neuromuscular junction.

Amyotropic Lateral Sclerosis

Amyotropic lateral sclerosis (ALS), or Lou-Gehrig's disease as it is sometimes called in the United States, is characterized by the progressive loss of cortical and spinal motor neurones. This leads to paralysis, respiratory depression and death. The incidence of the disease is 4–6 per 100,000 people, is more common in men and increases with age. Both sporadic and familial forms of the disease have been described, with the former accounting for ~90% of cases.

Despite intensive investigation, the aetiology of sporadic ALS is unknown and an effective therapy is lacking. An autoimmune origin is suggested by the increased incidence of associated autoimmune disorders in ALS patients and reports (albeit controversial) that ALS is associated with antibodies to presynaptic Ca^{2+} channels. In initial studies, it was found that Ig from ALS patients enhances ACh release at the neuromuscular junction (Uchitel et al., 1988). It also increases P-type Ca^{2+} currents in both cerebellar Purkinje cells and in artificial bilayers, by prolonging the single-channel open time (Llinás et al., 1993). Since P/Q-type Ca^{2+} channels primarily account for Ca^{2+}-dependent ACh release at the neuromuscular junction (Protti et al., 1996), this argues that ALS antibodies may interact with presynaptic P/Q-type channels to increase the presynaptic Ca^{2+} current and so enhance transmitter release. Indeed, 23% of ALS patients have antibodies to P/Q-type Ca^{2+} channels (Lennon et al., 1995).

There is also evidence that ALS antibodies interact with additional types of Ca^{2+} channel, because ~75% of patients have antibodies to L-type Ca^{2+} channels (Smith et al., 1992). Furthermore, ALS Ig increases Ca^{2+} currents in a motoneurone cell line, which, based on their pharmacology, are distinct from L-, N- and P-types (Mosier et al., 1995). It is not known whether this heterogeneity arises because different Ca^{2+} channel α_1-subunits share the same epitope, because ALS antibodies bind to a regulatory subunit common to several types of Ca^{2+} channel, or because ALS sera may contain more than one type of antibody. Nor is it known how the potentiation of the Ca^{2+} current produced by ALS IgG is involved in the pathogenesis of the disease. One suggestion, however, is that ALS antibodies contribute to neuronal cell death by stimulating Ca^{2+} currents and elevating intracellular Ca^{2+} levels. Further studies are required to determine if this idea is correct and to confirm the presence of antibodies to Ca^{2+} channels in ALS patients, because several groups have failed to find such antibodies and there have been some concerns raised about the Ig preparation used in the earlier studies (Vincent and Drachman, 1996).

Acquired Neuromyotonia

Acquired neuromyotonia (NMT) is a rare autoimmune disorder characterized by a hyperexcitability of the motor nerves that leads to spontaneous muscle twitching (myokymia), painful muscle cramps and muscle weakness. The myotonia differs from that exhibited by patients with mutations in voltage-gated Na^+ and Cl^- channels in that it is spontaneous, continuous and not triggered by exercise. As in the case of these other channel disorders, however, the increased muscular activity associated with NMT may lead to muscle hypertrophy. Excessive sweating is also often a problem, perhaps due to the involvement of the autonomic nervous system, and about 20% of patients have a small cell lung carcinoma or a thymus tumour.

Acquired neuromyotonia is associated with antibodies against voltage-gated K^+-channels (K_V channels) in the peripheral motor nerves. These are proposed to lead to a reduction of the membrane K^+ permeability and prolongation of the action potential at the nerve terminal. Consequently, neurotransmitter release will be enhanced, thus accounting for the myokymia. There is accumulating evidence in favour of this hypothesis. First, antibodies against K_V channels have been detected in 50% of NMT patients (Shillito et al., 1995; Hart et al., 1997). Second, injection of mice with IgG isolated from NMT patients caused a small increase in transmitter release at the neuromuscular junction, similar to that produced by low concentrations of K^+ channel blockers such as 4-aminopyridine. Repetitive firing of dorsal root ganglion neurones was also induced by one patient's IgG (Fig. 21.8). Finally, some patients show clinical improvements after plasma exchange.

The symptoms of acquired neuromyotonia are thus consistent with the presence of autoantibodies to voltage-gated K^+ channels in the motor nerve terminals. Although these antibodies also interact with brain K^+ channels, they are prevented from reaching their target antigen in vivo by the presence of the blood–brain barrier. Consequently, NMT is not usually associated with disturbance of the central nervous system (although some patients do suffer from hallucinations or personality changes). Questions that remain to be answered include which type(s) of K_V channel is the actual target antigen and what triggers the production of autoantibodies against this channel.

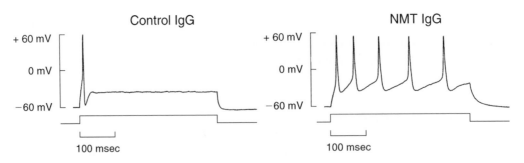

FIGURE 21.8 NMT ANTIBODIES REDUCE K^+ CURRENTS

Action potentials recorded in response to an injected current pulse from dorsal root ganglion cells after incubation for 24 hr with control IgG (A) or IgG isolated from a patient with acquired neuromyotonia (B). From Shillito et al. (1995).

AUTOANTIBODIES TO ION CHANNELS NOT INVOLVED IN NEUROMUSCULAR TRANSMISSION

Rasmussen's Encephalitis

Rasmussen's encephalitis (RE) is a (fortunately rare) progressive disorder that presents in childhood and is characterized by severe epilepsy, hemiplegia (paralysis of one side of the body), dementia and inflammation of the brain. The disease is confined to one cerebral hemisphere and causes its progressive atrophy. Similar symptoms were observed in two rabbits immunised with the GluR3 glutamate receptor subunit, which led Rogers and his colleagues (1994) to investigate whether Rasmussen's encephalitis might be an autoimmune disease. They detected GluR3 antibodies in the serum of affected children and further showed that in one of these patients plasma exchange reduced the antibody titre and concomitantly improved neurological function (Fig. 21.9). Subsequent studies confirmed these early data and provided additional evidence that Rasmussen's encephalitis is indeed an autoimmune disease (Andrews and McNamara, 1996).

The question of how GluR3 antibodies exert their deleterious effects is still unresolved. It has, however, been established that GluR3 antibodies are able to activate AMPA/kainate receptors in cortical neurones (Fig. 21.10) and that the region of GluR3 with which the RE antibodies interact lies within the agonist binding site (Tywman *et al.*, 1995). Neuronal excitation induced by receptor activation may therefore precipitate the epileptic seizures which characterise the disease and contribute to excitotoxic cell death. This is not the only cause of cell death, however, because other reports indicate that GluR3 antibodies can kill cortical cells in a complement-dependent manner even in the presence of GluR antagonists (Andrews and McNamara, 1996). Thus both receptor activation and complement-mediated cell lysis may be involved. Other issues that remain to be resolved include how the disease is triggered, why it remains confined to a single cerebral hemisphere and how circulating antibodies gain access to the brain.

Cardiomyopathy

Antibodies against the mitochondrial ATP/ADP carrier are found at high frequency in the sera of patients with dilated cardiomyopathy or myocarditis, presumably as a consequence of tissue damage. These antibodies cross-react with the β-subunit of the voltage-gated Ca^{2+} channel in cardiac myocytes and increase Ca^{2+} current amplitude and myocardial contractile force (Morad *et al.*, 1988). The enhanced Ca^{2+} permeability may lead to eventual Ca^{2+} overload and thereby contribute to the pathogenesis of dilated cardiomyopathy.

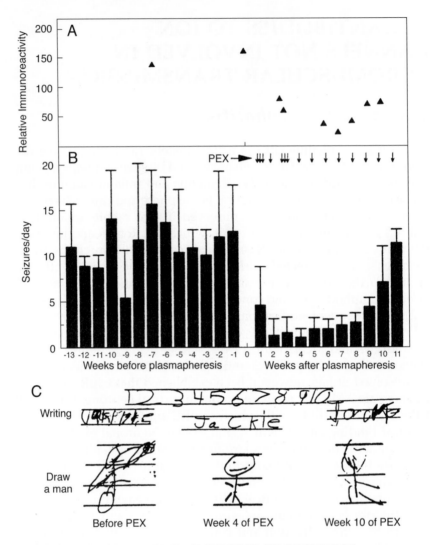

FIGURE 21.9 EFFECTS OF RASMUSSEN'S ENCEPHALITIS ANTIBODIES
Correlation of GluR3 immunoreactivity, seizure frequency and neurological function in a child with Rasmussen's encephalitis. Plasma exchange (PEX) was initiated at the arrow. During the first 7 weeks of PEX, antibody titre decreased (A) in parallel with seizure frequency (B). There was an associated increase in the ability of the child to write, draw and do simple maths problems (C). She now also spoke spontaneously. From Rogers *et al.* (1994), Fig. 4.

Type-1 Diabetes Mellitus

Type-1 diabetes mellitus is characterized by a progressive autoimmune destruction of the insulin-secreting pancreatic β-cells within the islets of Langerhans so that eventually the patient is dependent on exogenously supplied

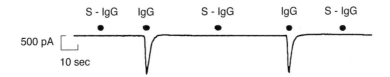

FIGURE 21.10 RASMUSSEN'S ENCEPHALITIS ANTIBODIES ACTIVATE GLUTAMATE RECEPTORS

IgG purified from a patient with Rasmussen's encephalitis (IgG) evokes an inward current in mouse fetal cortical neurones whereas serum from which the IgG fraction had been removed (S-IgG) was ineffective. IgG-activated currents were blocked by CNQX, suggesting that they were mediated by AMPA on kainate receptors. From Twyman *et al.* (1995).

insulin for survival. The aetiology of the disease is quite distinct from that of type-2 diabetes (Chapter 8) and it presents much earlier, usually between 11 and 14 years of age. Type-1 diabetes susceptibility shows a clear genetic component. Some of the susceptibility disease genes encode molecules involved in the regulation of immune responses, and antibodies to islet cell components are present in the sera of type-1 diabetic patients (Christie, 1992). Both the initial trigger and the mechanism of β-cell destruction in type-1 diabetes are unknown and are the subject of intensive investigation. However, the IgM fraction of serum from type-1 diabetic patients was shown to enhance the amplitude of the β-cell voltage-gated Ca^{2+} current (Juntti-Berggren *et al.*, 1993). It is therefore possible that an IgM-mediated Ca^{2+} influx may contribute to β-cell destruction by increasing cellular Ca^{2+} levels.

Stiff-Man Syndrome

Stiff-man syndrome is characterized by progressive rigidity of the skeletal muscles with intermittent painful muscle spasms. The clinical symptoms result from the simultaneous activation of both agonist and antagonist muscle groups (in normal individuals, activation of agonist muscles is accompanied by relaxation of the antagonist muscles). The condition is therefore believed to result from an imbalance between excitatory and inhibitory pathways controlling motoneurone activity, an idea which is supported by the ability of GABA agonists, such as diazepam, to alleviate the symptoms of stiff-man syndrome.

Stiff-man syndrome is associated with autoantibodies to glutamic acid decarboxylase (GAD) (Solimena *et al.*, 1988). This cytosolic enzyme is found in GABAergic neurones where it catalyses the formation of GABA from glutamate. It is not yet clear whether GAD antibodies produce the disease phenotype or if they are simply a consequence of the lysis of GAD-expressing cells. The fact that there are no obvious clinical differences between patients who are antibody positive and those who are antibody negative supports the latter hypothesis. However, the presence of additional autoimmune disorders, such as type-1 diabetes mellitus, Graves disease, or myasthenia gravis, in 60% of antibody-positive stiff-man patients suggests an autoimmune pathogenesis (Ellis and Atkinson, 1996). Further work is also required to determine whether GAD antibodies interfere with GABAergic neurotransmission.

CHAPTER 22

ION CHANNELS IN VIRUSES

INFLUENZA VIRUS

Influenza is far more serious than most people appreciate. Although we generally think of 'flu' as no more than an unpleasant (often yearly) experience, new and highly virulent forms arise regularly. Pandemic infections occurred in 1890, 1918, 1957 and 1968 and were associated with very high mortality rates. The influenza pandemic of 1918–1919, for example, killed more people than did the first and second World Wars combined. Within less than 5 months between 20 and 25 million people died, and in some areas entire villages were eliminated (Garrett, 1995). Even the less virulent varieties of influenza can cause considerable morbidity and mortality in the elderly and the young.

Three types of influenza viruses (A, B and C) can be distinguished by their nucleoproteins. They vary in the extent to which their surface proteins mutate, with the highest mutation rates being found for the type A virus. As individuals are unlikely to have developed resistance against mutant strains, most pandemics are caused by the type A influenza virus.

All viruses must enter a host cell in order to proliferate as they lack the necessary biochemical machinery to manufacture proteins themselves. Once inside the host cell, they multiply rapidly and new virus particles are then released by budding from the cell membrane. The type A influenza virus is surrounded by a lipid membrane, known as the viral envelope, which is derived from the plasma membrane of the host cell. Embedded within this viral envelope are three integral membrane proteins encoded by the virus RNA. These are haemaggluttinin, neuraminidase and the M2 protein. Haemaggluttinin and neuraminidase are involved in attachment of the virus to the surface of the host cell and subsequent virus internalization, while the M2 protein is a proton channel that plays a key role in viral replication (Lamb et al., 1994). There are only 4–16 M2 channels in a mature virus particle. The genetic information of the virus is contained within the viral envelope and consists of five separate strands of single-stranded RNA. These are entwined together in a spiral structure and associated with nucleocapsid proteins and RNA transcriptase.

Structure of the M2 Channel

The M2 protein is 97 amino acids long and has a molecular mass of 15 kDa. It has a single transmembrane domain of 19 amino acids (residues 25–43), with the N terminus being extracellular and the C terminus intracellular. The M2 monomer assembles into a homotetrameric channel, which is stabilised by disulphide bridges linking the N termini of adjacent subunits. The minimalistic nature of the channel structure lends itself to molecular modelling studies and to analysis of the relationship between channel structure and function by serial mutation of individual residues. These have suggested that the transmembrane domains cross the membrane as α-helices, which are symmetrically arranged around a central pore (Fig. 22.1).

When M2 is expressed in mouse erythroleukaemia (MEL) cells, it is directed to the plasma membrane and large whole-cell currents are induced by increasing the extracellular proton concentration (Chizhmakov *et al.*, 1996) (Fig. 22.2). Noise analysis suggests that the single-channel conductance, however, is very small. It appears that in MEL cells the M2 channel is selectively permeable to protons and does not conduct either Na^+ or K^+ (strictly speaking, it is not known whether the channel conducts protons or hydroxyl ions, as it is difficult to distinguish this in electrophysiological studies). By contrast, in *Xenopus* oocytes, and in artificial bilayers, a significant permeability to other monovalent cations has been reported (Pinto *et al.*, 1992). The reason for this variability in channel selectivity is unknown. In addition to permeating the M2 channel, protons also influence the channel open probability, and channel activation requires the binding of protons to an external site on the N terminus of the protein which has a pK of ~7.0 (at +60 mV). No currents are observed at pH 8.5 and channel activation is fully complete at pH 4.0. The proton-binding site may be histidine 37 (Wang *et al.*, 1995).

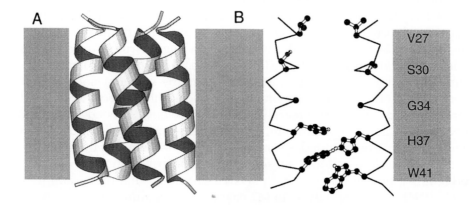

FIGURE 22.1 STRUCTURE OF THE TETRAMERIC M2 CHANNEL
(A) The M2 channel is formed by the coassembly of 4 momomers, each of which crosses the membrane as an α-helix. (B) Two opposing monomers of the channel viewed perpendicular to the pore. Residues important for amantadine binding are indicated. Figure supplied by Dr. Mark Sansom.

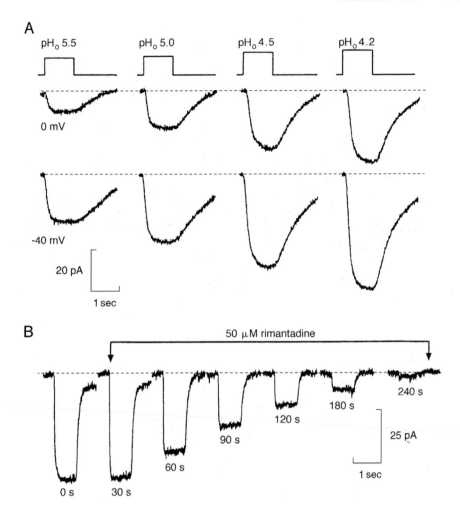

A

pH$_o$ 5.5   pH$_o$ 5.0   pH$_o$ 4.5   pH$_o$ 4.2

0 mV

-40 mV

20 pA

1 sec

B

50 μM rimantadine

240 s

180 s

120 s

90 s

25 pA

60 s

1 sec

0 s   30 s

FIGURE 22.2 MEMBRANE CURRENTS INDUCED BY LOW pH IN MEL CELLS TRANSFECTED WITH M2 cDNA
(A) Currents recorded at 0 mV or −40 mV in response to changing the external pH from 7.2 to the concentration indicated. The dashed line indicates the current level at pH 7.2. (B) Currents recorded at −70 mV in response to changing external pH from 7.2 to 5.2. The inhibitor rimantidine was added at the arrow. From Chizhmakov *et al.* (1996).

How Does the M2 Channel Influence Viral Replication?

The life cycle of the influenza virus begins with its entry into the host cell. Once inside, the virus commandeers the transcriptional and translation machinery of the host cell and directs it to make new viral RNA and proteins. These are then assembled into new viruses, packaged, and finally released into the circulation by viral budding. This results in the death of the host cell.

The virus enters the cell by receptor-mediated endocytosis (Fig. 22.3). This process is initiated by the binding of haemaggluttinin (HA) to sialic acid residues present on the surface of the host cell. Once the virus has entered the endosome compartment, the low pH triggers a conformational change in HA that leads to the fusion of the viral membrane with that of the endosome and enables the release of viral RNA complexes into the cell cytoplasm. These complexes consist of viral RNA bound to a matrix protein (M1) which anchors the RNA to the inner face of the membrane. The RNA must dissociate from the M1 protein in order for it to be transported to the nucleus, where replication takes place. The viral proteins are synthesized within the endoplasmic reticulum, combined with the RNA, and directed to the Golgi for release via the constitutive secretory pathway of the cell.

The M2 channel functions at two stages of viral replication: within the early endosome and in the *trans*-Golgi (Hay, 1992; Lamb *et al.*, 1994). The orientation of the M2 channel within the viral and endosomal membranes is related to its function. When the virus is endocytosed, the M2 channel is activated by the low pH of the endosome. The subsequent increase in proton conductance causes a fall in the internal pH of the virus particle and promotes the dissociation of RNA from the M1 protein. In certain avian influenza

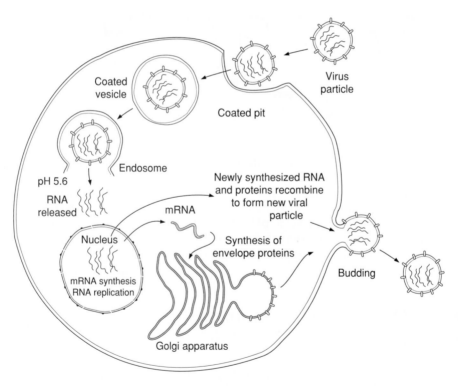

FIGURE 22.3 THE ROLE OF M2 IN THE LIFE CYCLE OF THE TYPE A INFLUENZA VIRUS
For description, see text. After Lamb *et al.* (1994).

viruses the M2 channel is also required to reduce the acidity of the *trans*-Golgi compartment, which allows the acid-sensitive haemagluttinins in the virus envelope to reach the plasma membrane without being proteolytically cleaved. The orientation of the M2 channel in the *trans*-Golgi membrane is such that the proton-sensitive site faces the acid interior so that channel activation permits proton efflux and increases the pH of the *trans*-Golgi compartment. This orientation also means that the M2 channel arrives at the plasma membrane the correct way round for release by viral budding.

Drug Therapy

The antiviral drugs amantadine and rimantidine are highly specific for the type A influenza virus and are ineffective against type B or C influenza viruses. These drugs block viral replication by inhibiting the M2 channel from the external side of the membrane (Pinto *et al.*, 1992). The mechanism is not clearly understood. One idea is that amantadine simply plugs the channel pore. However, the slow onset of the amantadine block suggests instead that the drug may influence the channel allosterically (Chizhmakov *et al.*, 1996). Amantadine is very effective at alleviating the symptoms of influenza and has been used clinically as a prophylactic in nursing homes, or for those particularly at risk from infection. Unfortunately, the virus mutates very rapidly and amantadine-resistant strains quickly develop. In addition, amantadine is a teratogen and therefore cannot be administered in pregnancy. Analysis of the mutations associated with amantadine resistance has yielded information not only about the function of the M2 channel (see earlier), but also about the residues involved in drug interactions or pore formation. These studies have shown that amantadine resistance results from mutations in the transmembrane domain of M2, including residues V27, A30, S31 and G34 (Fig. 22.1) (Hay, 1992; Holsinger *et al.*, 1994). These residues lie on the same face of an α-helix and it is therefore suggested that they may line the channel pore. The hope for the future is that it may be possible to design amantadine analogues with fewer side effects, that have higher affinity, and which interact with the channel at several different sites simultaneously so that they will be more resistant to the high viral mutation rate.

OTHER TYPES OF VIRUSES

The M2 channel appears to be specific for the type A influenza virus. Different proteins, however, play a similar role in the life cycle of other viruses. For example, the NB protein of the type B influenza virus also serves as an ion channel (Sunstom *et al.*, 1996). In artificial lipid membranes, these channels have a variable conductance (minimum 10 pS) and are cation selective at normal pH. They are also permeable to protons (Ewart *et al.*, 1996). The NB protein has no sequence homology to M2, but like M2, contains a hydrophobic sequence of 19 amino acids that might span the membrane. Other structurally analogous proteins include the SH proteins of the paramyxoviruses and human respiratory syncytial virus.

Vpu is an 80-amino acid protein encoded by the human immunodeficiency virus HIV-1. The N-terminal part of the molecule contains a single transmembrane-spanning domain (residues 6–28) and the C-terminal part is intracellular. Vpu forms weakly cation-selective channels when the mRNA is expressed in *Xenopus* oocytes or when the purified protein is incorporated into lipid bilayers (Ewart *et al.*, 1996). As is the case for the M2 channel, it is believed that the Vpu channel is formed from the association of a number of Vpu monomers. The role of the Vpu protein in the life cycle of the HIV1 virus is not fully understood. Its function differs from that of the M2 protein because Vpu is not found in the viral envelope and HIV-1 infects cells by direct fusion rather than by pH-dependent endocytosis. Instead, Vpu appears to be important for the budding and release of viral particles from HIV-1-infected cells, since this process is compromised when the transmembrane domain of Vpu is mutated. Replication is, however, only slowed and not completely abolished. It is to be expected that rapid progress will be made in understanding the functional role of the Vpu channel as it is an obvious therapeutic target.

CHAPTER 23

ION CHANNELS AS LETHAL AGENTS

Many ion channels are secreted by cells as offensive weapons. These channels insert into the membrane of the target cell, where they cause an increase in permeability that leads to lysis and cell death. Such channels include a range of bacterial and fungal toxins (many of which act as antibiotics), animal venoms, and proteins that play important roles in the human immune system.

ION CHANNELS OF THE VERTEBRATE IMMUNE SYSTEM

An important function of ion channels in the immune system is to kill foreign pathogens. This is achieved in two principal ways: by activation of the complement system and by cytotoxic T lymphocytes. Both mechanisms involve the self-association of proteins to form a macromolecular complex. This is accompanied by a change in the proteins' conformation, which leads to a transition from a hydrophilic state, in which they are soluble, to an amphiphilic state in which they are able to insert into the lipid bilayer and form a large pore (Peitsch and Tschopp, 1991).

Complement

The complement system has the capacity to lyse the cell membranes of many microorganisms. It consists of a number of proteins, C1–C9, which operate primarily as an enzyme cascade. These proteins are found in serum at variable, although relatively high, concentrations. Confusingly, they are named for the order in which they were discovered, rather than the order in which they are activated. There are two distinct pathways of complement activation, one triggered by immune complexes and the other by bacterial products. Complement activation has three functions. First, the attachment of specific complement proteins to target microorganisms, which enables them to be recognised and engulfed by phagocytes. This process, known as opsonization,

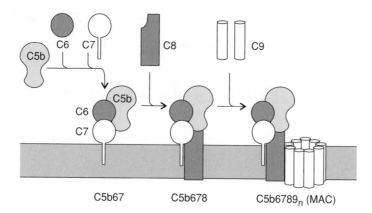

FIGURE 23.1 MECHANISM OF COMPLEMENT PORE FORMATION
C5b coassembles with circulating C6 and C7 and attaches to the membrane
of the target cell membrane. One C8 molecule then binds to a site on
C5b and inserts into the membrane where it facilitates the insertion and
subsequent polymerization of C9 molecules, which form a lytic pore
known as the membrane attack complex.

is probably the most important function of complement. Secondly, activation
of phagocytes by proteins of the complement system or their breakdown
products. Thirdly, lysis of foreign microorganisms by insertion of ion channels
into their cell membranes. We will consider only the last of these actions here.

The pore-forming molecules of the complement system consist of C6, C7,
C8, and C9. These proteins participate in the formation of the membrane attack
complex (MAC). The initial step in formation of the MAC is the enzymatic
activation of C5 by C5 convertase to give C5b (Fig. 23.1). This is followed
by the successive binding of C6 and C7 to form the C5bC6C7 complex, which
attaches to the membrane of the foreign target cell. Subsequently, C8 binds
to C5b and inserts into the membrane, where it supports the insertion and
self-assembly of a variable number of C9 monomers (usually between 12 and
18) to form a large pore (Fig. 23.1). Although insertion of C8 is associated with
a small amount of lysis, that produced by the addition of C9 is much greater.

The channel-forming properties of the various components of the MAC
have been investigated in artificial lipid bilayers. The C5b-8 complex enhances
bilayer permeability, but discrete single-channel currents have not been reported, suggesting that C5b-8 forms rather small pores. The C9 monomer
(M_r 71 kDa) forms a cation-selective, voltage-activated ion channel when
added to an artificial lipid bilayer, with a single-channel conductance of
~15 pS in 100 mM NaCl. Addition of C9 to bilayers primed with C5b-8
initially also causes the appearance of single-channel currents with ~15 pS
conductance, but this is followed by the development of much larger single-
channel currents of as much as 0.5 nS conductance (Shiver et al., 1991). These
results fit a model in which, during formation of the MAC, multiple C9
molecules self-assemble to form a pore whose diameter depends on the

number of C9 molecules that are incorporated. A major conformational change of C9 takes place during pore formation as the molecule unfolds from its globular form in solution to an elongated structure 16 nm in length, which can span the membrane. More than twelve C9 molecules are required to form the classical tubular structure observed for the MAC in the electron microscope (Fig. 23.2), which has a diameter of ~10 nm. The lytic activity of the complex is also determined by the number of C9 molecules: at least 3 C9 molecules must combine to form a pore large enough to induce lysis of *Escherichia coli*.

Attachment of C5b67 to the membrane occurs only in the close vicinity of the site at which the complex is formed because a number of soluble inhibitors in serum bind to free C5bC6C7 and inactivate it. Should this process fail and the C5bC6C7 complex become attached to autologous (host) cells, other mechanisms prevent the subsequent insertion of C8 and, if this fails, inactivate the MAC complex. One such mechanism involves a membrane-anchored protein known as a homologous restriction factor (HRF), which binds C8. These devices serve to reduce damage to autologous tissues by complement complexes. Under certain conditions, however, the MAC may attack the cells of the host, producing inflammation and tissue damage. This may occur in patients with autoimmune diseases, as when antigen–antibody complexes form at the motor endplates in myasthenia gravis (see Chapter 21). Complement activation may also cause tissue injury when immune complexes get lodged in the walls of blood vessels or in tissues, as in bacterial endocarditis or systemic lupus erythematosus. In addition, excessive activa-

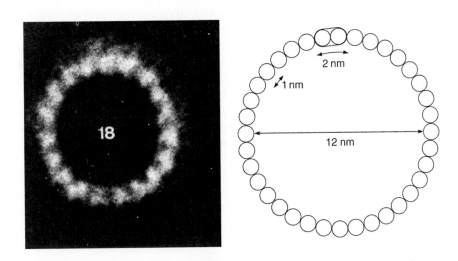

FIGURE 23.2 THE poly-C9 PORE
(A) Electron micrograph of a poly-C9 pore formed from 18 monomers, viewed from on top. (B) Geometry of the poly-C9 pore shown in A. Each C9 monomer is composed two α-helices. Thirty-six α-helices (two per monomer) can be arranged around a 12 nm diameter pore. From Peitsch and Tshcopp (1991).

TABLE 23.1   CHROMOSOMAL LOCATION OF COMPLEMENT GENES
INVOLVED IN THE MAC COMPLEX

Gene	Protein	Chromosome location (human)	Reference
C5B	C5b	9q32-34	Wetsel *et al.* (1998)
C6	C6	5p	Jeramiah *et al.* (1990); Setien *et al.* (1993)
C7	C7	5p	Jeramiah *et al.* (1990); Setien *et al.* (1993)
C8A	C8α	1	Kaufmann *et al.* (1989)
C8B	C8β	1	Kaufmann *et al.* (1989)
C8G	C8γ	9q	Kaufmann *et al.* (1989)
C9	C9	5p13	Abbott *et al.* (1989); Setien *et al.* (1993)

tion of the complement system by overwhelming bacterial infections can cause host cell lysis. Patients suffering from paroxysmal nocturnal haemoglobinuria type III are deficient in the HRF protein which removes the MAC from erythrocyte membranes (Hänsch *et al.*, 1987). These patients therefore suffer from an abnormal sensitivity to intrinsic complement activation.

Complement is an important defence against infection by bacteria and microorganisms such as *Mycoplasma*, trypanosomes, and fungi. This is nicely illustrated by the fact that the pathogenicity of many bacterial strains is related to their ability to resist complement-mediated destruction (Chapel, 1996). Patients in which any of the MAC components are lacking, or defective, are prone to recurrent infections by *Neisseria meningitidis* and *N. gonorrhoea*. Among the black population of South Africa, C6 deficiency is common. It has been suggested that the high prevalence of the deficiency may be associated with protection against the high infant mortality that follows gastroenteritis and endotoxic shock, because individuals with C6 deficiency do not succumb to endotoxic shock. C7 and C8 deficiency occur in Israeli and the Middle Eastern populations, where they are associated with recurrent, nonfatal meningitis from neisserial infections. C9 deficiency is relatively common among the Japanese (1 in 1000). Although the initial patients identified were asymptomatic, subsequent studies have shown that C9 deficiency also confers increased susceptibility to systemic meningococcal infection by *N. meningitidis*.

The chromosomal location of the different complement components involved in cell lysis in man is given in Table 23.1. The C6, C7, and C9 gene loci are closely linked on chromosome 5. The eighth component of complement is composed of three subunits encoded by different genes: two of these lie close together on chromosome 1, but the third is located on chromosome 9q. The mutations in these genes that are associated with the hereditary deficiency diseases are in the process of being identified.

Perforins

Perforin is a channel-forming protein secreted by cytotoxic T lymphocytes and natural killer cells that mediates the destruction of virally infected host cells and tumour cells (Kägi *et al.*, 1996). Within the T cell, it is contained in cytoplasmic vesicles (called granules). On T-cell activation these granules undergo exocytosis and release their contents close to the site of contact between the lymphocyte and its target cell. Unlike C9, perforin does not require a protein receptor to bind to the target membrane. Instead, the higher Ca^{2+} concentration it encounters when it contacts the extracellular solution induces a conformational change that enables perforin to insert directly into the phospholipid membrane. It then undergoes oligomerization to form a large polyperforin pore composed of 12–18 perforin monomers (Fig. 23.3). This results in the swelling and subsequent lysis of the target cell.

The human perforin gene is located on chromosome 10q22 (Fink *et al.*, 1992) and encodes a glycoprotein of 534 amino acids, with a molecular mass similar to that of the MAC complement protein C9. The N-terminal part of the perforin molecule is also highly homologous to that of C9. The first 34 amino acids of the N terminus form the cytolytic domain, and within this region 19 amino acids, arranged in an amphipathic β-sheet, are sufficient to form the ion channel pore (Persechini *et al.*, 1992). An additional α-helical domain within the central region of the perforin molecule (residues 189–218) is required for optimal lytic activity and may be involved in the subsequent self-assembly of the perforin molecule. A novel method has been used to record the activity of single perforin channels. This method, known as fluorescence microphotolysis, allows the researcher to resolve the flux of a fluorescently labeled macromolecule through a single perforin pore (Peters *et al.*, 1990). Using this method, the perforin pore was calculated to have a diameter of ~5 nm. A similar pore diameter was calculated from conventional single-channel recordings of perforin pores in artificial lipid bilayers (Bashford *et al.*, 1988).

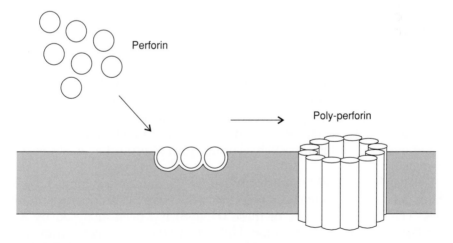

FIGURE 23.3 THE ACTION OF PERFORIN
In the presence of Ca^{2+}, perforin molecules bind to the target cell membrane and subsequently polymerize to form a polyperforin pore.

Perforin-deficient mice are unable to clear infection with lymphocytic choriomeningitis virus and their CD8[+] T cells show little or no antiviral activity (Kägi *et al.*, 1994). In addition, perforin-deficient mice fail to reject transplanted fibroblasts, suggesting that perforin-induced cell lysis may also be important in graft rejection. Cytotoxic T cells appear to be equipped with a mechanism that prevents the lysis of their own membranes by the perforins that they secrete, but the identity of the proteins involved is not known.

Defensins

The defensins are another family of channel-forming antimicrobial peptides which play an important role in the mammalian immune response by causing lysis of target cells (Kagan *et al.*, 1994; Boman, 1995). They are found in the granules of phagocytic blood cells, such as neutrophils and macrophages, in the Paneth cells of the small intestine and in the epithelia of the tongue, nasal passages and conducting airways. Epithelial cells of the intestine and airways secrete defensins directly onto their target cells. By contrast, neutrophils and macrophages do not secrete defensins into the extracellular solution, but into endocytotosed vesicles containing ingested microorganisms.

Defensins are effective against a variety of bacteria, mycobacteria, fungi and enveloped viruses. Their importance in innate immunity is exemplified by the fact that lack of neutrophil defensins gives rise to two disorders characterised by recurrent infections: Chediak–Higashi syndrome and specific granule deficiency (Ganz *et al.*, 1988). The antimicrobial activity of defensins is optimal at low ionic strength and is inactivated by high salt concentrations, a factor which may contribute to the recurrent bacterial infections characteristic of the airways of cystic fibrosis patients (see Chapter 12). In the airways, expression of defensins is upregulated after birth and after exposure to bacterial lipopolysaccharides.

The defensins form voltage-gated weakly anion-selective channels in artificial lipid membranes. A wide variety of single-channel current amplitudes have been observed and, as these tend to increase with time, it is believed that defensin molecules may aggregate to produce pores of larger diameter. Channel opening requires a negative potential (~ -70 mV) on the *trans* side of the membrane (the opposite side of bilayer to that which the defensin is added). This explains why a target cell is only killed when it is metabolically active and thus has a significant resting membrane potential and why depolarising metabolic inhibitors such as dinitrophenol protect against the action of defensins.

Defensins are small cationic peptides, 29–35 amino acids long. Six different human defensins have been identified which are the products of multiple genes. The structure of one of these (HNP-3) has been analysed by X-ray crystallography at 1.9 Å resolution (Hill *et al.*, 1991). It has overall dimensions of $26 \times 15 \times 15$ Å. In contrast to many other lytic peptides, HNP-3 has no α-helices. Instead it is composed of a three-stranded antiparallel β-sheet which is stabilised by three intramolecular disulphide bonds. The mechanism of channel formation is not clear, but since the disulphide bridges make the molecule fairly rigid, it seems unlikely that it will undergo a marked change on membrane insertion. HNP-3 cystallizes as a dimer, which resembles a basket with an apolar base and a polar top that includes the two amino and two carboxy termini. It therefore seems likely that such dimers are involved in pore formation.

A number of related defensins are found in other vertebrates and in plants. Lytic peptides (crecropins) also constitute part of the immune system of the silkmoth.

Magainins

The magainins are two closely related peptides secreted by the skin of the African clawed toad, *Xenopus laevis* (Boman, 1995). Each consists of 23 amino acids and forms weakly anion-selective ion channels with multiple conductance levels in lipid bilayers. The magainins have broad-spectrum antimicrobial activity: they act on both gram-negative and gram-positive bacteria, on fungi and on protozoa. They were first isolated because of the absence of infection observed following *Xenopus* surgery, despite nonsterile conditions. Derivatives of the magainins with improved potency and antimicrobial spectra are currently in clinical trials as putative topical antibiotics for skin and eye infections.

ION CHANNELS OF BACTERIA, FUNGI, AND PROTOZOANS

Bacterial Ion Channels

Bacteria indulge in incessant chemical warfare. Many secrete water-soluble proteins that kill other bacteria by virtue of their ability to form pores in the membranes of their target cells. This induces lysis and cell death. Some of these bacterial toxins have adverse effects in man, while others are used as antibiotics and antifungal agents. For the researcher, a particular advantage of such channel-forming proteins is that they are relatively small and often soluble. This means that their structure can be studied at high resolution by techniques such as X-ray crystallography and nuclear magnetic resonance. Their small size also means that it is possible to carry out scanning mutagenesis, systematically changing each residue in turn to examine its effect on channel function. Clearly, this is impossible with channels as large as the voltage-gated Na^+ channel, which is ~2000 amino acids long.

Antibiotics

Gramicidin
Gramicidin is an antibiotic secreted by the gram-negative bacterium *Bacillus brevis*. It is a short peptide, only 15 amino acids long, in which D- and L-

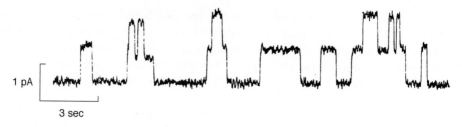

FIGURE 23.4 GRAMICIDIN
Single-channel currents recorded from an artificial lipid bilayer to which
gramicidin has been added. From Bamberg and Lauger (1974).

amino acids alternate (Fig. 23.4). All the residues are hydrophobic and this,
together with the addition of an ethanolamine to the C terminus, makes the
molecule highly nonpolar and thus insoluble in water. The peptide forms a
single-stranded, right-handed β-helix that spirals around a central pore. This
structure is stabilised by hydrogen bonding. Each gramicidin molecule acts
as a hemichannel and two molecules, one from each side of the membrane,
meet head to head to form the channel. The pore is 25 Å long and has a
diameter of 4 Å.

Gramicidin has the distinction of being the first ion channel from which
single-channel currents were recorded. In 1970, Hladky and Haydon observed
that following the addition of a small quantity of the antibiotic to the solution
on one side of an artificial lipid bilayer, the membrane current fluctuated in
regular square steps (Fig. 23.4). These current steps correspond to the opening
and closing of single gramicidin channels. The single-channel conductance
is ~30 pS in 100 mM RbCl (Neher *et al.*, 1978). Gramicidin channels are
selectively permeable to monovalent cations, with a selectivity sequence $H^+ >
Cs^+ \sim Rb^+ > NH_4^+ > Tl^+ > Li^+$ when the membrane is exposed to symmetrical
solutions. This is similar to the relative mobility of these ions in free solution
and indicates that the channel behaves as a water-filled pore. Because of the
simplicity of the molecule and the relative ease with which it can be studied,
an impressive amount of information is available concerning the properties
of the gramicidin channel. The peptide has also been used as a test bed for the
development of many of our ideas about the way in which ion channels work.

Gram-negative bacteria are surrounded by two membranes, an outer
envelope and an inner plasma membrane, which are separated by a periplasmic space. The outer membrane acts as an additional permeability barrier
and tends to exclude lipophilic substances, such as many of the naturally
occurring antibiotics. These antibiotics are effective, however, on gram-
positive bacteria which lack the outer envelope. Gramicidin selectively targets
gram-positive bacteria. It is used clinically, in conjunction with other antibiotics, to combat bacterial infections of the eye and ear.

Colicins

The colicins are a family of pore-forming toxins of ~600 residues produced by *E. coli* and related bacteria (Sansom, 1994). They kill gram-negative

bacteria. Three stages are involved in this process. The colicin molecule first binds to the external envelope of the target bacterium. It is then translocated across the periplasmic space to the inner bacterial membrane. Finally, it inserts into the inner membrane, forming an ion channel that produces membrane depolarization and cell lysis. Each of these stages is governed by a specific domain of the channel protein. In many colicins (e.g., A, E, Ia), the N terminal (T) domain is responsible for translocation across the outer bacterial membrane, the central R domain contains a receptor-binding site, and the ion channel is formed from the C terminus of the molecule (the C domain). Although colicins are bactericidal, their antibiotic activity is weak and they are not used clinically.

The structure of colicin Ia has been resolved at 3 Å resolution by X-ray crystallography (Wiener et al., 1997; Fig. 23.5, see color plate). The molecule is composed of three domains (T,R,C) which are connected by two extremely long α-helices. In the crystal structure, the T and C domains pack close to each other with the R domain at the tip of a 160-Å long hairpin loop. Resolution of the structure of colicin Ia suggests a model for how the different stages of colicin action are achieved (Fig. 23.6). The first step involves the binding of the R domain of colicin Ia to a protein (Cir) in the outer bacterial membrane. The Cir protein is normally anchored to the TonB protein which spans the periplasmic space. The T domain of the colicin molecule disrupts this interaction and binds to the TonB protein in place of Cir. This is followed by the translocation of the C domain across the periplasmic space to the inner bacterial membrane: the length of the α-helix connecting the R and C domains (160 Å) is just sufficient to span the periplasmic gap (150 Å). Once it reaches the inner membrane, the C domain undergoes a series of conformational

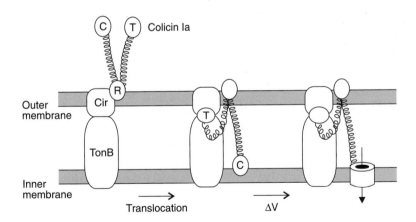

FIGURE 23.6 MODEL FOR COLICIN PORE FORMATION

Binding of the R domain of colicin to Cir is followed interaction of the T domain with the TonB protein and translocation of the C domain across the membrane. In the presence of a membrane potential (V), the helices of the C domain repack to form an ion channel, which precipitates cell lysis. Modified from Sansom (1997).

changes which result in channel formation. This involves reorientation of the 10 helices that are packed together in the C domain. A central helix hairpin (C8-9), which is highly hydrophobic, inserts spontaneously into the membrane. This is followed by a voltage-dependent insertion of other helices. The precise structure of the colicin channel is still unclear, but it is thought to be composed of two helix hairpins (C8-9 and C6-7) that insert into the membrane and form the walls of the pore.

The bacterium *Bacillus thuringiensis* secretes δ-endotoxin. Like the colicins, δ-endotoxin has a three-domain structure. In this case, however, the N terminus of the protein participates in channel formation while the central domain is involved in receptor binding. The toxin is used as an insecticide: it binds to receptors in the wall of the insect gut, lysing the cells and causing diarrhoea and death by dehydration. The bacterium itself is also used as a biological control agent to limit insect populations—for example, in greenhouses.

Antifungal agents

Amphotericin, from the bacterium *Streptomycetes nodosus* and nystatin, from *Streptomyces nouresi* and other *Streptomyces* species, are known as polyene antibiotics. They only form ion channels in membranes that contain sterols, such as cholesterol. Thus they induce pores in the membrane of fungi and other eukaryotic cells, but are ineffective against bacteria, whose membranes are sterol free. They also appear to have a higher affinity for the ergosterol-containing membranes of fungi than the cholesterol-rich membranes of mammalian cells. Hence they are used as antifungal drugs.

Amphotericin forms cation-selective channels. Surprisingly, nystatin forms cation-selective channels when it is added to one side of the membrane only, but anion-selective channels when it is added to both sides of the membrane. Amphotericin and nystatin have been used as tools to enable the study of whole-cell currents in intact cells using the patch-clamp method (Rae *et al.*, 1991). Inclusion of the antibiotic in the pipette solution increases the ionic permeability of the patch of membrane under the electrode, providing electrical access to the cell interior and enabling it to be voltage clamped. The pores formed by the antibiotic are too small, however, to allow the passage of molecules larger than ~500 kDa. Thus metabolic and second messenger systems in the cell remain intact.

Both amphotericin and nystatin are in routine clinical use. Amphotericin B has been used since the 1960s to treat serious systemic fungal infections. Indeed, because of the recent rise in the incidence of immune-compromised individuals (due to AIDS and the use of immunosupressants to treat transplant patients), it is becoming increasingly more important. It is active against most fungi and yeast, and is also used to treat leishmaniasis (unusually, the membranes of this protozoan parasite contain ergosterol precursors). Nystatin is principally used for the treatment of *Candida albicans* (thrush) and for infections of the skin, mucous membranes or intestinal tract. Both antifungal agents suffer from the fact that they are also effective against host tissues.

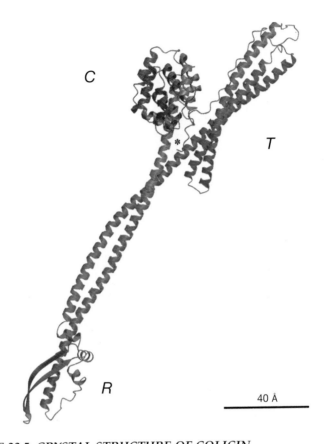

FIGURE 23.5 CRYSTAL STRUCTURE OF COLICIN
Ribbon representation of the structure of colicin 1A with the receptor-binding (R), translocation (T) and channel-forming (C) domains indicated. From Weiner *et al.* (1997).

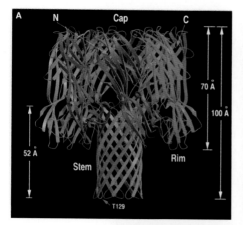
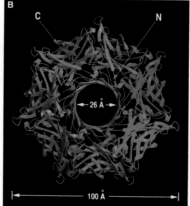
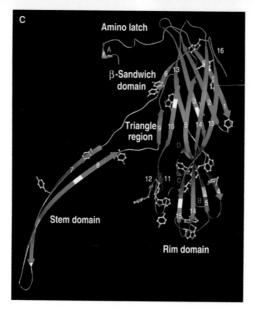

FIGURE 23.9 CRYSTAL STRUCTURE OF α-TOXIN (HAEMOLYSIN)
Ribbon representation of the heptameric α-toxin channel, with each protomer in a different colour. (A) Viewed parallel to the plane of the membrane. Approximate locations of the cap, rim, and stem are indicated. (B) Viewed from on top. The N and C terminus of a single protomer are indicated. (C) Isolated protomer viewed from the "outside" of the heptameric channel. The stem, rim, and amino latch domains are indicated. From Song *et al.* (1996).

For example, amphotericin B can cause decreased renal function, anaphylaxis, fever and nausea. Toxicity can be reduced, however, by administration of amphotericin in a lipid formulation (Hartsel and Bolard, 1996).

Bacterial toxins

Many bacteria secrete a cocktail of soluble toxins that have deleterious effects on eukaryotic cells. Some of these toxins act as ion channels and cause cell lysis; their relative contribution to the pathogenicity of the bacterium is, however, variable.

Many lytic toxins are secreted as inactive water-soluble pro-toxins (Fig. 23.7). This prevents damage to the bacterium itself during transit to the target cell, which is especially important in the case of gram-negative species, as the toxin has to pass across both inner and outer membranes. The toxin binds to specific receptors on the target cell surface and the presence or absence of such receptors is responsible for the cell specificity of toxin action. Activation of the toxin then takes place, usually by proteolytic removal of part of the molecule. This enables the individual toxin monomers to associate into an oligomeric complex. Some toxins, such as those of the gram-positive bacterium *Staphylococcus,* do not require proteolytic activation but associate directly with other monomers. Oligomerization is accompanied by a series of conformational changes which enable the complex to insert into the bilayer. Once inserted, the complex forms a large hole (1–12 nm in diameter). The leakage of cellular constituents through this pore leads to osmotic swelling, lysis and

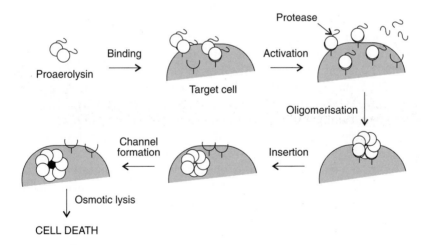

FIGURE 23.7 *MODEL FOR AEROLYSIN ACTION*
Proaerolysin is secreted by *Aeromonas hydrophila* as an inactive water-soluble dimer. This binds to a receptor on the surface of the target cell and subsequent proteolysis converts proaerolysin to the active form (aerolysin) by removal of part of the C terminus of the molecule. This is followed by oligomerization with other aerolysin monomers, insertion into the bilayer, and formation of a heptameric channel. After Buckley (1992).

the ultimate death of the cell. A challenging question is how a water-soluble molecule can reorientate to form an amphiphilic molecule with a hydrophobic domain capable of inserting into the membrane.

The large size of the pores produced by the bacterial lytic toxins has been exploited by research workers to permeabilise the plasma membrane and so introduce large molecules into cells (Lindau and Gomperts, 1991). Toxins such as α-toxin or streptolysin-O are excellent tools for cell permeabilization because they are stable, soluble in water and attach exclusively to the plasma membrane (they do not reach intracellular organelles). The cutoff size of the pore (~4 kDa) means that large macromolecules such as enzymes are retained within the cell while the cytosolic concentration of small regulatory molecules (ATP, cyclic AMP, etc.) can be readily manipulated by addition to the bath solution. Many studies on the mechanism of secretion have utilised this method.

Another class of bacterial toxins, to which diphtheria toxin, tetanus toxin, botulinum toxin and anthrax toxin belong, act on intracellular targets. A common feature of these toxins is that they consist of an enzymatically active polypeptide, which is translocated across the cell membrane into the cytosol where it inhibits key cellular enzymes. This translocation is associated with the induction of an ion channel, but as the pathogenic role of these ion channels is controversial, they are not discussed here. In the next section, we consider some of the more scientifically interesting and medically important bacterial toxins that form lytic pores.

Staphylococcus aureus

Staphylococci are gram-positive bacteria that cause skin infections, such as abscesses, or wound infections following surgery or trauma. On occasion, however, they may produce life-threatening septicaemia. They also secrete enterotoxins which can cause food poisoning when ingested, or give rise to toxic-shock syndrome. A great number of toxins, with diverse effects, are produced by S. aureus. Their pathogenicity was first recognised following a tragedy in the Australian town of Bundaberg in 1928 when 21 children who were vaccinated with diphtheria vaccine infected with S. aureus became severely ill. Twelve of them subsequently died. The most potent of the lytic toxins secreted by S. aureus is α-toxin (also known as α-haemolysin). It causes lysis of lymphocytes, platelets and endothelial cells (Bhakdi and Tranum-Jensen, 1991). The concentration required to produce lysis varies from 1 nM for rabbit erythrocytes to more than 1 μM for human platelets. The ability of α-toxin to form lytic pores is reduced considerably in the absence of cholesterol, which helps explain why it does not damage bacteria (whose membranes lack cholesterol).

Alpha-toxin is secreted as a hydrophilic soluble monomer of M_r, 34 kDa (Fig. 23.8). This binds to the membrane of the target cell and associates with other monomers to form a heptameric channel that inserts into the bilayer. The channel is slightly anion selective and has a single-channel conductance of ~90 pS. It permits the passage of molecules with a molecular mass of up to 2–3000 Da.

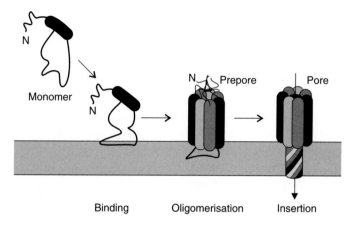

FIGURE 23.8 α-TOXIN FORMS LYTIC PORES IN MEMBRANES
Model of α-toxin action. α-toxin is secreted by *S. aureus* as a water-soluble
monomer. It binds to the membrane of the target cell and coassembles with
six other monomers to form a heptameric pore. After Engleman (1996).

The structure of the α-toxin channel has been determined to 1.9 Å resolution (Song *et al.*, 1996). It is mushroom shaped, with the stem spanning the
membrane and the cap resting on its outer surface (Fig. 23.9, see color plate),
and it measures 100 Å in height and up to 100 Å in diameter. Domains
protruding from the underside rim of the cap probably help anchor the
channel within the lipid bilayer. The channel is built up of seven monomers,
which, when they are incorporated into the complex, are known as protomers.
These are assembled around a central fluid-filled pore that varies from 14 to
46 Å in diameter and spans the length of the complex.

The stem is 52 Å long and is divided into an upper region, which pene-
trates the cap, and a lower region (28 Å long), which spans the membrane
(Fig. 23.9). The outer surface of the lower stem is hydrophobic and believed
to interact with the nonpolar portion of the lipid bilayer. There is a ring of
charged and polar residues at the very base of the stem, which presumably
contacts the cytoplasm of the target cell and helps to anchor the channel
within the bilayer. The pore is a right-handed β-barrel, composed of 14
antiparallel β-strands. Each protomer contributes a hairpin loop, formed from
two β-strands, which dips into the membrane and wraps around the barrel
by ~180°. The pore is narrowest at the junction of the stem with the cap (the
neck) and expands into a wider vestibule in the cap region (Fig. 23.9). Within
the stem, the pore diameter varies between 14 and 24 Å, depending on the
volume of the side chains that protrude into the lumen. Both ends of the
stem are defined by rings of positively charged and negatively charged amino
acids and the intervening region is lined by neutral residues. This arrangement of rings of charge separated by nonpolar residues is reminiscent of that
proposed for the acetylcholine receptor (see Chapter 15).

Studies have shown that the secondary structure of the α-toxin monomer
does not change markedly on oligomerization. However, two regions of the

protein, known as the amino latch (residues 1–20) and the glycine-rich region (residues 110–148), become buried within the molecule and are thus no longer susceptible to proteolytic cleavage. The glycine-rich region ends up lining the channel pore, while the amino latch is involved in linking adjacent protomers together. The amino latch is critical for heptamer formation and cell lysis, because mutations within this domain abolish lytic activity. A key question is how the hydrophobic residues which line the outer surface of the stem β-strands and facilitate insertion into the membrane, are protected in the water-soluble form of the monomer. One possibility is that they are screened by packing against a nonpolar part of the molecule. This would also prevent assembly of the heptamer until interaction with the membrane triggers β-strand formation.

Aeromonas

Aeromonads are gram-negative bacteria that are found in water, soil, and cold-blooded animals. Three species cause diarrhoea in man: *Aeromonas hydrophila*, *A. sobria* and *A. caviae*. They are also associated with deep wound infections. Such infections are frequent in tropical and subtropical regions, but are less common in temperate climates. It is believed that contaminated water is the main route of infection.

The major cause of the pathogenicity of *A. hydrophila* is the pore-forming toxin aerolysin produced by the bacterium (Fig. 23.7). It is secreted as a 470 amino acid pro-toxin and is subsequently activated by proteolytic removal of approximately 40 amino acids from the C terminus (van der Goot *et al.*, 1994). Activation is carried out both by the bacterium itself and by a variety of mammalian proteases. Both proaerolysin, and aerolysin itself, exist in solution as a dimer and bind with high affinity to glycophorin molecules in erythrocyte membranes. The high density of this glycoprotein in red cell membranes means that the toxin is concentrated on the surface of the target cell. The receptor on the cells of the digestive tract, where aerolysin exerts its major pathogenic effect, is unknown. It is believed that binding normally precedes proteolytic activation. Once activated, the aerolysin dissociates into monomers, which then self-assemble to form a heptameric complex. Insertion of this heptamer into the bilayer produces a large pore. In artificial lipid bilayers these pores are slightly anion selective, with a single channel conductance of 420 pS (van der Goot *et al.*, 1994) and they remain open between -70 and $+70$ mV. Pore formation makes the target cell permeable to molecules as large as 3000 Da and results in osmotic lysis and cell death.

In solution, proaerolysin forms crystals whose structure has been analysed at 2.6-Å resolution (Parker *et al.*, 1994). Electron microscopy of two-dimensional crystals in lipid bilayers has also been achieved (Wilmsen *et al.*, 1992). These studies suggest that the pore structure resembles that of the α-toxin of *S. aureus*, described earlier.

Pseudomonas

A great many other bacteria also secrete channel-forming molecules. Among these, the gram-negative bacterium *Pseudomonas aeruginosa* has

achieved notoriety as it is the cause of 'hospital' infection in immunosup-
pressed or debilitated patients. It is also a major cause of death in patients
with cystic fibrosis (Chapter 12). Unfortunately, *P. aeruginosa* is naturally
resistant to most antimicrobials and antiseptics. *Pseudomonas* produces a
29 kDa pore-forming toxin that causes the formation of pores with 2 nm
diameter. The toxin is secreted as an inactive precursor of 286 amino acids and
is activated by proteolytic cleavage of the last 20 amino acids. Subsequently it
binds to a receptor in the target cell: in erythrocyte membranes this has been
identified as the water channel AQP1 (see Chapter 19). Residues 197–202
play an important role in this binding (Struckmeier *et al.*, 1995). Following
binding, the protein adopts a conformation that facilitates insertion into the
membrane and self-assembly to form a channel.

Fungal Ion Channels

Alamethicin

Alamethicin is produced by the fungus *Trichoderma viridie*. It belongs to
a family of peptides that are known as peptaibols because they have a high
content of the unusual amino acid α-aminoisobutyric acid (Aib) and an α-
amino alcohol (ol) at the C terminus (see Sansom, 1993). Alamethicin is 20
amino acids long and contains eight α-aminoisobutyric acid residues. The C-
terminal residue is phenylalaninol, a derivative of phenylalanine, and the N-
terminal residue (α-aminoisobutyric acid) is acetylated. The high-resolution
crystal structure of the molecule has been determined and reveals that it is
predominantly a single α-helix with a kink induced by the proline residue
at position 14 (Fig. 23.10) (Fox and Richards, 1982). This kink may act as a
molecular hinge, enabling the N- and C-terminal segments to move relative
to one another.

When alamethicin is incorporated into planar lipid bilayers, single-
channel currents are observed. A notable feature of these currents is that the
amplitude of the current steps is not constant so that the single-channel
conductance varies (Fig. 23.11). This suggests that the alamethicin channel
is composed of a variable number of monomers and that increasing the
number of alamethicin molecules which contribute to the pore enlarges its
radius, so increasing its conductance. The alamethicin channel is therefore
conceived of as a central pore surrounded by a variable number of α-helices
(probably from 4 to 11) (Fig. 23.10B).

Alamethicin channels are selectively permeable to monovalent cations
with the selectivity sequence $K^+ > Rb^+ > Cs^+ > Na^+ > Li^+$. The origin of
this cation selectivity is not clear as the channel remains cation selective even
when the only negatively charged residue (a glutamate at position 18) is
neutralised (Hall *et al.*, 1984). As the alamethicin pores grow larger, indicated
by an increase in the size of the single-channel current steps, the cation
selectivity gets weaker. This is expected, as ions will pass through the pore
in a more hydrated state, and interact with its walls less, when the pore

A   B

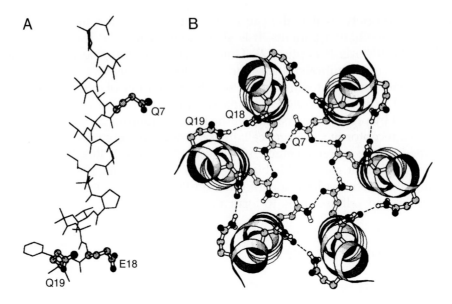

FIGURE 23.10 ALAMETHICIN
(A) Structure of a single alamethicin molecule. The molecule has a kink in its length due to the presence of a proline residue at position 14. Hydrophilic side chains are shown in 'ball and stick' format. (B) Alamethicin channel formed from six monomers viewed from the N-terminal mouth. From Sansom (1993).

diameter is larger. Alamethicin currents are activated by voltage and bursts of channel openings occur more frequently at *cis*-positive potentials (the *cis* side of the bilayer is the one to which alamethicin is added). The mechanism of this voltage-gated activation remains controversial. There are two favoured ideas (see Sansom, 1993). The first model (Fig. 23.12A) proposes that in the absence of a membrane potential the N-terminal segments of the alamethicin

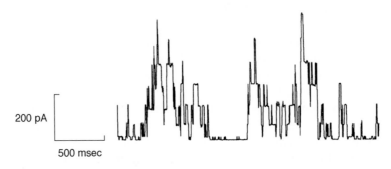

200 pA

500 msec

FIGURE 23.11 ALAMETHICIN SINGLE-CHANNEL CURRENTS
Single-channel currents recorded from a lipid bilayer doped with alamethicin. Individual events of different amplitudes can be resolved. From Sansom (1991).

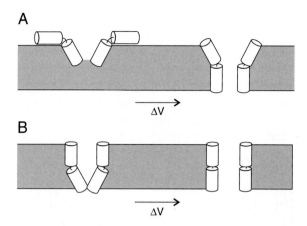

FIGURE 23.12 TWO MODELS FOR VOLTAGE-DEPENDENT
GATING OF ALAMETHICIN
In response to an applied voltage the channel is suggested to undergo a
conformational change which either (A) allows it to span the bilayer and
so form a channel or (B) reduces the kink angle and so opens the pore.

monomers may adopt an orientation that is either perpendicular or parallel
to the bilayer. Imposition of a voltage field causes all the monomers to reorient
in the perpendicular direction, thereby inserting into the bilayer, where they
self-assemble to form a channel. An alternative idea (Fig. 23.12B) proposes
that all the alamethicin molecules are fully inserted into the bilayer in the
absence of the voltage field, but that the molecules are bent at the proline
hinge so that the pore is blocked. Application of a voltage field across the
membrane straightens out this kink and opens the pore.

Alamethicin itself has only weak antibacterial action. However, a number
of alamethicin analogues—both natural and synthetic—are used as antibiotics.

Protozoan Ion Channels

Intracellular parasites

Intracellular protozoan parasites are phagocytosed by the mammalian
host cell and confined within acidified vacuoles known as phagosomes. They
then either multiply within the phagosome or escape into the cytoplasm,
where they proliferate. In either case, they must ultimately disrupt the vacuolar membrane in order to leave the phagosome. There is accumulating evidence that protozoan parasites secrete pore-forming proteins which insert
into the vacuolar membrane, causing osmotic lysis of the organelle and release
of the parasite. In this section, we consider one example of such a protozoan
pore, that secreted by *Trypanosoma cruzi.*

Trypanosoma cruzi is the causative agent of Chagas' disease, which affects around 20 million people in Central and South America. Infected individuals may develop cardiomyopathy, encephalitis, megacolon and megaoesophagus up to several years after the original infection. It is transmitted within and between human and animal populations by blood-sucking triatomine bugs such as the common bedbug, *Rhodnius prolixus*. Transmission is by contamination with bedbug faeces containing infective trypanosomes or by transfusion with infected blood. In the acute phase of infection the parasite is found in the blood, but it subsequently leaves the circulation and persists in the tissues, particularly in heart, skeletal and smooth muscles and in connective tissue.

The intracellular form of *T. cruzi* (known as an amastigote) secretes a 60- to 75-kDa protein, which, like perforin, cross-reacts with antibodies to C9, the pore-forming subunit of complement (Andrews *et al.*, 1990). At low pH, this protein has lytic activity. Furthermore, when fractions enriched in lytic activity were incorporated into lipid bilayers, single-channel currents were observed. Like the channels formed by single C9 molecules, the predicted pore diameter was too small to cause lysis, and argues that the lytic pore is formed by oligomerization. The predicted diameter of the lytic pore is ~10 nm. These observations support the view that the *T. cruzi* protein may form a pore similar to that of C9 and perforin. This pore will be activated by the low pH of the phagosome within which the parasite is confined, and thus lead to lysis of the vacuole and parasite release.

One wonders how these parasites escape damage to their own membranes by the haemolytic pores they secrete: perhaps they have a molecule analogous to that of mammalian cells which removes the lytic complex from the membrane (see page 387).

Amoebic pores

Pore-forming proteins are also produced by extracellular protozoan parasites, such as the amoeba *Entamoeba histolytica* (Lynch *et al.*, 1982) and the amoebo-flagellate *Naegleria fowleri* (Young and Lowrey, 1989). These can cause extensive tissue damage and account for the pathogenicity of the organism.

About one-tenth of the world's population is infected with *E. histolytica*. In most European countries and in North America the incidence is less than 2% but it may reach 30% in tropical countries. *E. histolytica* is an obligate parasitic gut amoeba that lives mainly in the proximal colon and caecum where it feeds on bacteria in the gut lumen. This commensal form of the organism is harmless. Clinical problems arise, however, when the amoebae invade the colonic mucosa. On contact with the mucosal cells the amoeba secretes a lytic protein (known as amoebopore), which induces lysis of the host cells and causes tissue necrosis. This favours the further penetration of the amoeba into the tissue. What triggers the change from the commensal form to the pathogenic invasive form is unknown. Invasive intestinal amoebiasis produces variable symptoms that range from mild changes in bowel habit to severe dysentery. In the more serious cases, large ulcerative lesions may occur along the whole of the colon. This results in diarrhoea, associated with the loss of blood and intestinal fluids. Perforation of the colon may occur leading to peritonitis. *Entamoeba histolytica* infection can also manifest as hepatic amoebiasis, a condition in which the amoebae invade the liver, where they form abscesses.

The amoebopore secreted by *E. histolytica* is a ~13-kDa protein that forms weakly cation-selective channels with a single-channel conductance of ~1.6 nS (in 1 M KCl) in artificial membranes (Lynch *et al.*, 1982).

Naegleria fowleri is a free-living amoebo-flagellate that causes potentially fatal inflammation of the meninges—the membranes that surround the brain—and necrosis of brain tissue (especially the olfactory bulbs). The amoeba enters the brain via the olfactory epithelium and infection usually occurs as a result of swimming in infected water. The amoebae grow readily in the warm water of health spas and in the air-cooling units of air-conditioning systems but the incidence of infection is extremely rare. *Naegleria fowleri* secretes a 66 kDa protein which functions as a lytic ion channel (Young and Lowrey, 1989). It is cytotoxic to erythrocytes and various cell lines. In lipid bilayers, the single-channel conductance ranges from 150 to 400 pS.

VENOMS

A range of channel-forming molecules are found in venoms (Sansom, 1991). For example, melittin, bombolitins, and mastoparan are produced by the honeybee, the bumblebee, and certain wasps, respectively.

Melittin is the active component of honeybee venom (*Apis mellifera*). In unsensitized individuals, a single sting is painful but not dangerous unless there is obstruction of the airways (as may occur with stings on the tongue). The venom from as few as 30 stings, however, can cause death in children—although some individuals have survived more than 1000 stings. Multiple stings by European honeybees are rare, but the aggressive African honeybee (*A. mellifera scutellata*) may mount mass attacks, which can be fatal in man. In the United States, there are around 50 deaths a year from bee stings. Most of these deaths are not due to the poisonous effects of the venom itself, but rather to the fact that it rapidly triggers a fatal allergic reaction in sensitized individuals.

Melittin is a 26 amino acid peptide that forms ion channels in lipid membranes. Its toxic effects appear to result principally from the activation of mast cells, pore formation leading to a calcium influx which triggers degranulation. This degranulation releases large amounts of histamine and initiates a typical inflammatory response consisting of local vasodilation, swelling and weal formation. Severe envenoming can produce generalised vasodilation, hypotension and bronchoconstriction.

The crystal structure of the melittin molecule reveals that it consists of two α-helical segments, comprising residues 1–10 and 13–26, with a central kink induced by a proline at position 14. The N-terminal region is largely amphipathic and there is a cluster of charged and polar residues at the C terminus. The channel is a tetramer formed from the aggregation of four melittin monomers. It is believed that the N-terminal helix crosses the bilayer, while the highly charged C terminus prevents it from diffusing right across and anchors the protein at the lipid/water interface. Melittin shows a strong voltage-dependent activation, with currents being observed only at *cis*-positive potentials. The channels are anion selective, show multiple conductance levels, and have shorter lifetimes than those of alamethicin.

CHAPTER 24

A RAGBAG OF CHANNELS

A number of ion channels do not fit easily into any of the other chapters, yet do not merit a chapter of their own. Despite having no structural or functional relationship to one another, they are collected together here. This chapter therefore resembles the leftover scraps of material that made up my mother's ragbag: full of small pieces of information that are too interesting to throw away.

ATP-GATED ION CHANNELS

ATP is an important neurotransmitter both within the central nervous system (CNS) and at nerve-smooth muscle synapses (North and Barnard, 1997). The term purinergic receptor was coined to describe receptors gated by ATP, and purinergic transmission proposed by Geoffrey Burnstock in 1972. Two classes of purinergic receptors have been identified: P2X and P2Y. The latter belong to a family of G-protein–coupled receptors and are not discussed here. The P2X receptors, however, comprise a family of ligand-gated cation channels (permeable to both Na^+ and Ca^{2+}) which are opened by the binding of extracellular ATP. No mutations in P2X receptors have been associated with any human disease to date, but as this does not necessarily imply a lack of involvement, a short description of these channels is included here for completeness.

Structure

Seven different genes encoding P2X receptors have been cloned, termed $P2X_1$ to $P2X_7$ (North and Barnard, 1997). They encode proteins of between 400 and 600 amino acids which share 35–48% sequence identity. Hydrophobicity plots suggest that, like the Kir and ENaC channels described in Chapters 8 and 13, P2X receptors have only two transmembrane domains with intracellular amino and carboxy termini (Fig. 24.1). The N terminus is always very short (<30 amino acids), but the C terminus varies in length from ~25 residues in the $P2X_6$ receptor to as long as 240 residues in the $P2X_7$ receptor. The long extracellular loop (~270 residues) linking the two transmembrane domains contains between 2–6 potential glycosylation sites, and 10 conserved cysteine residues which may potentially stabilise the structure by disulphide bond formation.

The stoichiometry of the P2X receptors remains unknown, but it is likely that several subunits are required to form the channel. Since ATP-gated ion channels can be observed by expression of a single type of P2X receptor, homomeric channels are clearly viable (Valera *et al.*, 1994). Heteromerization is also possible because coexpression of $P2X_2$ and $P2X_3$ receptor mRNAs generates currents with properties differing from those of the parent molecules (Lewis *et al.*, 1995). There is evidence that both homomeric and heteromeric channels occur *in vivo*, because the properties of homomeric $P2X_1$ channels are very similar to those observed for ATP-gated currents in native smooth muscle cells (Evans *et al.*, 1995), whereas those of other tissues can only be reproduced by coexpression of two different P2X receptors (Lewis *et al.*, 1995).

$P2X_1$ has been localised to chromosome 17p13.3, $P2X_3$ to chromosome 11q12, and $P2X_2$ to chromosome 12q24.32. The location of the other P2X receptors has not yet been published.

Functional Roles

ATP is cosecreted with other transmitters from the terminals of nerves that synapse onto various types of smooth muscle, including those of the vas deferens, blood vessels and gut. It mediates the fast component of the excitatory postsynaptic potential in these tissues. ATP was first shown to be a transmitter in the CNS by Edwards and colleagues (1992), but effects of ATP on many other types of neurone, including those of the hippocampus, hypothalamus and dorsal root ganglia, have also been found. The pharmacology of these responses is consistent with the involvement of P2X receptors, although the subunit(s) that mediates these responses has in most cases not been identified. ATP also stimulates trigeminal nociceptive (pain-sensitive) neurones and plays a key role in the sensation of tooth pain, probably by activation of $P2X_3$ and heteromeric $P2X_2/P2X_3$ receptors (Cook *et al.*, 1997).

P2X receptors show different expression patterns and functional properties

In situ hybridization and immunocytochemistry have shown that the different P2X receptors show distinct expression patterns (North and Barnard, 1997). $P2X_1$ is widely expressed throughout the brain, and in sensory ganglion, smooth muscle and heart. By contrast, $P2X_3$ receptors appear to be confined to sensory nerves. Unlike other P2X receptors, $P2X_7$ is not found in neurones or muscle but is expressed in white blood cells and brain microglia.

The different P2X receptors vary in their sensitivity to ATP, with $P2X_7$ being the most sensitive followed by $P2X_1$ and $P2X_3$. They also differ in their rates of desensitization and in their sensitivity to the antagonists suranim and pyridoxal 5-phosphate-6-azophenyl-2',4'-disulphonic acid (PPADS). The $P2X_7$ receptor has some remarkable properties. Prolonged ATP application leads to the opening of very large lytic pores, which results in cell death. Deletion of the C terminus of the molecule prevents this effect and reveals that the $P2X_7$ receptor also acts as a ATP-gated cation-selective channel of small conductance (Surprenant et al., 1996). The mechanism of lytic pore formation by the $P2X_7$ receptor remains a mystery, although it is known to be potentiated by the reduction of extracellular divalent cations. $P2X_7$ receptors are found on many different types of blood cells, including macophages, mast cells and eosinophils. The large holes they form have been utilised experimentally to permeabilize these cell types and load them with membrane-impermeant substances in order, for example, to examine their effect on secretion (Lindau and Gomperts, 1991). More recently, lytic pore formation by ATP has also been described in brain microglial cells (Ferrai et al., 1996).

Correlating Structure and Function

Little is known to date about the relationship between the primary structure and the functional properties of P2X receptors. The binding site for extracellular ATP has not yet been identified and there is no obvious sequence homology with known ATP-binding proteins. A similar situation is found for the intracellular ATP-binding site of Kir channels (Chapter 8). The cysteine-scanning method (Chapter 2) has been used to identify the amino acid residues that line the channel pore (Rassendren et al., 1997). Figure 24.1 indicates residues which, when substituted by cysteine, are able to interact with extracellular thiol reagents. It is noteworthy that several of these residues lie with the putative second transmembrane domain. In this respect, the P2X receptor resembles Kir and ENaC channels (see Chapters 8 and 13), where TM2 also contributes to the pore. One residue (D349) was only accessible to the extracellular solution when the channel was opened by ATP and must therefore lie on the cytosolic side of the channel gate.

VANILLOID RECEPTORS

Capsaicin is a common ingredient of Indian, Indonesian and Mexican food. Tucked innocuously away inside prettily coloured peppers it explodes within the mouth like a volcano, creating a severe burning sensation (some people actually like it!). Attempts to douse the fire with water only succeed in spreading it further around the mouth and the initial pain is often followed by an outbreak of sweating. I imagine most people are familiar with this unfortunate scenario. The capsaicin content differs between different varieties

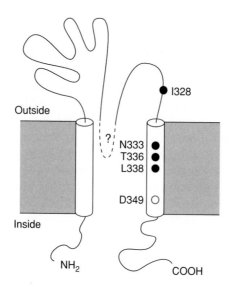

Outside

I328

? N333 ●
T336 ●
L338 ●

D349 ○

Inside

NH₂

COOH

FIGURE 24.1 PUTATIVE MEMBRANE TOPOLOGY OF THE P2X
RECEPTOR
Putative membrane topology of the P2X receptor, based on hydropathy
analysis. By analogy with ENaC channels it seems possible that a pore
loop is present, but as this is still only an hypothesis it has been indicated
by a dashed line. Residues that are believed to line the channel pore
are indicated.

of pepper. In 1912, Wilbur Scoville calibrated the potency of the spice by
measuring how much an extract of the pepper must be diluted until it was
only barely detectable when placed on the tongue. On the Scoville scale, the
relatively mild bell pepper has <1 heat units, the hotter jalapeno has 10^3, the
fiery habanero has 10^5 and pure capsaicin has a massive 10^7 heat units.

The target with which capsaicin interacts turns out to be a ligand-gated
ion channel and it was cloned by Caterina and colleagues from sensory
neurones (1997). Because the vanilloid moiety is an essential component of
capsaicin, they named their channel the vanilloid receptor (VR1). It is 838
amino acids long and has a predicted molecular mass of 95 kDa. Hydropathy
analysis suggests that VR1 has six putative transmembrane domains (TMs),
a pore loop between TMs 5 and 6, and intracellular amino and carboxy
termini (Fig. 24.2). The N terminus contains a proline-rich region followed
by three ankyrin repeats. The latter are domains which bind the structural
protein ankyrin and are presumably involved in localising and anchoring
VR1 at the nerve terminal. The vanilloid receptor shows greatest sequence
homology with a family of channels that are believed to be activated by
depletion of intracellular calcium stores (Montel and Rubin, 1989; Hardie
and Minke, 1993). These channels are known as *trp* channels because the first
member to be isolated caused a mutant phenotype in *Drosophila* known as
the transient receptor potential (*trp*). Unlike *trp* channels, however, VR1 does

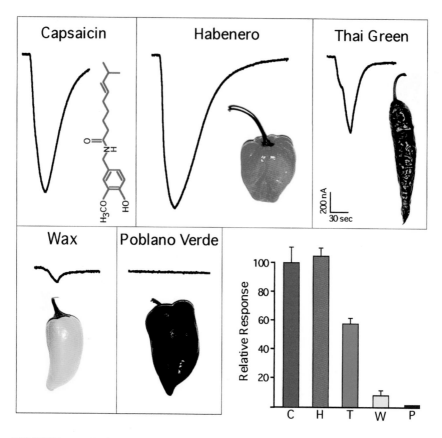

FIGURE 24.4 RESPONSES ELICITED BY DIFFERENT PEPPER VARIETIES
VR1 currents elicited by capsaicin and extracts derived from four different varieties of pepper. The potency of each pepper, relative to capsaicin, is plotted in the graph at the bottom right. The perceived hotness of each pepper in Scoville units are habanero (H) $1–3 \times 10^6$, Thai green (T) $5–10 \times 10^5$, wax (W) $5–10 \times 10^4$, and poblano verde (P) $1–1.5 \times 10^3$. Capsaicin (C) is rated as 1.6×10^6 units.

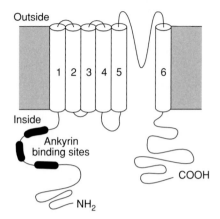

FIGURE 24.2
Putative membrane topology of VR1 based on hydropathy analysis.

not appear to be gated by store depletion. VR1 is expressed almost exclusively, and at high density, in trigeminal and dorsal root sensory ganglia. This matches the distribution of nociceptive (pain-sensitive) neurones and is consistent with a role for the receptor in pain sensation.

Expression studies have demonstrated that VR1 is a non-selective cation channel with a high calcium permeability (P_{Ca}/P_{Na} = ~10). Thus receptor activation not only causes neuronal excitation but also elevates intracellular calcium. This may explain why capsaicin acts as an excitotoxin and why continuous exposure to capsaicin for several hours kills nociceptive neurones *in vivo* and cells expressing VR1 *in vitro* (Fig. 24.3). People who regularly

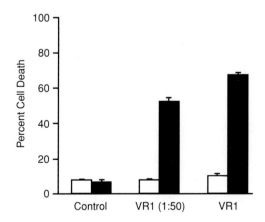

FIGURE 24.3 CAPSAICIN INDUCES THE DEATH OF CELLS EXPRESSING VR1
Percentage of dead cells observed 7 hr after addition of 3 μM capsaicin (black bars) or control solution (white bars) to control cells or cells expressing the vanilloid receptor VR1. From Caterina *et al.* (1997).

consume very spicy food become desensitized to the effects of capsaicin, as well as other noxious stimuli. The selective destruction of nociceptive neurons may be one reason for this desensitization. Alternatively, VR1 receptors might be downregulated in some way in response to prolonged capsaicin exposure. Whatever the reason, the desensitization of pain fibres by capsaicin is the basis for the use of the spice as an analgesic in arthritis and in viral and diabetic neuropathies (it is applied topically, as a cream).

VR1 is activated both by pure vanilloids, such as capsaicin, and by pepper extracts. The relative potency of the pepper extracts in gating VR1 correlates well with their perceived hotness on the Scoville scale (Fig. 24.4, see color plate). The steepness of the dose–response curve suggests that the binding of more than one capsaicin molecule is required for receptor activation. One possible explanation for this cooperativity is that the channel is multimeric, like other channels which have a six TM topology: by analogy with the K_V and CNG channels we may speculate that it is a tetramer (see Chapters 6 and 11). In addition to capsaicin, VR1 is activated by resiniferatoxin. This toxin is isolated from the spurge *Euphorbia resinifera* and is responsible for the powerful burning sensation and skin irritation induced by the milky sap of these plants. Ruthenium red and the synthetic antagonist capsazepine block capsaicin-induced responses. The location of the binding site for capsaicin is unknown, but because the spice is lipophilic, and therefore membrane permeant, it could act from either the outside or the inside of the membrane, or even within it. The lipophilicity of capsaicin is well known to anyone who has tried to quench the burning sensation it produces with water—all that happens is that the fire is spread around the mouth.

VR1 is not only activated by capsaicin. It also is gated by heat, which may in fact serve as the physiological stimulus. In HEK cells transfected with VR1, a rapid increase in temperature induces a large inward current, with properties resembling those of capsaicin-activated currents. A similar current is also induced in sensory neurones by noxious levels of heat (Cesare and McNaughton, 1996). The ability of high temperatures to activate VR1 may explain why capsaicin is perceived as hot. Whether an endogenous ligand for VR1 also exists is an interesting question.

CHANNELS INVOLVED IN PROGRAMMED CELL DEATH

Members of the Bcl-2 family are intracellular membrane-associated proteins that regulate apoptosis, otherwise known as programmed cell death. Some function as suppressors of apoptosis (e.g., Bcl-2 and Bcl-X$_L$), whereas others promote cell death (e.g., Bax).

Although they show no sequence homology, the three-dimensional structure of the anti-apoptotic protein Bcl-X$_L$ is strikingly similar to the pore-forming domains of the bacterial colicins and diphtheria toxins discussed in

Chapter 23 (Muchmore *et al.*, 1996). Like these toxins, Bcl-2 contains two core α-helices arranged in a hairpin structure surrounded by five to eight α-helices. Both Bcl-2 and Bcl-X$_L$ have been shown to have pore-forming activity in artificial planar lipid bilayers. The channels formed by Bcl-2 are cation selective. Like the colicins, Bcl-2 only forms channels at low pH and there is no detectable channel activity at neutral pH (Schendel *et al.*, 1997). In contrast, cation-selective Bcl-X$_L$ channels are observed at physiological pH (Minn *et al.*, 1997). What is the functional role of these proteins and how do they mediate their anti-apoptotic effects? Unfortunately, answers to these questions are still a matter for speculation. Bcl-2 and Bcl-X$_L$ possess a stretch of amino acids at their C termini that anchors the protein to intracellular membranes, including those of the outer mitochondrial membrane, the outer nuclear envelope and the endoplasmic reticulum. One suggestion, therefore, is that they influence metabolic processes by regulating ion permeability across these membranes.

Bax is a pro-apoptotic member of the Bcl-2 family. Its role in cell death is exemplified by the fact that neurones of Bax knockout mice are resistant to naturally occurring apoptosis and to deprivation of neurotrophic factors (Deckwerth *et al.*, 1996). Bax exhibits lytic properties and is able to lyse sheep erythrocytes and cultured sympathetic neurones within 3–6 hr. Presumably this is a consequence of its ability to form an ion channel in cell membranes, as demonstrated in artificial lipid bilayers by Antonsson and colleagues (1997). For reasons still not understood, Bcl-2 is able to prevent the lytic activity of Bax. This property may underlie the anti-apoptotic action of Bcl-2.

β-AMYLOID PEPTIDE

Alzheimer's disease (AD) is a devastating disease whose victims suffer from a progressive and insidious loss of memory. It generally, but not exclusively, occurs in the elderly and is becoming increasingly common in the Western world as more people survive into old age. Examination of the brain of AD patients reveals marked cortical atrophy, which particularly affects the temporal lobes and the hippocampus. This accounts for the symptoms of the disease. Histologically, the hallmark of AD is the presence of a large number of extracellular amyloid plaques and intraneuronal neurofibrillary tangles, which are particularly prominent in the neocortex and hippocampus. The major constituent of these plaques and tangles is β-amyloid, a 39–42 residue peptide derived by proteolytic cleavage of the much larger amyloid precursor protein (APP). The β-amyloid peptide tends to form insoluble fibrils because of its prominent β-pleated sheet structure. The APP gene is located on chromosome 21, and mutations within this gene have been identified in a few families with familial AD. An additional copy of chromosome 21 is found in patients with Down's syndrome, and it is noteworthy that these individuals usually develop symptoms of Alzheimer's disease, and amyloid deposition, within their thirties or forties. The extent to which β-amyloid contributes to the more common sporadic form of AD, however, is far from clear.

The β-amyloid peptide has been shown to have neurotoxic properties in various neuronal preparations. The mechanism of this toxicity is not completely clear and there are several contending views. One theory is that β-amyloid may contribute to neurodegeneration by inserting into the membrane to form an ion channel. This hypothesis is supported by the fact that β-amyloid forms ion channels in artificial lipid bilayers (Aripse *et al.*, 1993) and induces a novel ion channel in hypothalamic neuronal membranes (Kawahara *et al.*, 1997). Significantly, these channels are permeable to Ca^{2+}. They may therefore contribute to cytotoxicity by facilitating Ca^{2+} entry, both by their own intrinsic Ca^{2+} permeability and by causing depolarization and activation of voltage-gated Ca^{2+} channels. The mechanism of pore formation is unknown, but given the small size of the peptide it is likely that it involves oligomerization.

A LAST WORD

Predicting the future is notoriously difficult, and never more so than in science where the introduction of a new technique or theory can revolutionize a field overnight. Nevertheless, it seems safe to assume that the current rate of growth in ion channel studies will continue. Thus, we can look forward to further progress in unravelling the relationship between ion channel structure and function, and to the identification and characterization of many more diseases of man and animals that result from mutations in the genes encoding ion channel proteins. This flowering of molecular physiology is immensely exciting and drives the field forward at an ever faster pace. In this last chapter, we consider where rapid advances may be expected and those areas where large gaps in our understanding need to be filled.

Closing the Gaps

The Human Genome project is an international collaborative effort to map and characterise the ~75,000 genes in the human genome. Similar projects are under way to sequence the genomes of other organisms, including the mouse. Some genomes are already complete (*C. elegans,* the yeast *S. cerevisiae,* and many bacteria), and it seems probable that early in the next century we will also know the sequence of the entire human genome. This will be a major achievement. Yet it brings with it its own attendant problems. How do we make sense of the huge amount of information that is generated? And how does this knowledge benefit the patient with an inheritable disease?

For the researcher interested in ion channels, the first problem is to determine whether a given gene does, in fact, code for an ion channel. Initial clues as to whether this is the case may be obtained simply by comparing its nucleotide sequence with those of known ion channels. Confirmation, however, requires the demonstration that heterologous expression of the gene produces a novel ionic current. This is not always an easy task: some genes may look like ion channels but don't give rise to functional channels—the PKD2 gene, discussed in Chapter 9, is one such example. Does this mean the gene does not code for an ion channel but has a quite different function? Or is it that an essential subunit is lacking or that an, as yet unidentified, ligand is needed to activate the channel? In some cases, the answers to such questions may come from studies of transgenic animals or mouse models of disease. Despite the fact that the cloned δ_2 glutamate receptor subunit does not produce functional channels when heterologously expressed, it does code for an ion channel, since the *Lurcher* mouse phenotype, which results from mutations in the δ_2 gene, shows a novel constitutive inward current in Purkinje neurones (Chapter 16). Thus, it may take some time and much work to determine whether or not a given gene indeed codes for an ion channel.

The marriage of molecular biology and biophysics has given us some idea of which domains in ion channel proteins may be of functional importance. What is still lacking, however, is a detailed knowledge of how ion channel proteins are organised in three dimensions. This is critical for a real understanding of how ions permeate through the channel pore, of gating mechanisms and of drug-channel interactions. X-ray crystallography has provided such information for soluble proteins, such as haemoglobin and insulin. However, this type of study has proved notoriously difficult for ion channels, because of the problem of obtaining three-dimensional crystals of a protein that has evolved in the two dimensional environment of the lipid bilayer. The recent report of the crystal structure of a bacterial K^+ channel (Doyle *et al.*, 1998) is therefore a major achievement, and it heralds the beginning of a new age of structural channel studies. We now need to find ways of making crystals of more complex channels or new ways—such as solid-state nuclear magnetic resonance—of looking in detail at ion channel structure.

As I discovered when writing this book, there also remain major gaps in our understanding of how mutations in ion channels give rise to human and animal disease. Broadly speaking, these fall into two groups. First of all, it is not always clear how disease-causing mutations alter cellular function. Second, in many cases we do not understand how changes in function at the level of the single cell produce the disease phenotype in the whole organism. These are not simple questions to answer as they demand a detailed knowledge not only of the how the single cell works, but also of how the activity of the many different types of cell in the body are integrated. Thus, it is not always easy to determine how mutations in a given ion channel gene produce disease. The voltage-gated Ca^{2+} channel of skeletal muscle is a case in point. Although this gene has been cloned for several years, and much is known of its functional properties, the way in which mutations give rise to hypokalaemic periodic paralysis is still far from clear (see Chapter 9). Likewise, there remains considerable debate about the way in which mutations in the cystic fibrosis gene product (CFTR) cause airway disease (Chapter 12). What is needed is a much more detailed knowledge of both cellular and systems physiology. Indeed, while the human genome project is a major intellectual enterprise, once it is complete we face an even more daunting task—that of figuring out what all those genes actually do.

Benefits for the Patient?

How does identification of the genetic basis of a human disease benefit the individual patient? One obvious way is that it becomes possible to develop diagnostic tests for genetic screening and prenatal diagnosis. This is of particular value if strategies for preventing the subsequent development of disease in individuals carrying disease-causing mutations can be devised. A knowledge of the genetic cause of a disease may also enable its better clinical management. In the case of a genetically heterogeneous disease, for example, the ideal drug therapy may be markedly different depending on which gene is responsible; long QT syndrome is just one such example (Chapters 5 and 6). Ultimately, gene therapy—the replacement of the defective gene with a normal copy—may provide a means for treating genetic disease. It is wise to remember, however, that gene therapy requires not only a knowledge of the defective gene but also a detailed understanding of the pathogenesis of the disease, something that is not always available. Additionally, the introduction of the normal gene poses the problem of how to target it to the correct cells and how to avoid immune responses associated with the vector. Furthermore, in those patients who fail to produce any of the wild-type protein, the immune system may not recognise the introduced protein as 'self' and may mount an immune response to it. The upregulation of fetal genes, or closely related genes, which can substitute for the defective gene may be an alternative approach to classical gene therapy methods, as it avoids such immunological problems.

Searching for Further Information

It seems clear that we have not yet discovered all the ion channel genes that result in human or animal diseases, nor even all the disease-causing mutations in those genes that have been identified. Furthermore, as this book is not intended to be exhaustive, but rather to present the basic principles, not all those mutations known at the time of writing have been included. So where do you find those mutations that have not been mentioned—or even those that have not yet been identified as I write this? One of the best ways to discover them is to search the databases accessible via the World Wide Web. In this section, therefore, I provide the addresses of some of the more useful web sites for those who wish to explore further.

The most valuable web site is probably OMIM—Online Mendelian Inheritance in Man. This is a database of human genes and genetic disorders compiled by Dr. Victor A. McKusick and his colleagues at the Center for Medical Genetics, Johns Hopkins University in Baltimore (USA). For each gene, they supply the chromosomal location, a description of the gene structure, and details of any mutations which have been shown to cause heritable disease: references for all of this information are also provided. You can also search by disease to obtain information about its phenotype and its genetic basis: this is often a very useful way to proceed as many diseases are genetically heterogenous. OMIM provides extensive links to other relevant websites. Its address is http://www.ncbi.nlm.nih.gov/omim/.

Details of human mutations are also provided by the Human Gene Mutation Database at http://www.uwcm.ac.uk/uwcm/mg/hgmd0.html/ (Krawczak and Cooper, 1997). This provides a catalogue of all the mutations responsible for human inherited disease. Some individual websites for specific diseases also exist. For example, a database of mutations associated with cystic fibrosis is located at http://www.genet.sickkids.on.ca/cftr. And an excellent web site concerned with neurological disorders is maintained by the Neuromuscular Disease Center at Washington University School of Medicine in St. Louis, USA. This provides information about a range of neuromuscular and CNS disorders, including a clinical description of each syndrome, the associated pathology and the molecular basis of the disorder. It is located at http://www.neuro.wustl.edu/neuromuscular/. The section on ion channels, which is regularly updated, can be found at http://www.neuro.wustl.edu/neuromuscular/mother/chan.html.

Another important web site is maintained by The National Center for Biotechnology Information at http://www.ncbi.nlm.nih.gov/. Here you can obtain access to GenBank. This is the National Institutes of Health (USA) genetic sequence database, which comprises an annotated collection of all publicly available DNA sequences (Bensen *et al.*, 1998). As of December 1998, approximately 3,044,000 sequences had been registered. GenBank is part of the International Nucleotide Sequence Database Collaboration, which is made up of the DNA DataBank of Japan (DDBJ) and the European Molecular Biology Laboratory (EMBL), as well as GenBank. These three organizations collate and update sequence data on all cloned genes (daily) and exchange data on a daily basis. You can gain entry to any of these three databases directly, but in my experience, GenBank offers the easiest access. There you can find the sequence of the gene of your choice, enter new sequences (if you've cloned a new gene) and compare your sequence with those in the database. In the United Kingdom, help for anyone interested in cloning novel human genes themselves can be obtained from the human genome mapping project resource centre at http://www.hgmp.mrc.ac.uk.

Finally, MEDLINE (MEDlars onLINE) is the National Library of Medicine's bibliographic database covering the fields of medicine and the preclinical sciences. It contains the reference citations and abstracts from approximately 3,900 current biomedical journals published throughout the world. The database contains approximately 9 million records dating back to 1966. It may be accessed at http://www.ncbi.nlm.nih.gov/PubMed/.

Conclusions

Even a brief glance at this book shows that we have come a long way since the first ion channel was cloned in 1982. We now have a reasonable understanding of the relationship between the primary structure and the functional properties of several types of ion channel. And we have identified a large number of channelopathies. Nevertheless, it is also clear that there is still much to be done. The role of ion channels in the development of an organism is also an expanding research field. Most of the diseases described in this book arise from mutations in a single gene. The major killers in the western world today—heart disease, cancer, stroke, diabetes—are, however, almost certainly caused by a complex interaction between a number of different genes and the environment. Given the central role of ion channels in the physiology of the cell, it seems possible that defects in the pathways that regulate channel function (if not in the ion channel genes themselves) may also contribute to these diseases. Thus, despite the rapid progress that has been made in the field of ion channels during the past few years, there remain many puzzles and much to do. Sorting it all out is likely to take some time but promises to bring great intellectual and clinical rewards. It seems a fortunate time to be involved in ion channel research.

And Finally . . . . . .

Despite careful checking by many people, it is possible that some errors may have eluded us and remain in the published version of this book. If you discover any such errors, I would be grateful if you would inform the publishers at Academic Press so that they may be corrected in any future editions.

BIBLIOGRAPHY

Abbott, C., West, L., Povey, S., Jeremiah, S., Murad, Z., DiScipio, R., and Fey, G. (1989). The gene for human complement component C9 mapped to chromosome 5 by polymerase chain reaction. *Genomics* **4**, 606–609.

Adams, B. A., Tanabe, T, Mikami, A., Numa, S., and Beam, K. G. (1990). Intramembrane charge movement restored in dysgenic skeletal muscle by injection of dihydropyridine receptor cDNAs. *Nature* **346**, 569–572.

Adelman, J. P., Bond, C. T., Pessia, M., and Maylie, J. (1995). Episodic ataxia results from voltage-dependent potassium channels with altered functions. *Neuron* **15**, 1449–1454.

Adelman, J. P., Shen, K. Z., Kavanaugh, M. P., Warren, R. A., Wu, Y. N., Lagrutta, A., Bond, C. T., and North, R. A. (1992). Calcium-activated potassium channels expressed from cloned complementary DNAs. *Neuron* **9**, 209–216.

Adrian, R. H., and Bryant, S. H. (1974). On the repetitive discharge in myotonic muscle fibres. *J. Physiol.* **240**, 505–515.

Aguilar-Bryan, L., Nichols, C. G., Wechsler, S. W., Clement, J. P., Boyd, A. E., González, G., Herrera-Sosa, H., Nguy, K., Bryan, J., and Nelson, D. A. (1995). Cloning of the β-cell high-affinity sulphonylurea receptor: A regulator of insulin secretion. *Science* **268**, 423–426.

Ahern, G. P., Junankar, P. R., and Dulhunty, A. F. (1994). Single channel activity of the ryanodine receptor calcium release channel is modulated by FK-506. *FEBS Lett.* **352**, 369–374.

Aidley, D. J., and Stanfield. P. R. (1996). "Ion channels." Cambridge University Press.

Akabas, M. H., Kaufmann, C., Archdeacon, P., and Karlin, A. (1994). Identification of acetylcholine receptor channel-lining residues in the entire M2 segment of the α subunit. *Neuron* **13**, 919–927.

Akabas, M. H., Stauffer, D. A., Xu, M., and Karlin, A. (1992). Acetylcholine receptor channel structure probed in cysteine-substitution mutants. *Science* **258**, 307–310.

Amin, J., and Weiss, D. S. (1993). GABA$_A$ receptor needs two homologous domains of the β-subunit for activation by GABA but not by pentobarbital. *Nature* **366**, 565–569.

Anand, R., and Lindstrom, J. (1992). Chromosomal localization of seven neuronal nicotinic acetylcholine receptor subunit genes in humans. *Genomics* **13**, 962–967.

Anderson, C. R., and Stevens, C. F. (1973). Voltage-clamp analysis of acetylcholine-produced end-plate current fluctuations at the frog neuromuscular junction. *J. Physiol.* **235**, 655–691.

Anderson, M. P., Gregory, R. J., Thompson, S., Souza, D. W., Paul, S., Mulligan, R. C., Smith, A. E., and Welsh, M. J. (1991). Demonstration that CFTR is a chloride channel by alteration of its anion selectivity. *Science* **253**, 202–205.

Andrews, N. W., Abrams, C. K., Slatin, S. L., and Griffiths, G. (1990). A *T. cruzi*-secreted protein immunologically related to the complement component C9: Evidence for membrane pore-forming activity at low pH. *Cell* **61**, 1277–1287.

Andrews, P. I., and McNamara, J. O. (1996). Rasmussen's encephalitis: An autoimmune disorder? *Curr. Opin. Neurobiol.* **6**, 673–678.

Antonsson, B., Conti, F., Ciavatta, A., Montessuit, S., Lewis, S., Martinou, I., Bernasconi, L., Bernard, A., Mermod, J.-J., Mazzei, G., Maundrell, K., Gambale, F., Sadoul, R., and Martinou, J.-C. (1997). Inhibition of Bax channel-forming activity by Bcl-2. *Science* **277**, 370–372.

Ardell, M. D., Aragon, I., Oliveira, L., Porche, G. E., Burke, E., and Pittler, S. J.

(1996). The beta subunit of human rod photoreceptor cGMP-gated cation channel is generated from a complex transcription unit. *FEBS Lett.* **389**, 213–218.

Arispe, N., Rojas, E., and Pollard, H. B. (1993). Alzheimer disease amyloid β protein forms calcium channels in bilayer membranes: Blockade by tromethamine and aluminum. *Proc. Natl. Acad. Sci. USA* **90**, 567–571.

Armstrong, C. M. (1969). Inactivation of the potassium conductance and related phenomena caused by quaternary ammonium ion injection in squid axons. *J. Gen. Physiol.* **54**, 553–575.

Armstrong, C. M., and Bezanilla, F. (1977). Inactivation of the sodium channel. II. Gating current experiments. *J. Gen. Physiol.* **70**, 567–590.

Armstrong, C. M, Bezanilla, F., and Rojas, E. (1973). Destruction of sodium conductance inactivation in squid axons perfused with pronase. *J. Gen. Physiol.* **62**, 375–391.

Armstrong, N., Sun, Y., Chen, G. Q., and Gouaux, E. (1998). Structure of a glutamate-receptor ligand-binding core in complex with kainate. *Nature* **395**, 913–917.

Ashcroft, F. M. and Gribble, F. M. (1998). Correlating structure and function in ATP-sensitive K⁺ channels. *Trends in Neurosci.* **21**, 288–294.

Ashcroft, F. M., Harrison, D. E., and Ashcroft, S. J. H. (1984). Glucose induces closure of single potassium channels in isolated rat pancreatic β-cells. *Nature* **312**, 446–448.

Ashcroft, F. M. and Stanfield, P. R. (1982). Calcium inactivation in skeletal muscle fibres of the stick insect, *Carausius morosus*. *J. Physiol.* **330**, 349–372.

Ashcroft, S. J. H. and Ashcroft, F. M. (1990). Properties and functions of ATP-sensitive K-channels. *Cell Signal.* **2**, 197–214.

Atkinson, N. S., Roberston, G. A., and Ganetzky, B. (1991). A component of calcium-activated potassium channels encoded by the *Drosophila slo* locus. *Science* **253**, 551–555.

Atkison, P., Joubert, G., Barron, A., Grant, D., Paradis, K., Seidman, E., Wall, W.,

Rosenberg, H., Howard, J., Williams, S., and Stiller, C. (1995). Hypertrophic cardiomyopathy associated with tacrolimus in paediatric transplant patients. *Lancet* **345**, 894–896 (see also comments, pages 1644–1645).

Attree, O., Olivos, I. M., Okabe, I., Bailey, L. C., Nelson, D. L., Lewis, R. A., McInnes, R. R., and Nussbaum, R. L. (1992). The Lowe's oculocerebrorenal syndrome gene encodes a protein highly homologous to inositol polyphosphate-5-phosphatase. *Nature* **358**, 239–242.

Attwell, D., and Wilson, M. (1980). Behaviour of the rod network in the tiger salamander retina mediated by membrane properties of individual rods. *J. Physiol.* **309**, 287–315.

Auld, V. J., Goldin, A. L., Krafte, D. S., Catterall, W. A., Lester, H. A., Davidson, N., and Dunn, R. J. (1990). A neutral amino acid change in segment IIS4 dramatically alters the gating properties of the voltage-dependent sodium channel. *Proc. Natl. Acad. Sci. USA* 87, 323–327.

Baker, P. F., Hodgkin, A. L., and Shaw, T. I. (1961). Replacement of the protoplasm of a giant nerve fibre with artificial solutions. *Nature* **190**, 885–887.

Bamberg, E., and Lauger, P. (1974). Temperature-dependent properties of gramicidin A channels. *Biochim. Biophys. Acta* **367**, 127–133.

Barchi, R. L. (1995). Molecular pathology of the skeletal muscle sodium channel. *Annu. Rev. Physiol.* **57**, 355–385.

Barhanin, J., Lesage, F., Guillemare, E., Fink, M., Lazdunski, M., and Romey, G. (1996). KvLQT1 and IsK (minK) proteins associate to form the I_{Ks} cardiac potassium current. *Nature* **384**, 78–80.

Barker, P. M., Gowen, C. W., Lawson, E. E., and Knowles, M. R. (1997). Decreased sodium ion absorption across nasal epithelium of very premature infants with respiratory distress syndrome. *J. Pediatr.* **130**, 373–377.

Barker, P. M., Markiewicz, M., Parker, K. A., Walters, D. V., and Strang, L. B. (1990). Synergistic action of triiodothyronine and hydrocortisone on epinephrine-induced reabsorption of fetal lung liquid. *Pediatric Research* **27**, 588–591.

Barrio, L. C., Suchyna, T., Bargiello, T., Xu, L. X., Roginski, R. S., Bennett, M. V. L., and Nicholson, B. J. (1991). Gap junctions formed by connexins 26 and 32 alone and in combination are differently affected by applied voltage. *Proc. Natl. Acad. Sci. U.S.A.* **88,** 8410–8414.

Bashford, C. L., Menestrina, G., Henkart, P. A., and Pasternak, C. A. (1988). Cell damage by cytolysin: Spontaneous recovery and reversible inhibition by divalent cations. *J. Immunol.* **141,** 3965–3974.

Baukrowitz, T., Hwang, T-C., Nairn, A. C., and Gadsby, D. C. (1994). Coupling of CFTR Cl⁻ channel gating to an ATP hydrolysis cycle. *Neuron* **12,** 473–482.

Baukrowitz, T., Schulte, U., Oliver, D., Herlitze, S., Krauter, T., Tucker, S. J., Ruppersberg, J. P., and Fakler, B. (1998). PIP₂ and PIP as determinants for ATP inhibition of K_{ATP} channels. *Science* **282,** 1141–1144.

Beam, K. G., Knudson, C. M., and Powell, J. A. (1986). A lethal mutation in mice eliminates the slow calcium current in skeletal muscle cells. *Nature* **320,** 168–170.

Bear, C. E., Li, C. H., Kartner, N., Bridges, R. J., Jensen, T. J., Ramjeesingh, M., and Riordan, J. R. (1992). Purification and functional reconstitution of the cystic fibrosis transmembrane conductance regulator (CFTR). *Cell* **68,** 809–818.

Beck, C., Moulard, B., Steinlein, O., Guipponi, M., Vallee, L., Montpied, P., Baldy-Moulnier, M., and Malafosse, A. (1994). A nonsense mutation in the α_4 subunit of the nicotinic acetylcholine receptor (CHRNA4) cosegregates with 20q-linked benign neonatal familial convulsions (EBN1). *Neurobiol. Dis.* **1,** 95–99.

Beck, C. L., Fahlke, C., and George, A. L. (1996). Molecular basis for decreased muscle chloride conductance in the myotonic goat. *Proc. Natl. Acad. Sci. U.S.A.* **93,** 11248–11252.

Becker, C. M., Schmieden, V., Tarroni, P., Strasser, U., and Betz, H. (1992). Isoform-selective deficit of glycine receptors in the mouse mutant *spastic*. *Neuron* **8,** 283–289.

Beeson, D., Jeremiah, S., West, L. F., Povey, S., and Newsom-Davis, J. (1990). Assignment of the human nicotinic acetylcholine receptor genes: The α and δ subunit genes to chromosome 2 and the β subunit gene to chromosome 17. *Ann. Hum. Genet.* **54,** 199–208.

Behrens, M. I., Jalil, P., Serani, A., Vergara, F., and Alvarez, O. (1994). Possible role of apamin-sensitive K⁺ channels in myotonic dystrophy. *Muscle Nerve* **17,** 1264–1270.

Ben-Ari, Y., Khazipov, R., Leinekugel, X., Caillard, O., and Gaiarsa, J.-L. (1997). GABA_A, NMDA and AMPA receptors: A developmentally regulated "menage a trois." *Trends. Neurosci.* **20**(11), 523–529.

Benke, D., Mertens, S., Trzeciak, A., Gillessen, D., and Mohler, H. (1991). GABA_A receptors display association of γ_2 subunits with α_1 and β_2/β_3 subunits. *J. Biol. Chem.* **266,** 4478–4483.

Bennett, P. B., Yazawa, K., Makita, N., and George, A. L. (1995). Molecular mechanism for an inherited cardiac arrhythmia. *Nature* **376,** 683–685.

Benson, D. A., Boguski, M. S., Lipman, D. J., Ostell, J., and Ouellette, B. F. F. (1998). GenBank. *Nucl. Acids Res.* **26,** 1–7.

Bergoffen, J., Scherer, S. S., Wang, S., Scott, M. O., Bone, L. J., Paul, D. L., Chen, K., Lensch, M. W., Chance, P. F., and Fischbeck, K. H. (1993). Connexin mutations in X-linked Charcot-Marie-Tooth disease. *Science* **262,** 2039–2042.

Berridge, M. J. (1993). Inositol trisphosphate and calcium signalling. *Nature* **361,** 315–325.

Berridge, M. J., Downes, C. P., and Hanley, M. R. (1989). Neural and developmental actions of lithium: A unifying hypothesis. *Cell* **59,** 411–419.

Bezprozvanny, I., and Ehrlich, B. E. (1994). Inositol (1,4,5) trisphosphate (InsP₃)-gated Ca channels from cerebellum: Conduction properties for divalent cations and regulation by intraluminal calcium. *J. Gen. Physiol.* **104,** 821–856.

Bezprozvanny, I., and Ehrlich, B. E. (1995). The inositol 1,4,5-trisphosphate (InsP₃) receptor. *J. Membr. Biol.* **145,** 205–216.

Bezprozvanny, I., Watras, J., and Ehrlich, B. E. (1991). Bell-shaped calcium-response curves of Ins(1,4,5)P₃- and calcium-gated channels from endoplasmic reticulum of cerebellum. *Nature* **351,** 751–754.

Bhakdi, S., and Tranum-Jensen, J. (1991). Alpha-toxin of *Staphylococcus aureus*. *Microbiol. Rev.* **55**, 733–751.

Biervert, C., Schroeder, B. C., Kubisch, C., Berkovic, S. F., Propping, P., Jentsch, T. J., and Steinlein, O. K. (1998.) A potassium channel mutation in neonatal human epilepsy. *Science* **279**, 403–406.

Blair, L. A. C., Levitan, E. S., Marshall, J., Dionne, V. E., and Barnard, E. A. (1988). Single subunits of the GABA$_A$ receptor form ion channels with properties of the native receptor. *Science* **242**, 577–579.

Boman, H. G. (1995). Peptide antibiotics and their role in innate immunity. *Annu. Rev. Immunol.* **13**, 61–92.

Bone, L. J., Dahl, N., Lensch, M. W., Chance, P. F., Kelly, T., Le Guern, E., Magi, S., Parry, G., Shapiro, H., Wang, S., and Fischbeck, K. H. (1995). New connexin32 mutations associated with X-linked Charcot-Marie-Tooth disease. *Neurology* **45**, 1863–1866.

Bormann, J., Rundström, N., Betz, H., and Langosch, D. (1993). Residues within the transmembrane segment M2 determine chloride conductance of glycine receptor homo- and hetero-oligomers. *EMBO J.* **12**, 3729–3737.

Botero-Velez, M., Curtis, J. J., and Warnock, D. G. (1994). Brief Report: Liddle's Syndrome revisited—a disorder of sodium reabsorption in the distal tubule. *New Engl. J. Med.* **330**, 178–181.

Boucher, R. C., Stutts, M. J., Knowles, M. R., Cantley, L., and Gatzy, J. T. (1986). Na$^+$ transport in cystic fibrosis respiratory epithelia. Abnormal basal rate and response to adenylate cyclase activation. *J. Clin. Invest.* **78**, 1245–1252.

Brandt, S., and Jentsch, T. J. (1995). ClC-6 and ClC-7 are two novel broadly expressed members of the CLC chloride channel family. *FEBS Lett.* **377**, 15–20.

Brehm, P., and Eckert, R. (1978). Calcium entry leads to inactivation of calcium channel in *Paramecium*. *Science* **202**, 1203–1206.

Brightman, M. W., and Reese, T. S. (1969). Junctions between intimately apposed cell membranes in the vertebrate brain. *J. Cell Biol.* **40**, 648–677.

Brillantes, A.-M. B., Ondrias, K., Scott, A., Kobrinsky, E., Ondriasova, E., Moschella, M. C., Jayaraman, T., Landers, M., Ehrlich, B. E., and Marks, A. R. (1994). Stabilization of calcium release channel (ryanodine receptor) function by FK506 binding protein. *Cell* **77**, 513–523.

Britz-Cunningham, S. H., Shah, M. M., Zuppan, C. W., and Fletcher, W. H. (1995). Mutations of the Connexin43 gap-junction gene in patients with heart malformations and defects of laterality. *N. Engl. J. Med.*, **332**, 1323–1329.

Broillet, M.-C., and Firestein, S. (1997). β-subunits of the olfactory cyclic nucleotide-gated channel form a nitric oxide activated Ca^{2+} channel. *Neuron* **18**, 951–958.

Brown, D. A., and Adams, P. (1980). Muscarinic suppression of a novel voltage-sensitive K$^+$ current in a vertebrate neurone. *Nature* **283**, 673–676.

Brown, H. F., DiFrancesco, D., and Noble, S. J. (1979). How does adrenaline accelerate the heart? *Nature* **280**, 235–236.

Brown, M. J., Olver, R. E., Ramsden, C. A., Strang, L. B., and Walters, D. V. (1983). Effects of adrenaline and of spontaneous labour on the secretion and absorption of lung liquid in the foetal lamb. *J. Physiol.* **344**, 137–152.

Browne, D. L., Gancher, S. T., Nutt, J. G., Brunt, E. R. P., Smith, E. A., Kramer, P., and Litt, M. (1994). Episodic ataxia/myokymia syndrome is associated with point mutations in the human potassium channel gene KCNA1. *Nature Genetics* **8**, 136–140.

Brugnara, C., Gee, B., Armsby, C. C., Kurth, S., Sakamoto, M., Rifai, N., Alper, S. L., and Platt, O. S. (1996). Therapy with oral clotrimazole induces inhibition of the Gardos channel and reduction of erythrocyte dehydration in patients with sickle cell disease. *J. Clin. Invest.* **97**, 1227–1234.

Brunet, L. J., Gold, G. H., and Ngai, J. (1996). General anosmia caused by targeted disruption of the mouse olfactory cyclic nucleotide-gated cation channel. *Neuron* **17**, 681–693.

Brusa, R., Zimmermann, F., Koh, D.-S., Feldmeyer, D., Gass, P., Seeburg, P. H., and Sprengel, R. (1995). Early-onset epilepsy and postnatal lethality associated with an editing-deficient GluR-B allele in mice. *Science* **270,** 1677–1680.

Bruzzone, R., White, T. W., Scherer, S. S., Fischbeck, K. H., and Paul, D. L. (1994). Null mutations of connexin32 in patients with X-linked Charcot-Marie-Tooth disease. *Neuron* **13,** 1253–1260.

Bryant, S. H. (1969). Cable properties of external intercostal muscle fibres from myotonic and non-myotonic goats. *J. Physiol.* **204,** 539–550.

Bryant, S. H., and Morales-Aguilera, A. (1971). Chloride conductance in normal and myotonic muscle fibres and the action of monocarboxylic aromatic acids. *J. Physiol.* **219,** 367–383.

Buckle, V. J., Fujita, N., Ryder-Cook, A. S., Derry, J. M. J., Barnard, P. J., Lebo, R. V., Schofield, P. R., Seeburg, P. H., Bateson, A. N., Darlison, M. G., and Barnard, E. A. (1989). Chromosomal localization of GABA$_A$ receptor subunit genes: Relationship to human genetic disease. *Neuron* **3,** 647–654.

Buckley, J. T. (1992). Crossing three membranes: channel formation by aerolysin. *FEBS Lett.* **307,** 30–33.

Buckwalter, M. S., Cook, S. A., Davisson, M. T., White, W. F., and Camper, S. A. (1994). A frameshift mutation in the mouse α_1 glycine receptor gene (*Glra1*) results in progressive neurological symptoms and juvenile death. *Hum. Mol. Genet.* **3,** 2025–2030.

Burgess, D. L., Jones, J. M., Meisler, M. H., and Noebels, J. L. (1997). Mutation of the Ca^{2+} channel β subunit gene *Cchb4* is associated with ataxia and seizures in the lethargic (*lh*) mouse. *Cell* **88,** 385–392.

Burgess, D. L., Kohrman, D. C., Galt, J., Plummer, N. W., Jones, J. M., Spear, B. and Meisler, M. H. (1995). Mutation of a new sodium channel gene, Scn8a, in the mouse mutant "motor endplate disease" *Nature Genet.* **10,** 461–465.

Burnashev, N., Monyer, H., Seeburg, P. H., and Sakmann, B. (1992a). Divalent ion permeability of AMPA receptor channels is dominated by the edited form of a single subunit *Neuron* **8,** 189–198.

Burnashev, N., Schoepfer, R., Monyer, H., Ruppersberg, J. P., Günther, W., Seeburg, P. H., and Sakmann, B. (1992b). Control by asparagine residues of calcium permeability and magnesium blockade in the NMDA receptor. *Science* **257,** 1415–1419.

Burnashev, N., Villarroel, A., and Sakmann, B. (1996). Dimensions and ion selectivity of recombinant AMPA and kainate receptor channels and their dependence on Q/R site residues. *J. Physiol.* **496.1,** 165–173.

Burnstock, G. (1972). Purinergic nerves. *Pharmacol. Rev.* **24,** 509–581.

Butler, A., Tsunoda, S., McCobb, D. P., Wei, A., and Salkoff, L. (1993). *mSlo,* a complex mouse gene encoding "maxi" calcium-activated potassium channels. *Science* **261,** 221–224.

Cameron, A. M., Steiner, J. P., Sabatini, D. M., Kaplin, A. I., Walensky, L. D., and Snyder, S. H. (1995). Immunophilin FK506 binding protein associated with inositol 1,4,5-trisphosphate receptor modulates calcium flux. *Proc. Natl. Acad. Sci. USA* **92,** 1784–1788.

Canessa, C. M., Horisberger, J.-D., and Rossier, B. C. (1993). Epithelial sodium channel related to proteins involved in neurodegeneration. *Nature* **361,** 467–470.

Canessa, C. M., Schild, L., Buell, G., Thorens, B., Gautschi, I., Horisberger, J.-D., and Rossier, B. C. (1994). Amiloride-sensitive epithelial Na$^+$ channel is made of three homologous subunits. *Nature* **367,** 463–467.

Cannon, S. C. (1996). Sodium channel defects in myotonia and periodic paralysis. *Annu. Rev. Neurosci.* **19,** 141–164.

Cannon, S. C., Brown, R. H., and Corey, D. P. (1991). A sodium channel defect in hyperkalemic periodic paralysis: Potassium-induced failure of inactivation. *Neuron* **6,** 619–626.

Cannon, S. C., Brown, R. H. and Corey, D. P. (1993). Theoretical reconstruction of myotonia and paralysis caused by incomplete inactivation of sodium channels. *Biophys. J.* **65,** 270–288.

Cannon, S. C., Hayward, L. J., Beech, J., and Brown, R. H. (1995). Sodium channel inactivation is impaired in equine hyperkalemic periodic paralysis. *J. Neurophysiol.* **73,** 1892–1899.

Caplen, N. J., Alton, E. W. F., Middleton, P. G., Dorin, J. R., Stevenson, B. J., Gao, X., Durham, S. R., Jeffery, P. K., Hodson, M. E., Coutelle, C., Huang, L., Porteous, D. J., Williamson, R., and Geddes, D. M. (1995). Liposome-mediated CFTR gene transfer to the nasal epithelium of patients with cystic fibrosis. *Nature Med.* **1**(1), 39–46.

Carson, M. R., Travis, S. M., and Welsh, M. J. (1995). The two nucleotide-binding domains of cystic fibrosis transmembrane conductance regulator (CFTR) have distinct functions in controlling channel activity. *J. Biol. Chem.* **270**, 1711–1717.

Carter, E. P., Umenishi, F., Matthay, M. A., and Verkman, A. S. (1997). Developmental changes in water permeability across the alveolar barrier in the perinatal rabbit lung. *J. Clin. Invest.* **100**, 1071–1078.

Caterina, M. J., Schumacher, M. A., Tominaga, M., Rosen, T. A., Levine, J. D., and Julius, D. (1997). The capsaicin receptor: A heat-activated ion channel in the pain pathway. *Nature* **389**, 816–824.

Catterall, W. A. (1986). Molecular properties of voltage-sensitive sodium channels. *Ann. Rev. Biochem.* **55**, 953–985.

Catterall, W. A. (1993). Structure and function of voltage-gated ion channels. *Trends Neurosc.* **16**, 500–506.

Catterall, W. A. (1995). Structure and function of voltage-gated ion channels. *Annu. Rev. Biochem.* **64**, 493–531.

Cendes, F., Andermann, F., Carpenter, S., Zatorre, R. J., and Cashman, N. R. (1995). Temporal lobe epilepsy caused by domoic acid intoxication: Evidence for glutamate receptor–mediated excitotoxicity in humans. *Ann. Neurol.* **37**, 123–126.

Cesare, P., and McNaughton, P. (1996). A novel heat-activated current in nociceptive neurons and its sensitization by bradykinin. *Proc. Natl. Acad. Sci. USA* **93**, 15435–15439.

Chad, J. E., and Eckert, R. (1986). An enzymatic mechanism for calcium current inactivation in dialysed *Helix* neurones. *J. Physiol.* **378**, 31–51.

Champigny, G., Imler, J.-L., Puchelle, E., Dalemans, W., Gribkoff, V., Hinnrasky, J., Dott, K., Barbry, P., Pavirani, A., and Lazdunski, M. (1995). A change in gating mode leading to increased intrinsic Cl⁻ channel activity compensates for defective processing in a cystic fibrosis mutant corresponding to a mild form of the disease. *EMBO J.* **14**, 2417–2423.

Champigny, G., Voilley, N., Lingueglia, E., Friend, V., Barbry, P., and Lazdunski, M. (1994). Regulation of expression of the lung amiloride-sensitive Na⁺ channel by steroid hormones. *EMBO Journal* **13**, 2177–2181.

Chang, S. S., Grunder, S., Hanukoglu, A., Rösler, A., Mathew, P. M., Hanukoglu, I., Schild, L., Lu, Y., Shimkets, R. A., Nelson-Williams, C., Rossier, B. C., and Lifton, R. P. (1996). Mutations in subunits of the epithelial sodium channel cause salt wasting with hyperkalaemic acidosis, pseudohypoaldosteronism type 1. *Nature Genet.* **12**, 248–253.

Changeux, J. P., Galzi, J. L., Devillers-Thiery, A., and Bertrand, D. (1992). The functional architecture of the acetylcholine nicotinic receptor explored by affinity labelling and site-directed mutagenesis. *Q. Rev. Biophys.* **25**, 395–432.

Changeux, J. P., Kasai, M., and Lee, C. Y. (1970). The use of a snake venom toxin to characterise the cholinergic receptor protein. *Proc. Natl. Acad. Sci. USA* **67**, 1241–1247.

Chapel, H. M. (1996). Complement and disease. *In* "Oxford Textbook of Medicine" (D. J. Weatherall, J. G. G. Ledingham, and D. A. Warrell, eds.), 3rd Ed., pp. 175–182.

Charlier, C., Singh, N. A., Ryan, S. G., Lewis, T. B., Reus, B. E., Leach, R. J., and Leppart, M. (1998). A pore mutation in a novel KQT-like potassium channel gene in an idiopathic epilepsy family. *Nature Genet.* **18**, 53–55.

Chaudhari, N. (1992). A single nucleotide deletion in the skeletal muscle-specific calcium channel transcript of muscular dysgenesis (*mdg*) mice. *J. Biol. Chem.* **267**, 25636–25639.

Cheek, D. B., and Perry, J. W. (1957). A salt-wasting syndrome in infancy. *Arch. Dis. Child.* **33**, 252–256.

Chen, T.-Y., Peng, Y.-W., Dhallan, R. S., Ahamed, B., Reed, R. R., and Yau. K.-W.

(1993). A new subunit of the cyclic nucleotide-gated cation channel in retinal rods. *Nature* **362,** 764–767.

Cheng, A., van Hoek, A. N., Yeager, M., Verkman, A. S., and Mitra, A. K. (1997). Three-dimensional organisation of a human water channel. *Nature* **387,** 627–630.

Chepelinsky, A. B. (1994). The MIP transmembrane channel gene family. *In* "Handbook of Membrane Channels," pp. 413–432. Academic Press, San Diego.

Cheung, M., and Akabas, M. H. (1997). Locating the anion-selectivity filter of the cystic fibrosis transmembrane conductance regulator (CFTR) chloride channel. *J. Gen. Physiol.* **109,** 289–299.

Chizhmakov, I. V., Geraghty, F. M., Ogden, D. C., Hayhurst, A., Antoniou, M., and Hay, A. J. (1996). Selective proton permeability and pH regulation of the influenza virus M2 channel expressed in mouse erythroleukaemia cells. *J. Physiol.* **494**(2), 329–336.

Choi, K. L., Mossman, C., Aube, J., and Yellen, G. (1993). The internal quaternary ammonium receptor site of *Shaker* potassium channels. *Neuron* **10,** 533–541.

Choiunard, S. W., Wilson, G. F., Schlingen, A. K., and Ganetzky, B. (1995). A potassium channel subunit related to the aldoketo reductase superfamily is encoded by the *Drosophila Hyperkinetic* locus. *Proc. Natl. Acad. Sci. U.S.A.* **92,** 6763–6767.

Christie, M. J., North, R. A., Osborne, P. B., Douglass, J., and Adelman, J. P. (1990). Heteropolymeric potassium channels expressed in *Xenopus* oocytes from cloned subunits. *Neuron* **4,** 405–411.

Christie, M. R. (1992). Aetiology of type I diabetes: Immunological aspects. In "Insulin: Molecular Biology to Pathology" F. M. Ashcroft and S. J. H. Ashcroft, eds. pp 306–346.

Clement, J. P. IV, Kunjilwar, K., Gonzalez, G., Schwanstecher, M., Panten, U., Aguilar-Bryan, L., and Bryan, J. (1997). Association and stoichiometry of K_{ATP} channel subunits. *Neuron* **18,** 827–838.

Clements, J. D., and Westbrook, G. L. (1991). Activation kinetics reveal the number of glutamate and glycine binding sites on the N-methyl-D-aspartate receptor. *Neuron* **7,** 605–613.

Coburn, C. M., and Bargmann, C. I. (1996). A putative cyclic nucleotide-gated channel is required for sensory development and function in *C. elegans. Neuron* **17,** 695–706.

Cohen, N. A., Brenman, J. E., Snyder, S. H., and Bredt, D. S. (1996). Binding of the inward rectifier K^+ channel Kir2.3 to PSD-95 is regulated by protein kinase A phosphorylation. *Neuron* **17,** 759–767.

Cohen-Haguenauer, O., Barton, P. J., Buonanno, A., Cong, N. V., Masset, M., de Tand, M. F., Merlie, J., and Frezal, J. (1989). Localization of the acetylcholine receptor γ subunit gene to human chromosome 2q32-qter. *Cytogenet. Cell Genet.* **52,** 124–127.

Compton, S. J., Lux, R. L., Ramsey, M. R., Strelich, K. R., Sanguinetti, M. C., Green, L. S., Keating, M. T., and Mason, J. W. (1996). Genetically defined therapy of inherited long-QT syndrome. Correction of abnormal repolarization by potassium. *Circulation* **94,** 1018–1022.

Cook, S. P., Vulchanova, L., Hargreaves, K. M., Elde, R., and McCleskey, E. W. (1997). Distinct ATP receptors on pain-sensing and stretch-sensing neurons. *Nature* **387,** 505–508.

Cooper, E., Couturier, S., and Ballivet, M. (1991). Pentameric structure and subunit stoichiometry of a neuronal nicotinic acetylcholine receptor. *Nature* **350,** 235–238.

Corcos, I. A., Lafreniere, R. G., Begy, C. R., Loch-Caruso, R., Willard, H. F., and Glover, T. W. (1992). Refined localization of human connexin32 gene locus, GJB1, to Xq13.1. *Genomics* **13,** 479–480.

Cotten, J. F., Ostedgaard, L. S., Carson, M. R., and Welsh, M. J. (1996). Effect of cystic fibrosis-associated mutations in the fourth intracellular loop of cystic fibrosis transmembrane conductance regulator. *J. Biol. Chem.* **271,** 21279–21284.

Coulter, K. L., Perier, F., Radeke, C. M., and Vandenberg, C. A. (1995). Identification and molecular localisation of a pH sensing domain for the inward rectifier potassium channel HIR. *Neuron* **15,** 1157–1168.

Cui, J., Cox, D. H., and Aldrich, R. W. (1997). Intrinsic voltage dependence and Ca^{2+} regulation of *mslo* large conductance Ca-activated K$^+$ channels. *J. Gen. Physiol.* **109**, 647–673.

Culiat, C. T., Stubbs, L. J., Montgomery, C. S., Russell, L. B., and Rinchik, E. M. (1994). Phenotypic consequences of deletion of the $\gamma3$, $\alpha5$, or $\beta3$ subunit of the type A gamma-aminobutyric acid receptor in mice. *Proc. Natl. Acad. Sci. USA* **91**, 2815–2818.

Cummins, T. R., and Sigworth, F. J. (1996). Impaired slow inactivation in mutant sodium channels. *Biophys. J.* **71**, 227–236.

Curran, M. E., Splawski, I., Timothy, K. W., Vincent, G. M., Green, E. D., and Keating, M. T. (1995). A Molecular Basis for Cardiac Arrhythmia: *HERG* Mutations Cause Long QT Syndrome. *Cell* **80**, 795–803.

Curtis, B. M., and Catterall, W. A. (1984). Purification of the calcium antagonist receptor of the voltage-sensitive calcium channel from skeletal muscle transverse tubules. *Biochemistry* **23**, 2113–2118.

Cutting, G. R., Curristin, S., Zoghbi, H., O'Hara, B., Seldin, M. F. and Uhl, G. R. (1992). Identification of a putative gamma-aminobutyric acid (GABA) receptor subunit rho2 cDNA and colocalization of the genes encoding rho2 (GABRR2) and rho1 (GABRR1) to human chromosome 6q14–q21 and mouse chromosome 4. *Genomics* **12**, 801–806.

Cutting, G. R., Lu, L., O'Hara, B. F., Kasch, L. M., Montrose-Rafizadeh, C., Donovan, D. M., Shimada, S., Antonarakis, S. E., Guggino, W. B., Uhl, G. R., and Kazazian, H. H., Jr. (1991). Cloning of the gamma-aminobutyric acid (GABA) $\rho1$ cDNA: A GABA receptor subunit highly expressed in the retina. *Proc. Natl. Acad. Sci. USA* **88**, 2673–2677.

Czajkowski, C., and Karlin, A. (1991). Agonist binding site of *Torpedo* electric tissue nicotinic acetylcholine receptor: A negatively charged region of the δ subunit within 0.9 nm of the α subunit binding site disulfide. *J. Biol. Chem.* **266**, 22603–22612.

Czajkowski, C., and Karlin, A. (1995). Structure of the nicotinic receptor acetylcholine-binding site: Identification of acidic residues in the δ subunit within 0.9 nm of the α subunit-binding site disulphide. *J. Biol. Chem.* **270**, 3160–3164.

Czajkowski, C., Kaufmann, C., and Karlin, A. (1993). Negatively charged amino acid residues in the nicotinic receptor δ subunit that contribute to the binding of acetylcholine. *Proc. Natl. Acad. Sci. USA* **90**, 6285–6289.

Dani, J. A., and Heinemann, S. (1996). Molecular and cellular aspects of nicotine abuse. *Neuron* **16**, 905–908.

Davies, P. A., Hanna, M. C., Hales, T. G., and Kirkness, E. F. (1997). Insensitivity to anaesthetic agents conferred by a class of GABA$_A$ receptor subunit. *Nature* **385**, 820–823.

De Leon, M., Wang, Y., Jones, L., Perez-Reyes, E., Wei, X., Soong, T. W., Snutch, T. P., and Yue, D. T. (1995). Essential Ca^{2+}-binding motif for Ca^{2+}-sensitive inactivation of L-type Ca^{2+} channels. *Science* **270**, 1502–1506.

De Waard, M., Liu, H., Walker, D., Scott, V. E. S., Gurnett, C. A., and Campbell, K. P. (1997). Direct binding of G-protein $\beta\gamma$ complex to voltage-dependent calcium channels. *Nature* **385**, 446–450.

De Waard, M., Witcher, D. R., Pragnell, M., Liu, H., and Campbell, K. P. (1995). Properties of the $\alpha_1\beta$ anchoring site in voltage-dependent Ca^{2+} channels. *J. Biol. Chem.* **270**, 12056–12064.

Deckwerth, T. L., Elliott, J. L., Knudson, C. M., Johnson, E. M., Snider, W. D., and Korsmeyer, S. J. (1996). BAX is required for neuronal death after trophic factor deprivation and during development. *Neuron* **17**, 401–411.

Deen, P. M. T., Croes, H., van Aubel, R. A. M. H., Ginsel, L. A., and van Os, C. H. (1995). Water channels encoded by mutant aquaporin-2 genes in nephrogenic diabetes insipidus are impaired in their cellular routing. *J. Clin. Invest.* **95**, 2291–2296.

Deen, P. M. T., and van Os, C. H. (1998). Epithelial aquaporins. *Curr. Op. Cell Biol.* **10**, 435–442.

Deen, P. M. T., Verdijk, M. A. J., Knoers, N. V. A. M., Wieringa, B., Monnens, L. A. H., van Os, C. H., and van Oost, B. A. (1994a). Requirement of human renal water channel aquaporin-2 for vasopressin-dependent concentration of urine. *Science* **264**, 92–95.

Deen, P. M. T., Weghuis, D. O., van Kessel, G., Wieringa, B., and van Os, C. H. (1994b). The human gene for water channel aquaporin 1 (AQP1) is localized on chromosome 7p15–7p14. *Cytogenet. Cell Genet.* **65**, 243–246.

Denning, G. M., Anderson, M. P., Amara, J. F., Marshall, J., Smith, A. E., and Welsh, M. J. (1992). Processing of mutant cystic fibrosis transmembrane conductance regulator is temperature-sensitive. *Nature* **358**, 761–764.

Dennis, M., Giraudat, J., Kotzyba-Hibert, F., Goeldner, M., Hirth, C., Chang, J. Y., Lazure, C., Chretien, M., and Changeux, J. P. (1988). Amino acids of the *Torpedo marmorata* acetylcholine receptor alpha subunit labelled by a photoaffinity ligand for the acetylcholine binding site. *Biochemistry* **27**, 2346–2357.

Denoyelle, F., Weil, D., Maw, M. A., Wilcox, S. A., Lench, N. J., Allen-Powell, D. R., Osborn, A. H., Dahl, H. H., Middleton, A., Houseman, M. J., Dode, C., and Marlin, S. *et al.* (1997). Prelingual deafness: High prevalence of a 30delG mutation in the connexin 26 gene. *Hum. Mol. Genet.* **6**, 2173–2177.

Dermietzel, R., Traub, O., Hwang, T. K., Beyer, E., Bennett, M. V. L., Spray, D. C., and Willecke, K. (1989). Differential expression of three gap junction proteins in developing and mature brain tissues. *Proc. Natl. Acad. Sci. U.S.A.* **86**, 10148–10152.

Derst, C., Konrad, M., Köckerling, A., Károlyi, L., Deschenes, G., Daut, J., Karschin, A., and Seyberth, H. W. (1997). Mutations in the ROMK gene in antenatal Bartter syndrome are associated with impaired K^+ channel function. *Biochem. Biophys. Res. Commun.* **230**, 641–645.

Dhallan, R. S., Macke, J. P., Eddy, R. L., Shows, T. B., Reed, R. R., Yau, K.-W., and Nathans, J. (1992). Human rod photoreceptor cGMP-gated channel: Amino acid sequence, gene structure, and functional expression. *J. Neurosci.* **12**, 3248–3256.

DiFrancesco, D. (1986). Characterization of single pacemaker channels in cardiac sino-atrial node cells. *Nature* **324**, 470–473.

DiFrancesco, D., and Tortora, P. (1991). Direct activation of cardiac pacemaker channels by intracellular cyclic AMP. *Nature* **351**, 145–147.

Diriong, S., Lory, P., Williams, M. E., Ellis, S. B., Harpold, M. M. and Taviaux, S. (1995). Chromosomal localization of the human genes for alpha 1A, alpha 1B and alpha 1E voltage-dependent Ca^{2+} channel subunits. *Genomics* **30**, 605–609.

Dolly, J. O.. and Parcej, D. N. (1996). Molecular properties of voltage-gated K^+ channels. *J. Bioeng. Biomembr.* **28**, 231–253.

Dolphin, A. C. (1995). Voltage-dependent calcium channels and their modulation by neurotransmitters and G proteins. *Exp. Physiol.* **80**, 1–36.

Dolphin, A. C. (1998). Mechanisms of modulation of voltage-dependent calcium channels by G proteins. *J. Physiol.* **506**, 3–11.

Doupnik, C. A., Davidson, N., and Lester, H. A. (1995). The inward rectifier potassium channel family. *Curr. Op. Neurobiol.* **5**, 268–277.

Doyle, D. A., Cabral, J. M., Pfuetzner, R. A., Kuo, A., Gulbis, J. M., Cohen, S. L., Chait, B. T., and MacKinnon. R. (1998). The structure of the potassium channel: Molecular basis of K^+ conduction and selectivity. *Science* **280**, 69–77.

Drachman, D. B., Angus, C. W., Adams, R. N., Michelson, J. D., and Hoffman, G. J. (1978). Myasthenic antibodies cross-link acetylcholine receptors to accelerate degradation. *N. Engl. J. Med.* **298**, 1116–1122.

Drumm, M. L., Wilkinson, D. J., Smith, L. S., Worrell, R. T., Strong, T. V., Frizzell, R. A., Dawson, D. C., and Collins, F. S. (1991). Chloride conductance expressed by ΔF508 and other mutant CFTRs in *Xenopus* oocytes. *Science* **254**, 1797–1799.

Dryja, T. P., Finn, J. T., Peng, Y.-W., McGee, T. L., Berson, E. L., and Yau, K.-W. (1995). Mutations in the gene encoding the α-subunit of the rod cGMP-gated channel in autosomal recessive retinitis pigmentosa. *Proc. Natl. Acad. Sci. U.S.A.* **92**, 10177–10181.

Dumaine, R., Wang, Q., Keating, M. T., Hartmann, H. A., Schwartz, P. J., Brown, A. M., and Kirsch, G. E. (1996). Multiple mechanisms of Na^+ channel-linked Long QT syndrome. *Circul. Res.* **78**, 916–924.

Dunlap, K., Luebke, J. I., and Turner, T. J. (1995). Exocytotic Ca^{2+} channels in mammalian central neurons. *Trends Neurosci.* **18**, 89–98.

Eastwood, S. L., McDonald, B., Burnet, P. W. J., Beckwith, J. P., Kerwin, R. W., and Harrison, P. J. (1995). Decreased expression of mRNAs encoding non-NMDA glutamate receptors GluR1 and GluR2 in medial temporal lobe neurons in schizophrenia. *Mol. Brain Res.* **29**, 211–223.

Eaton, D. C., Becchetti, A., Ma, H., and Ling, B. N. (1995). Renal sodium channels: Regulation and single channel properties. *Kidney International* **48**, 941–949.

Ebers, G. C., George, A. L., Barchi, R. L., Ting-Passador, S. S., Kallen, R. G., Lathrop, G. M., Beckmann, J. S., Hahn, A. F., Brown, W. F., Campbell, R. D., and Hudson, A. J. (1991). Paramyotonia congenita and hyperkalemic periodic paralysis are linked to the adult muscle sodium channel gene. *Ann. Neurol.* **30**, 810–816.

Echevarría, M., Windhager, E. E., and Frindt, G. (1996). Selectivity of the renal collecting duct water channel aquaporin-3. *J. Biol. Chem.* **271**, 25079–25082.

Edwards, F. A., Gibb, A. J., and Colquhoun, D. (1992). ATP receptor–mediated synaptic currents in the central nervous system. *Nature* **359**, 144–147.

Egan, M., Flotte, T., Afione, S., Solow, R., Zeitlin, P. L., Carter, B. J., and Guggino, W. B. (1992). Defective regulation of outwardly rectifying Cl^- channels by protein kinase A corrected by insertion of CFTR. *Nature* **358**, 581–584.

Ek-Vitorin, J. F., Calero, G., Morley, G. E., Coombs, W., Taffet, S. M., and Delmar, M. (1996). pH regulation of connexin 43: Molecular analysis of the gating particle. *Biophys. J.* **71**, 1273–1284.

Elgoyhen, A. B., Johnson, D. S., Boulter, J., Vetter, D. E., and Heinemann, S. (1994). α_9: an acetylcholine receptor with novel pharmacological properties expressed in rat cochlear hair cells. *Cell* **79**, 705–715.

Ellis, T. M., and Atkinson, M. A. (1996). The clinical significance of an autoimmune response against glutamic acid decarboxylase. *Nature Med.* **2**, 148–153.

Elmqvist, D., Hofmann, W. W., Kugelberg, J., and Quastel, D. M. J. (1964). An electrophysiological investigation of neuromuscular transmission in myasthenia gravis. *J. Physiol.* **174**, 417–434.

Engel, A. G. (1991). Review of evidence for loss of motor nerve terminal calcium channels in Lambert–Eaton myasthenic syndrome. *Ann. N. Y. Acad. Sci.* **635**, 246–258.

Engel, A. G., Lambert, E. H., and Howard, F. M. (1977). Immune complexes (IgG and C3) at the motor endplate in myasthenia gravis: Ultrastructural and light microscopic localization and electrophysiologic correlations. *Mayo Clin. Proc.* **52**, 267–280.

Engel, A. G., Ohno, K., Milone, M., Wang, H. L., Nakano, S., Bouzat, C., Pruitt, J. N., Hutchinson, D. O., Brengman, J. M., Bren, N., Sieb, J. P., and Sine, S. M. (1996). New mutations in acetylcholine receptor subunit genes reveal heterogeneity in the slow-channel congenital myasthenic syndrome. *Hum Mol. Genet.* **5**, 1217–1227.

Engelman, D. M. (1996). Crossing the hydrophobic barrier: insertion of membrane proteins. *Science* **274**, 1850–1851.

Eubanks, J. H., Puranam, R. S., Kleckner, N. W., Bettler, B., Heinemann, S. F., and McNamara, J. O. (1993). The gene encoding the glutamate receptor subunit GluR5 is located on human chromosome 21q21.1–22.1 in the vicinity of the gene for familial amyotrophic lateral sclerosis. *Proc. Natl. Acad. Sci. USA* **90**, 178–182.

Evans, R. J., Lewis, C., Buell, G., Valera, S., North, R. A., and Surprenant, A. (1995). Pharmacological characterization of heterologously expressed ATP-gated cation channels (P2X purinoceptors). *Mol. Pharmacol.* **48**, 178–183.

Ewart, G. D., Sutherland, T., Gage, P. W., and Cox, G. B. (1996). The Vpu protein of the human immunodeficiency virus type I forms cation-selective ion channels. *J. Virol.* **70**, 7108–7115.

Fahlke, C., Beck, C. L., and George, A. L. (1997a). A mutation in autosomal dominant myotonia congenita affects the pore properties of the muscle chloride channel. *Proc. Natl. Acad. Sci. U.S.A.* **94**, 2729–2734.

Fahlke, C., Rhodes, T. H., Desai, R. R., and George, A. L. (1998). Pore stoichiometry of a voltage-gated chloride channel. *Nature* **394**, 687–690.

Fahlke, C., Knittle, T., Gurnett, C. A., Campbell, K. P., and George, A. L. (1997b). Subunit stoichiometry of human muscle chloride channels. *J. Gen. Physiol.* **109**, 93–104.

Fakler, B., Schulz, J. H., Yang, J., Schulte, U., Brändle, U., Zenner, H. P., Jan, L. Y., and Ruppersberg, J. P. (1996). Identification of a titratable lysine residue that determines sensitivity of kidney potassium channels (ROMK) to intracellular pH. *EMBO J.* **15**, 4093–4099.

Fambrough, D. M., Drachman, D. B., and Satyamurti, S. (1973). Neuromuscular junction in myasthenia gravis: Decreased acetylcholine receptors. *Science* **182**, 293–295.

Fatt, P., and Katz, B. (1951). An analysis of the endplate potential recorded with an intracellular microelectrode. *J. Physiol.* **115**, 320–370.

Ferguson, D. G., Schwartz, H. W., and Franzini-Armstrong, C. (1984). Subunit structure of junctional feet in triads of skeletal muscle: A freeze-drying, rotary-shadowing study. *J. Cell Biol.* **99**, 1735–1742.

Ferrari, D., Villalba, M., Chiozzi, P., Falzoni, S., Ricciardi-Castagnoli, P., and Di Virgilio, F. (1996). Mouse microglial cells express a plasma membrane pore gated by extracellular ATP. *J. Immunol.* **156**, 1531–1539.

ffrench-Constant, R. H., Rocheleau, T. A., Steichen, J. C., and Chalmers, A. E. (1993). A point mutation in a *Drosophila* GABA receptor confers insecticide resistance. *Nature* **363**, 449–451.

Fink, T. M., Zimmer, M., Weitz, S., Tschopp, J., Jenne, D. E., and Lichter, P. (1992). Human perforin (PRF1) maps to 10q22, a region that is syntenic with mouse chromosome 10. *Genomics* **13**, 1300–1302.

Firsov, D., Schild, L., Gautschi, I., Mérillat, A. M., Schneeberger, E., and Rossier, B. C. (1996). Cell surface expression of the epithelial Na channel and a mutant causing Liddle syndrome: A quantitative approach. *Proc. Natl. Acad. Sci. U.S.A.* **93**, 15370–15375.

Fisher, S. E., von Bakel, I., Lloyd, S. E., Pearce, S. H. S., Thakker, R. V., and Craig, I. W. (1995). Cloning and characterization of CLCN5, the human kidney chloride channel gene implicated in Dent Disease (an X-linked hereditary nephrolithiasis). *Genomics* **29**, 598–606.

Fletcher, C. F., Lutz, C. M., O'Sullivan, T. N., Shaughnessy, J. D., Hawkes, R., Frankel, W. N., Copeland, N. G., and Jenkins, N. A. (1996). Absence epilepsy in tottering mutant mice is associated with calcium channel defects. *Cell* **87**, 607–617.

Fontaine, B., Khurana, T. S., Hoffman, E. P., Bruns, G. A. P., Haines, J. L., Trofatter, J. A., Hanson, M. P., Rich, J., McFarlane, H., Yasek, D. M., Romano, D., Gusella, J. F., and Brown, R. H. (1990). Hyperkalaemic periodic paralysis and the adult muscle sodium channel α-subunit gene. *Science* **250**, 1000–1002.

Forman, S. A., Yellen, G., and Thiele, E. A. (1996). Alternative mechanism for pathogenesis of an inherited epilepsy by a nicotinic AChR mutation. *Nature Genet.* **13**, 396–397.

Fox, R. O., and Richards, F. M. (1982). A voltage-gated ion channel model inferred from the crystal structure of alamethicin at 1.5 Å resolution. *Nature* **300**, 325–330.

Franciolini, F., and Petris, A. (1990). Chloride channels of biological membranes. *Biochim. Biophys. Acta* **1031**, 247–259.

Franzini-Armstrong, C., and Protasi, F. (1997). Ryanodine receptors of striated muscles: A complex channel capable of multiple interactions. *Physiol. Rev.* **77**, 699–729.

Fu, Y. H., Pizzuti, A., Fenwick, R. G., Jr., King, J., Rajnarayan, S., Dunne, P. W., Dubel, J., Nasser, G. A., Ashizawa, T., de Jong, P., Wieringa, B., Korneluk, R., Perryman, M. B., Epstein, H. F., and Caskey, C. T. (1992). An unstable triplet repeat in a gene related to myotonic muscular dystrophy. *Science* **255**, 1256–1258.

Fujii, J., Otsu, K., Zorzato, F., de Leon, S., Khanna, V. K., Weiler, J. E., O'Brien, P. J., and MacLennan, D. H. (1991). Identification of a mutation in porcine ryanodine receptor associated with malignant hyperthermia. *Science* **253**, 448–451.

Fukunaga, H., Engel, A. G., Lang, B., Newsom-Davis, J., and Vincent, A. (1983). Passive transfer of Lambert–Eaton myasthenic syndrome with IgG from man to mouse depletes the presynaptic membrane active zones. *Proc. Natl. Acad. Sci. USA* **80**, 7636–7640.

Fukunaga, H., Engel, A. G., Osame, M., and Lambert, E. H. (1982). Paucity and disorganization of presynaptic membrane active zones in the Lambert–Eaton myasthenic syndrome. *Muscle Nerve* **5**, 686–697.

Furuichi, T., Yoshikawa, S., Miyawaki, A., Wada, K., Maeda, N., and Mikoshiba, K. (1989). Primary structure and functional expression of the inositol 1,4,5-trisphosphate binding protein P_{400}. *Nature* **342**, 32–38.

Fushimi, K., Uchida, S., Hara, Y., Hirata, Y., Marumo, F., and Sasaki, S. (1993). Cloning and expression of apical membrane water channel of rat kidney collecting tubule. *Nature* **361**, 549–552.

Galzi, J. L., Bertrand, D., Devillers-Thiery, A., Revah, F., Bertrand, S., and Changeux, J. P. (1991). Functional significance of aromatic amino acids from three peptide loops of the α_7 neuronal nicotinic receptor site investigated by site-directed mutagenesis. *FEBS Lett.* **294**, 198–202.

Galzi, J. L., Devillers-Thiery, A., Hussy, N., Bertrand, S., Changeux, J. P., and Bertrand, D. (1992). Mutations in the channel domain of a neuronal nicotinic receptor convert ion selectivity from cationic to anionic. *Nature* **359**, 500–505.

Galzi, J. L., Revah, F., Black, D., Goeldner, M., Hirth, C., and Changeux, J. P. (1990). Identification of a novel amino acid α-tyrosine 93 within the cholinergic ligand-binding sites of the acetylcholine receptor by photoaffinity labelling: Additional evidence for a three-loop model of the cholinergic ligand binding sites. *J. Biol. Chem.* **265**, 10430–10437.

Ganz, T., Metcalf, J. A., Gallin, J. I., Boxer, L. A., and Lehrer, R. I. (1988). Microbicidal/cytotoxic proteins of neutrophils are deficient in two disorders: Chediak–Higashi syndrome and "specific" granule deficiency. *J. Clin. Invest.* **82**, 552–556.

García-Anoveros, J., Derfler, B., Neville-Golden, J., Hyman, B. T., and Corey, D. P. (1997). BNaC1 and BNaC2 constitute a new family of human neuronal sodium channels related to degenerins and epithelial sodium channels. *Proc. Natl. Acad. Sci. U.S.A.* **94**, 1459–1464.

Gardos, G. (1958). The function of calcium in the potassium permeability of human erythrocytes. *Biochim. Biophys. Acta* **30**, 653–654.

Garrett, L. (1995). "The Coming Plague." Virago Press.

Gauss, R., Seifert, R., and Kaupp, U. B. (1998). Molecular identification of a hyperpolarization-activated channel in sea urchin sperm. *Nature* **393**, 583–587.

Gautam, M., Noakes, P. G., Mudd, J., Nichol, M., Chu, G. C., Sanes, J. R., and Merlie, J. P. (1995). Failure of post-synaptic specialization to develop at neuromuscular junctions of rapsyn deficient mice. *Nature* **377**, 232–236.

Geiger, J. R. P., Melcher, T., Koh, D.-S., Sakmann, B., Seeburg, P. H., Jonas, P., and Monyer, H. (1995). Relative abundance of subunit mRNAs determines gating and Ca^{2+} permeability of AMPA receptors in prinicpal neurons and interneurons in rat CNS. *Neuron* **15**, 193–204.

George, A. L., Crackower, M. A., Abdalla, J. A., Hudson, A. J., and Ebers, G. C. (1993). Molecular basis of Thomsen's disease (autosomal dominant myotonia congenita). *Nature Genet.* **3**, 305–310.

George, A. L., Komisarof, J., Kallen, R. G., and Barchi, R. L. (1992). Primary structure of the adult human skeletal muscle voltage-dependent sodium channel. *Ann. Neurol.* **31**, 131–137.

George, A. L., Ledbetter, D. H., Kallen, R. G., and Barchi, R. L. (1991). Assignment of a human skeletal muscle sodium channel α–subunit gene (SCN4A) to 17q23.1-25.3. *Genomics* **9**, 555–556.

George, A. L., Varkony, T. A., Drabkin, H. A., Han, J., Knops, J. F., Finley,

W. H., Brown, G. B., Ward, D. C., and Haas, M. (1995). Assignment of the human heart tetrodotoxin-resistant voltage-gated Na$^+$ channel α-subunit gene (SCN5A) to band 3p21. *Cytogenet. Cell Genet.* **68,** 67–70.

Gillard, E. F., Otsu, K., Fujii, J., Duff, C., de Leon, S., Khanna, V. K., Britt, B. A., Worton, R. G., and MacLennan, D. H. (1992). Polymorphisms and deduced amino acid substitutions in the coding sequence of the ryanodine receptor (RYR1) gene in individuals with malignant hyperthermia. *Genomics* **13,** 1247–1254.

Gillard, E. F., Otsu, K., Fujii, J., Khanna, V. K., de Leon, S., Derdemezi, J., Britt, B. A., Duff, C. L., Worton, R. G., and MacLennan, D. H. (1991). A substitution of cysteine for arginine 614 in the ryanodine receptor is potentially causative of human malignant hyperthermia. *Genomics* **11,** 751–755.

Goldman, M. J., Anderson, G. M., Stolzenberg, E. D., Kari, U. P., Zasloff, M., and Wilson, J. M. (1997). Human β-defensin-1 is a salt-sensitive antibiotic in lung that is inactivated in cystic fibrosis. *Cell* **88,** 553–560.

Gordon, S. E., and Zagotta, W. N. (1995). Localization of regions affecting an allosteric transition in cyclic nucleotide-activated channels. *Neuron* **14,** 857–864.

Gorter, J. A., Petrozzino, J. J., Aronica, E. M., Rosenbaum, D. M., Opitz, T., Bennett, M. V. L., Connor, J. A., and Zukin, R. S. (1997). Global ischaemia induces downregulation of Glur2 mRNA and increases AMPA receptor–mediated Ca^{2+} influx in hippocampal CA1 neurons of gerbil. *J. Neurosci.* **17,** 6179–6188.

Goulding, E. H., Tibbs, G. R., and Siegelbaum, S. A. (1994). Molecular mechanism of cyclic-nucleotide-gated channel activation. *Nature* **372,** 369–374.

Gowen, C. W., Lawson, E. E., Gingras, J., Boucher, R. C., Gatzy, J. T., and Knowles, M. R. (1988). Electrical potential difference and ion transport across nasal epithelium of term neonates: Correlation with mode of delivery, transient tachypnea of the newborn and respiratory rate. *J. Pediatr.* **113,** 121–127.

Gray, M. A., Harris, A., Coleman, L., Greenwell, J. R., and Argent, B. E. (1989). Two types of chloride channel on duct cells cultured from human fetal pancreas. *Am. J. Physiol.* **257,** C240–C251.

Greger, V., Knoll, J. H. M., Woolf, E., Glatt, K., Tyndale, R. F., DeLorey, T. M., Olsen, R. W., Tobin, A. J., Sikela, J. M., Nakatsu, Y., Brilliant, M. H., Whiting, P. J., and Lalande, M. (1995). The γ-aminobutyric acid receptor γ_3 subunit gene (GABRG3) is tightly linked to the α_5 subunit gene (GABRA5) on human chromosome 15q11-q13 and is transcribed in the same orientation. *Genomics* **26,** 258–264.

Gregg, R. G., Couch, F., Hogan, K., and Powers, P. A. (1993a). Assignment of the human gene for the α_1 subunit of the skeletal muscle DHP-sensitive Ca^{2+} channel (CACNL1A3) to chromosome 1q31-q32. *Genomics* **15,** 107–112.

Gregg, R. G., Powers, P. A., and Hogan, K. (1993b). Assignment of the human gene for the β subunit of the voltage-dependent calcium channel (CACNLB1) to chromosome 17 using somatic cell hybrids and linkage mapping. *Genomics* **15,** 185–187.

Gregory, R. J., Rich, D. P., Cheng, S. H., Souza, D. W., Paul, S., Manavalan, P., Anderson, M. P., Welsh, M. J., and Smith, A. E. (1991). Maturation and function of cystic fibrosis transmembrane conductance regulator variants bearing mutations in putative nucleotide-binding domains 1 and 2. *Mol. Cell. Biol.* **11**(8), 3886–3893.

Grenningloh, G., Rienitz, A., Schmitt, B., Methfessel, C., Zensen, M., Beyreuther, K., Gundelfinger, E. D., and Betz, H. (1987). The strychnine-binding subunit of the glycine receptor shows homology with nicotinic acetylcholine receptors. *Nature* **328,** 215–220.

Grenningloh, G., Schmieden, V., Schofield, P. R., Seeburg, P. H., Siddique, T., Mohandas, T. K., Becker, C. M., and Betz, H. (1990). Alpha subunit variants of the human glycine receptor: Primary structures, functional expression and chromosomal localization of the corresponding genes. *EMBO J.* **9,** 771–776.

Griffin, C. A., Ding, C. L., Jabs, E. W., Hawkins, A. L., Li, X., and Levine, M. A. (1993). Human rod cGMP-gated cation channel gene maps to 4p12-to-centromere by chromosomal *in situ* hybridization. *Genomics* **16,** 302–303.

Gronemeier, M., Condie, A., Prosser, J., Steinmeyer, K., Jentsch, T. J., and Jockusch, H. (1994). Nonsense and missense mutations in the muscular chloride channel gene ClC-1 of myotonic mice. *J. Biol. Chem.* **269,** 5963–5967.

Gründer, S., Thiemann, A., Pusch, M., and Jentsch, T. J. (1992). Regions involved in the opening of ClC-2 chloride channel by voltage and cell volume. *Nature* **360,** 759–762.

Gunderson, K. L., and Kopito, R. R. (1995). Conformational states of CFTR associated with channel gating: the role of ATP binding and hydrolysis. *Cell* **82,** 231–239.

Gundlach, A. L., Dodd, P. R., Grabara, C. S. G., Watson, W. E. J., Johnston, G. A. R., Harper, P. A. W., Dennis, J. A., and Healy, P. J. (1988). Deficit of spinal cord glycine/strychnine receptors in inherited myoclonus of Poll Hereford calves. *Science* **241,** 1807–1810.

Günther, U., Benson, J., Benke, D., Fritschy, J.-M., Reyes, G., Knoflach, F., and Crestani, F., Aguzzi, A., Arigoni, M., Lang, Y., Bluethmann, H., Mohler, H., and Lüscher, B. (1995). Benzodiazepine-insensitive mice generated by targeted disruption of the γ_2 subunit gene of γ-aminobutyric acid type A receptors. *Proc. Natl. Acad. Sci. USA* **92,** 7749–7753.

Hakamata, Y., Nakai, J., Takeshima, H., and Imoto, K. (1992). Primary structure and distribution of a novel ryanodine receptor/calcium release channel from rabbit brain. *FEBS Lett.* **312,** 229–235.

Hall, J. E., Vodyanoy, I., Balasubramanian, T. M., and Marshall, G. R. (1984). Alamethicin: A rich model for channel behavior. *Biophys. J.* **45,** 233–247.

Hamill, O. P., Marty, A., Neher, E., Sakmann, B., and Sigworth, F. J. (1981). Improved patch-clamp technique for high-resolution current recordings from cells and cell-free membrane patches. *Pflüg. Arch.* **391,** 85–100.

Hamill, O. P., and McBride, D. W. (1996). A supramolecular complex underlying touch sensitivity. *Trends Neurosci.* **19,** 258–261.

Han, J. Lu, C.-M., Brown, G. B., and Rado, T. A. (1991). Direct amplification of a single dissected chromosomal segment by polymerase chain reaction: a human brain sodium channel gene is on chromosome 2q22-q23. *Proc. Natl. Acad. Sci. USA* **88,** 335–339.

Handford, C. A., Lynch, J. W., Baker, E., Webb, G. C., Ford, J. H., Sutherland, G. R., and Schofield, P. R. (1996). The human glycine receptor β subunit: primary structure, functional characterisation and chromosomal location of the human and murine genes. *Mol. Brain Res.* **35,** 211–219.

Hanner, M., Schmalhofer, W. A., Munujos, P., Knaus, H. G., Kaczorowski, G. J., and Garcia, M. L. (1997). The β-subunit of the high conductance calcium-activated potassium channel contributes to the high-affinity receptor for charybdotoxin. *Proc. Natl. Acad. Sci. USA* **94,** 2853–2858.

Hänsch, G. M., Schonermark, S., and Roelcke, D. (1987). Paroxysmal nocturnal hemoglobinuria type III: lack of an erythrocyte membrane protein restricting the lysis by C5b-9. *J. Clin. Invest.* **80,** 7–12.

Hansson, J. H., Schild, L., Lu, Y., Wilson, T. A., Gautschi, I., Shimkets, R., Nelson-Williams, C., Rossier, B. C., and Lifton, R. P. (1995a). A *de novo* missense mutation of the β subunit of the epithelial sodium channel causes hypertension and Liddle syndrome identifying a proline-rich segment critical for regulation of channel activity. *Proc. Natl. Acad. Sci. U.S.A.* **92,** 11495–11499.

Hansson, J. H., Nelson-Williams, C., Suzuki, H., Schild, L., Shimkets, R., Lu, Y., Canessa, C., Iwasaki, T., Rossier, B. C., and Lifton, R. P. (1995b). Hypertension caused by a truncated epithelial sodium channel γ-subunit: genetic heterogeneity of Liddle syndrome. *Nature Genetics* **11,** 76–82.

Hardie, R. C., and Minke, B. (1993). Novel Ca^{2+} channels underlying transduction in

Drosophila photoreceptors: implications for phosphoinositide-mediated Ca^{2+} mobilization. *Trends Neurosci.* **16**(9), 371–376.

Hart, I. K., Waters, C., Vincent, A., Newland, C., Beeson, D., Pongs, O., Morris, C., and Newsom-Davis, J. (1997). Autoantibodies detected to expressed K$^+$ channels are implicated in neuromyotonia. *Ann. Neurol.* **41**, 238–246.

Hartenstein, B., Schenkel, J., Kuhse, J., Besenbeck, B., Kling, C., Becker, C. M., Betz, H., and Weiher, H. (1996). Low level expression of glycine receptor beta subunit transgene is sufficient for phenotype correction in *spastic* mice. *EMBO J.* **15**, 1275–1282.

Hartmann, H. A., Kirsch, G. E., Drewe, J. A., Taglialatela, M., Joho, R. H., and Brown, A. M. (1991). Exchange of conduction pathways between two related K$^+$ channels. *Science* **251**, 942–944.

Hartsel, S., and Bolard, J. (1996). Amphotericin B: New life for an old drug. *Trends Pharmacol. Sci.* **17**, 445–449.

Hay, A. J. (1992). The action of adamantanamines against influenza A viruses: Inhibition of the M2 ion channel protein. *Virology* **3**, 21–30.

Hayward, L. J., Brown, R. H., and Cannon, S. C. (1996). Inactivation defects caused by myotonia-associated mutations in the sodium channel III-IV linker. *J. Gen. Physiol.* **107**, 559–576.

Hayward, L. J., Brown, R. H., and Cannon, S. C. (1997). Slow inactivation differs among mutant Na channels associated with myotonia and periodic paralysis. *Biophys. J.* **72**, 1204–1219.

Hedblom, E., and Kirkness, E. F. (1997). A novel class of GABA$_A$ receptor subunit in tissues of the reproductive system. *J. Biol. Chem.* **272**, 15346–15350.

Heginbotham, L., Abramson, T., and MacKinnon, R. (1992). A functional connection between the pores of distantly related ion channels as revealed by mutant K$^+$ channels. *Science* **258**, 1152–1155.

Heginbotham, L., and MacKinnon, R. (1992). The aromatic binding site for tetraethylammonium ion on potassium channels. *Neuron* **8**, 483–491.

Heginbotham, L., Lu, Z., Abramson, T., and MacKinnon, R. (1994). Mutations in the K$^+$ channel signature sequence. *Biophys. J.* **66**, 1061–1067.

Heinemann, S. H., Terlau, H., Stühmer, W., Imoto, K., and Numa, S. (1992). Calcium channel characteristics conferred on the sodium channel by single mutations. *Nature* **356**, 441–443.

Herlitze, S., Hockerman, G. H., Scheuer, T., and Catterall, W. A. (1997). Molecular determinants of inactivation and G protein modulation in the intracellular loop connecting domains I and II of the calcium channel alpha 1A subunit. *Proc. Natl. Acad. Sci. USA* **94**, 1512–1516.

Heuser, J. E. (1977). Synaptic vesicle exocytosis revealed in quick-frozen frog neuromuscular junctions treated with 4-aminopyridine and given a single electrical shock. *Soc. Neurosci. Symp.* **2**, 215–239.

Heuser, J. E., and Salpeter, S. R. (1979). Organization of acetylcholine receptors in quick-frozen, deep-etched and rotary-replicated Torpedo postsynaptic membrane. *J. Cell Biol.* **82**, 150–173.

Higgins, C. F. (1992). ABC transporters: From microorganisms to man. *Annu. Rev. Cell Biol.* **8**, 67–113.

Higuchi, M., Single, F. N., Kohler, M., Sommer, B., and Seeburg, P. H. (1993). RNA editing of AMPA receptor subunit GluR-B: a base-paired intron–exon structure determines position and efficiency. *Cell* **75**, 1361–1370.

Hill, C. P., Yee, J., Selsted, M. E., and Eisenberg, D. (1991). Crystal structure of defensin HNP-3, an amphiphilic dimer: mechanisms of membrane permeabilization. *Science* **251**, 1481–1485.

Hille, B. (1992). "Ionic Channels of Excitable Membranes," 2nd Ed. Sinauer Associates, Sunderland, MA

Hirai, H., Kirsch, J., Laube, B., Betz, H., and Kuhse, J. (1996). The glycine binding site of the *N*-methyl-D-aspartate receptor subunit NR1: identification of novel determinants of co-agonist potentiation in the extracellular M3–M4 loop region. *Proc. Natl. Acad. Sci. USA* **93**, 6031–6036.

Hladky, S. B., and Haydon, D. A. (1970). Discreteness of conductance change in bimolecular lipid membranes in the presence of certain antibiotics. *Nature* **225,** 451–453.

Ho, K., Nichols, C. G., Lederer, W. J., Lytton, J., Vassilev, P. M., Kanazirska, M. V., and Hebert, S. C. (1993). Cloning and expression of an inwardly rectifying ATP-regulated potassium channel. *Nature* **362,** 31–38.

Hockerman, G. H., Peterson, B. Z., Johnson, B. D. and Catterall, W. A. (1997). Molecular determinants of drug binding and action on L-type calcium channels. *Annu. Rev. Pharmacol. Toxicol.* **37,** 361–396.

Hodgkin, A. L., and Huxley, A. F. (1952). A quantitative description of membrane current and its application to conduction and excitation in nerve. *J. Physiol.* **117,** 500–544.

Hollmann, M., Hartley, M., and Heinemann, S. (1991). Ca^{2+} permeability of KA-AMPA-gated glutamate receptor channels depends on subunit composition. *Science* **252,** 851–853.

Hollmann, M., and Heinemann, S. (1994). Cloned glutamate receptors. *Annu. Rev. Neurosci.* **17,** 31–108.

Hollmann, M., Maron, C., and Heinemann, S. (1994). *N*-glycosylation site tagging suggests a three transmembrane domain topology for the glutamate receptor GluR1. *Neuron* **13,** 1331–1343.

Hollmann, M., O'Shea-Greenfield, A., Rogers, S. W., and Heinemann, S. (1989). Cloning by functional expression of a member of the glutamate receptor family. *Nature* **342,** 643–648.

Holmgren, M., Jurman, M. E., and Yellen, G. (1996). N-type inactivation and the S4-S5 linker region of the *Shaker* K$^+$ channel. *J. Gen. Physiol.* **108,** 195–206.

Holsinger, L. J., Nichani, D., Pinto, L. H., and Lamb, R. A. (1994). Influenza A virus M2 ion channel protein: a structure–function analysis. *J. Virol.* **68,** 1551–1563.

Homanics, G. E., DeLorey, T. M., Firestone, L. L., Quinlan, J. J., Handforth, A., Harrison, N. L., Krasowski, M. D., Rick, C. E. M., Korpi, E. R., Mäkelä, R., Brilliant, M. H., Hagiwara, N., Ferguson, C., Snyder, K., and Olsen, R. W. (1997). Mice devoid of γ-aminobutyrate type A receptor β_3 subunit have epilepsy, cleft palate, and hypersensitive behavior. *Proc. Natl. Acad. Sci. USA* **94,** 4143–4148.

Hong, K., and Driscoll, M. (1994). A transmembrane domain of the putative channel subunit MEC-4 influences mechanotransduction and neurodegeneration in *C. elegans. Nature* **367,** 470–473.

Hoshi, T., Zagotta, W. N., and Aldrich, R. W. (1990). Biophysical and molecular mechanisms of *Shaker* potassium channel inactivation. *Science* **250,** 533–538.

Hoshi, T., Zagotta, W. N., and Aldrich, R. W. (1991). Two types of inactivation in *Shaker* K$^+$ channels: effects of alterations in the carboxy-terminal region. *Neuron* **7,** 547–556.

Hosie, A. M., Aronstein, K., Sattelle, D. B., and ffrench-Constant, R. H. (1997). Molecular biology of insect neuronal GABA receptors. *Trends Neurosci.* **20,** 578–583.

Hsaio, C. D., Sun, Y. J., Rose, J., and Wang, B. C. (1996). The crystal structure of glutamine-binding protein from Escherichia coli. *J. Mol. Biol.* **262,** 255–242.

Hsieh, C.-L., Kumar, N. M., Gilula, N. B., and Francke, U. (1991). Distribution of genes for gap junction membrane channel proteins on human and mouse chromosomes. *Somat. Cell Molec. Genet.* **17,** 191–200.

Huang, C.-L., Slesinger, P. A., Casey, P. J., Jan, Y. N. and Jan, L. Y. (1995). Evidence that direct binding of Gβγ to the GIRK1 G protein-gated inwardly rectifying K$^+$ channel is important for channel activation. *Neuron* **15,** 1133–1143.

Huang, L. Y. M., Catterall, W. A., and Ehrenstein, G. (1978). Selectivity of cations and nonelectrolytes for acetylcholine-activated channels in cultured muscle cells. *J. Gen. Physiol.* **71,** 397–410.

Huang, M., and Chalfie, M. (1994). Gene interactions affecting mechanosensory transduction in *Caenorhabditis elegans. Nature* **367,** 467–470.

Hughes, J., Ward, C. J., Peral, B., Aspinwall, R., Clark, K., San Millán, J. L., Gamble, V., and Harris, P. C. (1995). The polycystic kidney disease 1 (*PKD1*) gene encodes a novel protein with multiple cell recognition domains. *Nature Genet.* **10,** 151–160.

Hummler, E., Barker, P., Gatzy, J., Beermann, F., Verdumo, C., Schmidt, A., Boucher, R., and Rossier, B. C. (1996). Early death due to defective neonatal lung liquid clearance in αENaC-deficient mice. *Nature Genet.* **12**, 325–328.

Hwang, T. C., Nagel, G., Nairn, A. C., and Gadsby, D. C. (1994). Regulation of the gating of the cystic fibrosis transmembrane conductance regulator Cl channels by phosphorylation and ATP hydrolysis. *Proc. Natl. Acad. Sci. USA* **91**, 4698–4702.

Hyde, S. C., Gill, D. R., Higgins, C. F., Trezise, A. E. O., MacVinish, L. J., Cuthbert, A. W., Ratcliff, R., Evans, M. J., and Colledge, W. H. (1993). Correction of the ion transport defect in cystic fibrosis transgenic mice by gene therapy. *Nature* **362**, 250–255.

Ikeda, S., Nasrallah, J. B., Dixit, R., Preiss, S., and Nasrallah, M. E. (1997). An aquaporin-like gene required for the *Brassica* self-incompatibility response. *Science* **276**, 1564–1566.

Imoto, K., Busch, C., Sakmann, B., Mishina, M., Konno, T., Nakai, J., Bujo, H., Mori, Y., Fukuda, K., and Numa, S. (1988). Rings of negatively charged amino acids determine the acetylcholine receptor conductance. *Nature* **335**, 645–648.

Imoto, K., Konno, T., Nakai, J., Wang, F., Mishina, M., and Numa, S. (1991). A ring of uncharged polar amino acids as a component of channel constriction in the nicotinic acetylcholine receptor. *FEBS Lett.* **289**, 193–200.

Inagaki, N., Gonoi, T., Clement, IV, J. P., Namba, N., Inazawa, J., Gonzalez, G., Aguilar-Bryan, L., Seino, S., and Bryan, J. (1995). Reconstitution of IK$_{ATP}$: an inward rectifier subunit plus the sulphonylurea receptor. *Science* **270**, 1166–1170.

Inagaki, N., Gonoi, T., Clement, IV, J. P., Wang, C. Z., Aguilar-Bryan, L., Bryan, J. and Seino, S. (1996). A family of sulfonylurea receptors determines the pharmacological properties of ATP-sensitive K$^+$ channels. *Neuron* **16**, 1011–1017.

International Collaborative Study Group For Bartter-like Syndrome. (1997). Mutations in the gene encoding the inwardly rectifying renal potassium chan-nel ROMK cause the antenatal variant of Bartter syndrome: evidence for genetic heterogeneity. *Human Mol. Genet.* **6**, 17–26.

Ionasescu, V., Ionasescu, R., and Searby, C. (1996). Correlation between connexin 32 gene mutations and clinical phenotype in X-linked dominant Charcot-Marie-Tooth neuropathy. *Am. J. Med. Genetics* **63**, 486–491.

Isacoff, E. Y., Jan, Y. N., and Jan, L. Y. (1991). Putative receptor for the cytoplasmic inactivation gate in the *Shaker* K$^+$ channel. *Nature* **353**, 86–90.

Ishibashi, K., Sasaki, S., Saito, F., Ikeuchi, T., and Marumo, F. (1995). Structure and chromosomal localization of a human water channel (AQP3) gene. *Genomics* **27**, 352–354.

Ishii, T. M., Maylie, J., and Adelman, J. P. (1997a). Determinants of apamin and d-tubocurarine block in SK potassium channels. *J. Biol. Chem..* **272**, 23195–23200.

Ishii, T. M., Silvia, C., Hirschberg, B., Bond, C. T., Adelman, J. P., and Maylie, J. (1997b). A human intermediate conductance calcium-activated potassium channel. *Proc. Natl. Acad. Sci. USA* **94**, 11651–11656.

Ismailov, I. I., Awayda, M. S., Jovov, B., Berdiev, B. K., Fuller, C. M., Dedman, J. R., Kaetzel, M. A., and Benos, D. J. (1996). Regulation of epithelial sodium channels by the cystic fibrosis transmembrane conductance regulator. *J. Biol. Chem.* **271**, 4725–4732.

Isom, L. L., de Jongh, K. S., Patton, D. E., Reber, B. F. X., Offord, J., Charbonneau, H., Walsh, K., Goldin, A. L., and Catterall, W. A. (1992). Primary structure and functional expression of the β1-subunit of the rat brain sodium channel. *Science* **256**, 839–842.

Jansen, G., Groenen, P. J., Bachner, D., Jap, P. H., Coerwinkel, M., Oerlemans, F., van den Broek, W., Gohlsch, B., Pette, D., Plomp, J. J., Molenaar, P. C., Nederhoff, M. G., van Echteld, C. J., Dekker, M., Berns, A., Hameister, H., and Wieringa, B. (1996). Abnormal myotonic dystrophy protein kinase levels produce only mild myopathy in mice. *Nat. Genet.* **13**, 316–24.

Jarvi, K., Zielenski, J., Wilschanski, M., Durie, P., Buckspan, M., Tullis, E., Markiewicz, D., and Tsui, L. C. (1995). Cystic fibrosis transmembrane conductance regulator and obstructive azoospermia. *Lancet* **345,** 1578.

Jentsch, T. J., and Günther, W. (1997). Chloride channels: an emerging molecular picture. *Bioessays* **19,** 117–126.

Jentsch, T. J., Steinmeyer, K., and Schwarz, G. (1990). Primary structure of *Torpedo marmorata* chloride channel isolated by expression cloning in *Xenopus* oocytes. *Nature* **348,** 510–514.

Jeremiah, S. J., Abbott, C. M., Murad, Z., Povey, S., Thomas, H. J., Solomon, E., DiScipio, R. G., and Fey, G. H. (1990). The assignment of the genes coding for human complement components C6 and C7 to chromosome 5. *Ann. Hum. Genet.* **54,** 141–147.

Ji, S., George, A. L., Horn, R., and Barchi, R. L. (1996). Paramyotonia congenita mutations reveal different roles for segments S3 and S4 of domain D4 in hSkM1 sodium channel gating. *J. Gen. Physiol.* **107,** 183–194.

Ji, S., Sun, W., George, A. L., Horn, R., and Barchi, R. L. (1994). Voltage-dependent regulation of modal gating in the rat SkM1 sodium channel expressed in *Xenopus* oocytes. *J. Gen. Physiol.* **104,** 625–643.

Johnson, J. W., and Ascher, P. (1987). Glycine potentiates the NMDA response in cultured mouse brain neurons. *Nature* **325,** 529–531.

Johnston, I., Lang, B., Leys, K., and Newsom-Davis, J. (1994). Heterogeneity of calcium channel autoantibodies detected using a small-cell lung cancer line derived from a Lambert–Eaton myasthenic syndrome patient. *Neurology* **44,** 334–338.

Jonas, E. A., and Kaczmarek, L. K. (1996). Regulation of potassium channels by protein kinases. *Curr. Op. Neurobiol.* **6,** 318–323.

Jonas, E. A., Knox, R. J., and Kaczmarek, L. K. (1997). Giga-ohm seals on intracellular membranes: a technique for studying intracellular ion channels in intact cells. *Neuron* **19,** 7–13.

Joris, L., Dab, I., and Quinton, P. M. (1993). Elemental composition of human airway surface fluid in healthy and diseased airways. *Annu. Rev. Respir. Dis.* **148,** 1633–1637.

Joutel, A., Bousser, M. G., Biousse, V., Labauge, P., Chabriat, H., Nibbio, A., Maciazek, J., Meyer, B., Bach, M. A, Weissenbach, J., Lathrop, G. M., and Tournier-Lasserve, E. (1993). A gene for familial hemiplegic migraine maps to chromosome 19. *Nature Genetics* **5,** 40–45.

Jung, J. S., Bhat, R. V., Preston, G. M., Guggino, W. B., Baraban, J. M., and Agre, P. (1994a). Molecular characterization of an aquaporin cDNA from brain: candidate osmoreceptor and regulator of water balance. *Proc. Natl. Acad. Sci. USA* **91,** 13052–13056.

Jung, J. S., Preston, G. M., Smith, B. L., Guggino, W. B., and Agre, P. (1994b). Molecular structure of the water channel through aquaporin CHIP. *J. Biol. Chem.* **269,** 14648–14654.

Juntti-Berggren, L., Larsson, O., Rorsman, P., Ämmälä, C., Bokvist, K., Wåhlander, K., Nicotera, P., Dypbukt, J., Orrenius, S., Hallberg, A., and Berggren, P. O. (1993). Increased activity of L-type Ca^{2+} channels exposed to serum from patients with type I diabetes. *Science* **261,** 86–90.

Jurkat-Rott, K., Lehmann-Horn, F., Elbaz, A., Heine, R., Gregg, R. G., Hogan, K., Powers, P. A., Lapie, P., Vale-Santos, J. E., Weissenbach, J., and Fontaine, B. (1994). A calcium channel mutation causing hypokalemic period paralysis. *Hum. Mol. Genet.* **3,** 1415–1419.

Jurkat-Rott, K., Uetz, U., Pika-Hartlaub, U., Powell, J., Fontaine, B., Melzer, W., and Lehmann-Horn, F. (1998). Calcium currents and transients of native and heterologously expressed mutant skeletal muscle DHP receptor $\alpha 1$ subunits (R528H). *FEBS Lett.* **423,** 198–204.

Kaczmarek, L. K., and Levitan, I. B. (1987). "Neuromodulation: The Biochemical Control of Neuronal Excitability." Oxford University Press, New York.

Kagan, B. L., Ganz, T., and Lehrer, R. I. (1994). Defensins: a family of antimicrobial and cytotoxic peptides. *Toxicology* **87,** 131–149.

Kägi, D., Ledermann, B., Bürki, K., Seiler, P., Odermatt, B., Olsen, K. J., Podack, E. R., Zinkernagel, R. M., and Hengartner, H. (1994). Cytotoxicity mediated by T cells and natural killer cells is greatly impaired in perforin-deficient mice. *Nature* **369**, 31–37.

Kägi, D., Ledermann, B., Bürki, K., Zinkernagel, R. M., and Hengartner, H. (1996). Molecular mechanisms of lymphocyte-mediated cytotoxicity and their role in immunological protection and pathogenesis *in vivo*. *Annu. Rev. Immunol.* **14**, 207–232.

Kakei, M., Kelly, R. P., Ashcroft, S. J. H., and Ashcroft, F. M. (1986). The ATP-sensitivity of K⁺ channels in rat pancreatic β-cells is modulated by ADP. *FEBS Lett.* **208**, 63–66.

Kane, C., Shepherd, R. M., Squires, P. E., Johnson, P. R. V., James, R. F. L., Milla, P. J., Aynsley-Green, A., Lindley, K. J., and Dunne, M. J. (1996). Loss of functional K$_{ATP}$ channels in pancreatic β-cells causes persistent hyperinsulinemic hypoglycemia of infancy. *Nature Medicine* **2** (12), 1344–1347.

Kanno, K., Sasaki, S., Hirata, Y., Ishikawa, S., Fushimi, K., Nakanishi, S., Bichet, D. G., and Marumo, F. (1995). Urinary excretion of aquaporin-2 in patients with diabetes insipidus. *N. Engl. J. Med.* **332**, 1540–1545.

Karlin, A., and Akabas, M. H. (1995). Toward a structural basis for the function of nicotinic acetylcholine receptors and their cousins. *Neuron* **15**, 1231–1244.

Karp, S. J., Masu, M., Eki, T., Ozawa, K., and Nakanishi, S. (1993). Molecular cloning and chromosomal localization of the key subunit of the human N-methyl-D-aspartate receptor. *J. Biol. Chem.* **268**, 3728–3733.

Kartner, N., Hanrahan, J. W., Jensen, T. J., Naismith, A. L., Sun, S. Z., Ackerley, C. A., Reyes, E. F., Tsui, L. C., Rommens, J. M., Bear, C. E., and Riordan, J. R. (1991). Expression of the cystic fibrosis gene in non-epithelial invertebrate cells produces a regulated anion conductance. *Cell* **64**, 681–91.

Kaufman, K. M., Snider, J. V., Spurr, N. K., Schwartz, C. E., and Sodetz, J. M. (1989). Chromosomal assignment of genes encoding the α, β and γ subunits of human complement protein C8: identification of a close physical linkage between the α and β loci. *Genomics* **5**, 475–480.

Kaupp, U. B. (1995). Family of cyclic nucleotide gated ion channels. *Curr. Op. Neurobiol.* **5**, 434–442.

Kaupp, U. B., Niidome, T., Tanabe, T., Terada, S., Bönigk, W., Stühmer, W., Cook, N. J., Kangawa, K., Matsuo, H., Hirose, T., Miyata, T., and Numa, S. (1989). Primary structure and functional expression from complementary DNA of the rod photoreceptor cyclic GMP-gated channel. *Nature* **342**, 762–766.

Kawahara, M., Arispe, N., Kuroda, Y., and Rojas, E. (1997). Alzheimer's disease amyloid β-protein forms Zn²⁺-sensitive, cation-selective channels across excised membrane patches from hypothalamic neurons. *Biophys. J.* **73**, 67–75.

Keating, K. E., Quane, K. A., Manning, B. M., Lehane, M., Hartung, E., Censier, K., Urwyler, A., Klausnitzer, M., Muller, C. R., Heffron, J. J. A., and McCarthy, T. V. (1994). Detection of a novel *RYR1* mutation in four malignant hyperthermia pedigrees. *Hum. Mol. Genet.* **3**, 1855–1858.

Kelsell, D. P., Dunlop, J., Stevens, H. P., Lench, N. J., Liang, J. N., and Parry, G., Mueller, R. F., and Leigh, I. M. (1997). Connexin 26 mutations in hereditary non-syndromic sensorineural deafness. *Nature* **387**, 80–83.

Kemp, P. J., and Olver, R. E. (1996). G protein regulation of alveolar ion channels: implications for lung fluid transport. *Experimental Physiol.* **81**, 493–504.

King, L. S., and Agre, P. (1996). Pathophysiology of the aquaporin water channels. *Annu. Rev. Physiol.* **58**, 619–648.

King, L. S., Nielsen, S., and Agre, P. (1996). Aquaporin-1 water channel protein in the lung: Ontogeny steroid-induced expression and distribution in the rat. *J. Clin. Invest.* **97**, 2183–2191

Kingsmore, S. F., Giros, B., Suh, D., Bieniarz, M., Caron, M. G., and Seldin, M. F. (1994). Glycine receptor beta-subunit gene mutation in *spastic* mouse associated with

LINE-1 element insertion. *Nature Genet.* **7,** 136–141.

Kirsch, J., Kuhse, J., and Betz, H. (1995). Targeting of the glycine receptor subunits to gephyrin-rich domains in transfected human embryonic kidney cells. *Mol. Cell. Neurosci.* **6,** 450–461.

Kirsch, J., Wolters, I., Triller, A., and Betz, H. (1993). Gephyrin antisense oligonucleotides prevent glycine receptor clustering in spinal neurons. *Nature* **366,** 745–748.

Knaus, H. G., Folander, K., Garcia-Calvo, M., Garcia, M. L., Kaczorowski, G. J., Smith, M., and Swanson, R. (1994a). Primary sequence and immunological characterisation of the β-subunit of high conductance Ca^{2+}-activated K^+ channel from smooth muscle. *J. Biol. Chem.* **269,** 17274–17278.

Knaus, H. G., Garcia-Calvo, M., Kaczorowski, G. J., and Garcia, M. L. (1994b). Subunit composition of the high conductance Ca-activated K^+ channel from smooth muscle, a representative of the *mSlo* and *slowpoke* family of potassium channels *J. Biol. Chem.* **269,** 3921–3924.

Knoll, J. H. M., Sinnett, D., Wagstaff, J., Glatt, K., Wilcox, A. S., Whiting, P. M., Wingrove, P., Sikela, J. M., and Lalande, M. (1993). FISH ordering of reference markers and of the gene for the α_5 subunit of the γ-aminobutyric acid receptor (GABRA5) within the Angelman and Prader–Willi syndrome chromosomal regions. *Hum. Mol. Genet.* **2,** 183–189.

Ko, Y. H., and Pedersen, P. L. (1995). The first nucleotide binding fold of the cystic fibrosis transmembrane conductance regulator can function as an active ATPase. *J. Biol. Chem.* **270**(38), 22093–22096.

Koch, M. C., Steinmeyer, K., Lorenz, C., Ricker, K., Wolf, F., Otto, M., Zoll, B., Lehmann-Horn, F., Grzeschik, K.-H., and Jentsch, T. J. (1992). The skeletal muscle chloride channel in dominant and recessive human myotonia. *Science* **257,** 797–800.

Kofuji, P., Davidson, N., and Lester, H. A. (1995). Evidence that neuronal G-protein-gated inwardly rectifying K^+ channels are activated by $G\beta\gamma$ subunits and function as heteromultimers. *Proc. Natl. Acad. Sci. U.S.A.* **92,** 6542–6546.

Kofuji, P., Hofer, M., Millen, K. J., Millonig, J. H., Davidson, N., Lester, H. A., and Hatten, M. E. (1996). Functional analysis of the weaver mutant GIRK2 K^+ channel and rescue of weaver granule cells. *Neuron* **16,** 941–952.

Köhler, M., Burnashev, N., Sakmann, B., and Seeburg, P. H. (1993). Determinants of Ca^{2+} permeability in both TM1 and TM2 of high-affinity kainate receptor channels: diversity by RNA editing. *Neuron* **10,** 491–500.

Köhler, M., Hirschberg, B., Bond, C. T., Kinzie, J. M., Marrion, N. V., Maylie, J., and Adelman, J. P. (1996). Small-conductance calcium-activated potassium channels from mammalian brain. *Science* **273,** 1709–1714.

Köhr, G., and Seeburg, P. H. (1996). Subtype-specific regulation of recombinant NMDA receptor-channels by protein tyrosine kinases of the src family. *J. Physiol.* **492.2,** 445–452.

Kohrman, D. C., Smith, M. R., Goldin, A. L., Harris, J., and Meisler, M. H. (1996). A missense mutation in the sodium channel Scn8a is responsible for cerebellar ataxia in the mouse mutant *jolting*. *J. Neurosci.* **16,** 5993–5999.

Komatsu, H., Mori, I., Rhee, J. S., Akaike, N., and Ohshima, Y. (1996). Mutations in a cyclic nucleotide-gated channel lead to abnormal thermosensation and chemosensation in *C. elegans*. *Neuron* **17,** 707–718.

Konno, T., Busch, C., von Kitzing, E., Imoto, K., Wang, F., Nakai, J., Mishina, M., Numa, S., and Sakmann, B. (1991). Rings of anionic amino acids as structural determinants of ion selectivity in the acetylcholine receptor channel. *Proc. R. Soc. Lond. B* **244,** 69–79.

Kornau, H.-C., Schenker, L. T., Kennedy, M. B., and Seeburg, P. H. (1995). Domain interaction between NMDA receptor subunits and the postsynaptic density protein PSD-95. *Science* **269,** 1737–1740.

Korpi, E. R., Kleingoor, C., Kettenmann, H., and Seeburg, P. H. (1993). Benzodiazepine-induced motor impairment linked to point mutation in cerebellar GABA$_A$ receptor. *Nature* **361,** 356–359.

Körschen, H. G., Illing, M., Seifert, R., Sesti, F., Williams, A., Gotzes, S., Colville, C., Müller, F., Dosé, A., Godde, M., Molday, L., Kaupp, U. B., and Molday, R. S. (1995). A 240 kDa protein represents the complete β-subunit of the cyclic nucleotide-gated channel from rod photoreceptor. *Neuron* **15,** 627–636.

Kostrzewa, M., Kohler, A., Eppelt, K., Hellam, L., Fairweather, N. D., Levy, E. R., Monaco, A. P., and Muller, U. (1996). Assignment of genes encoding GABA$_A$ receptor subunits α_1, α_6, β_2, and γ_2 to a YAC contig of 5q33. *Eur. J. Hum. Genet.* **4,** 199–204.

Koyama, K., Sudo, K., and Nakamura, Y. (1994). Mapping of the human nicotinic acetylcholine receptor β_3 gene (CHRNB3) within chromosome 8p11.2. *Genomics* **21,** 460–461.

Krapivinsky, G., Gordon, E. A., Wickman, K., Velimirovic, B., Krapivinsky, L., and Clapham, D. E. (1995a). The G-protein-gated atrial K$^+$ channel I$_{KACh}$ is a heteromultimer of two inwardly-rectifying K$^+$ channel proteins. *Nature* **374,** 135–141.

Krapivinsky, G., Krapivinsky, L., Wickman, K., and Clapham, D. E. (1995b). G$_{\beta\gamma}$ binds directly to the G protein-gated K$^+$ channel, I$_{KACh}$. *J. Biol. Chem.* **270,** 29059–29062.

Krawczak, M., and Cooper D. N. (1997). The human gene mutation database. *Trends Genet.* **13,** 121–122.

Kubo, Y., Baldwin, T. J., Jan, Y. N., and Jan, L. Y. (1993a). Primary structure and functional expression of mouse inward rectifier potassium channel. *Nature* **362,** 127–133.

Kubo, Y., Reuveny, E., Slesinger, P. A., Jan, Y. N., and Jan L. Y. (1993b). Primary structure and functional expression of a rat G-protein-coupled muscarinic potassium channel. *Nature* **364,** 802–806.

Kuhse, J., Betz, H., and Kirsch, J. (1995). The inhibitory glycine receptor: architecture, synaptic localization and molecular pathology of a postsynaptic ion-channel complex. *Curr. Opin. Neurobiol.* **5,** 318–323.

Kuhse, J., Schmieden, V., and Betz, H. (1990). A single amino acid exchange alters the pharmacology of neonatal rat glycine receptor subunit. *Neuron* **5,** 867–873.

Kumar, N. M., and Gilula, N. B. (1996). The gap junction communication channel. *Cell* **84,** 381–388.

Kuner, T., Wollmuth, L. P., Karlin, A., Seeburg, P. H., and Sakmann, B. (1996). Structure of the NMDA receptor channel M2 segment inferred from the accessibility of substituted cysteines. *Neuron* **17,** 343–352.

Kuryatov, A., Gerzanich, V., Nelson, M., Olale, F., and Lindstrom, J. (1997). Mutation causing autosomal dominant nocturnal frontal lobe epilepsy alters Ca^{2+} permeability, conductance, and gating of human $\alpha4\beta2$ nicotinic acetylcholine receptors. *J. Neurosci.* **17,** 9035–9047.

Kuryatov, A., Laube, B., Betz, H., and Kuhse, J. (1994). Mutational analysis of the glycine-binding site of the NMDA receptor: structural similarity with bacterial amino acid–binding proteins. *Neuron* **12,** 1291–1300.

Kutsuwada, T., Kashiwabuchi, N., Mori, H., Sakimura, K., Kushiya, E., Araki, K., Meguro, H., Masaki, H., Kumanishi, T., Arakawa, M., and Mishina, M. (1992). Molecular diversity of the NMDA receptor channel. *Nature* **358,** 36–41.

Lamb, R. A., Holsinger, L. J., and Pinto, L. H. (1994). The influenza A virus M2 ion channel protein and its role in the influenza virus life cycle. *In* "Cellular Receptors for Animal Viruses," pp. 303–321. Cold Spring Harbor Laboratory Press, Cold Spring Harbor, NY.

Lang, B., Newsom-Davis, J., Prior, C., and Wray, D. (1983). Antibodies to motor nerve terminals: an electrophysiological study of a human myasthenic syndrome transferred to mouse. *J. Physiol.* **344,** 335–345.

Langosch, D., Laube, B., Rundström, N., Schmieden, V., Bormann, J., and Betz, H. (1994). Decreased agonist affinity and chloride conductance of mutant glycine receptors associated with human hereditary hyperekplexia. *EMBO J.* **13,** 4223–4228.

Larrson, H. P., Baker, O. S., Dhillon, D. S., and Isacoff, E. Y. (1996). Transmembrane movement of the *Shaker* K$^+$ channel S4. *Neuron* **16**, 387–397.

Laube, B., Hirai, H., Sturgess, M., Betz, H., and Kuhse, J. (1997). Molecular determinants of agonist discrimination by NMDA receptor subunits: analysis of the glutamate binding site on the NR2B subunit. *Neuron* **18**, 493–503.

Laurie, D. J., and Seeburg, P. H. (1994). Regional and developmental heterogeneity in splicing of the rat brain NMDAR1 mRNA. *J. Neurosci.* **14**(5), 3180–3194.

Lee, M. D., Bhakta, K. Y., Yonescu, R., Griffin, C. A., Copeland, N. G., Gilbert, D. J., Jenkins, N. A., Preston, G. M., and Agre, P. (1996). The human aquaporin-5 gene: Molecular characterization and chromosomal localization. *J. Biol. Chem.* **271**, 8599–8604.

Lee, W. S., and Hebert, S. C. (1995). ROMK inwardly rectifying ATP-sensitive K$^+$ channel. Expression in rat distal nephron segments. *Am. J. Physiol.* **268**, F1124–1131.

Lehmann-Horn, F., Rüdel, R., and Ricker, K. (1987). Membrane defects in paramyotonia congenita (Eulenburg). *Muscle Nerve* **10**, 633–641.

Lehmann-Horn, F., Rüdel, R., Ricker, K., Lorkovic, H., Dengler, R., and Hopf, H. C. (1983). Two cases of adynamic episodica hereditaria: in vitro investigation of muscle membrane and contractile parameters. *Muscle Nerve* **6**, 113–121.

Lennon, V. A., Kryser, T. J., Griesmann, G. E., O'Suilleabhain, P. E., Windebank, A. J., Woppman, A., Miljanich, G. P., and Lambert, E. H. (1995). Calcium channel antibodies in the Lambert–Eaton myasthenic syndrome and other paraneoplastic syndromes. *N. Engl. J. Med.* **332**, 1467–1474.

Letts, V. A., Felix, R., Biddlecome, G. H., Arikkath, J., Mahaffey, C. L., Valenzuela, A., Bartlett, F. S., Mori, Y., Campbell, K. P., and Frankel, W. N. (1998). The mouse stargazer gene encodes a neuronal Ca^{2+} channel γ-subunit. *Nature Genet.* **19**, 340–347.

Levin, G., Keren, T., Peretz, T., Chikvashvili, D., Thornhill, W. B., and Lotan, I. (1995).

Regulation of RCK1 currents with a cAMP analog via enhanced protein synthesis and direct channel phosphorylation. *J. Biol. Chem.* **270**, 14611–14618.

Lewis, C., Neidhart, S., Holy, C., North, R. A., Buell, G., and Surprenant, A. (1995). Coexpression of P2X$_2$ and P2X$_3$ receptor subunits can account for ATP-gated currents in sensory neurons. *Nature* **377**, 432–435.

Li, M., Jan, Y. N., and Jan, L. Y. (1992). Specification of subunit assembly by the hydrophilic amino-terminal domain of the *Shaker* potassium channel. *Science* **257**, 1225–1230.

Li, M., Unwin, N., Stauffer, K. A., Jan, Y. N., and Jan, L. Y. (1994). Images of purified *Shaker* potassium channels. *Curr. Biol.* **4**, 110–115.

Liddle, G. W., Bledsoe, T., and Coppage, W. S. (1963). A familial renal disorder simulating primary aldosteronism but with negligible aldosterone secretion. *Trans. Assoc. Am. Physicians* **76**, 199–213.

Lindau, M., and Gomperts, B. D. (1991). Techniques and concepts in exocytosis: focus on mast cells. *Biochim. Biophys. Acta* **1071**, 429–471.

Lingueglia, E., Champigny, G., Lazdunski, M., and Barbry, P. (1995). Cloning of the amiloride-sensitive FMRFamide peptide-gated sodium channel. *Nature* **378**, 730–733.

Linsdell, P., Tabcharani, J. A., Rommens, J. M., Hou, Y. X., Chang, X. B., Tsui, L. C., Riordan, J. R., and Hanrahan, J. W. (1997). Permeability of wild-type and mutant cystic fibrosis transmembrane conductance regulator chloride channels to polyatomic anions. *J. Gen. Physiol.* **110**, 355–364.

Lipicky, R. J., Bryant, S. H., and Salmon, J. H. (1971). Cable parameters, sodium, potassium, chloride and water content, and potassium efflux in isolated external intercostal muscle of normal volunteers and patients with myotonia congenita. *J. Clin. Invest.* **50**, 2091–2103.

Litt, M., Luty, J., Kwak, M., Allen, L., Magenis, R. E., and Mandel, G. (1989). Localization of a human brain sodium channel gene (SCN2A) to chromosome 2. *Genomics* **5**, 204–208.

Liu, Y., Jurman, M. E., and Yellen, G. (1996a). Dynamic rearrangement of the outer mouth of a K^+ channel during gating. *Neuron* **16**, 859–867.

Liu, D. T., Tibbs, G. R., and Siegelbaum, S. A. (1996b). Subunit stoichiometry of cyclic-nucleotide gated channels and the effects of subunit order on channel function. *Neuron* **16**, 983–990.

Llinás, R., Sugimori, M., Cherksey, B. D., Smith, R. G., Delbono, D., Stefani, E., and Appel, S. (1993). IgG from amyotrophic lateral sclerosis patients increases current through P-type calcium channels in mammalian cerebellar Purkinje cells and in isolated channel protein in lipid bilayer. *Proc. Natl. Acad. Sci. USA* **90**, 11743–11747.

Lloyd, S. E., Pearce, S. H. S., Fisher, S. E., Steinmeyer, K., Schwappach, B., Scheinman, S. J., Harding, B., Bolino, A., Devoto, M., Goodyer, P., Rigden, S. P. A., Wrong, O., Jentsch, T. J., Craig, I. W., and Thakker, R. V. (1996). A common molecular basis for three inherited kidney stone diseases. *Nature* **379** 445–449.

Lloyd, S. E., Günther, W., Pearce, S. H. S., Thomson, A., Bianchi, M. L., Bosio, M., Craig, I. W., Fisher, S. E., Scheinman, S. J., Wrong, O., Jentsch, T. J., and Thakker, R. V. (1997a). Characterisation of renal chloride channel, *CLCN5*, mutations in hypercalciuric nephrolithiasis (kidney stones) disorders. *Hum. Molec. Genet.* **6**, 1233–1239.

Lloyd, S. E., Pearce, S. H. S., Günther, W., Kawaguchi, H., Igarashi, T., Jentsch, T. J., and Thakker, R. V. (1997b). Idiopathic low molecular weight proteinuria associated with hypercalciuric nephrocalcinosis in Japanese children is due to mutations of the renal chloride channel (CLCN5). *J. Clin. Invest.* **99**, 967–974.

Lobos, E. A. (1993). Five subunit genes of the human muscle nicotinic acetylcholine receptor are mapped to two linkage groups on chromosomes 2 and 17. *Genomics* **17**, 642–650.

Lobos, E. A., Rudnick, C. H., Watson, M. S., and Isenberg, K. E. (1989). Linkage disequilibrium study of RFLPs detected at the human muscle nicotinic acetylcho-line receptor subunit genes. *Am. J. Hum. Genet.* **44**, 522–533.

Logan, J., Hiestand, D., Daram, P., Huang, Z., Muccio, D. D., Hartman, J., Haley, B., Cook, W. J., and Sorscher, E. J. (1994). Cystic fibrosis transmembrane conductance regulator mutations that disrupt nucleotide binding. *J. Clin. Invest.* **94**, 228–236.

Lomeli, H., Mosbacher, J., Melcher, T., Hoger, T., Geiger, J. R. P., Kuner, T., Monyer, H., Higuchi, M., Bach, A., and Seeburg, P. H. (1994). Control of kinetic properties of AMPA receptor channels by nuclear RNA editing. *Science* **266**, 1709–1713.

Lopatin, A. N., Makhina, E. N., and Nichols, C. G. (1994). Potassium channel block by cytoplasmic polyamines as the mechanism of intrinsic rectification. *Nature* **372**, 366–369.

Lorenz, C., Meyer-Kleine, C., Steinmeyer, K., Koch, M. C., and Jentsch, T. J. (1994). Genomic organization of the human muscle chloride channel ClC-1 and analysis of novel mutations leading to Becker-type myotonia. *Hum. Mol. Gent.* **3**, 941–946.

Lorenz, C., Pusch, M., and Jentsch, T. J. (1996). Heteromultimeric CLC chloride channels with novel properties. *Proc. Natl. Acad. Sci. U.S.A.* **93**, 13362–13366.

Lu, M., Lee, M. D., Smith, B. L., Jung, J. S., Agre, P., Verdijk, M. A. J., Merkx, G., Rijss, J. P. L., and Deen, P. M. T. (1996). The human AQP4 gene: definition of the locus encoding two water channel polypeptides in brain. *Proc. Natl. Acad. Sci. USA* **93**, 10908–10912.

Lu, Z., and MacKinnon, R. (1994). Electrostatic tuning of Mg^{2+} affinity in an inward-rectifier K^+ channel. *Nature* **371**, 243–246.

Lüddens, H., Korpi, E. R., and Seeburg, P. H. (1995). $GABA_A$/benzodiazepine receptor heterogeneity: neurophysiological implications. *Neuropharmacology* **34**, 245–254.

Ludewig, U., Pusch, M., and Jentsch, T. J. (1996). Two physically distinct pores in the dimeric ClC-0 chloride channel. *Nature* **383**, 340–343.

Ludewig, U., Jentsch, T. J., and Pusch, M. (1997). Inward rectification in ClC-0 chloride channels caused by mutations in several protein regions. *J. Gen. Physiol.* **110,** 165–171.

Ludwig, A., Zong, X., Jeglitsch, M., Hofmann, F., and Biel, M. (1998). A family of hyperpolarization-activated mammalian cation channels. *Nature* **393,** 587–91.

Lynch, E. C., Rosenberg, I. M., and Gitler, C. (1982). An ion-channel forming protein produced by *Entamoeba histolytica. EMBO J.* **1,** 801–804.

Lynch, J. W., Rajendra, S., Pierce, K. D., Handford, C. A., Barry, P. H., and Schofield, P. R. (1997). Identification of intracellular and extracellular domains mediating signal transduction in the inhibitory glycine receptor chloride channel. *EMBO J.* **16,** 110–120.

Ma, T., Yang, B., Gillespie, A., Carlson, E. J., Epstein, C. J., and Verkman, A. S. (1998). Severely impaired urinary concentrating ability in transgenic mice lacking aquaporin-1 water channels. *J. Biol. Chem.* **273,** 4296–4299.

Maassen, J. A., and Kadowaki, T. (1996). Maternally inherited diabetes and deafness: A new diabetes subtype. *Diabetologia* **39,** 375–382.

MacKenzie, A. E., Korneluk, R. G., Zorzato, F., Fujii, J., Phillips, M., Iles, D., Wieringa, B., Leblond, S., Bailly, J., Willard, H. F., Duff, C., Worton, R. G., and MacLennan, D. H. (1990). The human ryanodine receptor gene: its mapping to 19q13.1, placement in a chromosome 19 linkage group and exclusion as the gene causing myotonic dystrophy. *Am. J. Hum. Genet.* **46,** 1082–1089.

MacKinnon, R. (1991). Determination of the subunit stoichiometry of a voltage-activated potassium channel. *Nature* **350,** 232–235.

MacKinnon, R., Aldrich, R. W., and Lee, A. W. (1993). Functional stoichiometry of *Shaker* potassium channel inactivation. *Science* **262,** 757–759.

MacKinnon, R., Heginbotham, L., and Abrahamson, T. (1990). Mapping the receptor site for charybdotoxin, a pore-blocking potassium channel inhibitor. *Neuron* **5,** 767–771.

MacKinnon, R., and Yellen, G. (1990). Mutations affecting TEA blockade and ion permeation in voltage-activated K^+ channels. *Science* **250,** 276–279.

Majumder, K., De-Biasi, M., Wang, Z., and Wible, B. A. (1995). Molecular cloning and functional expression of a novel potassium channel β-subunit from human atrium *FEBS Lett.* **361,** 13–16.

Makita, N., Bennett, P. B., and George, A. L. (1994a). Voltage-gated Na^+ channel β_1 subunit mRNA expressed in adult human skeletal muscle, heart and brain is encoded by a single gene. *J. Biol. Chem.* **269,** 7571–7578.

Makita, N., Sloan-Brown, K., Weghuis, D. O., Ropers, H. H., and George, A. L. (1994b). Genomic organization and chromosomal assignment of the human voltage-gated Na^+ channel β_1 subunit gene (SCN1B). *Geonomics* **23,** 628–634.

Makowski, L., Caspar, D. L. D., Phillips, W. C., and Goodenough, D. A. (1977). Gap junction structures II. Analysis of X-ray diffraction data. *J. Cell. Biol.* **74,** 629–645.

Mall, M., Hipper, A., Greger, R., and Kunzelmann, K. (1996). Wild type but not ΔF508 CFTR inhibits Na^+ conductance when coexpressed in *Xenopus* oocytes. *FEBS Lett.* **381,** 47–52.

Malo, M. S., Srivastava, K., Andresen, J. M., Chen, X.-N., Korenberg, J. R., and Ingram, V. M. (1994). Targeted gene walking by low stringency polymerase chain reaction: assignment of a putative human brain sodium channel gene (SCN3A) to chromosome 2q24-31. *Proc. Natl. Acad. Sci. USA.* **91,** 2975–2979.

Malosio, M. L., Marquezè-Pouey, B., Kuhse, J., and Betz, H. (1991). Widespread expression of glycine receptor subunit mRNAs in the adult rat and developing rat brain. *EMBO J.* **10,** 2401–2409.

Mandich, P., Schito, A. M., Bellone, E., Antonacci, R., Finelli, P., Rocchi, M., and Ajmar, F. (1994). Mapping of the human NMDAR2B receptor subunit gene (GRIN2B) to chromosome 12p12. *Genomics* **22,** 216–218.

Marples, D., Christensen, S., Christensen, E. I., Ottosen, P. D., and Nielsen, S. (1995). Lithium-induced downregulation of aquaporin-2 water channel expression in rat kidney medulla. *J. Clin. Invest.* **95**, 1838–1845.

Marples, D., Frøkiaer, J., Dørup, J., Knepper, M. A., and Nielsen, S. (1996). Hypokalemia-induced downregulation of aquaporin-2 water channel expression in rat kidney medulla and cortex. *J. Clin. Invest.* **97**, 1960–1968.

Marrion, N. V. (1997). Control of M-current. *Annu. Rev. Physiol.* **59**, 483–504.

Mason, W. P., Graus, F., Lang, B., Honnorat, J., Delattre, J.-Y., Valldeoriola, F., Antoine, J. C., Rosenblum, M. K., Rosenfeld, M. R., Newsom-Davis, J., Posner, J. B., and Dalmau, J. (1997). Small-cell lung cancer, paraneoplastic cerebellar degeneration and the Lambert–Eaton myasthenic syndrome. *Brain* **120**, 1279–1300.

Matsuda, H., Saigusa, A., and Irisawa, H. (1987). Ohmic conductance through the inwardly rectifying K^+ channel and blocking by internal Mg^{2+}. *Nature* **325**, 156–159.

Matsumoto, M., Nakagawa, T., Inoue, T., Nagata, E., Tanaka, K., Takano, H., Minowa, O., Kuno, J., Sakakibara, S., Yamada, M., Yoneshima, H., Miyawaki, A., Fukuuchi, Y., Furuichi, T., Okano, H., Mikoshiba, K., and Noda, T. (1996). Ataxia and epileptic seizures in mice lacking type 1 inositol 1,4,5-trisphosphate receptor. *Nature* **379**, 168–171.

May, A., Ophoff, R. A., Terwindt, G. M., Urban, C., van Eijk, R., Haan, J., Diener, J. H., Lindhout, D., Frants, R. R., Sandkuijl, L. A., and Ferrari, M. D. (1995). Familial hemiplegic migraine locus on 19p13 is involved in the common forms of migraine with and without aura. *Hum. Genet.* **96**, 604–608.

McLean, P. J., Farb, D. H., and Russek, S. J. (1995). Mapping of the α_4 subunit gene (GABRA4) to human chromosome 4 defines an α_2-α_4-β_1-γ_1 gene cluster: further evidence that modern $GABA_A$ receptor gene clusters are derived from an ancestral cluster. *Genomics* **26**, 580–586.

McDonald, T. V., Yu, Z., Ming, Z., Palma, E., Meyers, M. B., Wang, K. W., Goldstein, S. A. N., and Fishman, G. I. (1997), A minK-HERG complex regulates the cardiac potassium current I_{Kr}. *Nature* **388**, 289–292.

McGehee, D. S., and Role, L. W. (1995). Physiological diversity of nicotinic acetylcholine receptors expressed by vertebrate neurons. *Annu. Rev. Physiol.* **57**, 521–546.

McKernan, M. G., and Shinnick-Gallagher, P. (1997). Fear conditioning induces a lasting potentiation of synaptic currents *in vitro*. *Nature* **390**, 607–611.

McKernan, R. M., Quirk, K., Prince, R., Cox, P. A., Gillard, N. P., Ragan, C. I., and Whiting, P. (1991). $GABA_A$ receptor subtypes immunopurified from rat brain with alpha subunit-specific antibodies have unique pharmacological properties. *Neuron* **7**, 667–676.

McManus, O. B., Helms, L. M. H., Pallanck, L., Ganetzky, B., Swanson, R., and Leonard, R. J. (1995). Functional role of the β-subunit of high conductance calcium-activated potassium channels. *Neuron* **14**, 645–650.

McNamara, J. O., Eubanks, J. H., McPherson, J. D., Wasmuth, J. J., Evans, G. A., and Heinemann, S. F. (1992). Chromosomal localization of human glutamate receptor genes. *J. Neurosci.* **12**, 2555–2562.

McNicholas, C. M., and Canessa, C. M. (1997). Diversity of channels generated by the different combinations of epithelial sodium channel subunits. *J. Gen. Physiol.* **109**, 681–692.

McNicholas, C. M., Guggino, W. B., Schwiebert, E. M., Hebert, S. C., Giebisch, G., and Egan, M. E. (1996a). Sensitivity of a renal K^+ channel (ROMK2) to the inhibitory sulfonylurea compound glibenclamide is enhanced by coexpression with the ATP-binding cassette transporter cystic fibrosis transmembrane regulator. *Proc. Natl. Acad. Sci. U.S.A.* **93**, 8083–8088.

McNicholas, C. M., Yang, Y., Giebisch, G., and Herbert, S. C. (1996b). Molecular site for nucleotide binding on an ATP-sensitive renal K^+ channel (ROMK2). *Am. J. Physiol.* **271**, F275–285.

Medbo, J. I., and Sejersted, O. M. (1990). Plasma potassium changes with high intensity exercise. *J. Physiol.* **421**, 105–122.

Meguro, H., Mori, H., Araki, K., Kushiya, E., Kutsuwada, T., Yamazaki, M., Kumanishi, T., Arakawa, M., Sakimura, K., and Mishina, M. (1992). Functional characterization of a heteromeric NMDA receptor channel expressed from cloned cDNAs. *Nature* **357**, 70–74.

Meyer, G., Kirsch, J., Betz, H., and Langosch, D. (1995). Identification of a gephyrin binding motif on the glycine receptor β-subunit. *Neuron* **15**, 563–572.

Meyer-Kleine, C., Steinmeyer, K., Ricker, K., Jentsch, T. J., and Koch, M. C. (1995). Spectrum of mutations in the major human skeletal muscle chloride channel gene (CLCN1) leading to myotonia. *Am. J. Hum. Genet.* **57**, 1325–1334.

Mickelson, J. R., and Louis, C. F. (1996). Malignant hyperthermia: excitation-contraction coupling, Ca^{2+} release channel, and cell Ca^{2+} regulation defects. *Physiol. Rev.* **76**, 537–592.

Middleton, R. E., Pheasant, D. J., and Miller, C. (1996). Homodimeric architecture of a ClC-type chloride ion channel. *Nature* **383**, 337–340.

Mignery, G. A., Sudhof, T. C., Takei, K., and De Camilli, P. (1989). Putative receptor for inositol 1,4,5-trisphosphate similar to ryanodine receptor. *Nature* **342**, 192–195.

Mignon, C., Fromaget, C., Mattaei, M.-G., Gros, D., Yamasaki, H., and Mesnil, M. (1996). Assignment of connexin 26 (GJB2) and 46 (GJA3) genes to human chromsosome 13q11-q12 and mouse chromosome 14D1-E1 by in situ hybridization. *Cytogenet. Cell. Genet.* **72**, 185–186.

Mihic, S. J., Ye, Q., Wick, M. J., Koltchine, V. V., Krasowski, M. D., Finn, S. E., Mascia, M. P., Valenzuela, C. F., Hanson, K. K., Greenblatt, E. P., Harris, R. A., and Harrison, N. L. (1997). Sites of alcohol and volatile anaesthetic action on $GABA_A$ and glycine receptors. *Nature* **389**, 385–389.

Miller, C. (1982). Open-state substructure of single chloride channels from *Torpedo* electroplax. *Phil. Trans. Roy. Soc. Lond. B.* **299**, 401–411.

Minn, A. J., Velez, P., Schendel, S. L., Liang, H., Muchmore, S. W., Fesik, S. W., Fill, M., and Thompson, C. B. (1997). Bcl-X_L forms an ion channel in synthetic lipid membranes. *Nature* **385**, 353–357.

Mishina, M., Takai, T., Imoto, K., Noda, M., Takahashi, T., Numa, S., Methfessel, C., and Sakmann, B. (1986). Molecular distinction between fetal and adult forms of muscle acetylcholine receptor. *Nature* **321**, 406–411.

Mitrovic, N., George, A. L., Lerche, H., Wagner, S., Fahlke, C., and Lehmann-Horn, F. (1995). Different effects on gating of three myotonia-causing mutations in the inactivation gate of the human muscle sodium channel. *J. Physiol.* **487(1)**, 107–114.

Mochizuki, T., Wu, G., Hayashi, T., Xenophontos, S. L., Veldhuisen, B., Saris, J. J., Reynolds, D. M., Cai, Y., Gabow, P. A., Pierides, A., Kimberling, W. J., Breuning, M. H., Deltas, C. C., Peters, D. J. M. and Somlo, S. (1996). *PKD2*, a gene for polycystic kidney disease that encodes an integral membrane protein. *Science* **272**, 1339–1342.

Monkawa, T., Miyawaki, A., Sugiyama, T., Yoneshima, H., Yamamoto-Hino, M., Furuichi, T., Saruta, T., Hasegawa, M., and Mikoshiba, K. (1995). Heterotetrameric complex formation of inositol 1,4,5-trisphosphate receptor subunits. *J. Biol. Chem.* **270**, 14700–14704.

Monnier, N., Procaccio, V., Stieglitz, P., and Lunardi, J. (1997). Malignant-hyperthermia susceptibility is associated with a mutation of the alpha1–subunit of the human dihydropyridine-sensitive L-type voltage-dependent calcium-channel receptor in skeletal muscle. *Am. J. Hum. Genet.* **60**, 1316–1325.

Montell, C., and Rubin, G. M. (1989). Molecular characterization of the *Drosophila trp* locus: a putative integral membrane protein required for phototransduction. *Neuron* **2**, 1313–1323.

Monyer, H., Burnashev, N., Laurie, D. J., Sakmann, B., and Seeburg, P. H. (1994). Developmental and regional expression in the rat brain and functional properties of four NMDA receptors. *Neuron* **12**, 529–540.

Monyer, H., Seeburg, P. H., and Wisden, W. (1991). Glutamate-operated channels: developmentally early and mature forms arise by alternative splicing. *Neuron* **6**, 799–810.

Monyer, H., Sprengel, R., Schoepfer, R., Herb, A., Higuchi, M., Lomeli, H., Burnashev, N., Sakmann, B., and Seeburg, P. H. (1992). Heteromeric NMDA receptors: molecular and functional distinction of subtypes. *Science* **256**, 1217–1221.

Moody, W. J., and Hagiwara, S. (1982). Block of inward rectification by intracellular H^+ in immature oocytes of the starfish *Mediaster aequalis. J. Gen. Physiol.* **79**, 115–30.

Morad, M., Davies, N. W., Ulrich, G., and Schultheiss, H. P. (1988). Antibodies against ADP–ATP carrier enhance Ca^{2+} current in isolated cardiac myocytes. *Am. J. Physiol.* **255**, H960–964.

Moreno, A. P., Saez, J. C., Fishman, G. I., and Spray, D. C. (1994). Human connexin 43 gap junction channels. Regulation of unitary conductances by phosphorylation. *Circ. Res.* **74**, 1050–1057.

Moriyoshi, K., Masu, M., Ishii, T., Shigemoto, R., Mizuno, N., and Nakanishi, S. (1991). Molecular cloning and characterization of the rat NMDA receptor. *Nature* **354**, 31–37.

Mosbacher, J., Schoepfer, R., Monyer, H., Burnashev, N., Seeburg, P. H., and Ruppersberg, J. P. (1994). A molecular determinant for submillisecond desensitization in glutamate receptors. *Science* **266**, 1059–1062.

Mosier, D. R., Baldelli, P., Delbono, O., Smith, R. G., Alexianu, M. E., Appel, S. H., and Stefani, E. (1995). Amyotrophic lateral sclerosis immunoglobulins increase Ca^{2+} currents in a motoneuron cell line. *Ann. Neurol.* **37**, 102–109.

Mossman, S., Vincent, A., and Newsom-Davis, J. (1986). Myasthenia gravis without acetylcholine-receptor antibody: A distinct disease entity. *Lancet* **i,** 116–119.

Motomura, M., Johnston, I., Lang, B., Vincent, A., and Newsom-Davis, J. (1995). An improved diagnostic assay for Lambert–Eaton myasthenic syndrome. *J. Neurol. Neurosurg. Psychiat.* **58**, 85–87.

Motomura, M., Lang, B., Johnston, I., Palace, J., Vincent, A., and Newsom-Davis, J. (1997). Incidence of serum anti–P/Q-type and anti–N-type calcium channel autoantibodies in the Lambert–Eaton myasthenic syndrome. *J. Neurol. Sci.* **147**, 35–42.

Muchmore, S. W., Sattler, M., Liang, H., Meadows, R. P., Harlan, J. E., Yoon, H. S., Nettesheim, D., Chang, B. S., Thompson, C. B., Wong, S.-L., Ng, S.-L., and Fesik, S. W. (1996). X-ray and NMR structure of human Bcl-X_L, an inhibitor of programmed cell death. *Nature* **381**, 335–341.

Mulders, S. M., Knoers, N. V. A. M., van Lieburg, A. F., Monnens, L. A. H., Leumann, E., Wuhl, E., Schober, E., Rijss, J. P. L., van Os, C. H., and Deen, P. M. T. (1997). New mutations in the AQP2 gene in nephrogenic diabetes insipidus resulting in functional but misrouted water channels. *J. Am. Soc. Nephrol.* **8**, 242–248.

Mülhardt, C., Fischer, M., Gass, P., Simon-Chazottes, D., Guénet, J. L., Kuhse, J., Betz, H., and Becker, C. M. (1994). The spastic mouse: aberrant splicing of glycine receptor beta subunit mRNA caused by intronic insertion of L1 element. *Neuron* **13**, 1003–1015.

Murray, N. M. F., and Newsom-Davis, J. (1981). Treatment with oral 4-aminopyridine in disorders of neuromuscular transmission. *Neurology* **31**, 265–271.

Nakai, J., Dirksen, R. T., Nguyen, H. T., Pessah, I. N., Beam, K. G., and Allen, P. D. (1996). Enhanced dihydropyridine receptor channel activity in the presence of ryanodine receptor. *Nature* **380**, 72–75.

Nakai, J., Imagawa, T., Hakamat, Y., Shigekawa, M., Takeshima, H., and Numa, S. (1990). Primary structure and functional expression from cDNA of the cardiac ryanodine receptor/calcium release channel. *FEBS Lett.* **271**, 169–177.

Nakamura, T., and Gold, G. H. (1987). A cyclic nucleotide-gated conductance in olfactory receptor cilia. *Nature* **325**, 442–444.

Nakatsu, Y., Tyndale, R. F., DeLorey, T. M., Durham-Pierre, D., Gardner, J. M., McDanel, H. J., Nguyen, Q., Wagstaff, J., Lalande, M., Sikela, J. M., Olsen, R. W., Tobin, A. J., and Brilliant, M. H., (1993). A cluster of three $GABA_A$ receptor subunit genes is deleted in a neurological mutant of the mouse *p* locus. *Nature* **364**, 448–450.

Naranjo, D., and Miller, C. (1996). A strongly interacting pair of residues on the contact surface of charybdotoxin and a *Shaker* K$^+$ channel. *Neuron* **16**, 123–130.

Navarro, B., Kennedy, M. E., Velimirovic, B., Bhat, D., Peterson, A. S., and Clapham, D. E., (1996). Nonselective and G$\beta\gamma$-insensitive weaver K$^+$ channels. *Science* **272**, 1950–1953.

Neher, E., and Sakmann, B. (1976). Single-channel currents recorded from the membrane of denervated frog muscle fibres. *Nature* **260**, 799–802.

Neher, E., and Sakmann, B. (1995). "Single Channel Recording," 2nd Ed. Plenum Press, New York.

Neher, E., Sandblom, J., and Eisenman, G. (1978). Ionic selectivity, saturation and block in gramicidin A channels. II. Saturation behaviour of single channel conductances and evidence for the existence of multiple binding sites in the channel. *J. Membr. Biol.* **40**, 97–116.

Nestorowicz, A., Wilson, B. A., Schoor, K. P., Inoue, H., Glaser, B., Landau, H., Stanley, C. A., Thornton, P. S., Clement, IV, J. P., Bryan, J., Aguilar-Bryan, L., and Permutt, M. A. (1996). Mutations in the sulfonylurea receptor gene are associated with familial hyperinsulinism in Ashkenazi Jews. *Human Molec. Genet.* **5**, 1813–1822.

Neyroud, N., Tesson, F., Denjoy, I., Lebovici, M., Donger, C., Barhanian, J., Fauré, S., Gary, F., Coumel, P., Petit, C., Schwartz, K., and Guicheney, P. (1997). A novel mutation in the potassium channel gene KvLQT1 causes the Jervel and Lange-Neilsen cardioauditory syndrome. *Nature Genet.* **15**, 186–189.

Nicholls, R. D. (1993). Genomic imprinting and candidate genes in the Prader–Willi and Angelman syndromes. *Curr. Opin. Genet. Dev.* **3**, 445–456.

Nichols, C. G., Shyng, S. L., Nestorowicz, A., Glaser, B., Clement, IV, J. P. Gonzalez, G., Aguilar-Bryan, L., Permutt, M. A., and Bryan, J. (1996). Adenosine diphosphate as an intracellular regulator of insulin secretion *Science* **272**, 1785–1787.

Nielsen, S., and Agre, P. (1995). The aquaporin family of water channels in kidney. *Kidney Int.* **48**, 1057–1068.

Nielsen, S., Nagelhus, E. A., Amiry-Moghaddam, M., Bourque, C., Agre, P., and Ottersen, O. P. (1997a). Specialised membrane domains for water transport in glial cells: high-resolution immunogold cytochemistry of aquaporin-4 in rat brain. *J. Neurosci.* **17**, 171–180.

Nielsen, S., Pallone, T., Smith, B. L., Christensen, E. I., Agre, P., and Maunsbach, A. B. (1995). Aquaporin-1 water channels in short and long loop descending thin limbs and in the descending vasa recta in rat kidney. *Am. J. Physiol.* **268**, F1023–F1037.

Nielsen, S., Terris, J., Andersen, D., Ecelbarger, C., Frøkiaer, J., Jonassen, T., Marples, D., Knepper, M. A., and Petersen, J. S. (1997b). Congestive heart failure in rats is associated with increased expression and targeting of aquaporin-2 water channel in collecting duct. *Proc. Natl. Acad. Sci. USA* **94**, 5450–5455.

Nilius, B., Hess, P., Lansman, J. B., and Tsien, R. W. (1985). A novel type of cardiac calcium channel in ventricular cells. *Nature* **316**, 443–446.

Noda, M., Shimizu, S., Tanabe, T., Takai, T., Kayano, T., Ikeda, T., Takahashi, H., Nakayama, H., Kanaoka, Y., Minamino, N., Kangawa, K., Matsuo, H., Raftery, M. A., Hirose, T., Inayama, S., Hayashida, H., Miyata, T., and Numa, S. (1984). Primary structure of *Electrophorus electricus* sodium channel deduced from cDNA sequence. *Nature* **312**, 121–127

Noda, M., Suzuki, H., Numa, S., and Stühmer, W. (1989). A single point mutation confers tetrodotoxin and saxitoxin insensitivity on the sodium channel II. *FEBS Lett.* **259**, 213–216.

Noda, M., Takahashi, H., Tanabe, T., Toyosato, M., Furutani, Y., Hirose, T., Asai, M., Inayama, S., Miyata, T., and Numa, S., (1982). Primary structure of α-subunit precursor of *Torpedo californica* acetylcholine receptor deduced from cDNA sequence. *Nature* **299**, 793–797.

Noda, M., Takahashi, H., Tanabe, T., Toyosato, M., Kikyotani, S., Hirose, T., Asai, M., Takashima, H., Inayama, S., Miyata, T., and Numa, S. (1983). Primary structures of β- and δ-subunit precursors of *Torpedo californica* acetylcholine receptor deduced from cDNA sequences. *Nature* **301**, 251–255.

North, R. A. (1996). Families of ion channels with two hydrophobic segments. *Curr. Op. Cell Biol.* **8,** 474–483.

North, R. A., and Barnard, E. A. (1997). Nucleotide receptors. *Curr. Opin. Neurobiol.* **7,** 346–357.

Nowak, L., Bregestovski, P., Ascher, P., Herbet, A., and Prochiantz, A. (1984). Magnesium gates glutamate-activated channels in mouse central neurones. *Nature* **307,** 462–465.

O'Brodovich, H., Hannam, V., Seear, M., and Mullen, J. B. M. (1990). Amiloride impairs lung water clearance in newborn guinea-pigs. *J. Appl. Physiol.* **68,** 1758–1762.

Odano, I., Anezaki, T., Ohkubo, M., Yonekura, Y., Onishi, Y., Inuzuka, T., Takahashi, M., and Tsuji, S. (1996). Decrease in benzodiazepine receptor binding in a patient with Angelman syndrome detected by iodine-123 iomazenil and single-photon emission tomography. *Eur. J. Nuclear Med.* **23,** 598–604.

Oh, B.-H., Pandit, J., Kang, C.-H., Nikaido, K., Gokcen, S., Ames, G. F.-L., and Kim SH (1993). Three dimensional structures of the periplasmic lysine/arginine/ornithine–binding protein with and without a ligand. *J. Biol. Chem.* **268,** 11348–11355.

Oh, S., Ri, Y., Bennett, M. V. L., Trexler, E. B., Verselis, V. K., and Bargiello, T. A. (1997). Changes in permeability caused by connexin 32 mutations underlie X-linked Charcot-Marie-Tooth disease. *Neuron* **19,** 927–938.

Ohno, K., Hutchinson, D. O., Milone, M., Brengman, J. M., Bouzat, C., Sine, S. M., and Engel, A. G. (1995). Congenital myasthenic syndrome caused by prolonged acetylcholine receptor channel openings due to a mutation in the M2 domain of the ε subunit. *Proc. Nat. Acad. Sci. USA* **92,** 758–762.

Ohno, K., Wang, H.-L., Milone, M., Bren, N., Brengman, J. M., Nakano, S., Quiram, P., Pruitt, J. N., Sine, S. M., and Engel, A. G. (1996). Congenital myasthenic syndrome caused by decreased agonist binding affinity due to a mutation in the acetylcholine receptor ε subunit. *Neuron* **17,** 157–170.

Ohta, T., Endo, M., Nakano, T., Morohoshi, Y., Wanikawa, K., and Ohga, A. (1989). Ca-induced Ca release in malignant hyperthermia-susceptible pig skeletal muscle. *Am. J. Physiol.* **256,** C358–C367.

Olney, J. W. (1994). Excitotoxins in foods. *Neurotoxicology* **15,** 535–544.

Olver, R. E., and Strang, L. B. (1974). Ion fluxes across the pulmonary epithelium and the secretion of lung liquid in the foetal lamb. *J. Physiol.* **241,** 327–357.

Omori, Y., Mesnil, M., and Yamasaki, H. (1996). Connexin 32 mutations from X-linked Charcot-Marie-Tooth Disease patients: functional defects and dominant negative effects. *Mol. Biol. Cell* **7,** 907–916.

Ophoff, R. A., Terwindt, G. M., Vergouwe, M. N., van Eijk, R., Oefner, P. J., Hoffman, S. M. G., Lamerdin, J. E., Mohrenweiser, H. W., Bulman, D. E., Ferrari, M., Haan, J., Lindhout, D., van Ommen, G.-J. B., Hofker, M. H., Ferrari, M. D., and Frants, R. R. (1996). Familial hemiplegic migraine and episodic ataxia type-2 are caused by mutations in the Ca^{2+} channel gene *CACNL1A4*. *Cell* **87,** 543–552.

Orr-Urtreger, A., Seldin, M. F., Baldini, A., and Beaudet, A. L. (1995). Cloning and mapping of the mouse α_7 neuronal nicotinic acetylcholine receptor. *Genomics* **26,** 399–402.

Otsu, K., Nishida, K., Kimura, Y., Kuzuya, T., Hori, M., Kamada, T., and Tada, M. (1994). The point mutation Arg615Cys in the Ca^{2+} release channel of skeletal sarcoplasmic reticulum is responsible for hypersensitivity to caffeine and halothane in malignant hyperthermia. *J. Biol. Chem.* **269,** 9413–9415.

Otsu, K., Willard, H. F., Khanna, V. K., Zorzato, F., Green, N. M., and MacLennan, D. H. (1990). Molecular cloning of cDNA encoding the Ca^{2+} release channel (ryanodine receptor) of rabbit cardiac muscle sarcoplasmic reticulum. *J. Biol. Chem.* **265,** 13472–13483.

Papazian, D. M., Schwarz, T. L., Tempel, B. L., Jan, Y. N., and Jan, L. Y. (1987). Cloning of genomic and complementary DNA from *Shaker*, a putative potassium channel gene from *Drosophila*. *Science* **237,** 749–753.

Papazian, D. M., Timpe, L. C., Jan, Y. N., and Jan, L. Y. (1991). Alteration of voltage-dependence of *Shaker* potassium channel by mutations in the S4 sequence. *Nature* **349,** 305–310.

Parcej, D. N., Scott, V. E. S., and Dolly, J. O. (1992). Oligomeric properties of alpha-dendrotoxin-sensitive potassium ion channels purified from bovine brain. *Biochemistry* **31,** 11084–11088.

Pardo, L. A., Heinemann, S. H., Terlau, H., Ludewig, U., Lorra, C., Pongs, O., and Stühmer, W. (1992). Extracellular K^+ specifically modulates a rat brain K^+ channel. *Proc. Natl. Acad. Sci.* **89,** 2466–2470.

Parker, M. W., Buckley, J. T., Postma, J. P. M., Tucker, A. D., Leonard, K., Pattus, F., and Tsernoglou, D. (1994). Structure of the *Aeromonas* toxin proaerolysin in its water-soluble and membrane-channel states. *Nature* **367,** 292–295.

Paschen, W., Blackstone, C. D., Huganir, R. L., and Ross, C. A. (1994). Human GluR6 kainate receptor (GRIK2): molecular cloning, expression, polymorphism, and chromosomal assignment. *Genomics* **20,** 435–440.

Patel, P. I., and Lupski, J. R. (1994). Charcot-Marie-Tooth disease: a new paradigm for the mechanism of inherited disease. *Trends in Genetics* **10,** 128–133.

Paterson, D. J., (1996). Antiarrhythmic mechanisms during exercise. *J. Appl Physiol.* **80,** 1853–1862.

Patil, N., Cox, D. R., Bhat, D., Faham, M., Myers, R. M., and Peterson, A. S. (1995). A potassium channel mutation in weaver mice implicates membrane excitability in granule cell differentiation. *Nature Genet.* **11,** 126–129.

Patrick, J., and Lindstrom, J. (1973). Autoimmune response to acetylcholine receptor. *Science* **180,** 871–872.

Patton, D. E., West, J. W., Catterall, W. A., and Goldin, A. L. (1992). Amino acid residues required for fast Na^+-channel inactivation: charge neutralizations and deletions in the III-IV linker. *Proc. Natl. Acad. Sci. USA* **89,** 10905–10909.

Paul, D. L. (1995). New functions for gap junctions. *Curr. Op. Cell Biol.* **7,** 665–672.

Peers, C., Johnston, I., Lang, B., and Wray, D. (1993). Cross-linking of presynaptic calcium channels: a mechanism of action for Lambert–Eaton myasthenic syndrome antibodies at the mouse neuromuscular junction. *Neurosci. Lett.* **153,** 45–48.

Peitsch, M. C., and Tschopp, J. (1991). Assembly of macromolecular pores by immune defense systems. *Curr. Opin. Cell Biol.* **3,** 710–716.

Pellegrini-Giampietro, D. E., Gorter, J. A., Bennett, M. V. L., and Zukin, R. S. (1997). The GluR2 (GluR-B) hypothesis: Ca^{2+}-permeable AMPA receptors in neurological disorders. *Trends Neurosci.* **20,** 464–470.

Pellegrini-Giampietro, D. E., Zukin, R. S., Bennett, M. V. L., Cho, S., and Pulsinelli, W. A. (1992). Switch in glutamate receptor subunit gene expression in CA1 subfield of hippocampus following global ischemia in rats. *Proc. Natl. Acad. Sci. USA* **89,** 10499–10503.

Perez-Reyes, E., Cribbs, L. L., Daud, A., Lacerda, A. E., Barclay, J., Williamson, M. P., Fox, M., Rees, M., and Lee, J. H. (1998). Molecular characterization of a neuronal low-voltage-activated T-type calcium channel. *Nature* **391,** 896–900.

Persechini, P. M., Ojcius, D. M., Adeodato, S. C., Notaroberto, P. C., Daniel, C. B., and Young, J. D. (1992). Channel-forming activity of the perforin N-terminus and a putative α-helical region homologous with complement C9. *Biochemistry* **31,** 5017–5021.

Peters, R., Sauer, H., Tschopp, J., and Fritzsch, G. (1990). Transients of perforin pore formation observed by fluorescence microscopic single channel recording. *EMBO J.* **9,** 2447–2451.

Peto, R., Lopez, A. D., Boreham, J., Thun, M., and Heath, C. (1992). Mortality from tobacco in developed countries: indirect estimation from national vital statistics. *Lancet* **339,** 1268–1278.

Pfeiffer, F., Graham, D., and Betz, H. (1982). Purification by affinity chromatography of the glycine receptor of rat spinal cord. *J. Biol. Chem.* **257,** 9389–9393.

Phillips, M. S., Fujii, J., Khanna, V. K., DeLeon, S., Yokobata, K., de Jong, P. J., and MacLennan, D. H. (1996). The structural organization of the human skeletal muscle ryanodine receptor RYR1) gene. *Genomics* **34,** 24–41.

Picciotto, M. R., Zoli, M., Lena, C., Bessis, A., Lallemand, Y., LeNovere, N., Vincent, P., Pich, E. M., Brulet, P., and Changeux, J.-P. (1995). Abnormal avoidance learning in mice lacking functional high-affinity nicotine receptor in the brain. *Nature* **374,** 65–67.

Picciotto, M. R., Zoli, M., Rimondini, R., Lena, C., Marubio, L. M., Pich, E. M., Fuxe, K., and Changeux, J.-P. (1998). Acetylcholine receptors containing the β2 subunit are involved in the reinforcing properties of nicotine. *Nature* **391,** 173–177.

Picco, C., and Menini, A. (1993). The permeability of cGMP-activated channel to organic cations in retinal rods of the tiger salamander. *J. Physiol.* **460,** 741–758.

Pinto, L. H., Holsinger, L. J., and Lamb, R. A. (1992). Influenza virus M_2 protein has ion channel activity. *Cell* **69,** 517–528.

Pinto, A., Gillard, S., Moss, F., Whyte, K., Brust, P., Williams, M., Stauderman, K., Harpold, M., Lang, B., Newsom-Davis, J., Bleakman, D., Lodge, D., and Boot, J. (1998). Human autoantibodies specific for the alpha1A calcium channel subunit reduce both P-type and Q-type calcium currents in cerebellar neurons. *Proc. Natl. Acad. Sci. USA* **95,** 8328–8333.

Plaitakis, A., Berl, S., and Yahr, M. D. (1982). Abnormal glutamate metabolism in an adult-onset degenerative neurological disorder. *Science* **216,** 193–196.

Plaitakis, A., Berl, S., and Yahr, M. D. (1984). Neurological disorders associated with deficiency of glutamate dehydrogenase *Ann. Neurol.* **15,** 144–153.

Powers, P. A., Scherer, S. W., Tsui, L. C., Gregg, R. G., and Hogan, K. (1994). Localization of the gene encoding the α_2/δ subunit (CACNL2A) of the human skeletal muscle voltage-dependent calcium channel to chromosome 7q21–q22 by somatic cell hybrid analysis. *Geomics* **19,** 192–193.

Powers, P. A., Liu, S., Hogan, K., and Gregg, R. G. (1993). Molecular characterization of the gene encoding the γ subunit of the human skeletal muscle 1,4-dihydropyridine-sensitive Ca^{2+} channel (CACNLG), cDNA sequence, gene structure, and chromosomal location. *J. Biol. Chem.* **268,** 9275–9279.

Pragnell, M., De-Waard, M., Mori, Y., Tanabe, T., Snutch, T. P., and Campbell, K. P. (1994). Calcium channel beta-subunit binds to a conserved motif in the I-II cytoplasmic linker of the alpha 1–subunit. *Nature* **368,** 67–70.

Preston, G. M., Carroll, T. P., Guggino, W. B., and Agre, P. (1992). Appearance of water channels in *Xenopus* oocytes expressing red cell CHIP28 protein. *Science* **256,** 385–387.

Preston, G. M., Smith, B. L., Zeidel, M. L., Moulds, J. J., and Agre, P. (1994). Mutations in aquaporin-1 in phenotypically normal humans without functional CHIP water channels. *Science* **265,** 1585–1587.

Pribilla, I., Takagi, T., Langosch, D., Bormann, J., and Betz, H. (1992). The atypical M2 segment of the beta subunit confers picrotoxinin resistance to inhibitory glycine receptor channels. *EMBO J.* **11,** 4305–4311.

Prior, P., Schmitt, B., Grenningloh, G., Pribilla, I., Multhaup, G., Beyreuther, K., Maulet, Y., Werner, P., Langosch, D., Kirsch, J., and Betz, H. (1992). Primary structure and alternative splice variants of gephyrin, a putative glycine receptor–tubulin linker protein. *Neuron* **8,** 1161–1170.

Pritchett, D. B., Lüddens, H., and Seeburg, P. H. (1989). Type I and type II GABA$_A$-benzodiazepine receptors produced in transfected cells. *Science* **245,** 1389–1392.

Pritchett, D. B., and Seeburg, P. H. (1991). Gamma-aminobutyric acid type A receptor point mutation increases the affinity of compounds for the benzodiazepine site. *Proc. Natl. Acad. Sci. USA* **88,** 1421–1425.

Protti, D. A., Reisin, R., Mackinley, T. A., and Uchitel, O. D. (1996). Calcium channel blockers and transmitter release at the normal human neuromuscular junction. *Neurology* **46,** 1391–1396.

Puckett, C., Gomez, C. M., Korenberg, J. R., Tung, H., Meier, T. J., Chen, X. N., and Hood, L. (1991). Molecular cloning and chromosomal localization of one of the human glutamate receptor genes. *Proc. Natl. Acad. Sci. USA* **88,** 7557–7561.

Puranam, R. S., Eubanks, J. H., Heinemann, S. F., and McNamara, J. O. (1993). Chromosomal localization of gene for human glutamate receptor subunit-7. *Somat. Cell Mol. Genet.* **19,** 581–588.

Pusch, M., Ludewig, U., Rehfeldt, A., and Jentsch, T. J. (1995a). Gating of the voltage-dependent chloride channel ClC-0 by the permeant anion. *Nature* **373,** 527–531.

Pusch, M., Noda, M., Stühmer, W., Numa, S., and Conti, F. (1991). Single point mutations of the sodium channel drastically reduce the pore permeability without preventing its gating. *Eur. Biophys. J.* **20,** 127–133.

Pusch, M., Steinmeyer, K., Koch, M. C., and Jentsch, T. J. (1995b). Mutations in dominant human myotonia congenita drastically alter the voltage dependence of the ClC-1 chloride channel. *Neuron* **15,** 1455–1463.

Quane, K. A., Healy, J. M. S., Keating, K. E., Manning, B. M., Couch, F. J., Palmucci, L. M., Doriguzzi, C., Fagerlund, T. H., Berg, K., Ording, H., Bendixen, D., Mortier, W., Linz, U., Muller, C. R., and McCarthy, T. V. (1993). Mutations in the ryanodine receptor gene in central core disease and malignant hyperthermia. *Nature Genet.* **5,** 51–55.

Quane, K. A., Keating, K. E., Healy, J. M. S., Manning, B. M., Krivosic-Horber, R., Krivosic, I., Monnier, N., Lunardi, J., and McCarthy, T. V. (1994). Mutation screening of the RYR1 gene in malignant hyperthermia: detection of a novel Tyr to Ser mutation in a pedigree with associated central cores. *Genomics* **23,** 236–239.

Quayle, J. M., Bonev, A. D., Brayden, J. E., and Nelson, M. T. (1994). Calcitonin gene-related peptide activated ATP-sensitive K^+ currents in rabbit arterial smooth muscle via protein kinase A. *J. Physiol.* **475,** 9–13.

Quinton, P. M. (1983). Chloride impermeability in cystic fibrosis. *Nature* **301,** 421–422.

Radermacher, M., Rao, V., Grassucci, R., Frank, J., Timerman, A. P., Fleischer, S., and Wagenknecht, T. (1994). Cryo-electron microscopy and three-dimensional reconstruction of the calcium release channel/ryanodine receptor from skeletal muscle. *J. Cell Biol.* **127,** 411–423.

Rae, J., Cooper, K., Gates, P., and Watsky, M. (1991). Low access resistance perforated patch recordings using amphotericin B. *J. Neurosci. Methods* **37,** 15–26.

Raimondi, E., Rubboli, F., Moralli, D., Chini, B., Fornasari, D., Tarroni, P., De Carli, L., and Clementi, F. (1992). Chromosomal localization and physical linkage of the genes encoding the human α_3, α_5, and β_4 neuronal nicotinic receptor subunits. *Genomics* **12,** 849–850.

Raina, S., Preston, G. M., Guggino, W. B., and Agre, P. (1995). Molecular cloning and characterization of an aquaporin cDNA from salivary, lacrimal and respiratory tissues. *J. Biol. Chem.* **270,** 1908–1912.

Rajendra, S., Lynch, J. W., Pierce, K. D., French, C. R., Barry, P. H., and Schofield, P. R. (1994). Startle disease mutations reduce the agonist sensitivity of the human inhibitory glycine receptor. *J. Biol. Chem.* **269,** 18739–18742.

Rajendra, S., Lynch, J. W., Pierce, K. D., French, C. R., Barry, P. H., and Schofield, P. R. (1995a). Mutation of an arginine residue in the human glycine receptor transforms beta-alanine and taurine from agonists into competitive antagonists. *Neuron* **14,** 169–175.

Rajendra, S., Vandenberg, R. J., Pierce, K. D., Cunningham, A. M., French, P. W., Barry, P. H., and Schofield, P. R. (1995b). The unique extracellular disulphide loop of the glycine receptor is a principal ligand binding element. *EMBO J.* **14,** 2987–2998.

Randle, P. J. (1993). Glucokinase and candidate genes for type 2 (non-insulin-dependent) diabetes mellitus. *Diabetologia* **36,** 269–275.

Ranganathan, R., Lewis, J. H., and MacKinnon, R. (1996). Spatial localization of the K^+ channel selectivity filter by mutant cycle-based structure analysis. *Neuron* **16,** 131–139.

Rassendren, F., Buell, G., Newbolt, A., North, R. A., and Surprenant, A. (1997).

Identification of amino acid residues contributing to the pore of a P2X receptor. *EMBO J.* **16**, 3446–3454.

Reaume, A. G., de Sousa, P. A., Kulkarni, S., Langille, B. L., Zhu, D., Davies, T. C., Juneja, S. C., Kidder, G. M., and Rossant, J. (1995). Cardiac malformation in neonatal mice lacking connexin 43. *Science* **267**, 1831–1834.

Reddy, M. M., Quinton, P. M., Haws, C., Wine, J. J., Grygorczyk, R., Tabcharani, J. A., Hanrahan, J. W., Gunderson, K. L., and Kopito, R. R. (1996). Failure of cystic fibrosis transmembrane conductance regulator to conduct ATP. *Science* **271**, 1876–1879.

Reinhart, P. H., and Levitan, I. B. (1995). Kinase and phosphatase activities intimately associated with a reconstituted calcium-dependent potassium channel. *J. Neuroscience* **15**, 4572–4579.

Renard, S., Voilley, N., Bassilana, F., Lazdunski, M., and Barbry, P. (1995). Localization and regulation by steroids of the α,β and γ subunits of the amiloride-sensitive Na$^+$ channel in colon, lung and kidney. *Pflügers Arch.* **430**, 299–307.

Renaud, J.-F., Desnuelle, C., Schmid-Antomarchi, H., Hugues, M., Serratrice, G., and Lazdunski, M. (1986). Expression of apamin receptor in muscles of patients with myotonic muscular dystrophy. *Nature* **319**, 678–680.

Rettig, J., Heinemann, S. H., Wunder, F., Lorra, C., Parcej, D. N., Dolly, J. O., and Pongs, O. (1994). Inactivation properties of voltage-gated K$^+$ channels altered by presence of β-subunit. *Nature* **369**, 289–294.

Rettig, J., Sheng, Z.-H., Kim, D. K., Hodson, C. D., Snutch, T. P., and Catterall, W. A. (1996). Isoform-specific interaction of the α1A subunits of brain Ca^{2+} channels with the presynaptic proteins syntaxin and SNAP-25. *Proc. Natl. Acad. Sci. USA* **93**, 7363–7368.

Rich, D. P., Gregory, R. J., Anderson, M. P., Manavalan, P., Smith, A. E., and Welsh, M. J. (1991). Effect of deleting the R domain on CFTR-generated chloride channels. *Science* **253**, 205–207.

Riecker, G., and Bolte, H. D. (1966). Membranpotentiale einzelner Skeletmuskel-zellen bei hypokaliämischer periodischer Muskelparalyse. *Klinische Wochenschrift* **44**, 804–807.

Riordan, J. R., Rommens, J. M., Kerem, B., Alon, N., Rozmahel, R., Grzelczak, Z., Zielenski, J., Lok, S., Plavsic, N., Chou, J.-L., Drumm, M. L., Iannuzzi, M. C., Collins, F. S., and Tsui, L.-C. (1989). Identification of the cystic fibrosis gene: cloning and characterization of complementary DNA. *Science* **245**, 1066–1073.

Roden, D. M., Lazzara, R., Rosen, M., Schwartz, P. J., Towbin, J., and Vincent, G. M. (1996). Multiple mechanisms in the long-QT syndrome. Current knowledge, gaps and future directions. SADS Foundation Task Force on LQTS. *Circulation* **94**, 1996–2012.

Rogers, S. W., Andrews, P. I., Gahring, L. C., Whisenand, T., Cauley, K., Crain, B., Hughes, T. E., Heinemann, S. F., and McNamara, J. O. (1994). Autoantibodies to glutamate receptor GluR3 in Rasmussen's encephalitis. *Science* **265**, 648–651.

Root, M. J., and MacKinnon, R. (1993). Identification of an external divalent cation-binding site in the pore of a cGMP-activated channel. *Neuron* **11**, 459–466.

Rosenfeld, M. R., Wong, E., Dalmau, J., Manley, G., Posner, J. B., Sher, E., and Furneaux, H. M. (1993). Cloning and characterization of Lambert–Eaton myasthenic syndrome antigen. *Ann. Neurol.* **33**, 113–120.

Rosenmund, C., Stern-Bach, Y., and Stevens, C. F. (1998). The tetrameric structure of a glutamate receptor channel. *Science* **280**, 1596–1599.

Rüdel, R., Lehmann-Horn, F., Ricker, K., and Küther, G. (1984). Hypokalemic periodic paralysis: in vitro investigation of muscle fiber membrane parameters. *Muscle Nerve* **7**, 110–120.

Rudolph, J. A., Spier, S. J., Byrns, G., Rojas, C. V., Bernoco, D., and Hoffman, E. P. (1992). Periodic paralysis in quarter horses: a sodium channel mutation disseminated by selective breeding. *Nature Genet.* **2**, 144–147.

Rudy, B., and Iversen, L. E., eds. (1992). Ion channels. *Methods Enzymol.* **207**. Academic Press, San Diego.

Ruiz, M. L., and Karpen, J. W. (1997). Single cyclic nucleotide-gated channels locked in different ligand-bound states. *Nature* **389,** 389–392.

Ruiz-Gómez, A., Morato, E., García-Calvo, M., Valdivieso, F., and Mayor, F. (1990). Localization of the strychnine binding site on the 48-kilodalton subunit of the glycine receptor. *Biochemistry* **29,** 7033–7040.

Russell, M. B., Rasmussen, B. K., Thorvaldsen, P., and Olesen, J. (1995). Prevalence and sex-ratio of the subtypes of migraine. *Intl. J. Epidemiol.* **24,** 612–618.

Ryan, S. G., Dixon, M. J., Nigro, M. A., Kelts, K. A., Markand, O. N., Terry, J. C., Shiang, R., Wasmuth, J. J., and O'Connell, P. (1992). Genetic and radiation hybrid mapping of the hyperekplexia region on chromosome 5q. *Am. J. Hum. Genet.* **51,** 1334–1343.

Saito, F., Sasaki, S., Chepelinsky, A. B., Fushimi, K., Marumo, F., and Ikeuchi, T. (1995). Human AQP2 and MIP genes, two members of the MIP family, map within chromosome band 12q13 on the basis of two-color FISH. *Cytogenet. Cell Genet.* **68,** 45–48.

Sakura, H., Ämmälä, C., Smith, P. A., Gribble, F. M., and Ashcroft, F. M. (1995). Cloning and functional expression of the cDNA encoding a novel ATP-sensitive potassium channel subunit expressed in pancreatic β-cells, brain, heart and skeletal muscle. *FEBS Lett.* **377,** 338–344.

Sakura, H., Ashcroft, S. J. H., Terauchi, Y., Kadowaki, T., and Ashcroft, F. M. (1998). Glucose modulation of ATP-sensitive K-currents in wild-type, homozygous and heterozygous glucokinase knock-out mice. *Diabetologia* **41,** 654–659.

Salata, J. J., Jurkiewicz, N. K., Wallace, A. A., Stupienski, R. F., Guinosso, P. J., and Lynch, J. J. (1995). Cardiac electrophysiological actions of the histamine H_1-receptor antagonists astemizole and terfenidine compared with chlorpheniramine and pyrilamine. *Circulation Research* **76,** 110–119.

Sambrook, J., Fritsch, E. F., and Maniatis, T. (1989). "Molecular Cloning: A Laboratory Manual," 2nd Ed. Cold Spring Harbor Laboratory Press, Cold Spring Harbor, N. Y.

Sanguinetti, M. C., Curran, M. E., Zou, A., Shen, J., Spector, P. S., Atkinson, D. L., and Keating, M. T. (1996). Coassembly of KVLQT1 and minK (IsK) proteins to form cardiac I_{Ks} potassium channel. *Nature* **384,** 80–83.

Sanguinetti, M. C., Jiang, C., Curran, M. E., and Keating, M. T. (1995). A mechanistic link between an inherited and an acquired cardiac arrhythmia: HERG encodes the I_{Kr} potassium channel. *Cell* **81,** 299–307.

Sansom, M. S. (1991). The biophysics of peptide models of ion channels. *Progr. Biophys. Mol. Biol.* **55,** 139–235.

Sansom, M. (1997). Structure of a molecular hole-punch. *Nature* **385,** 390–391.

Sansom, M. S. P. (1993). Structure and function of channel-forming peptaibols. *Q. Rev. Biophys.* **26**(4), 365–421.

Sansom, M. S. P. (1994). The fine art of pore formation. *Struct. Biol.* **1,** 563–567.

Sansom, M. S. P., and Kerr, I. D. (1993). Influenza virus M2 protein: a molecular modelling study of the ion channel. *Protein Eng.* **6,** 65–74.

Santoro, B., Liu, D. T., Yao, H., Bartsch, D., Kandel, E. R., Siegelbaum, S. A., and Tibbs, G. R. (1998). Identification of a gene encoding a hyperpolarization activated pacemaker channel of brain. *Cell* **93,** 717–729.

Santoro, B., Grant, S. G. N., Bartsch, D., and Kandel, E. R. (1997). Interactive cloning with the SH_3 domain of N-src identifies a new brain-specific ion channel protein with homology to Eag and cyclic nucleotide-gated channels. *Proc. Natl. Acad. Sci. U.S.A.* **94,** 14815–14820.

Sasaki, S., Fushimi, K., Saito, H., Saito, F., Uchida, S., Ishibashi, K., Kuwahara, M., Ikeuchi, T., Inui, K., Nakajima, K., Watanabe, T. X., and Marumo, F. (1994). Cloning, characterization and chromosomal mapping of human aquaporin of collecting duct. *J. Clin. Invest.* **93,** 1250–1256.

Satin, J., Kyle, J. W., Chen, M., Bell, P., Cribbs, L. L., Fozzard, H. A., and Rogart, R. B. (1992). A mutant of TTX-resistant cardiac sodium channels with TTX-sensitive properties. *Science* **256**, 1202–1205.

Saul, B., Schmieden, V., Kling, C., Mülhardt, C., Gass, P., Kuhse, J., and Becker, C. M. (1994). Point mutation of glycine receptor alpha 1 subunit in the spasmodic mouse affects agonist responses. *FEBS Lett.* **350**, 71–76.

Schendel, S. L., Xie, Z., Montal, M. O., Matsuyama, S., Montal, M., and Reed, J. C. (1997). Channel formation by antiapoptotic protein Bcl-2. *Proc. Natl. Acad. Sci. USA* **94**, 5113–5118.

Schild, L., Canessa, C. M., Shimkets, R. A., Gautschi, I., Lifton, R. P., and Rossier, B. C. (1995). A mutation in the epithelial sodium channel causing Liddle disease increases channel activity in the *Xenopus laevis* oocyte expression system. *Proc. Natl. Acad. Sci. U.S.A.,* **92**, 5699–5703.

Schild, L., Lu, Y., Gautschi, I., Schneeberger, E., Lifton, R. P., and Rossier, B. C. (1996). Identification of a PY motif in the epithelial Na channel subunits as a target sequence for mutations causing channel activation found in Liddle syndrome. *EMBO Journal* **15**, 2381–2387.

Schild, L., Schneeberger, E., Gautschi, I., and Firsov, D. (1997). Identifcation of amino acid residues in the α,β and γ subunits of the epithelial sodium channel (ENaC) involved in amiloride block and ion permeation. *J. Gen. Physiol.* **109**, 15–26.

Schmid-Antomarchi, H., Renaud, J.-F., Romey, G., Hugues, M., Schmid, A., and Lazdunski, M. (1985). The all-or-none role of innervation in expression of apamin receptor and of apamin-sensitive Ca^{2+}-activated K^+ channel in mammalian skeletal muscle. *Proc. Natl. Acad. Sci. USA* **82**, 2188–2191.

Schmieden, V., Grenningloh, G., Schofield, P. R., and Betz, H. (1989). Functional expression in *Xenopus* oocytes of the strychnine binding 48 kd subunit of the glycine receptor *EMBO J.* **8**, 695–700.

Schmieden, V., Kuhse, J., and Betz, H. (1993). Mutation of glycine receptor subunit creates beta-alanine receptor responsive to GABA. *Science* **262**, 256–258.

Schroeder, B. C., Kubisch, C., Stein, V., and Jentsch, T. J. (1998). Moderate loss of function of cyclic-AMP-modulated KCNQ2/KCNQ3 K^+ channels causes epilepsy. *Nature* **396**, 687–690.

Schultz, D., Mikala, G., Yatani, A., Engle, D. B., Iles, D. E., Segers, B., Sinke, R. J., Weghuis, D. O., Klockner, U., Wakamori, M., Wang, J.-I., Melvin, D., Varadi, G., and Schwartz, A. (1993). Cloning, chromosomal localization, and functional expression of the α_1 subunit of the L-type voltage-dependent calcium channel from normal human heart. *Proc. Natl. Acad. Sci. USA* **90**, 6228–6232.

Schwartz, P. J., Periti, M., and Malliani, A. (1975). Fundamentals of clinical cardiology: The long QT syndrome. *Am. Heart J.* **89**, 378–390.

Schwiebert, E. M., Egan, M. E., Hwang, T. H., Fulmer, S. B., Allen, S. S., Cutting, G. R., and Guggino, W. B. (1995). CFTR regulates outwardly rectifying chloride channels through an autocrine mechanism involving ATP. *Cell* **81**, 1063–1073.

Scoville, W. L. (1912). A note on capsicums. *J. Am. Pharm. Assoc.* **1**, 453–454.

Seino, S., Yamada, Y., Espinosa, R., Le Beau, M. M., and Bell, G. I. (1992). Assignment of the gene encoding the α_1 subunit of the neuroendocrine/brain-type calcium channel (CACNL1A2) to human chromosome 3, band p14.3. *Genomics* **13**, 1375–1377.

Serysheva, I. I., Orlova, E. V., Chiu, W., Sherman, M. B., Hamilton, S. L., and van Heel, M. (1995). Electron cryomicroscopy and angular reconstitution used to visualize the skeletal muscle calcium release channel. *Nature Struct. Biol.* **2**, 18–24.

Setien, F., Alvarez, V., Coto, E., DiScipio, R. G., and Lopez-Larrea, C. (1993). A physical map of the human complement component C6, C7 and C9 genes. *Immunogenetics* **38**, 341–344.

Sewing, S., Roeper, J., and Pongs, O. (1996). $K_V\beta1$ subunit binding specific for *Shaker*-related potassium channel α-subunits. *Neuron* **16**, 455–463.

Shamotienko, O. G., Parcej, D. N., and Dolly, J. O. (1997). Subunit combinations defined for K^+ channel K_V1 subtypes in synaptic membranes from bovine brain. *Biochemistry* **36**, 8195–8201.

Sheng, M., Liao, Y. J., Jan, Y. N., and Jan, L. Y. (1993). Presynaptic A-current based on heteromultimeric K^+ channels detected *in vivo*. *Nature* **365**, 72–75.

Sheng, M., Cummings, J., Roldan, L. A., Jan, Y. N., and Jan, L. Y. (1994a). Changing subunit composition of heteromeric NMDA receptors during development of rat cortex. *Nature* **368**, 144–147.

Sheng, Z. H., Rettig, J., Takahashi, M., and Catterall, W. A. (1994b). Identification of a syntaxin-binding site on N-type calcium channels. *Neuron* **13**, 1303–1313.

Sheppard, D. N., Rich, D. P., Ostedgaard, L. S., Gregory, R. J., Smith, A. E., and Welsh, M. J. (1993). Mutations in CFTR associated with mild-disease-form Cl^- channels with altered pore properties. *Nature* **362**, 160–164.

Shi, G., Nakahira, K., Hammond, S., Rhodes, K. J., Schechter, L. E., and Trimmer, J. S. (1996). β-subunits promote K^+ channel surface expression through effects early in biosynthesis. *Neuron* **16**, 843–852.

Shiang, R., Ryan, S. G., Zhu, Y.-Z., Hahn, A. F., O'Connell, P., and Wasmuth, J. J. (1993). Mutations in the α_1 subunit of the inhibitory glycine receptor cause the dominant neurologic disorder, hyperekplexia. *Nature Genet.* **5**, 351–358.

Shiels, A., and Bassnett, S. (1996). Mutations in the founder of the MIP gene family underlie cataract development in the mouse. *Nature Genet.* **12**, 212–215.

Shillito, P., Molenaar, P. C., Vincent, A., Leys, K., Zheng, W., van-den-Berg, R. J., Plomp, J. J., van-Kempen, G. T., Chauplannaz, G., Wintzen, A. R., van Dijk, J. G., and Newsom-Davis, J. (1995). Acquired neuromyotonia: evidence for autoantibodies directed against K^+ channels of peripheral nerves. *Ann. Neurol.* **38**, 714–722.

Shimkets, R. A., Warnock, D. G., Bositis, C. M., Nelson-Williams, C., Hansson, J. H., Schambelan, M., Gill, J. R., Ulick, S., Milora, R. V., Findling, J. W., Canessa, C. M., Rossier, B. C., and Lifton, R. P. (1994). Liddle's syndrome: heritable human hypertension caused by mutations in the β subunit of the epithelial sodium channel. *Cell* **79**, 407–414.

Shiver, J. W., Dankert, J. R., and Esser, A. F. (1991). Formation of ion-conducting channels by the membrane attack complex proteins of complement. *Biophys. J.* **60**, 761–769.

Shomer, N. H., Mickelson, J. R., and Louis, C. F. (1994). Caffeine stimulation of malignant hyperthermia-susceptible sarcoplasmic reticulum Ca^{2+} release channel. *Am. J. Physiol.* **267**, C1253–C1261.

Shyng, S. L., Ferrigni, T., Shepard, J. B., Nestorowicz, A., Glaser, B., Permutt, M. A., and Nichols, C. G. (1998). Functional analyses of novel mutations in the sulfonylurea receptor 1 associated with persistent hyperinsulinaemic hypoglycaemia of infancy. *Diabetes.* **47**, 1145–1151.

Sienaert, I., De Smedt, H., Parys, J. B., Missiaen, L., Vanlingen, S., Sipma, H., and Casteels, R. (1996). Characterization of a cytosolic and luminal Ca^{2+} binding site in the type 1 inositol 1,4,5-trisphosphate receptor. *J. Biol. Chem.* **271**, 27005–27012.

Signorini, S., Liao, Y. J., Duncan, A., Jan, L. Y., and Stoffel, M. (1997). Normal cerebellar development but susceptibility to seizures in mice lacking the G protein-coupled, inwardly rectifying K^+ channel GIRK2. *Proc. Natl. Acad. Sci. U.S.A.* **94**, 923–927.

Simon, A. M., Goodenough, D. A., Li, E., and Paul, D. L. (1997a). Female infertility in mice lacking connexin 37. *Nature* **385**, 525–529.

Simon, D. B., Bindra, R. S., Mansfield, T. A., Nelson-Williams, C., Mendonca, E., Stone, R., Schurman, S., Nayir, A., Alpay, H., Bakkaloglu, A., Rodriguez-Soriano, J., Morales, J. M., Sanjad, S. A., Taylor, C. M., Pilz, D., Brem, A., Trachtman, H., Griswold, W., Richard, G. A., John, E., and Lifton, R. P. (1997b). Mutations in the chloride channel gene, *CLCNKB*, cause Bartter's syndrome Type III. *Nature Genet.* **17**, 171–178.

Simon, D. B., Karet, F. E., Rodriguez-Soriano, J., Hamdan, J. H., DiPietro, A., Trachtman, H., Sanjad, S. A., and Lifton, R. P. (1996). Genetic heterogeneity of Bartter's syndrome revealed by mutations in the K^+ channel, ROMK. *Nature Genet.* **14**, 152–156.

Sine, S. M., Ohno, K., Bouzat, C., Auerbach, A., Milone, M., Pruitt, J. N., and Engel, A. G. (1995). Mutation of the acetylcholine receptor α-subunit causes a slow-channel myasthenic syndrome by enhancing agonist binding affinity. *Neuron* **15**, 229–239.

Singh, N. A., Charlier, C., Stauffer, D., DuPont, B. R., Leach, R. J., Melis, R., Ronen, G. M., Bjerre, I., Quattlebaum, T., Murphy, J. V., McHarg, M. L., Gagnon, D., Rosales, T. O., Peiffer, A., Anderson, V. E., and Leppert, M. (1998). A novel potassium channel gene, KCNQ2, is mutated in an inherited epilepsy of newborns. *Nature Genet.* **18**, 25–29.

Sipos, I., Jurkat-Rott, K., Harasztosi, C., Fontaine, B., Kovacs, L., Melzer, W., and Lehmann-Horn, F. (1995). Skeletal muscle DHP receptor mutations alter calcium currents in human hypokalaemic periodic paralysis myotubes. *J. Physiol.* **483**, 299–306.

Slesinger, P. A., Jan, Y. N., and Jan, L. Y. (1993). The S4-S5 loop contributes to the ion-selective pore of potassium channels. *Neuron* **11**, 739–749.

Slesinger, P. A., Patil, N., Liao, Y. J., Jan, Y. N., Jan, L. Y., and Cox, D. R. (1996). Functional effects of the mouse weaver mutation on G protein-gated inwardly rectifying K$^+$ channels *Neuron* **16**, 321–331.

Smith, J. J., Travis, S. M., Greenberg, E. P., and Welsh, M. J. (1996a). Cystic fibrosis airway epithelia fail to kill bacteria because of abnormal airway surface fluid. *Cell* **85**, 229–236.

Smith, J. S., Imagawa, T., Ma, J., Fill, M., Campbell, K. P., and Coronado, R. (1988). Purified ryanodine receptor from rabbit skeletal muscle is the calcium-release channel of sarcoplasmic reticulum. *J. Gen. Physiol.* **92**, 1–26.

Smith, L. A., Wang, X., Peixoto, A. A., Neumann, E. K., Hall, L. M., and Hall, J. C. (1996b). A *Drosophila* calcium channel α1 subunit gene maps to a genetic locus associated with behavioral and visual defects. *J. Neurosci.* **16**, 7868–7879.

Smith, P. L., Baukrowitz, T., and Yellen, G. (1996c). The inward rectification mechanism of the HERG cardiac potassium channel. *Nature* **379**, 833–836.

Smith, R. G., Hamilton, S., Hofmann, F., Schneider, T., Nastainczyk, W., Birnbaumer, L., Stefani, E., and Appel, S. H. (1992). Serum antibodies to L-type calcium channels in patients with amyotrophic lateral sclerosis. *N. Engl. J. Med.* **327**, 1721–1728.

Snutch, T. P., and Reiner, P. B. (1992). Ca^{2+} channels: diversity of form and function. *Curr. Opin. Neurobiol.* **2**, 247–253.

Snyder, P. M., Price, M. P., McDonald, F. J., Adams, C. M., Volk, K. A., Zeiher, B. G., Stokes, J. B., and Welsh, M. J. (1995). Mechanism by which Liddle's syndrome mutations increase activity of a human epithelial Na$^+$ channel. *Cell* **83**, 969–978.

Solimena, M., Folli, F., Denis-Donini, S., Corni, G. C., Pozza, G., De Camilli, P., and Vicari, A. M. (1988). Autoantibodies to glutamic acid decarboxylase in a patient with stiff-man syndrome, epilepsy and type-1 diabetes mellitus. *N. Engl. J. Med.* **318**, 1012–1020.

Sommer, B., Keinänen, K., Verdoorn, T. A., Wisden, W., Burnashev, N., Herb, A., Kohler, M., Takagi, T., Sakmann, B., and Seeburg, P. H. (1990a). Flip and flop: a cell-specific functional switch in glutamate-operated channels of the CNS. *Science* **249**, 1580–1585.

Sommer, B., Köhler, M., Sprengel, R., and Seeburg, P. H. (1991). RNA editing in brain controls a determinant of ion flow in glutamate-gated channels. *Cell* **67**, 11–19.

Sommer, B., Poustka, A., Spurr, N. K., and Seeburg, P. H. (1990b). The murine GABA$_A$ receptor delta-subunit gene: structure and assignment to human chromosome 1. *DNA Cell Biol.* **9**, 561–568.

Song, L., Hobaugh, M. R., Shustak, C., Cheley, S., Bayler, H., and Gouaux, J. E. (1996). Structure of staphylococcal α-hemolysin, a heptameric transmembrane pore. *Science* **274**, 1859–1866.

Sorrentino, V., Giannini, G., Malzac, P., and Mattei, M. G. (1993). Localization of a novel ryanodine receptor gene (RYR3) to human chromosome 15q14-q15 by *in situ* hybridization. *Genomics* **18**, 163–165.

Spector, P. S., Curran, M. E., Keating, M. T., and Sanguinetti, M. C. (1996a). Class III antiarrhythmic drugs block HERG, a human cardiac delayed rectifier K^+ channel. *Circ. Res.* **78**, 499–503.

Spector, P. S., Curran, M. E., Zou, A., Keating, M. T., and Sanguinetti, M. C. (1996b). Fast inactivation causes rectification of the I_{Kr} channel. *J. Gen. Physiol.* **107**, 611–619.

Spencer, P. S., Roy, D. N., Ludolph, A., Hugon, J., Dwivedi, M. P., and Schaumburg, H. H. (1986). Lathyrism: evidence for role of the neuroexcitatory amino acid BOAA. *Lancet* **2**, 1066–1067.

Spencer, P. S., Nunn, P. B., Hugon, J., Ludolph, A. C., Ross, S. M., Roy, D. N., and Robertson, R. C. (1987). Guam amyotrophic lateral sclerosis–Parkinsonism–dementia linked to a plant excitant neurotoxin. *Science* **237**, 517–522.

Splawski, I., Tristani-Firouzi, M., Lehmann, M. H., Sanguinetti, M. C., and Keating, M. T. (1997). Mutations in the hminK gene cause long QT syndrome and suppress IKs function. *Nature Genet.* **17**, 338–340.

Splitt, M. P., Burn, J., and Goodship, J. (1995). Connexin 43 mutations in sporadic and familial defects of laterality. *New Engl. J. Med.* **333**, 941–942.

Staub, O., Dho, S., Henry, P., Correa, J., Ishikawa, T., McGlade, J., and Rotin, D. (1996). WW domains of Nedd4 bind to the proline-rich PY motifs in the epithelial Na^+ channel deleted in Liddle's syndrome. *EMBO J.* **15**, 2371–2380.

Stefani, E., Ottolia, M., Noceti, F., Olcese, R., Wallner, M., Latorre, R., and Toro, L. (1997). Voltage-controlled gating in a large conductance Ca^{2+}-sensitive K^+ channel. *Proc. Natl. Acad. Sci. USA.* **94**, 5427–5431.

Steinlein, O., Smigrodzki, R., Lindstrom, J., Anand, R., Kohler, M., Tocharoentanaphol, C., and Vogel, F. (1994). Refinement of the localization of the gene for neuronal nicotinic acetylcholine receptor α_4 subunit (CHRNA4) to human chromosome 20q13.2-q13.3. *Genomics* **22**, 493–495.

Steinlein, O. K., Mulley, J. C., Propping, P., Wallace, R. H., Phillips, H. A., Suther-land, G. R., Scheffer, I. E., and Berkovic, S. F. (1995). A missense mutation in the neuronal nicotinic acetylcholine receptor α_4 subunit is associated with autosomal dominant nocturnal frontal lobe epilepsy. *Nature Genet.* **11**, 201–203.

Steinmeyer, K., Lorenz, C., Pusch, M., Koch, M. C., and Jentsch, T. J. (1994). Multimeric structure of ClC-1 chloride channel revealed by mutations in dominant myotonia congenita (Thomsen). *EMBO J.* **13**, 737–743.

Steinmeyer, K., Klocke, R., Ortland, C., Gronemeier, M., Jockusch, H., Gründer, S., and Jentsch, T. J. (1991). Inactivation of muscle chloride channel by transposon insertion in myotonic mice. *Nature* **354**, 304–308.

Stehno-Bittel, L., Lückoff, A., and Clapham, D. E. (1995). Calcium release from the nucleus by InsP$_3$ receptor channels. *Neuron* **14**, 163–167.

Stephenson, F. A., Duggan, M. J., and Pollard, S. (1990). The γ_2 subunit is an integral component of the γ-aminobutyric acid$_A$ receptor but the α_1 polypeptide is the principal site of the agonist benzodiazepine photoaffinity labeling reaction. *J. Biol. Chem.* **265**, 21160–21165.

Stern-Bach, Y., Bettler, B., Hartley, M., Sheppard, P. O., O'Hara, P. J., and Heinemann, S. F. (1994). Agonist selectivity of glutamate receptors is specified by two domains structurally related to bacterial amino acid-binding proteins. *Neuron* **13**, 1345–1357.

Strang, L. B. (1991). Fetal lung liquid: Secretion and reabsorption. *Physiol. Rev.* **71**, 991–1016.

Strautnieks, S. S., Thompson, R. J., Gardiner, R. M., and Chung, E. (1996a). A novel splice-site mutation in the gamma subunit of the epithelial sodium channel in three pseudohypoaldosteronism families. *Nature Genet.* **13**, 248–250.

Strautnieks, S. S., Thompson, R. J., Hanukoglu, A., Dillon, M. J., Hanokoglu, I., Kuhnle, U., Seckl, J., Gardiner, R. M., and Chung, E. (1996b). Localisation of pseudohypoaldosteronism genes to chromosome 16p12.2.-13.11 and 12p13.1-pter by homozygosity mapping. *Hum. Mol. Genet.* **5**, 293–299.

Street, V. A., Bosma, M. M., Demas, V. P., Regan, M. R., Lin, D. D., Robinson, L. C., Agnew, W. S., and Tempel, B. L. (1997). The type 1 inositol 1,4,5-trisphosphate receptor gene is altered in the *opisthotonos* mouse. *J. Neurosci.* **17,** 635–645.

Struckmeier, M., Xiong, G., and Lutz, F. (1995). *Pseudomonas aeruginosa* cytotoxin: the Asp[197]-Gly-Asp-Tyr-His-Tyr-His-Tyr[202] containing loop is critical for plasma membrane binding. *Naunyn-Schmiedeberg's Arch. Pharmacol.* **351,** 315–319.

Stühmer, W., Conti, F., Suzuki, H., Wang, X., Noda, M., Yahagi, N., Kubo, H., and Numa, S. (1989). Structural parts involved in activation and inactivation of the sodium channel. *Nature* **339,** 597–603.

Stutts, M. J., Canessa, C. M., Olsen, J. C., Hamrick, M., Cohn, J. A., Rossier, B. C., and Boucher, R. C. (1995). CFTR as a cAMP-dependent regulator of sodium channels. *Science* **269,** 847–850.

Sunada, Y., and Campbell, K. P. (1995). Dystrophin-glycoprotein complex: Molecular organization and critical roles in skeletal muscle. *Curr Opin. Neurol.* **8,** 379–384.

Sunstrom, N. A., Premkumar, L. S., Premkumar, A., Ewart, G., Cox, G. B., and Gage, P. W. (1996). Ion channels formed by NB, an influenza B virus protein. *J. Membr. Biol.* **150,** 127–132.

Supattapone, S., Worley, P. F., Baraban, J. M., and Snyder, S. H. (1988). Solubilization, purification and characterization of an inositol trisphosphate receptor. *J. Biol. Chem.* **263,** 1530–1534.

Surprenant, A., Rassendren, F., Kawashima, E., North, R. A., and Buell, G. (1996). The cytolytic P2Z receptor for extracellular ATP identified as a P2X receptor (P2X$_7$). *Science* **272,** 735–738.

Swenson, K. I., Jordan, J. R., Beyer, E. C., and Paul, D. L. (1989). Formation of gap junctions by expression of connexins in Xenopus oocyte pairs. *Cell* **57,** 145–155.

Szpirer, C., Molne, M., Antonacci, R., Jenkins, N. A., Finelli, P., Szpirer, J., Riviere, M., Rocchi, M., Gilbert, D. J., Copeland, N. G., and Gallo, V. (1994). The genes encoding the glutamate receptor subunits KA1 and KA2 (GRIK4 and GRIK5) are located on separate chromosomes in human, mouse, and rat. *Proc. Natl. Acad. Sci. USA* **91,** 11849–11853.

Tabcharani, J. A., Chang, X.-B., Riordan, J. R., and Hanrahan, J. W. (1991). Phosphorylation-regulated Cl$^-$ channel in CHO cells stably expressing the cystic fibrosis gene. *Nature* **352,** 628–631.

Takamori, M., Iwasa, K., and Komai, K. (1997). Antibodies to synthetic peptides of the α_{1A} subunit of the voltage-gated calcium channel in Lambert–Eaton myasthenic syndrome. *Neurology* **48,** 1261–1265.

Takano, H., Onodera, O., Tanaka, H., Mori, H., Sakimura, K., Hori, T., Kobayashi, H., Mishina, M., Tsuji, S. (1993). Chromosomal localization of the ε_1, ε_3, and ζ_1 subunit genes of the human NMDA receptor channel. *Biochem. Biophys. Res. Commun.* **197,** 922–926.

Takeshima, H., Iino, M., Takekura, H., Nishi, M., Kuno, J., Minowa, O., Takano, H., and Noda, T. (1994). Excitation-contraction uncoupling and muscular degeneration in mice lacking functional skeletal muscle ryanodine receptor gene. *Nature* **369,** 556–559.

Takeshima, H., Nishimura, S., Matsumoto, T., Ishida, H., Kangawa, K., Minamino, N., Matsuo, H., Ueda, M., Hanaoka, M., Hirose, T., and Numa, S. (1989). Primary structure and expression from complementary DNA of skeletal muscle ryanodine receptor. *Nature* **339,** 439–445.

Tanabe, T., Beam, K. G., Adams, B. A., Niidome, T., and Numa, S. (1990a). Regions of the skeletal muscle dihydropyridine receptor critical for excitation–contraction coupling. *Nature* **346,** 567–569.

Tanabe, T., Beam, K. G., Powell, J. A., and Numa, S. (1988). Restoration of excitation-contraction coupling and slow calcium current in dysgenic muscle by dihydropyridine receptor complementary DNA. *Nature* **336,** 134–139.

Tanabe, T., Mikami, A., Numa, S., and Beam, K. G. (1990b). Cardiac-type excitation-contraction coupling in dysgenic skeletal muscle injected with cardiac dihydropyridine receptor cDNA. *Nature* **344,** 451–453.

Tanabe, T., Takeshima, H., Mikami, A., Flockerzi, V., Takahashi, H., Kangawa, K., Kojima, M., Matsuo, H., Hirose, T., and Numa, S. (1987). Primary structure of the receptor for calcium channel blockers from skeletal muscle. *Nature* **328,** 313–318.

Tanaka, K., Watase, K., Manabe, T., Yamada, K., Watanabe, M., Takahashi, K., Iwama, H., Nishikawa, T., Ichihara, N., Kikuchi, T., Okuyama, S., Kawashima, N., Hori, S., Takimoto, M., and Wada, K. (1997). Epilepsy and exacerbation of brain injury in mice lacking the glutamate transporter GLT-1. *Science* **276,** 1699–1702.

Terauchi, Y., Sakura, H., Yasuda, K., Iwamoto, K., Takahashi, N., Ito, K., Kasai, H., Suzuki, H., Ueda, O., Kamada, N., Jishage, K., Komeda, K., Noda, M., Kanazawa, Y., Taniguchi, S., Miwa, I., Akanuma, Y., Kodama, T., Yazaki, Y., and Kadowaki, T. (1995). Pancreatic beta-cell specific targeted disruption of glucokinase gene. Diabetes mellitus due to defective insulin secretion to glucose. *J. Biol. Chem.* **270,** 30253–30256.

Terlau, H., Heinemann, S. H., Stühmer, W., Pusch, M., Conti, F., Imoto, K., and Numa, S. (1991). Mapping the site of block by tetrodotoxin and saxitoxin of the sodium channel-II. *FEBS Lett.* **293,** 93–96.

Thomas, P., Ye, Y., and Lightner, E. (1996). Mutation of the pancreatic islet inward rectifier Kir6.2 also leads to familial persistent hyperinsulinemic hypoglycemia of infancy. *Hum. Mol. Genet.* **5,** 1809–1812.

Thomas, P. M., Cote, G. J., Hallman, D. M., and Mathew, P. M. (1995a). Homozygosity mapping, to chromosome 11p, of the gene for familial persistent hyperinsulinemic hypoglycemia of infancy. *Am. J. Hum. Genet.* **56,** 416–421.

Thomas, P. M., Cote, G. J., Wohllk, N., Haddad, B., Mathew, P. M., Rabl, W., Aguilar-Bryan, L., Gagel, R. F., and Bryan, J. (1995b). Mutations in the sulphonylurea receptor gene in familial persistent hyperinsulinemic hypoglycemia of infancy. *Science* **268,** 426–429.

Tinker, A., Jan, T. N., and Jan, L. Y. (1996). Regions responsible for the assembly of inwardly-rectifying potassium channels. *Cell* **87,** 857–868.

Tinker, A., and Williams, A. J. (1993). Probing the structure of the conduction pathway of the sheep cardiac sarcoplasmic reticulum calcium release channel with permeant and impermeant organic cations. *J. Gen. Physiol.* **102,** 1107–1129.

Toyka, K. V., Drachman, D. B., Griffin, D. E., Pestronk, A., Winkelstein, J. A., Fischbeck, K. H., and Kao, I. (1977). Myasthenia gravis. Study of humoral immune mechanisms by passive transfer to mice. *N. Engl. J. Med.* **296,** 125–131.

Treinin, M., and Chalfie, M. (1995). A mutated acetylcholine receptor subunit causes neuronal degeneration in *C. elegans*. *Neuron* **14,** 871–877.

Tretter, V., Ehya, N., Fuchs, K., and Sieghart, W. (1997). Stoichiometry and assembly of a recombinant GABA$_A$ receptor subtype. *J. Neurosci.* **17,** 2728–2737.

Tsien, R. W. (1998). Key clockwork component cloned. *Nature* **391,** 839–841.

Tsiokas, L., Kim, E., Arnould, T., Sukhatme, V. P., and Walz, G. (1997). Homo- and heterodimeric interactions between the gene products of *PKD1* and *PKD2*. *Proc. Natl. Acad. Sci. USA* **94,** 6965–6970.

Tsui, L.-C. (1992). The spectrum of cystic fibrosis mutations. *Trends Genet.* **8,** 392–398.

Tucker, S. J., Gribble, F. M., Zhao, C., Trapp, S., and Ashcroft, F. M. (1997). Truncation of Kir6.2 produces ATP-sensitive K$^+$ channels in the absence of the sulphonylurea receptor. *Nature* **387,** 179–183.

Twyman, R. E., Gahring, L. C., Spiess, J., and Rogers, S. W. (1995). Glutamate receptor antibodies activate a subset of receptors and reveal an agonist binding site. *Neuron* **14,** 755–762.

Tzartos, S. J., Barkas, T., Cung, M. T., Kordossi, A., Loutrari, H., Marraud, M., Papadouli, I., Sakarellos, C., Sophianos, D., and Tsikaris, V. (1991). The main immunogenic region of the acetylcholine receptor. Structure and role in myasthenia gravis. *Autoimmunity* **8,** 259–270.

Uchitel, O. D., Appel, S. H., Crawford, F., and Sczcupak, L. (1988). Immunoglobulins from amyotrophic lateral sclerosis patients enhance spontaneous transmitter release from motor-nerve terminals. *Proc. Natl. Acad. Sci. USA* **85,** 7371–7374.

Umenishi, F., Carter, E. P., Yang, B., Oliver, B., Matthay, M. A., and Verkman, A. S. (1996). Sharp increase in rat lung water channel expression in the perinatal period. *Am. J. Respir. Mol. Biol.* **15,** 673–679.

Unger, V. M., Kumar, N. M., Gilula, N. B., and Yeager, M. (1997). Projection structure of a gap junction membrane channel at 7 Å resolution. *Nature Struct. Biol.* **4,** 39–43.

Unwin, N. (1993). Neurotransmitter action: opening of ligand-gated ion channels. *Cell* **72**/*Neuron* **10** Suppl. 31–41.

Unwin, N. (1995). Acetylcholine receptor channel imaged in the open state. *Nature* **373,** 37–43.

Unwin, P. N. T., and Ennis, P. D. (1984). Two configurations of a channel-forming membrane protein. *Nature* **307,** 609–613.

Unwin, P. N. T., and Zampighi, G. (1980). Structure of the junction between communicating cells. *Nature* **283,** 545–549.

Valera, S., Hussy, N., Evans, R. J., Adami, N., North, R. A., Surprenant, A., and Buell, G. (1994). A new class of ligand-gated ion channel defined by P2X receptor for extracellular ATP. *Nature* **371,** 516–519.

Vallet, V., Chraibi, A., Gaeggeler, H. P., Horisberger, J. D., and Rossier, B. C. (1997). An epithelial serine protease activates the amiloride-sensitive sodium channel. *Nature* **389,** 607–610.

van der Goot, F. G., Pattus, F., Parker, M., and Buckley, J. T. (1994). The cytolytic toxin aerolysin: from the soluble form to the transmembrane channel. *Toxicology* **87,** 19–28.

van Os, C. H., Deen, P. M. T., and Demspter, J. A. (1994). Aquaporins: water selective channels in biological membranes. Molecular structure and tissue distribution. *Biochim. Biophys. Acta* **1197,** 291–309.

Vance, J. M. (1991). Hereditary motor and sensory neuropathies. *J. Med. Genet.* **28,** 1–5.

Vandenberg, C. A. (1987). Inward rectification of a potassium channel in cardiac ventricular cells depends on internal magnesium ions. *Proc. Natl. Acad. Sci. U.S.A.* **84,** 2560–2564.

Vandenberg, R. J., French, C. R., Barry, P. H., Shine, J., and Schofield, P. R. (1992). Antagonism of ligand-gated ion channel receptors: two domains of the glycine receptor α subunit form the strychnine-binding site. *Proc. Natl. Acad. Sci. USA* **89,** 1765–1769.

Varnum, M. D., and Zagotta, W. N. (1997). Interdomain interactions underlying activation of cyclic nucleotide-gated channels. *Science* **278,** 110–113.

Vassilev, P. M., Scheuer, T., and Catterall, W. A. (1988). Identification of an intracellular peptide segment involved in sodium channel inactivation. *Science* **241,** 1658–1661.

Veenstra, R. D., Wang, H.-Z., Beblo, D. A., Chilton, M. G., Harris, A. L., Beyer, E. C., and Brink, P. R. (1995). Selectivity of connexin-specific gap junctions does not correlate with channel conductance. *Circ. Res.* **77,** 1156–1165.

Verdoorn, T. A., Burnashev, N., Monyer, H., Seeburg, P. H., and Sakmann, B. (1991). Structural determinants of ion flow through recombinant glutamate receptor channels. *Science* **252,** 1715–1718.

Verselis, V. K., Ginter, C. S., and Bargiello, T. A. (1994). Opposite voltage gating polarities of two closely related connexins. *Nature* **368,** 348–351.

Vetter, D. E., Mann, J. R., Wangemann, P., Liu, J., McLaughlin, K. J., Lesage, F., Marcus, D. C., Lazdunski, M., Heinemann, S. F., and Barhanin, J. (1996). Inner ear defects induced by null mutation of the *isk* gene. *Neuron* **17,** 1251–1264.

Villarroel, A., Burnashev, N., and Sakmann, B. (1995). Dimensions of the narrow portion of a recombinant NMDA receptor channel. *Biophys. J.* **68,** 866–875.

Villarroel, A., Herlitze, S., Witzemann, V., Koenen, M., and Sakmann, B. (1992). Asymmetry of the rat acetylcholine receptor subunits in the narrow region of the pore. *Proc. Roy. Soc. Lond.* **249,** 317–324.

Vincent, A. (1980). Immunology of acetylcholine receptors in relation to myasthenia gravis. *Physiol. Rev.* **60,** 756–824.

460   Bibliography

Vincent, A. (1997). Disorders of the Human Neuromuscular Junction. *Adv. Organ Biol.* **2,** 315–349.

Vincent, A., and Drachman, D. B. (1996). Amyotrophic lateral sclerosis and antibodies to voltage-gated calcium channels: new doubts. *Ann. Neurol.* **40,** 691–693.

Vincent, A., Newland, C., Brueton, L., Beeson, D., Riemersma, S., Huson, S. M., and Newsom-Davis, J. (1995). Arthrogryposis multiplex congenita with maternal autoantibodies specific for a fetal antigen. *Lancet* **346,** 24–25.

Vincent, A., Newland, C., Croxen, R., and Beeson, D. (1997). Genes at the junction: candidates for congenital myasthenic syndromes. *Trends Neurosci.* **20,** 15–22.

Vincent, A., and Newsom-Davis, J. (1985). Acetylcholine receptor antibody as a diagnostic test for myasthenia gravis: results in 153 validated cases and 2967 diagnostic assays. *J. Neurol. Neurosurg. Psychiat.* **48,** 1246–1252.

Virgo, L., Samarasinghe, S., and de Belleroche, J. (1996). Analysis of AMPA receptor subunit mRNA expression in control and ALS spinal cord. *Neuroreport* **7,** 2507–2511.

Voilley, N., Bassilana, F., Mignon, C., Merscher, S., Mattéi, M. G., Carle, G. F., Lazdunski, M., and Barbry, P. (1995). Cloning, chromosomal localization and physical linkage of the β and γ subunits (SCNN1B and SCNN1G) of the human epithelial amiloride-sensitive sodium channel. *Genomics* **28,** 560–565.

Voilley, N., Lingueglia, E., Champigny, G., Mattéi, M. G., Waldmann, R., Lazdunski, M., and Barbry, P. (1994). The lung amiloride-sensitive Na^+ channel: biophysical properties, pharmacology, ontogenesis and molecular cloning. *Proc. Natl. Acad. Sci. U.S.A.* **91,** 247–251.

Von-Brederlow, B., Hahn, A. F., Koopman, W. J., Ebers, G. C., and Bulman, D. E. (1995). Mapping the gene for acetazolamide responsive hereditary paroxysmal cerebellar ataxia to chromosome 19p. *Hum. Mol. Genet.* **4,** 279–284.

Wada, K., Ballivet, M., Boulter, J., Connolly, J., Wada, E., Deneris, E. S., Swanson, L. W., Heinemann, S., and Patrick, J. (1988). Functional expression of a new pharmacological subtype of brain nicotinic acetylcholine receptor. *Science* **240,** 330–334.

Wagstaff, J., Knoll, J. H. M., Fleming, J., Kirkness, E. F., Martin-Gallardo, A., Greenberg, F., Graham, J. M., Menninger, J., Ward, D., Venter, J. C., and Lalande, M. (1991). Localization of the gene encoding the $GABA_A$ receptor $\beta3$ subunit to the Angelman/Prader–Willi region of human chromosome 15. *Am. J. Hum. Genet.* **49,** 330–337.

Waldmann, R., Champigny, G., Bassilana, F., Voilley, N., and Lazdunski, M. (1995a). Molecular cloning and functional expression of a novel amiloride-sensitive Na^+ channel. *J. Biol. Chem.* **270,** 27411–27414.

Waldmann, R., Champigny, G., and Lazdunski, M. (1995b). Functional degenerin-containing chimeras identify residues essential for amiloride-sensitive Na^+ channel function. *J. Biol. Chem.* **270,** 11735–11737.

Waldmann, R., Champigny, G., Voilley, N., Lauritzen, I., and Lazdunski, M. (1996). The mammalian degenerin MDEG, an amiloride-sensitive cation channel activated by mutations causing neurodegeneration in *Caenorhabditis elegans*. *J. Biol. Chem.* **271,** 10433–10436.

Waldmann, R., and Lazdunski, M. (1998). H^+-gated cation channels: neuronal acid sensors in the NaC/DEG familiy of ion channels. *Curr. Op. Neurobiol.* **8,** 418–424.

Wallace, R. H., Wang, D. W., Singh, R., Scheffer, I. E., George, A. L., Phillips, H. A., Saar, K., Reis, A., Johnson, E. W., Sutherland, G. R., Berkovic, S. F., and Mulley, J. C. (1998). Febrile seizures and generalized epilepsy associated with a mutation in the Na^+-channel β_1-subunit gene SCN1B. *Nature Genet.* **19,** 366–370.

Wallner, M., Meera, P., and Toro, L. (1996). Determinant for β-subunit regulation in high-conductance voltage-activated and Ca^{2+}-sensitive K^+ channels: an additional transmembrane region at the N-terminus. *Proc. Natl. Acad. Sci. USA* **93,** 14922–14927.

Walz, T., Hirai, T., Murata, K., Heymann, J. B., Mitsuoka, K., Fujiyoshi, Y., Smith, B. L., Agre, P., and Engel, A. (1997). The three-dimensional structure of aquaporin-1. *Nature* **387,** 624–627.

Wang, C., Lamb, R. A., and Pinto, L. H. (1995a). Activation of the M2 ion channel of influenza virus: a role for the trans-membrane domain histidine residue. *Biophys. J.* **69,** 1363–1371.

Wang, H., Kunkel, D. D., Schwartzkroin, P. A., and Tempel, B. L. (1994). Localization of Kv1.1 and Kv1.2, two channel proteins, to synaptic terminals, somata and dendrites in the mouse brain. *J. Neurosci.* **14,** 4588–4599.

Wang, H.-L., Auerbach, A., Bren, N., Ohno, K., Engel, A. G., and Sine, S. M. (1997). Mutation in the M1 domain of the acetylcholine receptor α-subunit decreases the rate of agonist dissociation. *J. Gen. Physiol.* **109,** 757–766.

Wang, H.-S., Pan, Z., Shi, W., Brown, B. S., Wymore, R. S., Cohen, I. S., Dixon, J. E., and McKinnon, D. (1998). KCNQ2 and KCNQ3 potassium channel subunits: molecular correlates of the M-channel. *Science* **282,** 1890–1893.

Wang, Q., Curran, M. E., Splawski, I., Burn, T. C., Millholland, J. M., VanRaay, T. J., Shen, J., Timothy, K. W., Vincent, G. M., de Jager, T., Schwartz, P. J., Towbin, J. A., Moss, A. J., Atkinson, D. L., Landes, G. M., Connors, T. D., and Keating, M. T. (1996). Positional cloning of a novel potassium channel gene: KVLQT1 mutations cause cardiac arrhythmias. *Nature Genet.* **12,** 17–23.

Wang, Q., Shen, J., Li, Z., Timothy, K., Vincent, G. M., Priori, S. G., Schwartz, P. J., and Keating, M. T. (1995b). Cardiac sodium channel mutations in patients with long QT syndrome, an inherited cardiac arrhythmia. *Hum. Mol. Genet.* **4,** 1603–1607.

Wang, Q., Shen, J., Splawski, I., Atkinson, D., Li, Z., Robinson, J. L., Moss, A. J., Towbin, J. A. and Keating, M. T. (1995c). SCN5A mutations associated with an inherited cardiac arrhythmia, long QT syndrome. *Cell* **80,** 805–811.

Wang, X. J., Reynolds, E. R., Déak, P., and Hall, L. M. (1997). The *seizure* locus encodes the *Drosophila* homolog of the HERG potassium channel. *J. Neurosci.* **17,** 882–890.

Warmke, J. W., Drysdale, R., and Ganetzky, B. (1991). A distinct potassium channel polypeptide encoded by the *Drosophila* *eag* locus. *Science* **252,** 1560–1562.

Warmke, J. W., and Ganetsky, B. (1994). A family of potassium channel genes related to *eag* in *Drosophila* and mammals. *Proc. Natl. Acad. Sci. U.S.A.* **91,** 3438–3442.

Warner, A. E., Guthrie, S. C., and Gilula, N. B. (1984). Antibodies to gap-junctional protein selectively disrupt junctional communication in the early amphibian embryo. *Nature* **311,** 127–131.

Waterman, S. A., Lang, B., and Newsom-Davis, J. (1997). Effect of Lambert–Eaton myasthenic syndrome antibodies on autonomic neurons in the mouse. *Ann. Neurol.* **42,** 147–156

Wei, A., Jegla, T., and Salkoff, L. (1996). Eight potassium channel familes revealed by the *C. elegans* genome project. *Neuropharmacol.* **35,** 805–829.

Weiland, S., Witzemann, V., Villarroel, A., Propping, P., and Steinlein, O. (1996). An amino acid exchange in the second transmembrane segment of a neuronal nicotinic receptor causes partial epilepsy by altering its desensitization kinetics. *FEBS Lett.* **398,** 91–96.

Welsh, M. J., and Smith, A. E. (1993). Molecular mechanisms of CFTR chloride channel dysfunction in cystic fibrosis. *Cell* **73,** 1251–1254.

West, J. W., Patton, D. E., Scheuer, T., Wang, Y., Goldin, A. L., and Catterall, W. A. (1992). A cluster of hydrophobic amino acid residues required for fast Na^{+}-channel inactivation. *Proc. Natl. Acad. Sci. USA* **89,** 10910–10914.

Wetsel, R. A., Lemons, R. S., Le Beau, M. M., Barnum, S. R., Noack, D., and Tack, B. F. (1998). Molecular analysis of human complement component C5: localization of the structural gene to chromosome 9. *Biochemistry* **27,** 1474–1482.

Weyand, I., Godde, M., Frings, S., Weiner, J., Müller, F., Altenhofen, W., Hatt, H., and Kaupp, U. B. (1994). Cloning and functional expression of a cyclic-nucleotide-gated channel from mammalian sperm. *Nature* **368,** 859–863.

White, T. W., and Bruzzone, R. (1996). Multiple connexin proteins in single intercellular channels: connexin compatibility and functional consequences. *J. Bioenerg. Biomembr.* **28,** 339–350.

White, T. W., Bruzzone, R., Wolfram, S., Paul, D. L., and Goodenough, D. A. (1994). Selective interactions among the multiple connexin proteins expressed in the vertebrate lens: The second extracellular domain is a determinant of compatability between connexins. *J. Cell. Biol.* **125,** 879–892.

Wickman, K., and Clapham, D. E. (1995). Ion channel regulation by G proteins. *Physiol. Rev.* **75,** 865–885.

Wieland, H. A., Luddens, H., and Seeburg, P. H. (1992). A single histidine in GABA$_A$ receptors is essential for benzodiazepine agonist binding. *J. Biol. Chem.* **267,** 1426–1429.

Wiener, M., Freymann, D., Ghosh P., and Stroud, R. M. (1997). Crystal structure of colicin Ia. *Nature* **385,** 461–464.

Wilcox, A. S., Warrington, J. A., Gardiner, K., Berger, R., Whiting, P., Altherr, M. R., Wasmuth, J. J., Patterson, D., and Sikela, J. M. (1992). Human chromosomal localization of genes encoding the γ1 and γ2 subunits of the gamma-aminobutyric acid receptor indicates that members of this gene family are often clustered in the genome. *Proc. Natl. Acad. Sci USA* **89,** 5857–5861.

Wilke, K., Gaul, R., Klauck, S. M., and Poustka, A. (1997). A gene in human chromosome band Xq28 (*GABRE*) defines a putative new subunit class of the GABA$_A$ neurotransmitter receptor. *Genomics* **45,** 1–10.

Willecke, K., Jungbluth, S., Dahl, E., Hennemann, H., Heynkes, R., and Grzeschik, K. H. (1990). Six genes of the human connexin gene family coding for gap junctional proteins are assigned to four different human chromosomes. *Eur. J. Cell Biol.* **53,** 275–280.

Wilmsen, H. U., Leonard, K. R., Tichelaar, W., Buckley, J. T., and Pattus, F. (1992). The aerolysin membrane channel is formed by heptamerization of the monomer. *EMBO J.* **11,** 2457–2463.

Wingrove, P. B., Thompson, S. A., Wafford, K. A., and Whiting, P. J. (1997). Key amino acids in the γ subunit of the γ-aminobutyric acid$_A$ receptor that determine ligand binding and modulation at the benzodiazepine site. *Mol. Pharmacol.* **52,** 874–881.

Winter, M. C., and Welsh, M. J. (1997). Stimulation of CFTR activity by its phosphorylated R domain. *Nature* **389,** 294–296.

Wischmeyer, E., and Karschin, A. (1996). Receptor stimulation causes slow inhibition of IRK1 inwardly rectifying K$^+$ channels by direct protein kinase A-mediated phosphorylation. *Proc. Natl. Acad. Sci. U.S.A.* **93,** 5819–5823.

Witzemann, V., Schwarz, H., Koenen, M., Berberich, C., Villarroel, A., Wernig, A., Brenner, H. R., and Sakmann, B. (1996). Acetylcholine receptor ε-subunit deletion causes muscle weakness and atrophy in juvenile and adult mice. *Proc. Natl. Acad. Sci. USA* **93,** 13286–13291.

Wo, Z. G., and Oswald, R. E. (1995). Unraveling the modular design of glutamate-gated ion channels. *Trends Neurosci.* **18,** 161–168.

Wollmuth, L. P., Kuner, T., Seeburg, P. H., and Sakmann, B. (1996). Differential contribution of the NR1- and NR2A-subunits to the selectivity filter of recombinant NMDA receptor channels. *J. Physiol.* **491.3,** 779–797.

Xia, X.-M., Fakler, B., Rivard, A., Wayman, G., Johnson-Pais, T., Keen, J. E., Ishii, T., Hirschberg, B., Bond, C. T., Lutsenko, S., Maylie, J., and Adelman, J. P. (1998). Mechanism of calcium channel gating in small-conductance calcium-activated potassium channels. *Nature* **395,** 503–507.

Xu, M., and Akabas, M. H. (1996). Identification of channel-lining residues in the M2 membrane-spanning segment of the GABA$_A$ receptor α_1 subunit. *J. Gen. Physiol.* **107,** 195–205.

Xu, M., Covey, D. F., and Akabas, M. H. (1995). Interaction of picrotoxin with GABA$_A$ receptor channel-lining residues probed in cysteine mutants. *Biophys. J.* **69,** 1858–1867.

Xu, Z. C., Yang, Y.. and Hebert, S. C. (1996). Phosphorylation of the ATP-sensitive, inwardly rectifying K⁺ channel, ROMK, by cyclic AMP-dependent protein kinase. *J. Biol. Chem.* **271,** 9313–9319.

Yamada, M., Miyawaki, A., Saito, K., Nakajima, T., Yamamoto-Hino, M., Ryo, Y., Furuichi, T., and Mikoshiba, K. (1995). The calmodulin-binding domain in the mouse type 1 inositol 1,4,5-trisphosphate receptor. *Biochem. J.* **308,** 83–88.

Yamada, N., Makino, Y., Clark, R. A., Pearson, D. W., Mattei, M. G., Guenet, J. L., Ohama, E., Fujino, I., Miyawaki, A., Furuichi, T., and Mikoshiba, K. (1994). Human inositol 1,4,5-trisphosphate type-1 receptor, InsP3R1: structure, function, regulation of expression and chromosomal localization. *Biochem. J.* **302,** 781–790.

Yamamoto-Hino, M., Sugiyama, T., Hikichi, K., Mattei, M. G., Hasegawa, K., Sekine, S., Sakurada, K., Miyawaki, A., Furuichi, T., Hasegawa, M., and Mikoshiba, K. (1994). Cloning and characterization of human type 2 and type 3 inositol 1,4,5-trisphosphate receptors. *Recept. Channels* **2,** 9–22.

Yang, B., and Verkman, A. S. (1997). Water and glycerol permeabilities of aquaporins 1-5 and MIP determined quantitatively by expression of epitope-tagged constructs in *Xenopus* oocytes. *J. Biol. Chem.* **272,** 16140–16146.

Yang, J., Ellinor, P. T., Sather, W. A., Zhang, J. F., and Tsien, R. W. (1993). Molecular determinants of Ca²⁺ selectivity and ion permeation in L-type Ca²⁺ channels. *Nature* **366,** 158–161.

Yang, J., Jan, Y. N., and Jan, L. Y. (1995). Control of rectification and permeation by residues in two distinct domains in an inward rectifier K⁺ channel. *Neuron* **14,** 1047–1054.

Yang, N., George, A. L., and Horn, R. (1996). Molecular basis of charge movement in voltage-gated sodium channels. *Neuron* **16,** 113–122.

Yang, T., and Roden, D. M. (1996). Extracellular potassium modulation of drug block of I_{Kr}. Implications for *torsade de pointes* and reverse use-dependence. *Circulation* **93,** 407–411.

Yang, W. P., Levesque, P. C., Little, W. A., Condor, M. L., Ramakrishnan, P., Neubauer, M. G., and Blanar, M. A. (1998). Functional expression of two KvLQT1-related potassium channels responsible for an inherited idiopathic epilepsy. *J. Biol. Chem.* **273,** 19419–19423.

Yang, W. P., Levesque, P. C., Little, W. A., Condor, M. L., Shalaby, F. Y., and Blanar, M. A. (1997). KvLQT1, a voltage-gated potassium channel responsible for human cardiac arrhythmias. *Proc. Natl. Acad. Sci. U.S.A.* **94,** 4017–4021.

Yarom, Y., Sugimori, M., and Llinás, R. (1985). Ionic currents and firing patterns of mammalian vagal motorneurons *in vitro*. *Neuroscience* **16,** 719–737.

Yellen, G., Jurman, M. E., Abramson, T., and MacKinnon, R. (1991). Mutations affecting internal TEA blockade identify the probable pore-forming region of a K⁺ channel. *Science* **251,** 939–942.

Yoshikawa, F., Morita, M., Monkawa, T., Michikawa, T., Furuichi, T., and Mikoshiba, K. (1996). Mutational analysis of the ligand-binding site of the inositol 1,4,5-trisphosphate receptor. *J. Biol. Chem.* **271,** 18277–18284.

Young, J. D.-E., and Lowrey, D. M. (1989). Biochemical and functional characterization of a membrane-associated pore-forming protein from the pathogenic ameboflagellate *Naegleria fowleri*. *J. Biol. Chem.* **264,** 1077–1083.

Zagotta, W. N., Hoshi, T., and Aldrich, R. W. (1990). Restoration of inactivation of *Shaker* potassium channels by a peptide derived from ShB. *Science* **250,** 568–571.

Zeidel, M. L., Ambudkar, S. V., Smith, B. L., and Agre, P. (1992). Reconstitution of functional water channels in liposomes containing purified red cell CHIP28 protein. *Biochemistry* **31,** 7436–7440.

Zhang, J. F., Ellinor, P. T., Aldrich, R. W, and Tsien, R. W. (1994). Molecular determinants of voltage-dependent inactivation in calcium channels. *Nature* **372,** 97–100.

Zhang, Y., Chen, H. S., Khanna, V. K., De Leon, S., Phillips, M. S., Schappert, K., Britt, B. A., Browell, A. K., and MacLennan, D. H. (1993). A mutation in the human ryanodine receptor gene associated with central core disease. *Nature Genet.* **5,** 46–50.

Zhou, X.-W., Pfahnl, A., Werner, R., Hudder, A., Llanes, A., Luebke, A., and Dahl, G. (1997). Identification of a pore lining segment in gap junction hemichannels. *Biophys. J.* **72,** 1946–1953.

Zhuchenko, O., Bailey, J., Bonnen, P., Ashizawa, T., Stockton, D. W., Amos, C., Dobyns, W. B., Subramony, S. H., Zoghbi, H. Y., and Lee, C. C. (1997). Autosomal dominant cerebellar ataxia (SCA6) associated with small polyglutamine expansions in the alpha 1A-voltage-dependent calcium channel. *Nature Genet.* **15,** 62–69.

Zimmerman, A. L. (1995). Cyclic nucleotide gated channels. *Curr. Op. Neurobiol.* **5,** 296–303.

Zucchi, R., and Ronca-Testoni, S. (1997). The sarcoplasmic reticulum Ca^{2+} channel/ryanodine receptor: modulation by endogenous effectors, drugs and disease states. *Pharmacol. Rev.* **49,** 1–51.

Zukin, R. S., and Bennett, M. V. L. (1995). Alternatively spliced isoforms of the NMDAR1 receptor subunit. *Trends Neurosci.* **18,** 306–313.

Zuo, J., De Jager, P. L., Takahashi, K. A., Jiang, W., Linden, D. J., and Heintz, N. (1997). Neurodegeneration in *Lurcher* mice caused by mutation in delta2 glutamate receptor gene. *Nature* **388,** 769–773.

INDEX

Absolute refractory period, definition of, 39
Acetazolamide, 111, 175, 179
Acetylcholine (ACh),
  binding site in nAChR, 278
  inhibition of heart by, 144–145
  See also Nicotinic acetylcholine receptors
    (nAChR)
Acetylcholine receptors (AChR)
  muscarinic AChR, 117, 144, 269
  See also Nicotinic acetylcholine receptors
Acetylcholinesterase, 271, 283, 287
ACh, *see* Acetylcholine
Acid-sensing ion channels, 234–235, 247
Acquired neuromyotonia, *see*
    Neuromyotonia, acquired
Action potential,
  cardiac, and LQT syndrome, 93, 114,
    121
  description of, 38–39
  measurement of, 44
Activation, voltage-dependent
  general description of, 29, 34
  of K_V channel, 106
  of sodium channel, 73–76
Adenylate cyclase, in olfactory cilia, 206
ADNFLE, *see* Autosomal dominant
    nocturnal frontal lobe epilepsy
Adrenaline, 164, 176, 223, 245
adr mouse mutant, 191
Aerolysin, 395, 398
Aeromonas, 398
After-hyperpolarization, 39, 125–126
AHP, *see* After-hyperpolarization
Alamethicin, 399–401
β-Alanine, 313, 315, 316, 320–321
Alcohol and $GABA_B$ receptor, 326, 330,
    334–335
Alcohol intolerance in rat, 334–335
Aldosterone, 149, 236, 240, 243
Allele, definition of, 13–14
α-helix, definition of, 10–11
α-Toxin, *Staphylococcus aureus*, 396–398
Alternative splicing
  definition of, 6
  of glutamate receptors, 299, 303

of Kir1.1, 136, 150
of K_V channels, 100
Alzheimer's disease, β-amyloid peptide in,
    411
Amantadine, 383
Amiloride, 231, 237, 240, 246
Amino acid residues, properties of, 8–9
γ-amino butyric acid, *see* GABA
4-Aminopyridine, 369
Amoebic pores, 402–403
AMPA, as glutamate receptor agonist,
    291, 292, 293
AMPA receptors, 291–301
  alternative splicing, 299
  basic properties, 291–293
  chromosomal location of, 293
  ligand-binding site, 300
  mRNA editing, 295–299
  pore, 300
  See also Glutamate receptors
Amphotericin B, 49, 394–395
β-Amyloid peptide, 411–412
Amyotropic lateral sclerosis, 309, 372–373
Anaesthetics
  general, 176–177, 256, 257, 316–317,
    330–331
  local, 256
Angelman syndrome, 332–333
Anion channels, *see* Chloride channels;
    Glycine receptors; GABA receptors
Antiarrhythmic drugs, 94, 120, 122
Antibiotics
  gramicidin A, 391–392
  as ionophores and pores, 391–394, 396
Antibodies to ion channels, 361–378
  to adult nAChR, 364, 366–367
  to Ca^{2+} channels, 369–372
  to fetal nAChR, 367–368
  to GluR receptors, 375–378
  to glutamic acid decarboxylase, 377–378
  to K_V channels, 373–374
  use in defining membrane topology, 58
Anticholinesterase, 365–366
Anticodon, in RNA translation, 7